PHYSICAL CONSTANTS

(See Appendix J for more precise values)

	Symbol	Value
Gravitational constant	G	6.67×10^{-11} N-m^2/kg^2
Acceleration of gravity	g	9.81 m/s^2
Avogadro's number	N_0	6.02×10^{23} atoms/mole
Gas constant	R	8.31 J/mole-K $= 1.99$ cal/mole-K
Boltzmann's constant	k	1.38×10^{-23} J/K
Triple point of water	T_t	273.16 K
Electron mass	m_e	9.11×10^{-31} kg
Proton mass	m_p	1.67×10^{-27} kg
Quantum of electric charge	e	1.60×10^{-19} C
Permeability of free space	μ_0	$4\pi \times 10^{-7}$ H/m
Permittivity of free space	ϵ_0	8.85×10^{-12} F/m
	$\dfrac{1}{4\pi\epsilon_0}$	8.99×10^9 m/F
Planck's constant	h	6.63×10^{-34} J-s
Speed of light	c	3.00×10^8 m/s
Bohr radius	a_0	5.29×10^{-11} m
Electron Compton wavelength	$\dfrac{h}{m_e c}$	2.43×10^{-12} m

PHYSICAL DATA

Sun

Mass	1.99×10^{30} kg
Mean radius	6.96×10^{8} m
Mean surface temperature	5800 K
Mean density	1.410×10^{3} kg/m^3

Earth

Mass	5.98×10^{24} kg
Mean radius	6.37×10^{6} m
Mean distance from sun	1.496×10^{11} m
Mean orbital speed	29.8 km/s
Mean density	5.517×10^{3} kg/m^3

Moon

Mass	7.35×10^{22} kg
Mean radius	1.738×10^{6} m
Mean distance from earth	3.844×10^{8} m
Mean density	3.342×10^{3} kg/m^3

Terrestial

Atmospheric pressure (at sea level and 0°C)	76 cm Hg, 1.013×10^{5} N/m^2
Mass density of air (at sea level and 0°C)	1.293 kg/m^3
Mass density of water (at sea level and 0°C)	1.000×10^{3} kg/m^3
Mean molecular weight of dry air	29.0

USEFUL CONVERSION FACTORS

Angular measure:	57.3 degrees/rad
	0.0175 rad/deg
Area	9.29×10^{-2} m²/ft²
	6.45×10^{-4} m²/in.²
Distance	0.305 m/ft
	10^{-10} m/Å
	2.54 cm/in.
	1.61 km/mi
	1.50×10^{8} km/AU
Energy	1.60×10^{-19} J/eV
	4.186 J/cal
	10^{-7} J/erg
Force	4.45 N/lb
	10^{-5} N/dyne
Mass	14.6 kg/slug
	1.661×10^{-27} kg/amu
Power	746 watts/hp
Pressure	1.01×10^{5} (N/m²)/(atm)
Speed	0.305 (m/s)/(ft/s)
	0.447 (m/s)/(mi/hr)
	0.278 (m/s)/(km/hr)
Time	3.156×10^{7} s/yr
Volume	2.83×10^{-2} m³/ft³
	10^{-3} m³/liter

Physics
Basic principles

Without my work in natural science I should never have known human beings as they really are. In no other activity can one come so close to direct perception and clear thought or realize so fully the error of the senses, the mistakes of the intellect, the weakness and greatness of the human character.

GOETHE

Physics
Basic principles
volume II

Solomon Gartenhaus
Purdue University

HOLT, RINEHART AND WINSTON, INC.
New York   Chicago   San Francisco   Atlanta
Dallas   Montreal   Toronto   London

Library of Congress Cataloging in Publication Data
Gartenhaus, Solomon, 1929–
Physics: basic principles.
1. Physics. I. Title.
QC21.2.G33   530   74-19388
ISBN 0-03-088080-7
Printed in the United States of America
5 6 7 8 038 1 2 3 4 5 6 7 8 9

This book was set in Times Roman by European Printing

Editor:	**Dorothy Garbose Crane**
Designer:	**Scott Chelius**
Production supervisor:	**Robert Ballinger**
Printer and Binder:	**The Maple Press Co.**
Cover:	**Edmée Froment**

To Johanna

Preface

This book is the second volume of a text designed for a one-year introductory physics course for students of science and engineering. It deals with electromagnetism, including electromagnetic waves (Chapters 19–29), with geometrical and physical optics (Chapters 30–33), and concludes with an introduction to quantum phenomena. The companion Volume I treats mechanics (Chapters 2–12), heat and thermodynamics (Chapters 13–17), and wave motion (Chapter 18).

Although most of the subject matter of both volumes deals with the laws and the phenomena of classical physics, a substantial number of references to, and illustrations of, nonclassical phenomena have been distributed throughout the text. These are intended to give the student some feeling for recent developments, as well as to alert him to the limitations of certain of the classical laws, and of the seemingly universal applicability of others. Thus, he sees the ideas of the conservation of energy and of momentum applied not only to laboratory-sized bodies but to the motions and the interactions of nuclear, atomic, planetary, and stellar systems as well. In addition, a number of chapters contain optional sections (marked by a dagger †), which generalize or extend the text material in some way or describe related but more recent discoveries. For example, Chapter 1 includes a qualitative description of nuclei and of subnuclear particles, and in Chapter 19 the charge structure of the neutron and the proton is briefly discussed. Other material of this nature includes discussions of: the Lorentz transformation, as the high velocity generalization of the Galilean transformation (Section †3-12); the relativistic forms of the laws of energy and momentum conservation (Section †9-10); and Einstein's equivalence principle and the bending of light rays in a gravitational field (Section †5-9).

To a considerable extent, this book reflects my view that the student for whom the book is intended is preparing for a career for which some in-depth knowledge of physics is essential. With today's accelerating tempo of scientific discovery and technological innovation foremost in mind, I have endeavored, throughout the text, to emphasize the basic physical laws and to present them in as succinct and complete a manner as it is possible to do at the introductory level. Included therefore are not only extensive discussions of the scope of the laws and their interrelations but, where appropriate, their ranges of validity as well. By placing the focus on the basic physical laws in this way, it is hoped that the student will come to understand some of the physicists' empirical-logical approach to the world and will acquire thereby a realistic perception of these laws and a secure intellectual base from which to pursue further studies.

Specific features of the book which are directed toward implementing this approach include:

1. All basic physical laws, such as Newton's laws of motion, Coulomb's law, and Faraday's law, are first introduced by use of certain conceptually simple and easily visualized experiments. The laws themselves are then abstracted from these experiments and reformulated in quantitative mathematical terms. Finally, the important features of the laws and their relation with other laws are brought out and reinforced in the student's mind by applying them to a variety of physical situations.

2. Because of their conceptual importance, both as a starting point in physical reasoning and as a tool for codifying diverse physical phenomena, the various conservation laws receive particularly heavy emphasis. More than a third of the material on mechanics, for example, deals with these laws. Extensive discussions of these laws are presented and the student is able to see, at several levels, their connection with other physical laws, their interrelations, and their limits of validity.

3. Wherever appropriate, all physical quantities are first introduced by use of operational definitions. My view here is that in the student's subsequent capacity as a professional he may well be called upon to apply physical laws in regions bordering on their limits of validity. Therefore, he should be able to reexamine critically and systematically all basic definitions and physical laws in light of their underlying assumptions and for this purpose operational definitions are indispensable.

4. A considerable effort has been made to present the material in a way

to make it dovetail more closely with that found in more advanced courses. Thus, the language, the basic approach, and the attitude, but not the level, are those found more often in texts at the intermediate and higher level.

Some other features of the book are:

1. As is traditional in introductory physics courses of this type, it is assumed that the student is taking concurrently a course in calculus. Nevertheless, many of the basic ideas of the differential and the integral calculus are introduced as needed and integrated with the text. Detailed derivations, however, are usually relegated to one of the brief mathematical appendices. Vectors are introduced in Chapter 3 and used throughout the text. The dot product is defined in Chapter 7 and the cross product in Chapter 10.

2. Each chapter contains a wide selection of questions and problems. The more difficult problems have been marked by an asterisk (*) and those which require some knowledge of material covered in optional sections have been similarly marked by a dagger (†). As a general rule, the order of the problems follows closely the ordering of the text material. The questions generally call for answers of a qualitative or semiquantitative nature, and are intended mainly to help the student test his physical understanding of the text material. Also included are some questions whose main purpose is to be thought provoking.

3. A sizable number of worked examples (averaging more than ten per chapter) are distributed throughout the text. These span all levels of difficulty, from the simple formula-substitution type to the more complex ones whose purpose is to extend some aspect of a topic treated in the text. Also, many of these examples serve as models which the student can use as aid in problem solving.

4. The SI or International System of Units (often called the metric or the MKSA system) is used exclusively throughout the text. Included, however, are definitions of and conversion tables to the CGS and the English system of units. With only rare exception, the answers to all problems and worked examples that call for a numerical value are in metric units.

5. A broad spectrum of applications has been included in the text. Thus there is available considerable latitude in adapting it to courses of varying length or specialized needs.

In writing this book, I have received help in many forms and from many more sources than it is possible to acknowledge individually. I should like to thank particularly Orland E. Johnson, who has been a valuable source of advice on all aspects of the book throughout the long years of writing. I am also greatly indebted to David J. Ennis and Nicholas J. Giordano, who with skill and patience worked out solutions to all problems and were helpful in a variety of other ways; to John J. Brehm, Gary S. Kovener, Earl W. McDaniel, and Walter W. Wilson, who critically reviewed the manuscript and made many valuable suggestions; to the many of my colleagues at Purdue, particularly Edward Akeley, Irving Geib, Don Schlueter, Isadore Walerstein and Lonnie Van Zandt for much encouragement and for help when needed; and to the members of the editorial staff at Holt, Rinehart and Winston through whose patience, persistence, and labors this book ultimately came into being. Finally, I should like to thank Debbie Carr, Coleen Flanagan, and my wife Johanna for their invaluable secretarial assistance and my sons Mike and Kevin for their help with the galleys.

January 1975   Solomon Gartenhaus
                                                           Purdue University

Contents

xi

33 Polarization   1015

34 Epilogue   1043

Appendices

List of tables

Physics
Basic principles

19 Coulomb's law

From a long view of the history of mankind—seen from say ten thousand years from now—there can be little doubt that the most significant event of the 19th century will be judged as Maxwell's discovery of the laws of electrodynamics.

R. P. FEYNMAN

19-1 General introduction

For the next eleven chapters we shall be studying the branch of physics known as *electromagnetism*. As the name implies, this discipline deals with electric and magnetic phenomena and the relations between them. Although some of the qualitative observations on electricity and magnetism date back to antiquity, most of the key experimental and theoretical ideas were discovered by researchers who lived during the nineteenth century. Very prominent among these is James C. Maxwell (1831–1879), who succeeded in a synthesis of all experimental facts on electromagnetism known up to about 1860—and since discovered—in terms of a set of four equations, which today are known as *Maxwell's equations.* These very fundamental relations play a role in electromagnetism which is very analogous to that played by Newton's laws in mechanics. By use of these equations we are able to understand not only many aspects of the forces between atoms and molecules but also the principles underlying the operation of a variety of devices, including electric motors, radio and television transmitters and receivers, and high-energy particle accelerators such as cyclotrons and synchrotrons. The scope of these laws is obviously enormous.

For the earliest recorded observations of electric and magnetic phenomena we are indebted to the Greeks. The fact that rubbed amber acquires electrical properties, as evidenced by the fact that it attracts small pieces of straw, was recorded by Thales of Miletus circa 600 B.C. Similarly, the existence of lodestone (that is, "leading stone" or compass) appeared in Greek writings as early as 800 B.C., and the magnetic properties of magnetite (the iron ore consisting mainly of $FeO-Fe_2O_3$) was known to Pliny. From these very early and primitive observations, it took man more than 2000 years to establish the fact that electricity and magnetism are but different aspects of the same phenomenon. For this breakthrough it was necessary to await the development of the voltaic cell, or battery, which made possible the production of steady electric currents. In 1820, Hans Oersted (1777–1851) discovered that a wire through which such an electric current flows has properties similar to that of a magnet. Shortly thereafter, Michael Faraday (1791–1867) reported on his observations of the related effect that if a wire is moved near a magnet—or, equivalently, if a magnet is moved near a loop of wire—an electric current flows in the wire. Finally, Maxwell showed that these and a variety of other experimental facts could be correlated in terms of a small number of simple relations.

One of the very interesting by-products of Maxwell's formulation was his deduction that if the current flow in a wire varies in time, then waves would be radiated. Twenty years after Maxwell's enunciation of his theory, the existence of these *electromagnetic waves* was established experimentally by Heinrich Hertz (1857–1894). From these early observations of Hertz it was a relatively small step to develop "wireless" transmission of electromagnetic signals and to establish the fact that ordinary light is but one example of an electromagnetic wave. Indeed, as we shall see, it is possible to deduce the speed of light by making measurements of electromagnetic phenomena in the laboratory!

As a final note, toward the end of the nineteenth century it became apparent that there were certain logical inconsistencies between Maxwell's equations and Newton's laws of motion. This matter was unambiguously resolved by Albert Einstein (1879–1955) in 1905 when he enunciated his theory of relativity. Remarkably enough, Maxwell's equations withstood the test. Newton's laws did not and required modification.

19-2 Electric charge

One of the simplest ways to produce electric charge involves a hard-rubber object, such as a pocket comb. If you comb your hair vigorously with such a comb you will find that it is able to attract small objects such as pieces of paper. This is analogous to the ancient Greek observation that rubbed amber attracts bits of straw. More generally, experiment shows that two hard-rubber rods which have been rubbed with animal fur will repel each other. Similarly, if two glass rods are rubbed with silk, they will also be found to repel each other. However, if a hard-rubber rod that has been rubbed with fur is brought near a glass rod that has been rubbed with silk we find that they attract each other. In all three of these cases, the force between the rods decreases as the separation distance is increased.

A convenient way to demonstrate the existence of these forces is shown in Figure 19-1. A glass rod that has been rubbed with silk is suspended by a nonmetallic thread in such a way that it is free to rotate in a horizontal plane. If a hard-rubber rod which has been rubbed with fur is brought near the glass rod, it will be observed to rotate, thus demonstrating the existence of an attractive force between the rubbed parts of the rods. If the suspended rod in the figure is replaced by a rubber rod, the force is found to be repulsive.

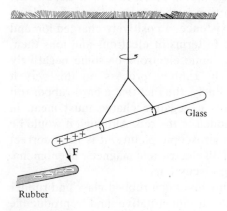

Glass

$+ + + +$

$\downarrow$ F

Rubber

Figure 19-1

More generally, experiment shows that when rubbed in an appropriate way, many solid substances will behave either as the glass rod or as the hard-rubber rod. That is, they will be either attracted by the glass rod and repelled by the hard-rubber rod, or vice versa. We characterize this property of attraction or of repulsion of a rubbed solid relative to a glass or a rubber rod by saying that these bodies have an *electric charge*. It is apparent from the above discussion that there exist at least two types of electric charge. Following the convention originally introduced by Benjamin Franklin, we call the type of electric charge on a glass rod rubbed with silk as *positive* and that on a rubber rod rubbed with fur as *negative*. In operational terms, then, any substance that repels a glass rod (and thus attracts a hard-rubber rod) is said to have a positive charge, and conversely, any substance that repels a hard-rubber rod (and thus attracts a rubbed glass rod) is said to have a negative charge. The plus and minus signs on the rods in Figure 19-1 reflect this universally accepted convention.

These results on the electric forces between charged solids may be conveniently summarized by the statement:

Positively charged bodies repel other positively charged bodies and negatively charged bodies repel other negatively charged bodies, whereas a body carrying either sign of electric charge will attract a body carrying charge of the opposite sign.

Let us now reexamine the above experimental results involving macroscopic bodies in light of our current understanding of atoms and molecules. Studies in chemistry and atomic physics have established that an atom consists of a very massive, positively charged nucleus, about which orbit a number of relatively light, negatively charged electrons. Under ordinary circumstances the atom is electrically neutral in that it carries no net charge. If, however, a negatively charged electron is torn from its parent nucleus, then two electric charges come into existence: a positively charged ion and the negatively charged electron itself. In terms of electrons and ions then, when a glass rod is rubbed with silk, some electrons or some negatively charged ions must come off during the rubbing process. In this way it acquires a positive charge. Correspondingly, the fact that a hard-rubber rod when rubbed with animal fur acquires a negative charge must mean in microscopic terms that electrons are added to the rod. Although it would be difficult, at this point, to justify this microscopic picture, it is fully correct and is a convenient way for thinking of all electric and magnetic phenomena. It will be freely used in the following discussions.

Unfortunately, the above experiments involving rubbed glass and rubber rods do not lend themselves readily to quantitative and reproducible experiments. The precise strength of the force between various electrically charged, macroscopic bodies depends sensitively on the detailed geometry of the bodies, on the length of time they are rubbed, and so forth. Thus to go from these qualitative observations to a quantitative measure of the forces between charged bodies it is convenient to make use of a class of materials known as conductors.

19-3 Charge induction and conductors

In a discussion of electric phenomena, it is convenient to characterize matter as being either a *conductor* or an *insulator*.[1] A conductor is a substance through which electric charge is readily transported, whereas an insulator is

---

[1]A third class of substances, known as *semiconductors*, lies somewhere between these two. In the following we shall be concerned only with conductors and insulators.

one which will not easily conduct charge. For example, the hard-rubber rods and the glass rods considered in Section 19-2 are insulators, for experiment shows that only the rubbed portions of the rods acquire an electric charge; there is no tendency for this charge to leave the surface and to diffuse into the interior of the rods. By contrast, if charge is placed on a metallic conductor, such as copper or silver, the electric charge appears not only on that portion of the conducting surface on which it is placed, but at various other parts of the conductor as well.

In addition to the familiar metallic solids, such as copper or silver, conductors also occur naturally in the form of liquids and gases. A *plasma* is a gas that consists of mobile charged particles and is thus a gaseous conductor. Plasmas can be produced in the laboratory; they also occur naturally in the upper regions of our atmosphere (the ionosphere), in the region surrounding the sun, and generally throughout much of intergalactic space. As exemplified by mercury, which at room temperature is a liquid, conductors can also exist in the liquid state. In addition to liquid metals, water containing various salts in solution, such as AgCl and Na_2CO_3, are also very good conductors. They are known as *electrolytes*. The human body itself is a relatively good conductor by virtue of the body fluids. The problem of describing the electrical behavior of gaseous and liquid conductors is generally much more complex than is the corresponding description of solid ones, and thus from now on we shall deal exclusively with solid conductors.

In connection with a discussion of electric charge in solid conductors, it is helpful to have available a microscopic picture of these materials. For this purpose we may think of a conductor as consisting of a rigid array of immobile and positively charged ions, which are arranged in a regular lattice of some type. Figure 19-2 shows a two-dimensional version of such a lattice. Interspersed between the fixed ions is an isotropically distributed gas of negatively charged electrons, with typically one or two electrons for each

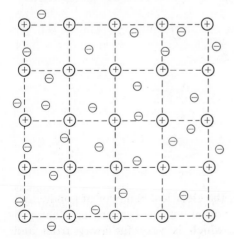

Figure 19-2

ion. In the normal situation of electrical neutrality, these electrons are distributed uniformly, so that in any small, but macroscopic, volume element of the lattice there is just as much positive charge associated with the ions as there is negative charge on the electrons. The lattice is in this case said to be *electrically neutral.* Note the important point that the positively charged ions are "rigidly" held in place but that the electrons are very mobile. This means that if electric charge is brought near a conductor, even though both the electrons and the ions will experience an additional force, only the electrons are free to move under its action. For simplicity we shall always think of the ions as maintaining fixed relative positions, although in actual fact they also undergo thermal motions about their equilibrium positions.

Figure 19-3a shows a charged glass rod near a conductor. According to the above physical picture, some of the electrons will be attracted by the electric charge on the rod and will move toward that portion of the conducting surface closest to the glass rod. As a result of this migration, a positive charge will appear on the more distant surface of the conductor. (It will be established in the next chapter that charge can only reside on the outer surface of a solid conductor.) Similarly, if a rubbed hard-rubber rod with its negative charge is brought near a conductor, some of the mobile electrons will be repelled by the rod, and the situation will then be as shown in Figure 19-3b. Hence, if charge is brought near a conductor an *induced charge* will appear on the surface of that conductor. Further, since the force between charged bodies decreases as the distance between them increases, the force between the conducting sphere and the rod, in Figure 19-3, is attractive. In both cases, the sign of the induced charge on that portion of the conducting surface nearest to the rod is opposite to the sign of the charge on the rod; thus there is a net attractive force between them.

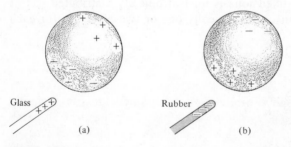

(a) (b)

Figure 19-3

19-4 Grounds

A concept of importance in a study of the electrical behavior of conductors is that of a *ground.* To introduce this notion imagine bringing up a charged hard-rubber rod to a conductor *A,* which is very far away from and connected to a second one *B* by a long conducting wire (Figure 19-4). As

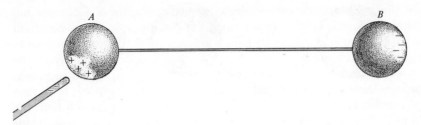

Figure 19-4

before, some of the electrons on *A* will respond to the repulsive force exerted by the negatively charged rod and will attempt to travel as far away from this charge as possible. Unlike the situation in Figure 19-3, however, where the electrons could not cross the boundary surface of the isolated conductor, in the present case, because of the connecting wire, these electrons can very easily leave *A* and travel to conductor *B*. We say that conductor *A* has been *grounded* provided that the second conductor *B* is so far away that the charge on it no longer influences the situation prevailing in the neighborhood of *A*. The symbol

will be used to denote the fact that a conductor has been grounded. If the conducting wire in Figure 19-4 is sufficiently long so that *B* is very far away from *A* and is thus grounded, then we represent this situation as in Figure 19-5.

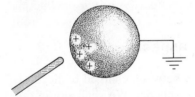

Figure 19-5

Mainly because of its metallic content and the existence of dissolved salts in the oceans, the earth itself is a relatively good conductor. Furthermore, because of its immense size it is in effect an infinite source as well as a sink of charge and is thus also a ground! Indeed, the name *ground* literally means the connecting of a conductor, by means of a conducting wire, to the earth. If we think of conductor *B* in Figure 19-4 as being the earth itself, then the negative charge appearing on it is always far away from conductor *A* and thus, even if the conducting wire connecting them is not particularly long, *A* will invariably be grounded under these circumstances.

As an application of the notion of grounding let us see how it can be used to place charge on a conductor. This process is known as *inducing charge* on a conductor or *charging it by induction*. Suppose, in Figure 19-6a, that a negatively charged rubber rod is brought up to an isolated conductor so that,

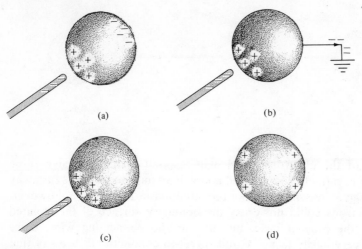

Figure 19-6

as shown, some of the electrons migrate to that side of the conductor farthest from the rod. If this conductor is now grounded, then some of the electrons will leave the original conductor and will no longer be of any direct interest. If the connection with the ground is then cut, the conductor is again isolated. Finally, as shown in Figures 19-6c and 19-6d, by removing the hard rubber rod, we obtain a conductor with a net positive charge on it.

Similarly, by use of a rubbed glass rod, a conductor containing a net negative charge can be produced.

19-5 Coulomb's law—qualitative aspects

To obtain a quantitative measure of the force between charged bodies, it is convenient to make use of small, conducting spheres. We shall say in this connection that two spheres are *small*, provided their radii are negligible compared to the distance separating them. In effect, then, small, conducting spheres may be thought of as charged particles, and in the following the term "particle" will always be used in this sense.

We shall now describe certain experiments that establish the following properties of the force between two charged particles:

1. The direction of the force lies along the line joining them.
2. The magnitude of the force varies inversely as the square of the distance between them.
3. The force is directly proportional to the charge on each particle.
4. For a given separation distance, the magnitude of the force, although not its sense, is independent of the sign of the charges.

Consider, in Figure 19-7, two charged particles, A and B, which have the respective masses m_1 and m_2 and are suspended by two insulating strings of

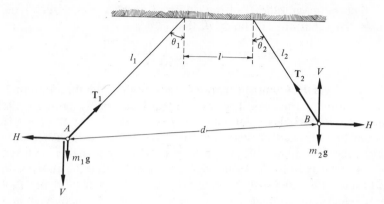

Figure 19-7

lengths l_1 and l_2 from a horizontal ceiling. Assuming, to be specific, that the particles have the same sign of charge, there is a repulsive force between them, and at equilibrium the strings will hang at certain angles, θ_1 and θ_2, with respect to the vertical. Besides their weights, m_1g and m_2g, each particle also experiences a force due to the tension in the strings (T_1 and T_2 in the figure) as well as a force **F** due to the electric repulsion between them. For convenience, only the horizontal and vertical components, H and V, respectively, of this electric force **F** have been drawn in the figure. Also, we have made use of Newton's third law, according to which the electric force on one of the particles is equal and opposite to that on the other.

Since both particles A and B are in static equilibrium under the combined action of these forces, it follows from the principles of mechanics that

$$-V + T_1 \cos \theta_1 - m_1g = 0$$
$$T_1 \sin \theta_1 - H = 0$$
$$T_2 \cos \theta_2 + V - m_2g = 0 \qquad \textbf{(19-1)}$$
$$H - T_2 \sin \theta_2 = 0$$

where the first two equations are the respective sums of the vertical and the horizontal components of the force acting on A, and the second two represent the corresponding quantities for B. If we eliminate the unknown tensions T_1 and T_2, the resultant two equations for the components V and H of the electric force may be solved in terms of the masses m_1 and m_2 and the two angles θ_1 and θ_2. Hence, since the four parameters m_1, m_2, θ_1, and θ_2 are easily measured independently, it follows that the strength of the electric force between the two particles can be obtained directly.

In terms of the setup in Figure 19-7, it is a straightforward matter to confirm the above properties of the electric force between two charged particles. By varying the distance l separating the points of suspension of the supporting strings, we can in effect vary the separation distance d between the particles. Carrying out a sequence of such experiments we find in this way that the magnitude F of the electric force **F** varies inversely as the square of the separation distance d; that is,

$$F = (V^2 + H^2)^{1/2} \propto \frac{1}{d^2}.$$

Further, since V and H are also separately measurable, the direction of **F** can also be measured in this way. The result is that **F** invariably lies along the line joining the two particles. Thus the first two properties of the electric force between two charged particles, as listed above, are established.

To confirm the fact that **F** is also proportional to the charge on each particle, we proceed as follows. Assuming, to be specific, that the charges on A and B in Figure 19-7 are positive, let us take, say, B and as in Figure 19-8 place it into electrical contact with a second originally uncharged and *identical* conducting sphere B'. It follows from symmetry that these two identical spheres must share the available electric charge equally; thus after the spheres are separated, the charge on B will be precisely half of its original value. If the experiment in Figure 19-7, with A still having its original charge, is now repeated, but with the charge on B decreased by a factor of two, we find that for any fixed value for d, the electric force **F** between them is halved. In this way, then, by varying the amount of charge on both spheres we may confirm that, in general, the force between two charged particles varies directly as the product of their charges. It is interesting to note that it is possible by use of this symmetry argument to deduce this direct proportionality without having specified a unit of electric charge. The question of units will be considered in Section 19-6.

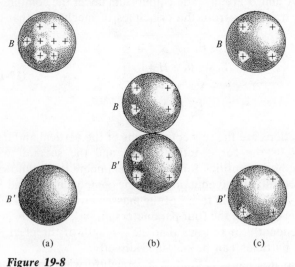

(a)   (b)   (c)

Figure 19-8

Finally, to confirm the fact that the magnitude of the force is independent of the sign of the charge, let us place equal, but opposite, charges on two

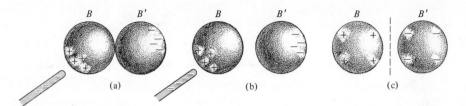

Figure 19-9

identical conducting spheres, B and B'. This may be done by placing the two conducting spheres into electrical contact, and then bringing up a charged rubber rod so that, as in Figure 19-9a, electrons migrate from B to B'. If the two spheres are separated and the rod is removed, then as shown in Figures 19-9b and c, the two conducting spheres will have equal and opposite charges. With a fixed charge on sphere A in Figure 19-7 we now carry out two successive experiments: one with sphere B, with its positive charge, and one with sphere B', with its equal but opposite negative charge. In this way, the fact that the magnitude of the force is independent of the sign of the charge is readily confirmed.

It should be noted that there are difficulties of a practical nature associated with the usage of the apparatus in Figure 19-7, and the above experiments should be thought of only as "thought experiments." Indeed, in ascertaining the properties of the force between charged particles, Coulomb (1736–1806) did not use this apparatus. Instead he used an apparatus similar to that shown in Figure 4-13, which was used by Cavendish to measure the gravitational constant G and is known as a torsion balance. Nevertheless, the experiments above are easy to understand and in principle may be carried out and used to justify quantitatively the stated properties of the electric force between small, charged bodies.

19-6 Coulomb's law—quantitative aspects

Based on the experiments of Section 19-5, we know that the electric force between two charged particles varies directly with the product of the charges and inversely with the square of the distance between them. In quantitative terms this means that if two particles of charges q and Q (to be defined below) are separated by a distance r, then the force **F** between them has the magnitude

$$F = k \frac{qQ}{r^2} \tag{19-2}$$

and its direction lies along the line joining the particles. As shown in Figure 19-10, if the charges have the same sign—that is, if they are either both positive or both negative, so that $Qq > 0$—then the force is repulsive. Correspondingly, if they are of the opposite sign, so that $Qq < 0$, then the

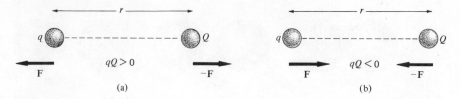

Figure 19-10

force is attractive. The proportionality constant k in (19-2) is arbitrary, and its detailed specification depends on the system of units adopted.

Although it is possible to do so, for technical reasons having to do mainly with the accuracy of experiments, it is *not* convenient to define a unit of charge in terms of (19-2) directly. As will be discussed in Chapter 26, it is more convenient first to define the SI unit of electric current of the *ampere* and then to define the associated unit of electric charge of the *coulomb* (C) as the amount of charge that is transported in 1 second along a wire through which flows a current of 1 ampere. A precise definition of the coulomb must therefore await our discussion of electric currents. In the interim, however, we shall make free use of this unit of electric charge as well as the related unit of the microcoulomb (μC), which is defined to be 10^{-6} coulomb. For purposes of orientation, let us note that the magnitude of the charge on a proton or an electron is 1.6×10^{-19} coulomb = 1.6×10^{-13} μC.

Once the unit of electric charge is established, the proportionality constant k in (19-2) is uniquely determined by experiment. It is customary in the SI system of units to introduce the symbol $1/(4\pi\epsilon_0)$ for this constant. Thus Coulomb's law in (19-2) becomes

$$F = \frac{1}{4\pi\epsilon_0} \frac{qQ}{r^2} \tag{19-3}$$

The constant ϵ_0 is called the *permittivity of free space*, and the factor 4π in this formula is for future convenience. In SI units, where F is measured in newtons, r is measured in meters, and q and Q are in coulombs, the constant $1/(4\pi\epsilon_0)$ has the experimental value[2]

$$\frac{1}{4\pi\epsilon_0} = 8.9875 \times 10^9 \frac{\text{N-m}^2}{\text{C}^2} \tag{19-4}$$

and the associated value for ϵ_0 itself is

$$\epsilon_0 = 8.8542 \times 10^{-12} \frac{\text{C}^2}{\text{N-m}^2}$$

For our purposes the approximate values

$$\frac{1}{4\pi\epsilon_0} = 9.0 \times 10^9 \frac{\text{N-m}^2}{\text{C}^2} \qquad \epsilon_0 = 8.9 \times 10^{-12} \frac{\text{C}^2}{\text{N-m}^2}$$

are usually adequate.

[2] More accurately, $10^7/4\pi\epsilon_0 \equiv c^2$, where c is the speed of light in vacuum.

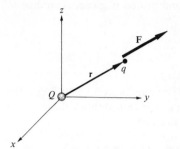

Figure 19-11

Even though force is a vector, in (19-3) the direction of action of this force is not explicitly stated. When making use of (19-3) it is always understood that this force is either attractive or repulsive, depending on the relative signs of the charges, and acts along the line joining the two particles.

For some purposes, it is convenient to rewrite Coulomb's law in its vector form. Suppose, in Figure 19-11, that the origin of a coordinate system is selected at the position of the particle of charge Q and that $\mathbf{r}$ represents the vector which describes the location of the other particle. In terms of these quantities, Coulomb's law for the force $\mathbf{F}$ acting on the particle of charge q is

$$\mathbf{F} = \frac{1}{4\pi\epsilon_0} qQ \frac{1}{r^3} \mathbf{r} \qquad (19\text{-}5)$$

since $\mathbf{r}/r$ is a unit vector parallel to the direction of $\mathbf{r}$. If the particles have charges of the same sign, then they repel each other, and the force $\mathbf{F}$ in this formula points along the direction of $\mathbf{r}$ as in the figure. On the other hand, if the charges are of opposite sign, then, according to (19-5), $\mathbf{F}$ points in the direction $-\mathbf{r}$. This is consistent with the fact that the particles attract each other in this case.

Example 19-1 Calculate the magnitude of the force of attraction between an electron and a proton separated by a distance of 0.5×10^{-10} meter, as they are in a hydrogen atom. Compare this with the force of attraction due to the action of gravity.

Solution We are given the data

$$q = -1.6 \times 10^{-19}\,\text{C} \qquad Q = +1.6 \times 10^{-19}\,\text{C} \qquad r = 0.5 \times 10^{-10}\,\text{m}$$

Since the charges are of opposite sign, the force is attractive, and its magnitude F according to (19-3) is

$$F = \frac{|qQ|}{4\pi\epsilon_0}\frac{1}{r^2} = 9.0 \times 10^9\,\frac{\text{N-m}^2}{\text{C}^2} \times \frac{(1.6 \times 10^{-19}\,\text{C})^2}{(0.5 \times 10^{-10}\,\text{m})^2}$$

$$= 9.2 \times 10^{-8}\,\text{N}$$

The gravitational force F_G between the electron and proton is given, according to Newton's law of universal gravitation, by

$$F_G = \frac{GMm}{r^2}$$

where $G = 6.67 \times 10^{-11}$ N-m^2/kg^2 is the gravitational constant, $m = 9.1 \times 10^{-31}$ kg is the electron mass, and $M = 1.67 \times 10^{-27}$ kg is the proton's mass. Substituting these values and using $r = 0.5 \times 10^{-10}$ meter, we obtain for F_G the value

$$F_G = 6.67 \times 10^{-11} \frac{\text{N-m}^2}{\text{kg}^2} \times \frac{(9.1 \times 10^{-31} \text{ kg}) \times (1.67 \times 10^{-27} \text{ kg})}{(0.5 \times 10^{-10} \text{ m})^2}$$

$$= 4.1 \times 10^{-47} \text{ N}$$

so the electric force between these particles is larger than the gravitational force by a factor of about 10^{40}! It is for this reason that the gravitational force between elementary particles can usually be neglected.

Example 19-2 Suppose the charge on the electron were not precisely the same in magnitude as that on the proton but, say, had the value $-(1 - 10^{-6}) = -0.999999$ times that of the proton.

(a) What would be the net charge associated with 1 mole of monatomic hydrogen?

(b) What would be the force of repulsion between 2 moles of hydrogen at a separation distance of 1 meter under these circumstances?

Solution

(a) In terms of Avogadro's number $N_0 = 6.0 \times 10^{23}$ atoms/mole, the charge on the protons is

$$N_0 \times 1.6 \times 10^{-19} \text{ C}$$

whereas that on the electrons would be

$$-N_0 \times (1 - 10^{-6}) \times 1.6 \times 10^{-19} \text{ C}$$

Thus the net charge Q_0 on 1 mole of hydrogen under these circumstances is

$$Q_0 = N_0 \times 1.6 \times 10^{-19} (1 - 1 + 10^{-6}) \text{ C}$$

$$= 6.0 \times 10^{23} \times 1.6 \times 10^{-19} \times 10^{-6} \text{ C}$$

$$= 9.6 \times 10^{-2} \text{ C}$$

(b) According to Coulomb's law, the force F of repulsion between two charges $Q_0 = 9.6 \times 10^{-2}$ coulomb at a separation distance of 1.0 meter is

$$F = \frac{Q_0^2}{4\pi\epsilon_0} \frac{1}{r^2} = 9.0 \times 10^9 \frac{\text{N-m}^2}{\text{C}^2} \times \frac{(9.6 \times 10^{-2} \text{ C})^2}{(1.0 \text{ m}^2)}$$

$$= 8.3 \times 10^7 \text{ N}$$

This is an enormous force and corresponds to approximately 10^4 tons!

19-7 Coulomb's law for collections of charged particles

Experiments analogous to those described in Figure 19-7, but involving a collection of more than two charged particles, show that the electric force acting on any one particle is the vector sum of the forces that would act on it if each of the other particles were the only other one present. For the case of three particles, for example, this means that the force on particle 1 is the vector sum of the forces produced on it if only particles 1 and 2 were present plus the force produced on it if only 1 and 3 were present. Thus to calculate the electric force on a charged particle in the presence of two or more others, we simply calculate the force due to each one by use of (19-5) and then add the results together vectorially.

As a special case consider the situation in Figure 19-12 of three particles of charges q_1, q_2, and q_3 of the same sign and separated by the distances r_{12}, r_{13}, and r_{23}, respectively. The force $\mathbf{F}_{12}$ on q_1 due to the presence of q_2 is directed as shown in the figure and has the magnitude

$$F_{12} = \frac{1}{4\pi\epsilon_0} \frac{q_1 q_2}{r_{12}^2}$$

Correspondingly, the force $\mathbf{F}_{13}$ on q_1 due to q_3 lies along the line joining q_1 and q_3 and has the magnitude

$$F_{13} = \frac{1}{4\pi\epsilon_0} \frac{q_1 q_3}{r_{13}^2}$$

The total force $\mathbf{F}_1$ on q_1 is, as shown in the figure, the vector sum of these:

$$\mathbf{F}_1 = \mathbf{F}_{12} + \mathbf{F}_{13} \tag{19-6}$$

The generalization of this formula to more than three charged particles is obvious and is illustrated in the examples which follow.

The relation in (19-6), which states that the force on a given particle is the vector sum of those produced by the other particles, is known as the *superposition principle*. Strictly speaking, this principle is valid only if the

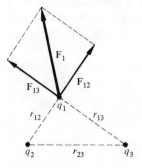

Figure 19-12

"particles" are actually point particles. For if they were, say, charged conducting spheres, then, as will be illustrated below, the electric force acting between spheres 1 and 2 will, in general, be different depending on the precise location of the third sphere. This follows from the fact that the presence of sphere number 3 will, in general, cause a redistribution of the electrons in spheres 1 and 2 and thus modify the electric force that the latter two exert on each other. Since in writing down (19-6) we have not made allowance for this possibility, this formula is of necessity restricted to point charges.

To illustrate this limitation on Coulomb's law, we shall now show that the force between two conducting spheres, only one of which carries an electric charge, is not zero as it would be for point particles. Consider, in Figure 19-13a, two conducting spheres A and B, which initially are separated by a distance much greater than their radii, and suppose that A has zero electric charge and B has a positive charge $Q \neq 0$. Since the spheres are separated by a very large distance, we may treat them effectively as point particles. Thus it follows from Coulomb's law that there is no force between them.

Let us now bring them closer together. As illustrated in Figure 19-13b, because of the positive charge on B some of the electrons in A will migrate to the side of the sphere closest to B. In other words, the presence of charged sphere B causes a redistribution of the charge in A. Because of the resultant charge separation and the inverse-square nature of Coulomb's law, it follows that there will now be a certain force of attraction $\mathbf{F}$ between the two spheres. Thus, even though A is overall electrically neutral, since B has a net charge and the separation distance between the spheres is comparable to their radii, they will attract each other. Of course, there is no real contradiction with (19-6), since it is valid only for point particles.

A similar failure of the superposition principle for large bodies can be seen in the following setup. Consider first the two uncharged conducting spheres, A and B, in Figure 19-14a. Since neither carries a charge, no charge redistribution takes place in either one, and there is no force between them. If, now, as in Figure 19-14b, a particle of, say, positive charge q is brought near, a rearrangment of charge within the two spheres takes place and they repel each other. That is, not only will each sphere experience a force of attraction due to the charged particle, but in addition each sphere will experience a force of repulsion due to the charge separation in the other

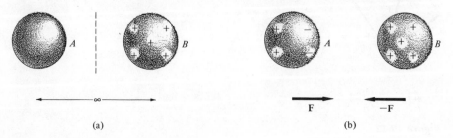

(a)   (b)

Figure 19-13

Figure 19-14

sphere. In other words, the presence of the charged particle causes an alteration in the force between the conducting spheres, again in seeming violation of the superposition principle. Note that this failure of superposition is due to the usage of bodies of finite dimensions; no such difficulties arise for point particles. In the following discussions we shall continue to restrict ourselves to a consideration of charged point particles; for these, Coulomb's law and the principle of superposition are always valid.

Example 19-3 Two negatively charged electrons, each of charge $-q$, are arranged as in Figure 19-15 near a positively charged alpha particle of charge $2q$. Assuming that $q = 1.6 \times 10^{-19}$ coulomb and $a = 1.0 \times 10^{-10}$ meter, calculate:
(a) The force on one of the electrons.
(b) The force on the positively charged alpha particle.

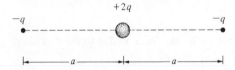

Figure 19-15

Solution Since the three particles lie along a straight line, it follows that all electric forces must also act along this direction. The force of repulsion F_r between the two electrons is given, according to Coulomb's law, by

$$F_r = \frac{q^2}{4\pi\epsilon_0}\left(\frac{1}{2a}\right)^2 = 9.0 \times 10^9 \frac{\text{N-m}^2}{\text{C}^2} \times \frac{(1.6 \times 10^{-19}\,\text{C})^2}{(2 \times 10^{-10})^2}$$
$$= 5.8 \times 10^{-9}\,\text{N}$$

The attractive force F_a between the alpha particle and either electron is

$$F_a = \frac{(+q)(2q)}{4\pi\epsilon_0}\frac{1}{a^2} = 9.0 \times 10^9 \frac{\text{N-m}^2}{\text{C}^2} \times \frac{2 \times (1.6 \times 10^{-19}\,\text{C})^2}{(1.0 \times 10^{-10}\,\text{m})^2}$$
$$= 4.6 \times 10^{-8}\,\text{N}$$

(a) The force on the electron on the right in Figure 19-15 is equal to the repulsive force $\mathbf{F}_r$ acting to the right plus the attractive force $\mathbf{F}_a$ due to the α particle acting to the left. Thus, the net force acting on it is to the *left* and has the magnitude

$$F = F_a - F_r = 4.6 \times 10^{-8}\,\text{N} - 5.8 \times 10^{-9}\,\text{N} = 4.0 \times 10^{-8}\,\text{N}$$

The force acting on the other electron has the same magnitude but acts to the right.

(b) The force on the α particle vanishes! This follows since the attractive force on it due to the electron on the right is just compensated for by the attractive force due to the other electron.

Example 19-4 Four particles, each of charge q, are located at the vertices of a square of side a. Calculate the force on one of these particles.

Solution The situation is shown in Figure 19-16. To calculate the force on, say, the particle located at the upper right-hand vertex, let us set up a Cartesian coordinate system as shown. Then the force F_1 due to the particle at the lower right-hand vertex is

$$F_1 = \frac{q^2}{4\pi\epsilon_0}\frac{1}{a^2}j$$

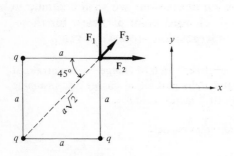

Figure 19-16

where j is a unit vector along the y-axis. Similarly, the force F_2 due to the particle at the upper left-hand vertex is

$$F_2 = \frac{q^2}{4\pi\epsilon_0}\frac{1}{a^2}i$$

Finally, the force F_3 due to the particle at the third vertex, which is at a distance $\sqrt{2}a$ from the given particle, is

$$F_3 = \frac{q^2}{4\pi\epsilon_0}\frac{1}{(\sqrt{2}a)^2}(i\cos 45° + j\sin 45°)$$

since the vector $(i\cos 45° + j\sin 45°)$ is a unit vector making equal angles with the x- and y-axes.

Collecting these results together, we find that the total force F is

$$F = F_1 + F_2 + F_3 = \frac{q^2}{4\pi\epsilon_0}\frac{1}{a^2}\left[i\left(1+\frac{\sqrt{2}}{4}\right)+j\left(1+\frac{\sqrt{2}}{4}\right)\right]$$

since

$$\cos 45° = \sin 45° = \frac{1}{2}\sqrt{2}$$

The magnitude of F is

$$F = \left(\sqrt{2}+\frac{1}{2}\right)\frac{q^2}{4\pi\epsilon_0}\frac{1}{a^2}$$

and its direction is the same as that of F_3. This latter feature is evident from the figure and the fact that the magnitudes of F_1 and F_2 are equal.

Example 19-5 A particle of charge q is on the perpendicular bisector and at a distance a away from a line charge of length $2l$ and of charge per unit length λ. Calculate the force on the particle.

Solution Consider this situation as shown in Figure 19-17. Let dF_x and dF_y represent the horizontal and vertical components of the force on the particle due to the charge $\lambda\,dx$ in an element of length dx at a distance x away from the y-axis. According to Coulomb's law,

$$dF_x = \frac{-q}{4\pi\epsilon_0}\,\lambda\,dx\,\frac{1}{a^2+x^2}\sin\theta$$

$$dF_y = \frac{q}{4\pi\epsilon_0}\,\lambda\,dx\,\frac{1}{a^2+x^2}\cos\theta \qquad\qquad \textbf{(19-7)}$$

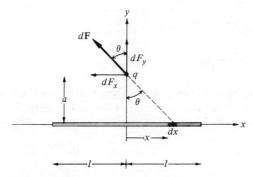

Figure 19-17

where the angle θ is as defined in the figure and the minus sign in the first equation reflects the fact that $d\mathbf{F}$ has a negative component along the x-axis. Reference to the figure shows that

$$\cos\theta = \frac{a}{(a^2+x^2)^{1/2}} \qquad \sin\theta = \frac{x}{(a^2+x^2)^{1/2}}$$

and thus (19-7) may be reexpressed in the form

$$dF_x = -\frac{\lambda q}{4\pi\epsilon_0}\frac{x}{(a^2+x^2)^{3/2}}\,dx$$

$$dF_y = \frac{\lambda q}{4\pi\epsilon_0}\frac{a}{(x^2+a^2)^{3/2}}\,dx \qquad\qquad \textbf{(19-8)}$$

The total force $\mathbf{F}$ and its two components F_x and F_y are obtained by adding together contributions of the form in (19-8) for all elements of the line charge. Expressing this in the form of an integral, we find that

$$F_x = \int_{-l}^{l} dF_x = -\frac{\lambda q}{4\pi\epsilon_0}\int_{-l}^{l}\frac{x}{(a^2+x^2)^{3/2}}\,dx$$

$$F_y = \int_{-l}^{l} dF_y = \frac{\lambda q a}{4\pi\epsilon_0}\int_{-l}^{l}\frac{dx}{(a^2+x^2)^{3/2}} \qquad\qquad \textbf{(19-9)}$$

where all constants have been taken out from under the integral signs. Reference to a table of integrals shows that

$$\int \frac{x}{(x^2 + a^2)^{3/2}} \, dx = -\frac{1}{(a^2 + x^2)^{1/2}}$$

$$\int \frac{dx}{(x^2 + a^2)^{3/2}} = \frac{x}{a^2} \frac{1}{(a^2 + x^2)^{1/2}}$$

and making use of these values, the integrals in (19-9) are readily evaluated as follows:

$$F_x = -\frac{\lambda q}{4\pi\epsilon_0} \left[-\frac{1}{(x^2 + a^2)^{1/2}} \right] \bigg|_{-l}^{l} = 0$$

$$F_y = \frac{\lambda q a}{4\pi\epsilon_0} \frac{1}{a^2} \left[\frac{x}{(x^2 + a^2)^{1/2}} \right] \bigg|_{-l}^{l}$$

$$= \frac{\lambda q}{4\pi\epsilon_0 a} \left[\frac{l}{(a^2 + l^2)^{1/2}} - \frac{-l}{(a^2 + l^2)^{1/2}} \right]$$

$$= \frac{2\lambda q l}{4\pi\epsilon_0 a (a^2 + l^2)^{1/2}}$$

The fact that $F_x = 0$ follows also by symmetry, since the x-component of the force due to the charge at the distance x cancels out that contribution due to the charge at $-x$. Or, equivalently, $F_x = 0$ follows since the integrand in the first equation of (19-9) is an odd function.

19-8   Limitations on charge

In the above discussion of electric charge and Coulomb's law, we have implicitly assumed that the electric charge q on a conductor can assume arbitrary positive and negative values. This is not correct! There is both an upper and a lower limit on the magnitude of the net amount of charge that can exist on a conductor.

The impossibility of having a very large amount of charge on a conductor is easy to see. The strong repulsive forces which would act between its components would produce severe stresses within the conductor and would ultimately cause it to explode. Normally this is not a problem, since, for a variety of practical reasons, it is not feasible to add enough charge to a rigid conductor to do this.

At the other end of the scale, we find an unambiguous lower limit to the smallest amount of charge that can be placed on a body. Experiment shows that there exists in nature a smallest subdivision of electric charge, and that all charge is an integral multiple of this smallest subdivision. We describe this by saying that electric charge is *quantized*, and refer to the smallest unit of charge as a *quantum of charge*. This quantum, for which the symbol e will be used, has the experimental value

$$e = 1.60219 \times 10^{-19} \, C \qquad\qquad \text{(19-10)}$$

Note that no one in the laboratory or elsewhere has ever observed an amount of charge smaller in magnitude than this value. Nor has anyone ever observed an amount of charge that is not a positive or negative integral multiple of this quantum e. The charge of the electron (e), the negative pion (π^-), the negatively charged kaon (K^-), the omega-minus (Ω^-), the antiproton ($\bar{p}$), and all other negatively charged elementary particles have precisely the charge $-e$. Similarly, the charge on the proton (p), the positron (e^+), the positive pion (π^+), the positively charged kaon (K^+), the sigma-plus (Σ^+), and all other positively charged elementary particles have the precise value $+e$. No deviations from the values $\pm e$ or integral multiples of $\pm e$ have ever been observed.

Despite this fact that no one has ever observed a particle the magnitude of whose charge is less than e, certain theoretical proposals have been put forth, mainly by M. Gell-Mann, the 1969 Nobel Prize winner, according to which all observed particles are composites made up of certain fundamental particles whose charges are multiples of $\pm\frac{1}{3}e$. These particles have been given the name "quarks." According to these ideas, there are six distinct quarks (three ordinary ones, of charges $-\frac{1}{3}e$, $-\frac{1}{3}e$, and $+\frac{2}{3}e$, and three antiquarks, with the opposite signs of charge). For example, the positive pion (π^+) is a composite consisting of a quark of charge $+\frac{2}{3}e$ and an antiquark with charge $+\frac{1}{3}e$. By their very nature, quarks would be exceedingly difficult to detect, and as of this writing, and despite many efforts, they have not yet been observed. Until such time when they are, quarks must remain in the nature of a theoretical speculation, and for the present we can assume that the quantum of charge in (19-10) is the smallest subdivision of electric charge.

A question related to the existence of the quantum of charge is whether or not the positive quantum of charge has precisely the same magnitude as the negative one. That is, are the charges on the electron and proton exactly equal and opposite? A partial answer to this question has been previously given in Example 19-2, where it was shown that if the charge on an electron differed by only 1 part per million from the value $-e$, then extraordinarily large repulsive forces would exist between electrically isolated bodies. Since forces of such a large magnitude would be easily observable, from the fact that they have not been seen it follows that the charge on the electron and the proton must be very nearly, if not precisely, equal and opposite.

If it is assumed then that the charge on the electron is precisely equal to that on the proton, the results of many experiments may be summarized by a conservation law known as *the law of the conservation of charge*. According to this law, the algebraic sum of the electric charges involved in any process is conserved. If, for example, an atom is ionized, the initial electric charge is zero, as is the algebraic sum of the charge of the electron and ion $e + (-e) = 0$ afterward. Or equally, if a certain positive electric charge is placed on a conductor, then an equal amount of negative charge must appear somewhere else. This principle of charge conservation, which has been implicitly assumed to be valid in the above discussions, is very firmly established experimentally and no one has ever observed a violation of it under any circumstances.

†19-9   Limitations on the concept of a point charge

It has been noted previously that two small charged conducting spheres will satisfy Coulomb's law only if their separation distance is very large compared to their radii. For if they come close together, the charge distribution in each one will be influenced by that on the other, and the force between them will be modified accordingly. Nevertheless, it used to be believed not very long ago that for microscopic particles, such as electrons, protons, and nuclei, Coulomb's law would be applicable for arbitrarily small separation distances.

After the Stanford linear accelerator became operational in the 1950s it became possible for the first time to probe very small separation distances by use of energetic electrons. It was discovered by Robert Hofstadter and collaborators that the nucleus of an atom is not at all a point particle, as seemed to be implied by other experiments. But rather they found that the nucleus of each atom has an extended charge distribution; that is, that the positive charge on a nucleus is not concentrated at a point but is distributed more or less uniformly within a sphere whose radius is directly proportional to the cube root of the atomic number. Figure 19-18 shows the charge density—that is, the charge per unit volume—of several nuclei as a function of the distance r from their centers. Note that in the central region of the nucleus the charge density is constant but that it drops off very rapidly toward the outer edge of the nucleus.

It might be argued that since a heavy nucleus is actually a compound structure consisting of uncharged neutrons and positively charged protons,

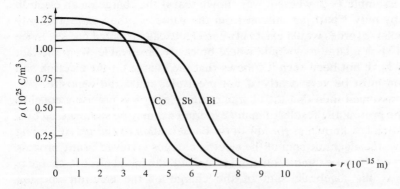

Figure 19-18   *The charge density $\rho(r)$ of the nuclei Co, Sb, and Bi as a function of the radial distance r from the center. (Abstracted from the article by Robert Hofstadter, Ann. Rev. Nuc. Sci., 7, 231 (1957).)*

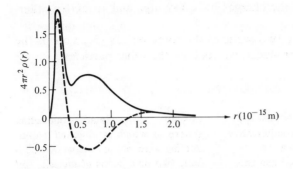

Figure 19-19 *The weighted charge densities $4\pi r^2\rho(r)$ of the proton (solid curve) and the neutron (dashed curve) as a function of radial distance. The total area under the dashed curve is zero since the neutron is electrically neutral. (Abstracted from the article by D. N. Wilson, H. F. Shopper, and R. R. Wilson, Phys. Rev. Lett. 6, 286 (1961).)*

results such as those in Figure 19-18 are not surprising. If on the other hand one were to probe, in the same way, the charge structure of a single proton or a neutron, then indeed they would be found to be *bona fide* point particles. Again, experiments carried out by Hofstadter demonstrated that even this is not true. But rather, as shown by the solid curve in Figure 19-19, the proton itself has a charge density $\rho(r)$, which extends over a distance $r \cong 10^{-15}$ meter. Also included on the same graph by a dashed curve is the charge density for the neutron. Even though the neutron is electrically neutral, it was discovered by scattering energetic electrons from deuterium that the neutron also has a charge structure. Specifically, for distances $r \lesssim 0.5 \times 10^{-15}$ meter, it was found to have a positive charge density, while for larger distances this charge density is negative by an appropriate amount so that the total charge Q_0

$$Q_0 = \int_0^\infty 4\pi r^2 dr\rho(r) \qquad \text{(neutron)}$$

vanishes.

At the present time it is generally believed that the electron, the muon, and the neutrino are point particles but that all other elementary particles have an extended structure of some type. It may well turn out that electrons and muons themselves also have an extended charge structure, but as yet no one has been able to suggest an experiment to resolve this very basic problem.

19-10 Summary of important formulas

The force **F** between two particles of charges q and Q separated by a distance r lies along the line joining the particles and has a magnitude given by Coulomb's law

$$F = \frac{1}{4\pi\epsilon_0}\frac{qQ}{r^2} \qquad (19\text{-}3)$$

The force **F** is repulsive if the charges have like sign and attractive otherwise.

For the case of more than two particles the force on any given one is the vector sum of the forces produced by each of the other particles.

QUESTIONS

1. A rubbed glass rod with its positive charge is brought near a spherical conductor suspended from an insulated string. It is observed that the sphere moves toward the rod, but that after making contact it jumps away and appears to be repelled by it. Explain these observations in terms of electron migrations within the conductor.

2. Consider the same situation as in Question 1, but suppose this time that the sphere is suspended by a conducting wire so that it is grounded. What is the behavior of the sphere now? Explain your answer and confirm your prediction by carrying out an experiment.

3. The essential components of an elec-troscope are a small, conducting sphere to which is attached a conducting wire, at the bottom of which hang two light strips of metallic foil (Figure 19-20). The strips are called leaves and are presumed to be far away from the sphere. It is observed as depicted in Figure 19-20b that if a rubbed glass rod is brought near the sphere, then the leaves of the electroscope diverge. Explain this effect in terms of electron motions.

4. In Figure 19-20b, what is the sign of the charge on the conducting sphere? What is the sign of the charge on the leaves?

5. A rubbed hard-rubber rod is brought near the conducting sphere of an electroscope. Explain what happens.

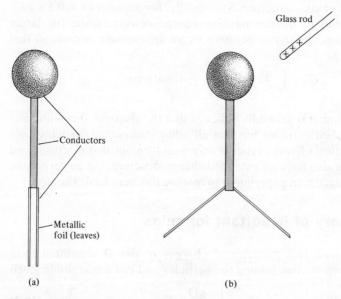

Glass rod

Conductors

Metallic
foil (leaves)

(a)   (b)

Figure 19-20

What is the sign of the charge on the sphere and on the leaves?

6. Consider again the situation in Figure 19-20b, but suppose that while the leaves are extended we ground the electroscope by touching the conducting sphere. Explain why the leaves now will collapse even though the rubbed glass rod is still near the sphere.

7. Subsequent to the collapse of the leaves in Question 6, suppose that the sphere is again disconnected from the ground. What will happen to the leaves? If the glass rod is removed, why will the leaves now diverge?

8. In Example 19-1 it was shown that the electric force between an electron and a proton is of the order of 10^{40} times greater than is their gravitational force. Explain why, despite this, it was necessary to include the weights of the two spheres in analyzing the setup in Figure 19-7.

9. Show that if the magnitudes of the charges on the electron and the proton differed by one part in 10^{18}, then the resultant electric force between two hydrogen atoms is comparable to their gravitational force. Why is this *not* a possible explanation for gravitational forces?

10. Suppose that an electroscope acquires an electric charge by the process outlined in Question 6. If the electroscope is allowed to remain in this state for some time, it is observed that the leaves collapse. What must have happened? Explain.

11. Explain why the leaves of a charged electroscope will collapse more readily when it is at a high elevation, say on the top of a mountain, than when it is at sea level. (*Note*: The first indications of the existence of cosmic rays came about as a result of this observation.)

12. In the construction of electroscopes, it is customary to surround the region below the sphere (see Figure 19-20a) by a conductor that is insulated from the metallic parts of the electroscope and is grounded. Can you think of a reason why this is desirable?

13. A particle of charge q is at a certain distance a away from a particle also of charge q. Where must a third charged particle be placed so that it experiences no electric force? Will this third particle exert a force on either of the others?

14. A very small insulating sphere is suspended from an insulating thread. Devise one experiment to determine whether the sphere is charged and one to ascertain the sign of this charge.

15. Repeat Question 14, but suppose this time that a *conducting* sphere is suspended from an insulating thread.

16. Present an argument that establishes that the net electric charge in the universe is "essentially" zero.

17. Two conducting spheres are found to attract each other with a certain electric force. Assuming that other charged bodies may be nearby, need the spheres necessarily have an electric charge? What about the case in which the spheres repel each other electrically?

PROBLEMS

In the following, use for the basic quantum of charge the value

$$\pm 1.6 \times 10^{-19}\,\text{C}$$

and for the constant $1/4\pi\epsilon_0$ the value

$$9.0 \times 10^9\,\frac{\text{N-m}^2}{\text{C}^2}$$

In specifying a force, remember that both a magnitude and a direction are required.

1. (a) Calculate the amount of positive charge contained on all of the protons of 1 mole of monatomic hydrogen! (*Note*: This electric charge, equal to about 9.65×10^4 coulombs, is called the *faraday*. It represents the electric charge on 1 mole of any singly ionized substance.) (b) What is the amount of charge on the electrons in this gas?

2. Suppose that the protons and the electrons in 1 mole of monatomic hydrogen are separated and placed on opposite ends of the earth. What would be the force between them? (Assume that $R_E = 6.5 \times 10^6$ meters.)

3. Calculate the electric force between the two protons in a helium nucleus, assuming that their separation distance is 2×10^{-15} meter. Based on this result, what can you say about the strength of the nuclear forces between two protons when they are at this separation distance?

4. Two particles, each of charge $2 \mu C$, are suspended at a separation distance of 30 cm by insulating strings each of length 0.5 meter (Figure 19-7). It is found that each of the strings make an angle of 30° with the vertical. Calculate (a) the mass of each particle, assuming that they are the same and (b) the tensions in the strings.

5. A nucleus of ^3H β-decays to ^3He according to the reaction

$$^3H \rightarrow {}^3He + e^- + \nu_e$$

where ν_e stands for the electrically neutral neutrino. If the separation distance between the electron and the ^3He nucleus immediately after the decay is 2×10^{-15} meter, calculate the force between them at this instant.

6. How much positive electric charge must be added to the sun and to the earth so as to nullify the gravitational force of attraction between them? Assume the following values: $M_S = 2.0 \times 10^{30}$ kg, $M_E = 6.0 \times 10^{24}$ kg, and $G = 6.67 \times 10^{-11}$ N-m²/kg².

7. What would the charge on a proton have to be so that the electric and the gravitational forces between two protons cancel one another?

8. Two particles of charges q_1 and q_2 ($q_1 > q_2 > 0$) are separated by a certain distance d. Suppose that an amount of charge q is transferred from q_1 and q_2 so that the resultant charges are $(q_1 - q)$ and $(q_2 + q)$. For what value of q is the force of repulsion between the particles a maximum?

9. Show that if (x, y, z) are the coordinates of q in Figure 19-11, then the x-component of the force **F** is

$$F_x = \frac{qQ}{4\pi\epsilon_0} \frac{x}{[x^2 + y^2 + z^2]^{3/2}}$$

and find the corresponding forms for F_y and F_z.

10. Three particles, each of charge $3.0 \mu C$, lie along a straight line at intervals of 0.5 meter. (a) Calculate the force on the middle particle. (b) Calculate the force on one of the end particles.

11. Particles having charges $-1.0 \mu C$, $+2.0 \mu C$, $+4.0 \mu C$ are located at the vertices of an equilateral triangle having sides 1 meter long. Calculate the magnitude of the force on (a) the $-1.0 \mu C$ particle; (b) the $+4.0 \mu C$ particle; and (c) the $+2.0 \mu C$ particle. (d) Show that the vector sum of these three forces is zero.

12. Two particles, of respective charges $-3.0 \mu C$ and $+4.0 \mu C$, are at a separation distance of 0.3 meter. Where, along the line joining them, can a third charge be placed so that it experiences no electric force?

13. Two particles, each of charge $+q$, are placed on diagonally opposite vertices of a square of side a, and

two particles each of charge $-q$ are placed at the remaining vertices. Calculate **(a)** the force on one of the particles of charge q and **(b)** the force on one of the negatively charged particles.

14. A particle of charge q and mass m is attached to a spring of constant k and of equilibrium length l_0. Suppose that a second particle of charge $-q$ is brought within a distance a of the place where the spring is attached (Figure 19-21). Show that when equilibrium is reached, the spring has been stretched by an amount d, which satisfies the equation

$$\frac{q^2}{4\pi\epsilon_0}\frac{1}{(a - l_0 - d)^2} = kd$$

What is the period of oscillations for small displacement from this equilibrium configuration? Assume that $a > l_0 + 3d$.

Figure 19-21

15. A particle of charge Q lies midway between two fixed identical charges of the same sign, each of magnitude q and separated by a distance $2b$.
   (a) What is the force on Q?
   (b) Suppose this particle is displaced from its original position by an amount y as shown in Figure 19-22. What is now the force acting on the particle?
   (c) Show that if $y \ll b$, then if the

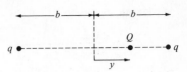

Figure 19-22

particle is released from its position in Figure 19-22, it will oscillate with simple harmonic motion at angular frequency ω, given by

$$\omega = \left(\frac{qQ}{\pi\epsilon_0 m}\frac{1}{b^3}\right)^{1/2}$$

where m is the mass of the particle.

16. Repeat (b) and (c) of Problem 15, but this time assume that the particle is displaced along the perpendicular bisector of the line joining the two fixed particles. Assume also that $Qq < 0$.

17. Two particles of charges $+q$ and $-q$ are placed at the respective points $(a, 0, 0)$, $(-a, 0, 0)$ in a certain coordinate system. Show that the force on a charged particle of charge $Q (> 0)$ located somewhere on the y-z plane (with coordinates $(0, y, z)$) points along the negative x-axis. What is the strength of this force as a function of y and z?

18. A particle of charge q lies along the axis of a uniform line charge of length l and of charge per unit length λ. Assume that the particle is at a distance a from the nearer end of the line charge as shown in Figure 19-23.

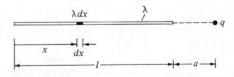

Figure 19-23

   (a) Show that the force dF on the particle due to an element of length dx is

$$dF = \frac{q\lambda \, dx}{4\pi\epsilon_0} \frac{1}{(a + l - x)^2}$$

(b) By integrating over all values of x from 0 to l calculate the total force on the particle.

*19. Eight particles, each of charge q, are distributed at relative angles of $2\pi/8 = \pi/4$ around a circle of radius a. A particle of charge Q is located on the axis of the circle and at a distance b from its center. Show that the magnitude of the force on Q is

$$\frac{2qQ}{\pi\epsilon_0} \frac{b}{(a^2 + b^2)^{3/2}}$$

and find the direction of this force (*Hint:* Recall that force is a vector quantity and thus vector addition must always be used.)

20. Calculate the force on the particle in Figure 19-23, but assume now that λ is given by

$$\lambda = \lambda_0 \left(1 - \frac{2x}{l}\right)$$

where λ_0 is a constant.

*21. A particle of charge Q is on the axis of a circular loop, of radius a and carrying a uniform charge per unit length λ.

(a) Calculate the horizontal and vertical components of the force $d\mathbf{F}$ due to an element of charge of length $a \, d\theta$ (Figure 19-24).

(b) Calculate the total force acting on Q.

*22. Calculate the magnitude of the force on the particle in Figure 19-24, but

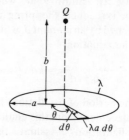

Figure 19-24

assume this time that the charge per unit length λ is

$$\lambda = \lambda_0 \sin\frac{\theta}{2}$$

where $\lambda_0 = $ constant.

20 The electrostatic field and Gauss' law

I have yet to see any problem, however complicated, which when you looked at it the right way did not become still more complicated.

P. ANDERSON

20-1 Introduction

Electrostatics is the study of charged particles at rest. Since the electric force between two charged particles is given directly by Coulomb's law, it might appear at first that with the statement of this law in Chapter 19 our study of electrostatics would be at an end. In actual fact nothing is farther from the truth. Coulomb's law, though basic to electrostatics, is only a first doorway to knowledge of this field.

Now the reason why Coulomb's law is only the beginning of electrostatics is due mainly to the fact that in the laboratory we are usually concerned with the behavior of electric charge in the presence of matter. Experiment shows that, in general, whenever electric charge is brought near matter, additional electric charges are induced in it. And since the distribution of this induced charge is not known *a priori*, it is not possible to apply Coulomb's law directly to these physically more interesting situations. If, for example, a charged particle is brought near an uncharged conducting body, the electric charges—that is, the ions and electrons—in the conductor will experience electric forces which cause a redistribution of charge within the body. In order to be able to apply Coulomb's law to this case it is necessary to know precisely how this charge is distributed within the conductor. Only after this information has been obtained can Coulomb's law be used to calculate the force on the given particle.

Consider, for example, the problem of a particle of charge q (>0) near an electrically neutral but isolated conducting sphere (Figure 20-1). The presence of the positively charged particle causes some of the free electrons inside to migrate to the nearer side of the sphere and to leave a compensating positive charge on the opposite side. Now if the charge distribution on the sphere were known in detail, it would be possible, in principle, to calculate by use of Coulomb's law, the force between the sphere and the particle. We would multiply each element of charge in the sphere by $q/4\pi\epsilon_0$, divide by the square of the distance between the particle and the given element, and carry out a vector summation over all such charge elements. Unfortunately, the charge distribution on the sphere is not known and thus it is not possible to proceed this way.

Similarly, if a charged particle is brought near an insulator a redistribution of electric charge within the insulator also takes place. Again, until the nature of the resultant charge has been ascertained, we cannot calculate the force of attraction between the particle and the insulating body.

Now mainly for this purpose of describing in a quantitative way how electric charge in matter responds to an external charged body, it is convenient to introduce the concept of the *electric field*. As we shall see, not only is this physical construct of decisive utility in the solution of problems in electrostatics, but it also has a certain physical reality of its own. For the moment, let us focus attention on problems of electrostatics and think of the electric field simply as a convenient way of characterizing the distribution of charge in ordinary matter when it is subjected to an electric force. The more profound aspects of the electric field, and in particular its indisputable reality, will become apparent only after we have studied time-dependent phenomena.

20-2 The electric field

Consider, in Figure 20-2, a collection of particles of charges, q_1, q_2, ... Let us bring up to these charges a *test particle* of (positive) charge q_0 and measure the electric force $\mathbf{F}$ that it experiences. In general, this force will be due to the charge on the given particles plus that induced in any matter that may be nearby. We define, provisionally, the *electric* or the *electrostatic field* $\mathbf{E}_0$ (due to the given charges) at the location of the test particle to be the force per unit charge on the test particle; that is,

$$\mathbf{E}_0 = \frac{1}{q_0}\mathbf{F} \tag{20-1}$$

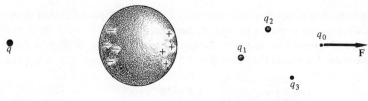

Figure 20-1   Figure 20-2

Now because of the fact that charge may be induced in matter, this provisional definition of the electric field is not completely satisfactory. The difficulty is that E_0 depends not only on the given charge distribution but also on the strength of the test particle and its associated ability to induce charge in nearby matter. This means, for example, that E_0 depends on the strength of the charge q_0 of the test particle and is not associated solely with the given collection of charged particles and the charge that these induce on nearby matter. Accordingly, we shall define the electric field E for the given charge distribution not by (20-1) but by its limiting form as the charge q_0 of the test particle becomes vanishingly small. Thus

$$E = \lim_{q_0 \to 0} \frac{1}{q_0} F \qquad (20\text{-}2)$$

where F is the force on the test particle. The advantage of the definition in (20-2) is that the presence of the test particle with its very small charge will not cause any redistribution of charge in nearby matter. Thus the electric field as here defined is a property *only of the given charge distribution*. Note that, in general, the force F on the test particle varies with position, so the electric field E will also vary from point to point in space.

Consider a region in space in which there exists an electric field E. If a particle of charge q is introduced into this region, it will experience the force qE, provided that q is sufficiently small that no additional charges are induced in nearby matter. Otherwise the force on it will not be qE but rather qE', with E' the electric field due to the original charge plus that induced on nearby matter by the charge q on the particle. Unless a statement is made explicitly to the contrary, it will always be assumed that the charge q is sufficiently small that the force on it is qE.

The unit of the electric field is that of force per unit charge. Hence the unit of E is the newton per coulomb (N/C).

Example 20-1 An electron of charge $q = -1.6 \times 10^{-19}$ coulomb is located in a region of space where there exists a uniform electric field of strength $E = 10^5$ N/C. Calculate:

(a) The force on the electron.

(b) Its acceleration.

Solution

(a) The force F on the electron is

$$F = qE = -1.6 \times 10^{-19} \, C \times 10^5 \, N/C$$
$$= -1.6 \times 10^{-14} \, N$$

The minus sign reflects the fact that this force is in a direction opposite to that of the electric field owing to the negative charge of the electron.

(b) According to Newton's second law, the acceleration a of the electron is F/m. Using the above value for F and the value $m = 9.1 \times 10^{-31}$ kg for the electron, we find that

$$a = \frac{1.6 \times 10^{-14} \text{ N}}{9.1 \times 10^{-31} \text{ kg}} = 1.8 \times 10^{16} \text{ m/s}^2$$

This acceleration is directed along the force and is thus opposite to that of the electric field.

Example 20-2 Two particles of charges q and $-q$ are attached to a rigid rod of length $2b$, with the entire structure in a uniform electric field **E** (Figure 20-3). Calculate:

(a) The force on the structure.

(b) The torque about the center of the rod, assuming that it is at an angle θ with respect to the direction of the electric field.

Solution

(a) Reference to the figure shows that for $q > 0$, the force on the upper particle, $q\mathbf{E}$, points along the direction of the field and that on the lower particle, $-q\mathbf{E}$, is equal and opposite. Hence the net force on the structure is zero.

(b) To calculate the torque τ on the structure about its center, recall from our studies in mechanics that the torque τ produced by a force **F** about a given origin is

$$\boldsymbol{\tau} = \mathbf{r} \times \mathbf{F}$$

where **r** is a vector from the origin to the point at which the force acts. According to this definition, the force on the upper particle produces a torque τ directed down into the plane of the diagram and with magnitude

$$\tau = bqE \sin \theta$$

since $|\mathbf{r}| = b$, $|\mathbf{F}| = qE$ and the angle between these vectors is θ. Applying the definition a second time but to the lower particle, we find precisely the same torque. Hence the torque τ acting on the structure is directed perpendicularly down into the plane of Figure 20-3 and has the magnitude

$$\tau = 2bqE \sin \theta$$

20-3 The electric field of discrete particles

In this section and the next we make use of the definition in (20-2) to derive formulas for the electric field **E** associated with certain charge distributions.

Consider first the case of a single particle of charge q. To determine the electric field **E** at a *field point*, which is at a displacement **r** from the particle, let us bring up a small test particle of charge q_0 (> 0) (Figure 20-4). According to Coulomb's law, the force **F** on the test particle is

$$\mathbf{F} = \frac{1}{4\pi\epsilon_0} \frac{qq_0}{r^2} \hat{\mathbf{r}}$$

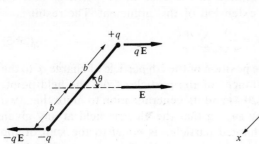

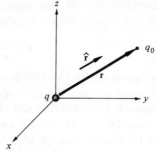

Figure 20-3   **Figure 20-4**

where $\hat{\mathbf{r}}$ is a unit vector along the direction from q to the test particle. Substitution into (20-2) yields

$$\mathbf{E} = \lim_{q_0 \to 0} \frac{\mathbf{F}}{q_0} = \lim_{q_0 \to 0} \frac{1}{q_0} \frac{1}{4\pi\epsilon_0} \frac{qq_0}{r^2} \hat{\mathbf{r}}$$

$$= \frac{q}{4\pi\epsilon_0} \frac{\hat{\mathbf{r}}}{r^2}$$

(20-3)

so for $q > 0$ the electric field is directed radially outward from the particle and varies inversely with the square of the distance from it. For a negatively charged particle the spatial variation is the same, but the field is directed radially inward toward the particle.

In a similar way we can obtain a formula for the electric field associated with more than one particle. Figure 20-5 shows a test particle of charge q_0 at a point that is at the distances r_1 and r_2 from two fixed particles of respective

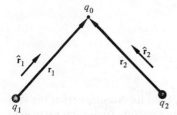

Figure 20-5

charges q_1 and q_2. Defining $\hat{\mathbf{r}}_1$ and $\hat{\mathbf{r}}_2$ to be the respective unit vectors from q_1 and q_2 to q_0, and making use of the superposition principle, we obtain

$$\mathbf{F} = \frac{1}{4\pi\epsilon_0} \frac{q_0 q_1}{r_1^2} \hat{\mathbf{r}}_1 + \frac{1}{4\pi\epsilon_0} \frac{q_0 q_2}{r_2^2} \hat{\mathbf{r}}_2$$

and thus, according to the definition in (20-2), the electric field $\mathbf{E}$ at the position of q_0 is

$$\mathbf{E} = \frac{1}{4\pi\epsilon_0} \left[\frac{q_1}{r_1^2} \hat{\mathbf{r}}_1 + \frac{q_2}{r_2^2} \hat{\mathbf{r}}_2 \right]$$

(20-4)

The generalization of this formula to the case of N particles may be obtained by a straightforward extension of this argument. The result is

$$\mathbf{E} = \frac{1}{4\pi\epsilon_0} \sum_{i=1}^{N} \frac{q_i \hat{\mathbf{r}}_i}{r_i^2} \tag{20-5}$$

where $\hat{\mathbf{r}}_i$ is a unit vector from the position of the ith particle of charge q_i to the field point, and $\{r_i\}$ are the distances of the particles from the field point.

It is interesting to note that (20-4) and its generalization to more than two particles may be interpreted by saying that the electric field at any given point due to a collection of charged particles is equal to the sum of the electric fields produced by each particle separately. This property, which is a direct consequence of the corresponding feature of Coulomb's law and (20-2), is usually described by saying that the *superposition principle* is also valid for the electric field. That is, the electric field due to a collection of particles is the sum of the fields produced by each particle separately.

Example 20-3 Two identical particles, each of charge q, are separated by a distance $2a$. Calculate the electric field at a point that lies:
(a) On the line joining the particles but is not between them.
(b) On the perpendicular bisector of the line joining them.

Solution Let us set up a coordinate system with the two particles located along the x-axis so that their coordinates are, respectively, $(a, 0)$ and $(-a, 0)$; see Figure 20-6.

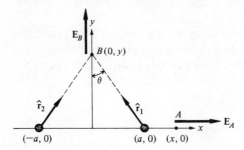

Figure 20-6

(a) If A is the field point with coordinates $(x, 0)$, then its distances from the two particles are $(x - a)$ and $(x + a)$, respectively. The unit vectors from each particle to the field point are parallel to the x-axis and thus to the unit vector $\mathbf{i}$ along this axis. Substituting these data into (20-5) and assuming that $x > a$, we obtain for the electric field $\mathbf{E}_A$ at point A

$$\mathbf{E}_A = \mathbf{i}\frac{q}{4\pi\epsilon_0}\left(\frac{1}{(x-a)^2} + \frac{1}{(x+a)^2}\right)$$

which as shown in the figure lies along the line joining the two particles.

(b) In a similar way, the electric field $\mathbf{E}_B$ at the point B with coordinates $(0, y)$ is

$$\mathbf{E}_B = \frac{q}{4\pi\epsilon_0}\left(\hat{\mathbf{r}}_1\frac{1}{y^2+a^2} + \hat{\mathbf{r}}_2\frac{1}{y^2+a^2}\right)$$

$$= \frac{q}{4\pi\epsilon_0}\frac{1}{y^2+a^2}(\hat{\mathbf{r}}_1 + \hat{\mathbf{r}}_2)$$

since the distance of each particle from the field point is $(a^2 + y^2)^{1/2}$. The unit vectors $\hat{\mathbf{r}}_1$ and $\hat{\mathbf{r}}_2$ are as defined in the figure, and hence the sum $(\hat{\mathbf{r}}_1 + \hat{\mathbf{r}}_2)$ is a vector parallel to the unit vector $\mathbf{j}$ along the y-axis. Since each of the unit vectors $\hat{\mathbf{r}}_1$ and $\hat{\mathbf{r}}_2$ has the component $\cos\theta$ along the y-axis, it follows that

$$\hat{\mathbf{r}}_1 + \hat{\mathbf{r}}_2 = 2\mathbf{j}\cos\theta$$

Finally, reference to the figure shows that $\cos\theta = y/(y^2 + a^2)^{1/2}$, so the final formula for $\mathbf{E}_B$ is

$$\mathbf{E}_B = \mathbf{j}\frac{q}{2\pi\epsilon_0}\frac{y}{(y^2 + a^2)^{3/2}}$$

Thus, $\mathbf{E}_B$ lies along the perpendicular bisector and decreases as $1/y^2$ for large y.

20-4   The electric field for continuous charge distributions

Although all electric charge in matter is localized on very small particles, that is, on electrons and ions, for many applications to problems involving macroscopic bodies it is convenient to think of these charge distributions as varying in a continuous way. This means that the actual discrete charge distribution is replaced by a smoothed-out, continuous *charge density* that represents, on the average, the charge per unit volume at any point in the body under consideration. The calculation of the electric field associated with these charge distributions is very similar to that of the discrete case considered in Section 20-3. The basic difference is that the sum in (20-5) must now be replaced by an appropriate integral. Let us illustrate the method by reference to several examples.

Example 20-4   Calculate the electric field at a perpendicular distance a from an infinite line charge of charge per unit length λ.

Solution   As in Figure 20-7, let us set up a coordinate system with the x-axis along the line charge and with the y-axis through the field point. The electric field $d\mathbf{E}$ at this

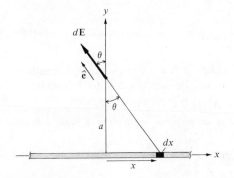

Figure 20-7

point due to the element of charge λdx at a distance x from the y-axis is

$$d\mathbf{E} = \frac{\lambda \, dx}{4\pi\epsilon_0} \frac{1}{(a^2 + x^2)} \hat{\mathbf{e}}$$

with $\hat{\mathbf{e}}$ a unit vector parallel to the line from dx to the field point.

Now, by symmetry, the component of $d\mathbf{E}$ along the x-axis due to the charge element at x will be canceled out by a contribution to the *total* electric field by the charge element of strength $\lambda \, dx$ located at the point $-x$. Thus, the total electric field will have only a component along the y-axis and we need only consider the component of $d\mathbf{E}$ along this axis. Calling this component dE, we have

$$dE = \mathbf{j} \cdot d\mathbf{E} = \frac{\lambda \, dx}{4\pi\epsilon_0} \frac{1}{a^2 + x^2} \cos\theta = \frac{\lambda \, dx}{4\pi\epsilon_0} \frac{a}{(a^2 + x^2)^{3/2}}$$

where in the final equality we have used $\cos\theta = a/(a^2 + x^2)^{1/2}$. The total electric field E is obtained by summing up these contributions for all values of x. Expressing this in the form of an integral, we have

$$E = \int dE = \frac{\lambda a}{4\pi\epsilon_0} \int_{-\infty}^{\infty} \frac{dx}{(a^2 + x^2)^{3/2}}$$

where the constant $\lambda a/4\pi\epsilon_0$ has been taken out from under the integral sign.

To evaluate the integral let us make in the integrand the substitution $x = a \tan\theta$. The upper and lower limits of the integral become $\pi/2$ and $-\pi/2$, respectively, and the infinitesimal dx becomes $dx = a \sec^2\theta \, d\theta$. Making these changes we obtain

$$E = \frac{\lambda a}{4\pi\epsilon_0} \int_{-\infty}^{\infty} \frac{dx}{(a^2 + x^2)^{3/2}} = \frac{\lambda a}{4\pi\epsilon_0} \int_{-\pi/2}^{\pi/2} \frac{a \sec^2\theta \, d\theta}{[a^2 + a^2 \tan^2\theta]^{3/2}}$$

$$= \frac{\lambda a}{4\pi\epsilon_0} \int_{-\pi/2}^{\pi/2} \frac{a \sec^2\theta \, d\theta}{a^3 \sec^3\theta} = \frac{\lambda}{4\pi\epsilon_0 a} \int_{-\pi/2}^{\pi/2} \cos\theta \, d\theta$$

$$= \frac{\lambda}{4\pi\epsilon_0 a} \sin\theta \Big|_{-\pi/2}^{\pi/2} = \frac{\lambda}{2\pi\epsilon_0 a}$$

where the third equality follows by use of the identity $1 + \tan^2\theta = \sec^2\theta$, and the fourth since $\cos\theta = 1/\sec\theta$.

Thus we conclude that the field at a distance a from an infinite line charge of charge per unit length λ is perpendicular to the line charge and has a strength E given by

$$E = \frac{\lambda}{2\pi\epsilon_0 a} \tag{20-6}$$

Example 20-5 A thin, circular loop of radius a has a charge per unit length λ. Calculate the electric field at a point on the axis of the loop at a distance b from it.

Solution Consider, in Figure 20-8a, the electric field $d\mathbf{E}$ due to an element of charge of strength $\lambda a \, d\theta$ and which lies at an angular separation θ from a fixed reference line. According to (20-3),

$$d\mathbf{E} = \frac{\lambda a \, d\theta}{4\pi\epsilon_0} \frac{\hat{\mathbf{e}}}{a^2 + b^2}$$

where ê is a unit vector oriented as shown. Now as indicated in Figure 20-8b, the component of $d\mathbf{E}$ parallel to the plane of the loop and due to the charge element located at θ is equal and opposite to the corresponding component produced by the charge element at $(\pi + \theta)$. Thus, as in Example 20-4, the electric field due to the entire loop will have a nonvanishing component only along the axis of the loop and only the component dE of the vector $d\mathbf{E}$ along this direction is of interest. Since the component of the unit vector ê along this direction is $\cos \phi$ and since $\cos \phi = b/(a^2 + b^2)^{1/2}$, it follows that

$$dE = \frac{\lambda a \, d\theta}{4\pi\epsilon_0} \frac{b}{(a^2 + b^2)^{3/2}}$$

Integrating this formula over all values of θ from 0 to 2π, we obtain for the total electric field E

$$E = \int dE = \int_0^{2\pi} \frac{\lambda a \, d\theta}{4\pi\epsilon_0} \frac{b}{(a^2 + b^2)^{3/2}} = \frac{\lambda ab}{4\pi\epsilon_0(a^2 + b^2)^{3/2}} \int_0^{2\pi} d\theta$$

$$= \frac{\lambda ab}{4\pi\epsilon_0(a^2 + b^2)^{3/2}} 2\pi$$

Thus the electric field is directed along the axis of the loop and has the magnitude

$$E = \frac{\lambda ab}{2\epsilon_0(a^2 + b^2)^{3/2}} \tag{20-7}$$

Example 20-6   A nonconducting circular disk of radius R has a charge per unit area σ. Calculate the electric field at a point on the axis and at a distance b from the disk (Figure 20-9).

Solution   Consider, as shown in the figure, an infinitesimal circular ring of the disk of radius r and of thickness dr. The area of this ring is $2\pi r \, dr$, and, since the charge per unit area on the disk is σ, the total charge on the ring is $\sigma 2\pi r dr$. According to (20-7), the electric field dE due to the charged ring in Figure 20-9 is

$$dE = \frac{r b \sigma \, dr}{2\epsilon_0(r^2 + b^2)^{3/2}}$$

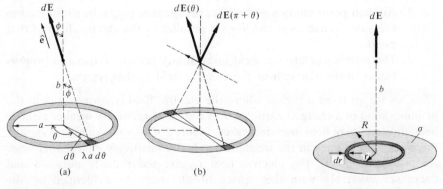

Figure 20-8   **Figure 20-9**

since the radius of the ring is r and its charge per unit length is now $2\pi\sigma r\,dr/2\pi r = \sigma\,dr$. As shown in the figure, this contribution to the electric field lies along the axis of the disk. The total electric field $\mathbf{E}$ also lies along this axis and its magnitude is obtained by summing up the contributions due to rings with radii from 0 to R. Expressing this sum as an integral we find that

$$E = \int dE = \int_0^R \frac{\sigma b}{2\epsilon_0} \frac{r\,dr}{(r^2+b^2)^{3/2}}$$

$$= \frac{\sigma b}{2\epsilon_0} \int_0^R \frac{r\,dr}{(r^2+b^2)^{3/2}} = \frac{\sigma b}{2\epsilon_0}\left[-\frac{1}{(b^2+r^2)^{1/2}}\right]\Big|_0^R$$

$$= \frac{\sigma b}{2\epsilon_0}\left[\frac{1}{b} - \frac{1}{(R^2+b^2)^{1/2}}\right]$$

so that the electric field E due to the entire disk is directed along the axis of the disk and has the magnitude

$$E = \frac{\sigma}{2\epsilon_0}\left[1 - \frac{b}{(R^2+b^2)^{1/2}}\right] \tag{20-8}$$

For the special case of an infinite plane of charge—that is, for $R \gg b$—the relation in (20-8) reduces to

$$E = \frac{\sigma}{2\epsilon_0} \tag{20-9}$$

and we note that this is independent of the distance b from the plane. In other words, the electric field due to an infinite nonconducting plane of charge is directed perpendicular to—and away from—the plane and has the *constant value* $\sigma/2\epsilon_0$ everywhere.

20-5   Geometrical representation of the electric field

Given any distribution of charge it is convenient to represent its associated electric field by a series of *directed lines*, which are called *electric field lines*. These lines are constructed so that they have the following two properties:

1. At each point along a given line, the tangent to the line (directed in the same sense as is the line) is parallel to the electric field at that point.
2. The number of electric field lines in any region of space is proportional to the strength of the electric field in that region.

Thus, as we go from a region where the electric field is strong—say in the neighborhood of a charged particle—to a region of relatively weaker electric field, the density of field lines decreases. Figure 20-10a shows a planar view of the electric field lines in the neighborhood of a positively charged particle. According to (20-3), the electric field is directed radially outward and decreases inversely with the square of distance. As evidenced by the arrowheads, the field lines in Figure 20-10a are also directed radially outward

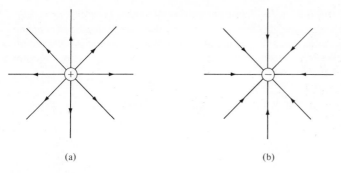

(a)   (b)

Figure 20-10

and their density decreases as we recede from the particle. Figure 20-10b shows the field lines for a negative particle. Note that here the field lines are oriented radially inward but that again in accordance with (20-3), the density of these lines increases as we approach the particle.

Because of the inverse square nature of Coulomb's law, it follows that in the immediate neighborhood of a charged particle the electric field is essentially determined by that of the particle alone. Thus, regardless of what other matter may be present, the electric field lines very close to a charged particle must always have the form in Figure 20-10. But what about the field lines in regions of space where the electric field is not dominated by a single particle? The following two properties of electric field lines are fundamental in this connection and are useful not only as an aid in visualizing field lines but are also indispensable in their construction. These properties are:

1. No two field lines can ever cross each other except at a point where there exists a charged particle.
2. All field lines are continuous in all regions of space containing no electric charge. Thus a field line must originate on a positively charged particle and terminate on a negatively charged particle, but no field line can ever originate or terminate at any point where there is no electric charge.

According to these properties, for example, the field lines near a charged body *cannot* have the structure depicted in Figure 20-11. The lines *A*, *B*, and *C* do not represent possible field lines since they either cross other field lines or else originate or terminate at a point at which there is no electric charge.

To justify the first of the above two properties, that in a charge-free region two field lines cannot cross, we may argue as follows. Suppose two field lines did cross. Then at such a point of crossing there would *not* be a unique tangent. This in turn would contradict the defining property of field lines that the tangent to a field line at a given point must be directed along the electric field at that point.

With regard to the second property, namely that all field lines in empty space must be continuous, it is unfortunately *not* possible to give a clean

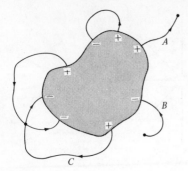

Figure 20-11

argument at an elementary level. Suffice it to say that in deriving this very crucial property of the continuity of the field lines, the fact that the electric field varies precisely with the inverse square of the distance from the particle plays a decisive role. If, for example, the exponent "2" in the inverse-square Coulomb law differed by any small amount from its established value of 2, then indeed electric field lines would *not* be continuous.

To illustrate the matter, consider, in Figure 20-12, three spherical surfaces concentric with a charged particle. Because of the continuity of electric field lines it is necessary that the same number of these lines cross the surface of each of the three spheres. For if this were not so, it would follow that at least one field line must originate or terminate in the region between the spheres and this would violate the continuity property of electric field lines in a charge-free region. (See Example 20-8 for a more detailed discussion of this matter.)

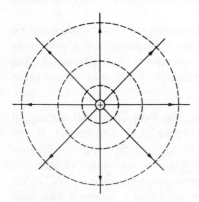

Figure 20-12

Figure 20-13 shows a planar view of the field lines associated with certain charge configurations. Parts (a) and (b) portray the field lines associated with two particles of opposite and of the same charge, respectively. The particular charge configuration shown in Figure 20-13a of equal and opposite

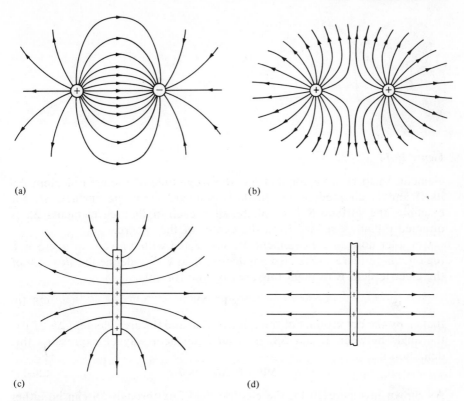

Figure 20-13

charges, q and $-q$, is known as an *electric dipole*, and the associated field as a *dipole field*. Note in each case that the field lines originate on the positive charge and terminate on the negative charge. For the case of the two positive charges, the endpoints of all field lines could not be included in the figure, and these lines can be thought of as terminating on negative charge at "infinity." Figure 20-13c shows the field lines associated with an edge-on view of a uniformly charged, nonconducting disk. Consistent with (20-8), the field on the axis of the disk is parallel to the axis and decreases away from the surface of the disk. Finally, Figure 20-13d shows the field lines associated with an infinite plane of charge. Consistent with (20-9), the field lines are uniform and perpendicular to the plane. Note that each of the diagrams in Figure 20-13 should be viewed in a three-dimensional perspective. The totality of field lines associated with the *dipole* in Figure 20-13a, for example, is obtained by rotating the figure about the line joining the two particles.

20-6 Electric flux

Consider, in Figure 20-14, a certain closed surface S. Let ΔS represent an infinitesimal element of area of S and define an associated vectorial area

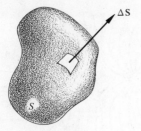

Figure 20-14

element **ΔS** to be a vector that has the magnitude of the area element ΔS itself and is directed perpendicularly outward from the surface. If, for example, the surface S is a sphere, then each of its area elements **ΔS** is directed radially outward from the center of the sphere.

Consider now an area element **ΔS** associated with a given surface S in a region where there exists also an electric field **E**. We define the *flux* $\Delta\Phi$ of the electric field **E** through this area by the formula

$$\Delta\Phi = \mathbf{E} \cdot \mathbf{\Delta S} \tag{20-10}$$

that is, by the dot product of the electric field and the area element **ΔS**. If θ is the angle between **E** and **ΔS**, then an equivalent way of expressing this definition is

$$\Delta\Phi = E \,\Delta S \cos\theta \tag{20-11}$$

As shown in Figure 20-15, the electric field flux through **ΔS** can be either positive, negative, or zero. The third possibility occurs if the electric field line lies in the surface element itself so that $\theta = \pi/2$.

More generally, consider a surface S composed of area elements **ΔS** to each of which a unique sense for the outward normal has been assigned. The total flux Φ_S through S is a sum of contributions of the type in (20-10); that is,

$$\Phi_S = \sum \mathbf{E} \cdot \mathbf{\Delta S} \tag{20-12}$$

where the sum extends over all area elements comprising the given surface

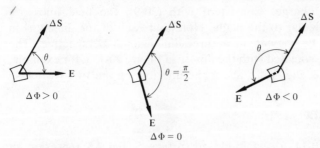

Figure 20-15

S. In the limit as these area elements tend to zero, the sum in (20-12) is defined to be the "surface integral"

$$\Phi_S = \int_S \mathbf{E} \cdot d\mathbf{S} \tag{20-13}$$

which may be viewed as simply another way of writing the limit of the sum in (20-12). For the special case that the surface S is closed and $d\mathbf{S}$ is directed outward, it is customary to write (20-13) in the form

$$\Phi_S = \oint_S \mathbf{E} \cdot d\mathbf{S} \qquad \text{(closed surface } S\text{)} \tag{20-14}$$

where the circle superimposed on the integral sign serves as a reminder that the surface S over which the integral is to be carried out is closed. Except for the very special cases to be considered below we shall *not* be concerned with evaluating surface integrals directly. However, we do need to give a meaning to these integrals and for this purpose it suffices to think of them as the limit of the sum in (20-12). From a physical point of view, the electric flux Φ_S is a measure of the number of field lines that go through the given surface.

Example 20-7 Calculate the flux through a planar surface of area A, which lies near an infinite, nonconducting plane carrying a charge per unit area σ. Assume that, as in Figure 20-16, α is the angle between the two planes and that $\hat{\mathbf{n}}$ is a unit vector normal to the surface.

Solution According to (20-9), the electric field is constant and normal to the charged plane and has magnitude $\sigma/2\epsilon_0$. Thus, since the angle between $\mathbf{E}$ and $\Delta\mathbf{S}$ is α, it follows from (20-11) that

$$\Phi = E\,\Delta S \cos\theta = \frac{\sigma}{2\epsilon_0} A \cos\alpha$$

Example 20-8 Calculate the flux through a spherical surface of radius R due to a particle of charge q at its center (Figure 20-17).

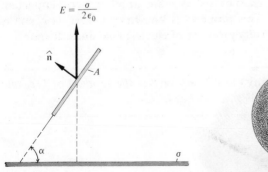

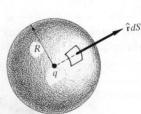

Figure 20-16

Figure 20-17

Solution According to (20-3), the electric field **E** at the distance R is

$$\mathbf{E} = \frac{q}{4\pi\epsilon_0}\frac{1}{R^2}\hat{\mathbf{r}}$$

where $\hat{\mathbf{r}}$ is a unit vector directed along a radius. For the sphere, the normal to its surface is along the radius and thus

$$d\mathbf{S} = \hat{\mathbf{r}}\, dS$$

By use of these formulas, the integrand in (20-14) becomes

$$\mathbf{E} \cdot d\mathbf{S} = \frac{q}{4\pi\epsilon_0}\frac{1}{R^2}\hat{\mathbf{r}} \cdot \hat{\mathbf{r}}\, dS$$

$$= \frac{q}{4\pi\epsilon_0}\frac{1}{R^2}\, dS$$

where in the second equality we have used the fact that $\hat{\mathbf{r}} \cdot \hat{\mathbf{r}} = 1$, which is valid for any unit vector. Furthermore, since the quantity $q/4\pi\epsilon_0 R^2$ is constant everywhere on the surface of the sphere it follows that Φ_S in (20-14) may be written

$$\Phi_S = \oint_S \mathbf{E} \cdot d\mathbf{S} = \frac{q}{4\pi\epsilon_0}\frac{1}{R^2}\oint dS = \frac{q}{4\pi\epsilon_0}\frac{1}{R^2}4\pi R^2$$

$$= \frac{q}{\epsilon_0}$$

where the third equality follows since for a sphere $\oint dS = 4\pi R^2$. Thus we have in this case

$$\Phi_S = \frac{q}{\epsilon_0}$$

which is independent of the radius of the sphere! In the next section we shall see that this result is true not only for a spherical surface but for *any* closed surface.

20-7 Gauss' law

The notion of electric flux enables us to state another very important property of the electric field. This property is known as *Gauss' law*, and is intimately related to the continuity property of electric field lines. It states:

The total flux out of any *closed surface S is the product of* $1/\epsilon_0$ *and the algebraic sum of* all *the charges inside S.*

In mathematical terms, Gauss' law is

$$\oint_S \mathbf{E} \cdot d\mathbf{S} = \frac{1}{\epsilon_0}Q \qquad\qquad \textbf{(20-15)}$$

where Q is the net charge, due allowance being made for sign, inside S and where in evaluating the surface integral each of the area elements dS is directed perpendicularly outward. Note that (20-15) is valid only for a *closed* surface S, but is applicable regardless of its shape. Thus in Figure 20-18 the flux out of the closed surfaces S_1 and S_2 are both zero since the total charge inside each one vanishes. Correspondingly, the flux out of S_3 is $+q/\epsilon_0$ and that out of S_4 is $-q/\epsilon_0$. We shall refer to the closed surface S implied in the integral in (20-15) as a *Gaussian surface*. Thus S_1, S_2, S_3, and S_4 are all Gaussian surfaces.

By making use of the concepts of the vector calculus, it is possible to derive Gauss' law by starting with the definition for the electric field in (20-2). Unfortunately, these methods are largely beyond our means. Suffice it to say that just as for the continuity property of the electric field lines, the fact that the Coulomb force law varies precisely with the inverse square of the distance plays a decisive role in such a proof. If the exponent in Coulomb's law were not precisely equal to 2, then Gauss' law would *not* be valid.

To confirm this fact that Gauss' law is valid only for an inverse-square force, suppose that this exponent were actually $(2+\eta)$, where η is some nonzero number. Then the "electric field" $\boldsymbol{\epsilon}$ associated with a particle of charge q would be

$$\boldsymbol{\epsilon} = \frac{q}{4\pi\epsilon_0} \frac{1}{r^{2+\eta}} \hat{\mathbf{r}}$$

with $\hat{\mathbf{r}}$ a unit vector directed radially outward from the particle. This time if we calculate the flux Φ_S through a spherical Gaussian surface of radius R and centered at the particle, then following the same steps as in Example 20-8, we find that

$$\Phi_S = \oint_S \boldsymbol{\epsilon} \cdot d\mathbf{S} = \frac{q}{4\pi\epsilon_0} \oint \frac{dS}{R^{2+\eta}} = \frac{q}{4\pi\epsilon_0} \frac{1}{R^{2+\eta}} 4\pi R^2$$

$$= \frac{q}{\epsilon_0} \frac{1}{R^{\eta}}$$

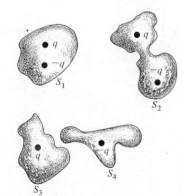

Figure 20-18

This varies with the radius R of the sphere, and hence the outward flux depends not only on the charge inside the surface but on the details of the surface as well. Thus, Gauss' law is *not* valid in this case.

Gauss' law can also be made plausible for the case where there is no charge inside the Gaussian surface. Consider, for example, the Gaussian surface and the field lines in Figure 20-19. Since, by hypothesis, there is no charge inside, it follows from the continuity property of field lines that every field line that enters the surface, such as at A, must leave the surface at some other point, such as B. Moreover, at A the angle θ between the electric field and the outward normal is obtuse and thus, according to (20-11), the flux at this point is negative. Correspondingly, at the point B the flux is positive, so it is at least possible for a cancellation to take place. Equivalently, we may argue that the fact that the total flux out of a closed surface containing no charge vanishes is due to the fact that no field lines can be created in its interior; thus every field line that enters the surface must leave it at some other point. If, on the other hand, some positive charge is inside the surface, then additional field lines will originate here and the flux would be positive. Correspondingly, if there is an excess of negative charge inside the surface, then additional field lines will enter it and the flux out of the surface will be negative.

Example 20-9   Consider the Gaussian surface $ABCDFGHJ$ in Figure 20-20. Assuming that there is no charge in its interior, confirm the fact that the flux out of this surface vanishes. Assume the existence of a uniform electric field $\mathbf{E}$ perpendicular to face $ABCD$.

Solution   Because of the given direction of the electric field, the only contribution to the flux integral in (20-14) comes from the two faces $ABCD$ and $FGHJ$. The outward normals of the remaining four faces are always perpendicular to the electric field and thus there is no contribution to the flux along these.

The flux Φ_1 through the lower face $ABCD$ is

$$\Phi_1 = -ES$$

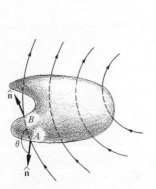

Figure 20-19

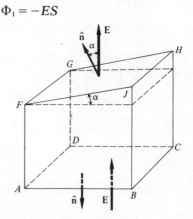

Figure 20-20

where S is the area of this face and the minus sign is due to the fact that the outward normal $\hat{n}$ to this face is directed opposite to **E**. Thus the angle θ in (20-11) is π and since $\cos \pi = -1$, the negative sign follows.

On the upper surface, *FGHJ*, the angle between the unit normal $\hat{n}$ and **E** is α. According to (20-11), the flux Φ_2 is thus

$$\Phi_2 = E \cos \alpha \, S_2$$

where S_2 is the area of this upper surface. It is apparent from the figure that S_2 is related to the area S of the lower face by

$$S_2 = \frac{S}{\cos \alpha}$$

since, for example, the lengths *AB* and *FJ* are related by $FJ = AB/\cos \alpha$. Substituting this value for S_2 into the formula for Φ_2, the factor $\cos \alpha$ cancels and we obtain

$$\Phi_2 = ES$$

Finally, combining this formula with the formula for Φ_1 we conclude that, in accordance with Gauss' law, the total outward flux $(\Phi_1 + \Phi_2)$ through the surface vanishes.

20-8   The electric field in the presence of a conductor

If a charged body, such as the glass rod in Figure 20-21, is brought near a conductor, induced charge will appear on it. The resultant total electric field **E** is due to both charges: the original charge on the rod plus that induced on the conductor. Because of the fact that the strength and the location of the induced charge are not generally known, we cannot make use of the methods of Sections 20-3 and 20-4 to calculate the total field **E**. In this section we shall show that by use of Gauss' law it is possible, nevertheless, to establish the validity of the following properties concerning the distribution of induced charge on a conductor and its effect on **E**.

1. **E** = 0 everywhere *inside* the conductor.
2. There is no net charge *inside* the conductor.
3. **E** is everywhere perpendicular to the bounding surface of the conductor.

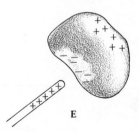

Figure 20-21

4. The induced charge per unit area σ on the conducting surface is related to the electric field E at the surface by

$$E = \frac{\sigma}{\epsilon_0} \qquad (20\text{-}16)$$

Property 1 may be established as follows. Suppose that the electric field inside the conductor were not zero. Then the electrons there would experience an unbalanced force and accelerate in a direction opposite to this electric field. This would imply, contrary to our hypothesis, that the situation is not static. Thus it follows that under conditions of electrostatic equilibrium—our only concern for the moment—the electric field must vanish everywhere in the interior of a conductor.

We may understand this physically in the following way. When the conductor is first introduced into the electric field, the field lines penetrate the conductor and exert forces on the mobile electrons in its interior. In response to this force, the electrons tend to travel in a direction opposite to the field and thus accumulate on a surface of the conductor. As more electrons collect at this surface, they repel other electrons, thereby counteracting the effect of the original field. Ultimately, these surface electrons produce an electric field which entirely cancels the original field. By way of example, in Figure 20-22 we present the initial and final situations when a rectangular slab of metal is introduced into a uniform electric field **E**. Figure 20-22a shows the initial nonequilibrium situation in which the field lines penetrate the conductor so that the electrons experience a force acting to the left. Finally, after a sufficient amount of negative charge has accumulated on the left face of the conductor (thereby leaving a positive charge on the opposite face) the electric field in the interior becomes zero. Only then does this migration of electrons stop. The time involved in going from the initial to the final configuration of zero electric field is typically of the order of 10^{-14} second. This time is so short that for practical purposes this charge rearrangement can be thought of as taking place instantaneously.

To establish property 2, that the interior of a conductor is charge free, consider in Figure 20-23 an arbitrary closed Gaussian surface S contained

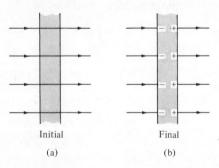

Initial

(a)

Final

(b)

Figure 20-22

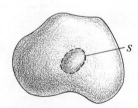

Figure 20-23

wholly within the interior of a conductor. According to property 1, the electric field vanishes everywhere on S and hence so must the flux Φ_S out of S. Applying Gauss' law to S, we conclude that the total charge inside S is also zero. Property 2 now follows from the fact that S is arbitrary. Note that this argument does *not* rule out the possibility of a surface charge on the conductor.

To establish property 3, that the electric field must be directed perpendicularly outward at each point of the conducting surface, we argue as follows. Suppose that at some point on the conducting surface the electric field had a component along the surface. Then the electrons near this point would experience a force and tend to move along the surface of the conductor. This contradicts the hypothesis that static conditions prevail. It follows then that the electric field at the surface of a conductor must be perpendicular to the surface.

Finally, let us establish property 4, according to which the electric field and the surface charge on a conductor are proportional to each other. To this end, consider, in Figure 20-24, a portion of the surface of a conductor in an electrostatic field E and suppose that the conductor has a surface charge σ.

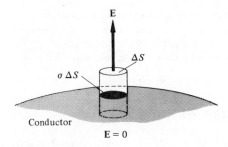

Figure 20-24

Let us construct a Gaussian surface in the form of a very small cylinder of cross-sectional area ΔS and with axis perpendicular to the surface at the given point. The total charge inside this Gaussian surface is $\sigma \Delta S$. Because of the facts that E is perpendicular to the surface of the conductor and vanishes inside, it follows by use of (20-12) that the total flux out of the end faces of the cylinder is simply $E \Delta S$. The flux out of the sides vanishes since E and ΔS are perpendicular along here. Applying Gauss' law we obtain

$$ E\ \Delta S = \frac{1}{\epsilon_0}\, \sigma\ \Delta S $$

and this simplifies to

$$ E = \frac{1}{\epsilon_0}\, \sigma $$

The validity of property 4 is thereby established.

Example 20-10 Show that the electric field in the interior of a hollow, charge-free, and closed conductor vanishes.

Solution To establish that the interior of the hollow, charge-free conductor A, in Figure 20-25a, has no electric field, consider first in Figure 20-25b a second conductor B, which is identical to A in all respects except that it is solid. According to the above arguments, regardless of the external charge, the electric field and the charge density vanish everywhere inside B. Thus, we may cut out any part of the interior of B without affecting either the external field lines or the surface charge. Since in this process of removing an interior portion of B we can produce the hollow conductor A, the desired result is established.

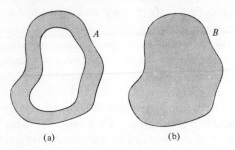

(a) (b)

Figure 20-25

This feature that the electric field in the interior of a hollow conductor vanishes is known as the principle of *electrostatic shielding*. If for any reason it is desired to shield out from a given region of space any stray electric fields that might exist, we may simply surround the region with a conducting shell and in this way obtain the desired shielding.

20-9 Experimental confirmation of the inverse-square law

Suppose a charge q is placed on a solid conductor. According to the arguments of Section 20-8, this charge must distribute itself on the outer surface of the conductor in such a way that the electric field inside vanishes. For if it did not, electrons would continue to migrate, and this is impossible under conditions of electrostatic equilibrium. Moreover, since E vanishes inside, it follows from Gauss' law that there can be no net charge anywhere in the interior of the conductor either. Thus we conclude that any charge placed on a conductor can reside only on its outer surface.

In a similar way, if a charge q is placed on a hollow closed conductor, it will also distribute itself on its *outer* surface. The electric field in this case must vanish not only in the interior of the conductor but in the cavity enclosed by the conductor as well. Thus by use of arguments similar to those presented in Example 20-10, it is easy to see that there can be no charge whatsoever on the inner surface of the conductor either.

Let us emphasize that in these arguments which establish that all charge placed on a conductor must reside on its outer surface, Gauss' law plays a decisive role. Now, as noted previously, Gauss' law is valid only for an inverse-square law, so if the exponent "2" in Coulomb's law did not have precisely this value, Gauss' law would be invalid. Hence if charge were placed on a hollow conductor, and if any of this charge were ever detected anywhere but on the outer surface, then a violation of Gauss' law would have been observed. In turn, this would imply that the exponent in Coulomb's law could not have the precise value of 2 that it does indeed have.

Even before the enunciation of Gauss' law in the nineteenth century, experiments designed to show that charge can reside only on the outer surface of a conductor were carried out by many people, including such notables as Benjamin Franklin, Henry Cavendish, Michael Faraday, and later on even Maxwell himself. The most accurate measurement in the nineteenth century, of the value for this exponent, was made by Maxwell, who established to an accuracy of 50 parts per million that indeed it had the value 2. That is, Maxwell showed that to the accuracy of his experiments the value for the exponent must fall in the range 1.99995–2.00005. More recently, Plimpton and Lawton (1936) improved the accuracy of these experiments even more and concluded that to 2 parts per billion, the exponent in Coulomb's law has the precise value of 2. At the present time, therefore, this exponent is bounded by the values 1.999999998 and 2.000000002, and the precise value of 2 is very plausible.

The experimental techniques underlying precision measurements of the exponent in Coulomb's law are all basically very similar and relatively easy to understand qualitatively. Figure 20-26a shows a hollow, spherical conduc-

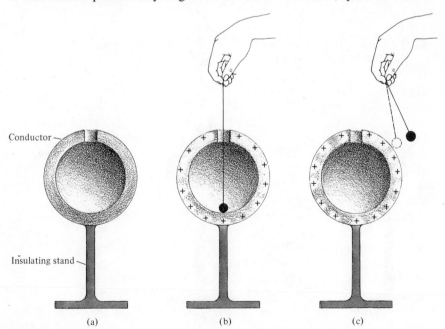

Conductor

Insulating stand

(a)   (b)   (c)

Figure 20-26

tor mounted on an insulating stand and with a small opening at the top so that small objects can be placed in the hollow part of the sphere. Suppose, as in Figure 20-26b, that we place a charge on the spherical conductor, lower into it a small, electrically neutral, conducting sphere, and let it come into contact with the inner surface of the hollow conductor. No measurable amount of charge ever accumulates on the small sphere. This shows that there is no charge on the inner surface of the hollow conductor. By contrast, if, as in Figure 20-26c, the small sphere is touched to the outer surface of the hollow conductor, it will immediately acquire an electric charge and will be repelled away. This confirms that all excess charge placed on a conductor resides entirely on its outer surface.

In a similar way we can establish that there is no electric field in the cavity. If, as before, a charge is placed on the hollow sphere, and a particle of very small charge is brought near its outer surface, then as shown in Figure 20-27a it will experience an electric force. This shows that outside the conducting shell there exists an electric field. By contrast, if this same charged conducting body is lowered into the cavity, it experiences no electric force, thus demonstrating that the electric field in the cavity is zero. By refinements of experiments of this sort the validity of Gauss' law and the fact that the exponent in Coulomb's must have the value of 2 has been confirmed to a very high degree of precision.

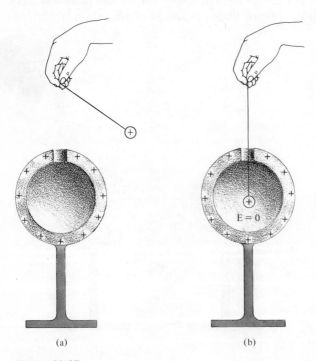

(a)   (b)

Figure 20-27

20-10 Applications of Gauss' law

The electric fields associated with certain charge distributions having a high degree of symmetry may sometimes be obtained by use of Gauss' law. In this section we illustrate the method—and incidentally derive a number of useful formulas—by reference to three examples. The procedure used in each case involves the construction of a Gaussian surface of a nature such that the electric flux out can be expressed algebraically in terms of the electric field at the surface. An application of Gauss' law then yields an explicit formula for the electric field.

Example 20-11 Calculate the electric field due to an infinite nonconducting plane which has a uniform charge per unit area σ.

Solution Consider, in Figure 20-28, a Gaussian surface in the form of a very small cylinder of cross-sectional area ΔS and with its axis perpendicular to the plane. Assume the plane bisects the cylinder. The total charge inside the cylinder is $\sigma \Delta S$.

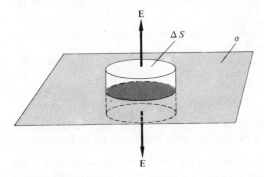

Figure 20-28

To calculate the electric flux, note first that, by symmetry, the electric field must be perpendicular to the plane. For if at some point the field had a component along the plane, this would imply more charge to one side of this point than the other and this is not possible for an infinite plane. Hence there are no contributions to the electric flux from the sides of the cylinder in the figure. On the other hand, there is a contribution to this flux at both the upper and lower faces. Again, by symmetry, the electric field **E** is the same on both sides of the plane, so the electric flux out of each of the end faces of the cylinder is $E\Delta S$. Hence the total flux out of the cylinder is $2E\,\Delta S$ and, by Gauss' law,

$$2E\,\Delta S = \frac{\sigma\,\Delta S}{\epsilon_0}$$

Solving for E we obtain

$$E = \frac{\sigma}{2\epsilon_0} \qquad\qquad (20\text{-}17)$$

which is identical to (20-9), as it must be.

Example 20-12 A very long nonconducting cylinder of radius a contains a uniform charge per unit volume ρ_0. Calculate the electric field at a point outside the cylinder.

Solution Let us construct a Gaussian surface in the form of a cylinder of radius r ($>a$) of length l and coaxial with the charged cylinder (the dashed surface in Figure 20-29).

Again, because of the symmetry, the electric field **E** must be directed along the radial direction and can vary only with the distance r from the axis of the cylinder. Since for a radial field of this type the electric flux out of the end faces vanishes, it follows that the total flux Φ out of the Gaussian surface is

$$\Phi = \oint \mathbf{E} \cdot d\mathbf{S} = E \int dS = 2\pi r l E$$

The total charge inside is $\rho_0 \pi a^2 l$ and thus by Gauss' law

$$2\pi r l E = \frac{1}{\epsilon_0} \pi a^2 l \rho_0$$

Solving for E we obtain

$$E = \frac{a^2 \rho_0}{2\epsilon_0} \frac{1}{r} \qquad (r \geqslant a) \qquad \textbf{(20-18)}$$

If we think of the charged cylinder as a line charge, then its charge per unit length λ is $\pi a^2 \rho_0$. Making this substitution in (20-18), we regain (20-6).

In the problems it is established that the field *inside* the charged cylinder in Figure 20-29 is

$$E = \frac{r\rho_0}{2\epsilon_0} \qquad (r \leqslant a) \qquad \textbf{(20-19)}$$

Example 20-13 Show that the electric field outside of a uniformly charged sphere is the same as if the total charge of the sphere were concentrated at its center.

Solution Consider, in Figure 20-30, a sphere of uniform charge of radius a and thus of total charge $Q_0 = (4\pi/3)a^3\rho_0$, with ρ_0 the charge density. By symmetry, the electric field must be radial and can vary only with the radial coordinate r. Hence the flux out of the spherical Gaussian surface of radius $r(>a)$ is $4\pi r^2 E$, where E is the electric field at a distance r from the sphere. By Gauss' law we have

$$4\pi r^2 E = \frac{1}{\epsilon_0} Q_0$$

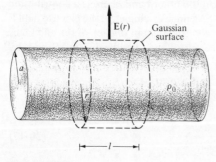

Figure 20-29

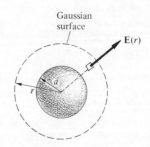

Figure 20-30

and thus we obtain the sought-for result

$$E = \frac{Q_0}{4\pi\epsilon_0} \frac{1}{r^2} \qquad (r \geq a) \qquad\qquad \text{(20-20)}$$

If we take a spherical Gaussian surface with radius $r(<a)$, it is shown in the problems that inside the sphere the electric field is

$$E = \frac{\rho_0}{3\epsilon_0} r \qquad (r \leq a) \qquad\qquad \text{(20-21)}$$

These results are plotted in Figure 20-31.

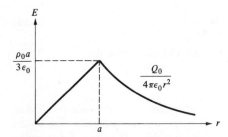

Figure 20-31

20-11 Summary of important formulas

The electric field **E** due to a distribution of charged particles is defined by

$$\mathbf{E} = \lim_{q_0 \to 0} \frac{1}{q_0} \mathbf{F} \qquad\qquad \text{(20-2)}$$

where **F** is the force on a test particle of charge q_0 placed at the given field point. For a single particle of charge q, the electric field is

$$\mathbf{E} = \frac{1}{4\pi\epsilon_0} \frac{q}{r^2} \hat{\mathbf{r}} \qquad\qquad \text{(20-3)}$$

where r is the distance of the particle from the field point and $\hat{\mathbf{r}}$ is a unit vector directed from the particle to the field point. For a collection of N particles of charges $q_1, q_2, \ldots, q_N$, **E** is

$$\mathbf{E} = \frac{1}{4\pi\epsilon_0} \sum_{i=1}^{N} \frac{q_i}{r_i^2} \hat{\mathbf{r}}_i \qquad\qquad \text{(20-5)}$$

where r_i is the distance between the ith particle and the field point and $\hat{\mathbf{r}}_i$ is a unit vector directed along the line joining the ith particle to the field point.

The electric flux $\Delta\Phi$ through a small vectorial area element $\Delta\mathbf{S}$ is defined by

$$\Delta\Phi = \mathbf{E} \cdot \Delta\mathbf{S} = E \, \Delta S \cos\theta \qquad\qquad \text{(20-10)}$$

where θ is the angle between the normal to the area element and the direction of the electric field. For a finite surface S, the flux Φ_S is

$$\Phi_S = \sum \mathbf{E} \cdot \Delta \mathbf{S} \qquad (20\text{-}12)$$

where the sum is to be carried out over every element $\Delta \mathbf{S}$ of the surface. In the limit as these area elements tend to zero, this sum approaches the surface integral

$$\Phi_S = \int_S \mathbf{E} \cdot d\mathbf{S} \qquad (20\text{-}13)$$

Gauss' law states that

$$\oint_S \mathbf{E} \cdot d\mathbf{S} = \frac{1}{\epsilon_0} Q \qquad (20\text{-}15)$$

where Q is the total charge enclosed by the closed surface S.

QUESTIONS

1. Why is it useful to introduce the notion of the electric field?
2. Explain why the definitions in (20-1) and (20-2) are equivalent if no conductors or insulators are present.
3. Consider an electric field line that lies entirely in the x-y plane. Why is the slope dy/dx at each point of this line given by

$$\frac{dy}{dx} = \frac{E_y}{E_x}$$

where E_y and E_x are the components of the electric field at the given point? How could you get the equations for the field lines by use of this property?
4. Consider the electric field $\mathbf{E}$ due to a positively charged point particle. Is the magnitude of the force on a nearby conducting sphere which carries a charge q_0 greater than or smaller than $|q_0|E$? Consider both cases of positive and negative q_0.
5. In what way, if any, is your answer to Question 4 modified if the field $\mathbf{E}$ is due to a uniformly charged, nonconducting sphere?
6. Consider an experiment for measuring the electric field $\mathbf{E}$ in the region surrounding a charged, conducting

sphere by probing with a particle of charge q_0. Explain why you would obtain different answers depending on whether the definition in (20-1) or (20-2) is used. What is the basic advantage of using the latter definition?
7. Let $\mathbf{E}^*$ represent the electric field associated with an isolated charged conducting sphere. Is the magnitude of the force on a nearby particle of charge q_0 greater or less than $|q_0 E|$? Consider all possibilities for the signs of the two charges.
8. Using the fact that the electric field outside of a uniformly charged, nonconducting sphere is the same as if all of the charge were concentrated at its center, explain why the force on such a sphere when placed into an external field is also the same as if all of the sphere's charge is concentrated at its center. (*Hint*: Use Newton's third law.)
9. Explain in physical terms why the strength of the electric field due to an infinite plane of charge is independent of distance from the plane.
10. Show by use of symmetry arguments that the electric field at the upper face of the cylinder in Figure 20-28 is

equal and opposite to that on the lower face.

11. A charged particle is originally at rest in a uniform electric field. Explain why it will follow a field line in its subsequent motion. Will it follow a field line if it originally had a component of velocity at right angles to a field line?

12. Repeat Question 11, but suppose that this time the electric field lines are curved.

13. Consider the electric field at a given point on the surface of a uniformly charged sphere. Show by symmetry arguments that every other point on the surface of the sphere has the same value for the electric field. (*Hint*: Consider an arbitrary rotation of the sphere about its center and use the fact that the physical situation is not altered in this process.)

14. Repeat Question 13, but this time consider the field at those points that are at a distance *r* from the center of the sphere.

15. Repeat Question 13, but this time assume that the charge distribution in the sphere is not uniform but is still spherically symmetric.

16. Consider the field produced by a particle of charge *q*. What is the flux out of each of the following surfaces: (**a**) A sphere surrounding the charge? (**b**) A cube surrounding the charge? (**c**) The surface bounded by two spheres of radii r_1 and r_2 ($r_1 < r_2$) concentric with the particle?

17. Consider the field produced by a dipole. What is the flux out of a cubical surface that completely surrounds the dipole? Does the normal component of the electric field vanish on the surface of the cube?

18. Explain why it is *not* possible to calculate the field of a dipole by use of Gauss' law alone.

19. A sphere contains a charge distribution that is everywhere positive and spherically symmetric. Show by use of Gauss' law that the field must be a maximum at the surface of the sphere. Compare with Figure 20-31.

20. A spherical hole is cut out of the center of a sphere that contains a uniform charge distribution. Why must the electric field in the cavity vanish? (*Hint*: Use Gauss' law and symmetry arguments.)

21. A particle of charge *q* is placed in the interior of a hollow conductor, which is grounded. What is the *total* charge on the inner surface of the conductor? What is the charge on the outer surface of the conductor? Is there an electric field outside?

22. Why is it necessary that the charge on the particle in Figure 20-27b be vanishingly small? Why must there be a force on it if the charge is not small? Would this imply that the electric field in the cavity surrounded by a charged conducting shell is not zero? Explain.

23. Present an argument to show that Gauss' law implies that electric field lines must be continuous.

24. A particle of charge *q* is inside the cavity of a grounded conducting shell. Explain why the total charge induced on the conductor is $-q$. (*Hint*: What is **E** inside the conductor?)

PROBLEMS

1. Calculate the strength of the electric field produced by a particle of charge 5.0 μC at distances of 10^{-2}, 10^{-1}, and 10 meters from the particle.

2. What is the force, magnitude and direction, exerted on a particle of charge $-2.0\ \mu$C when it is in a uniform field of strength 10^3 N/C?

3. Two particles, each of charge

+3.0 μC are located at the points $(2, 0, 0)$ and $(-2, 0, 0)$ in a certain coordinate system. Calculate the magnitude and direction of the field at the following points, assuming that all distances are measured in meters: (a) $(0, 0, 0)$; (b) $(1, 0, 0)$; and (c) $(3, 0, 0)$.

4. Repeat Problem 3, but this time calculate the field at the points (a) $(0, 1, 0)$; (b) $(0, -1, 0)$; and (c) $(2, 2, 0)$.

5. A dipole of dipole moment **p** consists of two particles of charges $+q$ and $-q$ and located in a certain coordinate system at the respective points $(a, 0, 0)$ and $(-a, 0, 0)$. In terms of these parameters the dipole moment **p** is defined by

$$\mathbf{p} = 2aq\,\mathbf{i}$$

where **i** is a unit vector along the positive x-axis. Calculate the electric field **E** of this dipole at the following points: (a) $(2a, 0, 0)$ and (b) $(-2a, 0, 0)$.

6. Show by use of the results of Problem 5 and Example 20-2 that the torque τ on a dipole of moment **p** in a uniform field **E** may be expressed in the form

$$\tau = \mathbf{p} \times \mathbf{E}$$

*7. Use the result of Example 20-2 to show that:
(a) The dipole is in stable equilibrium when $\theta = 0$, that is, when it is lined up with the field.
(b) The dipole is in unstable equilibrium when $\theta = \pi$. (*Note:* Stability occurs, if the forces and torques are of such a nature that the system tends to return to its original equilibrium state whenever it is slightly perturbed. Otherwise, it is unstable.)

8. As measured in a certain coordinate system, $2N$ particles, each of

charge q, are located at the points with coordinates $\{(na, 0, 0)\}$, where $n = \pm 1, \pm 2, \ldots, \pm N$. Show that the electric field **E** at the point $(0, d, 0)$ is

$$\mathbf{E} = \mathbf{j}\,\frac{qd}{2\pi\epsilon_0}\,\sum_{n=1}^{N}\frac{1}{[d^2 + n^2 a^2]^{3/2}}$$

9. What charge per unit area σ must be placed on a very large plane so that the associated electric field has a strength of 100 N/C?

10. A particle of charge q and mass m is suspended from an insulating thread and hangs near a vertical infinite plane, which has a charge per unit area σ (Figure 20-32).

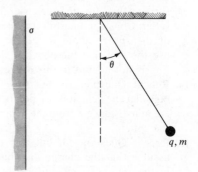

Figure 20-32

(a) Show that the angle θ which the thread makes with the vertical is given by

$$\tan \theta = \frac{q\sigma}{2\epsilon_0 mg}$$

(b) Calculate the tension T in the string.

11. Consider a proton in a certain uniform electric field **E**. (a) What must be the strength of this electric field so that it just balances the earth's gravitational field? (b) At what distance from a particle of charge $10\,\mu$C is a field of this strength achieved?

12. A straight line charge of length l carries a uniform charge per unit

length λ. Show that the magnitude of the electric field at a point on the axis of the line charge and at a distance x from the nearer end is

$$E = \frac{\lambda l}{4\pi\epsilon_0} \frac{1}{x(l+x)}$$

13. A line charge of length $2l$ consists of two parts: one half carrying a charge per unit length $+\lambda$ and the other half the charge per unit length $-\lambda$. Calculate the electric field at a distance y from the line charge and along its perpendicular bisector.

14. Calculate the magnitude and direction of the electric field at a perpendicular distance of 50 cm from a very long wire that carries a uniform charge of 2.0 μC/m.

15. An electron is at a distance of 2.0 cm from a very long wire and approaching it with an acceleration of 1.5×10^{13} m/s^2. What is the charge per unit length on the wire?

*16. A circular hole of radius a is cut out of an infinite plane that carries a charge per unit area σ. Calculate the electric field at a point along the axis of the hole and at a distance b from the plane. (*Hint:* Use Example 20-6 and the principle of superposition, according to which this field is the sum of the fields due to an infinite plane of charge $+\sigma$ plus that due to a disk of radius a carrying a charge $-\sigma$.)

17. A line charge of charge per unit length λ is in the form of a semicircle of radius a. Calculate the electric field at the center of the circle.

18. Calculate the electric field for the line charge in Problem 17, but this time take the field point at a distance b vertically above the center of the circle.

19. A particle of charge $-q_0$ $(q_0 > 0)$ and mass m is constrained to motion along the axis of a circular line

charge of radius a and of charge per unit length λ (>0).
(a) What is the force acting on the particle when it is at the center of the ring?
(b) Show that if it is displaced from this position (along the axis of the ring), then it executes simple harmonic motion with period

$$T = 2\pi \left(\frac{2\epsilon_0 m a^2}{q_0 \lambda}\right)^{1/2}$$

20. A disk of radius a carries a surface charge per unit area σ, which varies with radius r as

$$\sigma = \frac{\sigma_0}{a} r$$

where σ_0 is a positive constant.
(a) What is the total charge on the disk?
(b) Calculate the electric field at a distance b from the plane of the disk and along its axis. You may use the indefinite integral

$$\int \frac{x^2 dx}{(x^2 + b^2)^{3/2}} = -\frac{x}{(x^2 + b^2)^{1/2}} + \ln[x + (x^2 + b^2)^{1/2}]$$

21. A disk of radius R_0 carries a charge per unit area σ and has a hole of radius a cut out of its center. By an appropriate modification of (20-8) calculate the electric field at a point on the axis of the disk and at a distance b away from its center.

22. Show by use of (20-7) that the field on the axis of a uniformly charged circular loop of radius a has the magnitude

$$E = \frac{Qb}{4\pi\epsilon_0[a^2 + b^2]^{3/2}}$$

where Q is the total charge on the loop and b is the distance from the center of the loop. Does this reduce to the expected formula in the limit $b \gg a$?

*23. A cylinder of length l and radius R has a uniform charge per unit volume ρ_0. Show that the electric field at a point on the axis and at a distance b from the nearer end face is directed along the axis and has the magnitude

$$E = \frac{\rho_0}{2\epsilon_0}\left\{l + (R^2 + b^2)^{1/2} - [R^2 + (b + l)^2]^{1/2}\right\}$$

(*Hint*: Find the field $d\mathbf{E}$ due to a thin disk and integrate.)

24. Calculate the net charge inside a closed surface if the flux out of the surface is 5.0×10^4 N-m²/C.

*25. A particle of charge q is at the point $(-a, 0, 0)$ in a certain coordinate system (Figure 20-33).

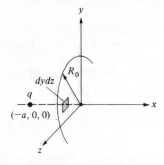

Figure 20-33

(a) Show that the flux $d\Phi$ through the element of area $dy\,dz$ at the point $(0, y, z)$ in the y-z plane is

$$d\Phi = \frac{qa}{4\pi\epsilon_0}\frac{1}{(a^2 + y^2 + z^2)^{3/2}}\,dy\,dz$$

(b) Calculate the total flux through a circle of radius R_0, centered at the origin and lying in the y-z plane. *Hint*: Transform to polar coordinates and evaluate

$$\Phi = \frac{qa}{4\pi\epsilon_0}\int_0^{R_0} r\,dr \int_0^{2\pi} d\theta \frac{1}{(a^2 + r^2)^{3/2}}$$

(c) What is the total flux through the entire y-z plane?

26. Two spheres of radii R_1 and R_2 $(< R_1)$ are concentric with a particle of charge q.
(a) What is the flux through the smaller sphere?
(b) What is the flux through the larger sphere?
(c) By use of your results to (a) and (b) calculate the flux through the closed surface consisting of these two spheres. Is your answer consistent with Gauss' law?

27. A small body carrying a charge of $2.5\ \mu$C is suspended in a cavity inside an uncharged conducting body.
(a) What is the electric field inside the conducting medium?
(b) By constructing an appropriate Gaussian surface, calculate the total charge induced on the inner walls of the cavity.
(c) What is the total charge on the outer surface of the conductor?

28. Consider an infinite cylinder of radius a, which carries a uniform charge per unit volume ρ_0. Construct a Gaussian surface as in Figure 20-29, but this time with $r < a$.
(a) Why must the electric field $\mathbf{E}$ be everywhere perpendicular to the curved part of the Gaussian surface?
(b) What is the total charge inside the Gaussian surface?
(c) Calculate the flux through the Gaussian surface and thus derive (20-19).

29. An infinitely long cylinder of radius a carries a uniform charge per unit volume $-\rho_0$ $(\rho_0 > 0)$ and is surrounded by a grounded conducting shell of radius b that is coaxial with the cylinder; see Figure 20-34. Calculate, by use of suitable Gaussian surfaces: (a) the electric field for $r \leqslant a$; (b) the electric field for $a \leqslant r \leqslant b$; and (c) the electric field for $r \geqslant b$.

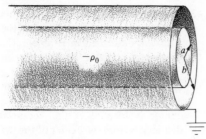

Figure 20-34

30. For the physical system of Problem 29, calculate the charge per unit area on the surface of the grounded conducting cylinder.

*31. An infinitely long cylinder of radius *a* has a charge per unit volume

$$\rho = \rho_0\left(\frac{1}{2} - \frac{r^2}{a^2}\right)$$

where ρ_0 is a positive constant and *r* is the radial distance. To calculate the electric field for $r \leq a$, construct a Gaussian surface in the form of a cylinder of radius *r* and length *l*.
(a) Show that the total charge inside the Gaussian surface is

$$Q(r) = \frac{\pi l \rho_0}{a^2} r^2 (a^2 - r^2)$$

(b) Calculate the electric field inside the cylinder.

32. Calculate the field outside of the cylinder of Problem 31 by constructing an appropriate Gaussian surface and using Gauss' law.

33. Derive (20-21), which gives the electric field at a point inside of a uniformly charged sphere of radius *a*. (*Hint*: Apply Gauss' law to a spherical surface of radius r ($< a$).)

34. By use of (20-20) and (20-21) prove that the electric field at a distance *r* from the center of a uniformly charged sphere is due to all charge no farther than *r* from the center of the sphere.

35. A nucleus of mass number *A* and containing *Z* protons may be thought of as a uniformly charged sphere of radius R_0 given (in meters) by

$$R_0 = 1.1 \times 10^{-15} A^{1/3}$$

(a) In terms of *A* and *Z* and the quantum of charge *e*, calculate the electric field outside of this nucleus, at a distance *r* from its center.
(b) Calculate the electric field inside the nucleus.
(c) Make a plot of the electric field as a function of *r* for the special nucleus $^{235}_{92}$U.

*36. A very small cylindrical hole is made along a diameter through a uniformly charged sphere of density ρ_0 (> 0) of radius *a*. Assume that the electric field inside this hole is the same as it would be if the hole were not present. Suppose now that a particle of charge $-q_0$ ($q_0 > 0$) and mass *m* is dropped into the hole. Assume no friction. (a) Show by use of (20-21) that the particle will move back and forth with simple harmonic motion. (b) Calculate the period of this motion.

*37. A sphere of radius *a* contains a uniform charge density ρ_0 and is surrounded by a concentric spherical shell of inner radius *b* ($> a$) and outer radius *c* and containing a uniform charge density $-\rho_0$. See Figure 20-35. By constructing suita-

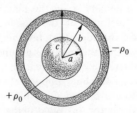

Figure 20-35

ble Gaussian surfaces calculate the electric field for values of *r* satisfying:
(a) $r \leq a$;
(b) $a \leq r \leq b$;

(c) $b \leqslant r \leqslant c$;

(d) $r \geqslant c$.

*38. Consider a spherically symmetric distribution of charge characterized by the charge density $\rho(r)$.

(a) Show by arguments of symmetry that the electric field must be purely radial and that it can depend only on r.

(b) Show that the electric field at the radial distance r is

$$E = \frac{1}{4\pi\epsilon_0}\frac{Q(r)}{r^2}$$

where $Q(r)$ is the total charge contained within a sphere of radius r:

$$Q(r) = 4\pi \int_0^r r^2 \, dr \, \rho(r)$$

39. Apply the result of Problem 38 to show that the electric field vanishes in a spherical cavity cut out of the center of a uniform distribution of charge.

40. A particle of charge q is placed at the center of a spherical conducting shell of inner radius a and outer radius b. Calculate the electric field for the following distances:

(a) $r < a$;

(b) $a \leqslant r \leqslant b$;

(c) $r > b$.

41. Consider the same situation as in Problem 40. (a) What is the surface charge density σ on the inner and outer surfaces of the conducting sphere? (b) Repeat (a), but this time assume that the sphere is grounded.

42. Consider the same situation as in Problem 40, but suppose that, in addition, a charge Q is placed on the conductor. (a) What is the total charge on the inner surface of the conducting sphere? (b) What is the total charge on the outer surface?

43. Calculate the electric field (in terms of q and Q) everywhere for the situation described in Problem 42.

*44. Consider, in Figure 20-36, a sphere of radius a containing a uniform charge density ρ_0 out of which is cut a spherical hole of radius c whose center is located at the vectorial distance **b** from the center of the big sphere.

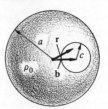

Figure 20-36

(a) Why can we express the total field **E** in the form

$$\mathbf{E} = \mathbf{E}_1 + \mathbf{E}_2$$

where $\mathbf{E}_1$ is the field due to a uniformly charged sphere of density ρ_0 and radius a, and $\mathbf{E}_2$ is the field produced by a uniformly charged sphere of radius c located at **b** and carrying a charge density $-\rho_0$?

(b) Show that if **r** is any point in the cavity, then

$$\mathbf{E}_1 = \frac{\rho_0}{3\epsilon_0}\mathbf{r}$$

(c) Show that $\mathbf{E}_2$ is

$$\mathbf{E}_2 = \frac{-\rho_0}{3\epsilon_0}(\mathbf{r} - \mathbf{b})$$

if **r** is any point in the cavity.

(d) By combining these results, show that the electric field **E** in the cavity is uniform and has the value

$$\mathbf{E} = \frac{\rho_0}{3\epsilon_0}\mathbf{b}$$

21 The electrostatic potential

In every explanation of natural phenomena we are compelled to leave the sphere of sense perception and to pass to things which are not the objects of sense and are defined only by abstract conceptions.

H. L. F. VON HELMHOLTZ (1821–1894)

21-1 Introduction

The electrostatic field has two very striking and useful mathematical properties. One of these we studied in Chapter 20, and is that it satisfies Gauss' law. The other has to do with a way of characterizing it in terms of a certain scalar quantity called the *electrostatic potential.* This important quantity is the subject matter of this chapter.

Because of the fact that the electric field associates with each point in space a vector—that is, both a magnitude and a direction—it is known as a *vector field.* By contrast, the electrostatic potential is a *scalar field* by virtue of the fact that it associates with each point in space a scalar quantity. We might expect therefore that if a quantitative description of electrostatic phenomena in terms of such a potential function were possible it would be considerably simpler than that in terms of the electric field itself. This is indeed the case, as we shall confirm in this chapter.

21-2   Work and the electric field

Consider, in Figure 21-1, a region of space in which there exists an electric field $\mathbf{E}$ and suppose that an external agent moves a particle of charge q_0 very slowly along a curve C from a point A to a point B. To achieve this, he must exert on the particle a certain force $\mathbf{F}$, which is equal and opposite to the electric force $q_0\mathbf{E}$ on it at each point of its path.

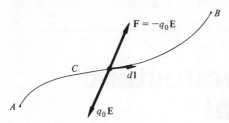

Figure 21-1

According to the definition in Chapter 7, the work W_{AB} that the agent carries out in moving the particle from A to B along a curve C is obtained by dividing the path into a sequence of infinitesimal displacements $\{\Delta l\}$ and adding together the works $\mathbf{F} \cdot \Delta l$ carried out along each of these. The work W_{AB}, which is the limit of this sum as the magnitudes of the displacements $|\Delta l|$ tend to zero, is them defined to be the line integral

$$W_{AB} = \int_A^B \mathbf{F} \cdot d\mathbf{l} \tag{21-1}$$

where $d\mathbf{l}$ is an infinitesimal element of path along C, and the integral is along the curve C from A to B. For the situation in Figure 21-1, since $\mathbf{F} = -q_0\mathbf{E}$ it follows that the work W_{AB} carried out may be expressed as

$$W_{AB} = -q_0 \int_A^B \mathbf{E} \cdot d\mathbf{l} \tag{21-2}$$

where the constant q_0 has been taken out from under the integral sign. Recall from the discussion in Chapter 7 that, in general, the value of the integral in (21-2) depends not only on the endpoints A and B of the path, but also on the particular curve C connecting them. In anticipation of a fact to be discussed in Section 21-3 that the integral in (21-2) is actually independent of the path, all reference to the particular path C has been omitted from this formula.

For the special case that the path C joining A and B is a straight line, the line integral in (21-2) reduces to an ordinary integral. The situation here is precisely the same as that considered in Chapter 7.

Example 21-1   A particle of charge q_0 is near an infinite nonconducting plane of charge per unit area σ. Calculate the work required to move it from A to B in Figure 21-2 along:

(a) The straight-line path of length a connecting A and B, assuming this line makes an angle α with the line $\overline{AD}$, which is perpendicular to the plane.

(b) The straight-line paths $\overline{AD}$ and $\overline{DB}$.

Solution

(a) In Example 20-6 it was established that the electric field associated with an infinite, nonconducting plane is directed perpendicular to the plane and has the value $\sigma/2\epsilon_0$. Substitution into (21-2) yields

$$W_{AB} = -q_0 \int_A^B \mathbf{E} \cdot d\mathbf{l} = -q_0 \int_A^B \frac{\sigma}{2\epsilon_0} \, dl \cos \alpha$$

$$= \frac{-q_0\sigma}{2\epsilon_0} \cos \alpha \int_A^B dl = -\frac{q_0\sigma a \cos \alpha}{2\epsilon_0}$$

since $\mathbf{E}$ is constant, as is the angle α between $\mathbf{E}$ and $d\mathbf{l}$.

(b) In this case note first that along $\overline{DB}$, $\mathbf{E}$ and $d\mathbf{l}$ are perpendicular, and hence no work is carried out along this portion of the path. Along $\overline{AD}$, $\mathbf{E}$ and $d\mathbf{l}$ are parallel, and thus

$$W_{AB} = -q_0 \int_A^D \mathbf{E} \cdot d\mathbf{l} = -q_0 \int_A^D \frac{\sigma}{2\epsilon_0} \, dl = -\frac{q_0\sigma}{2\epsilon_0} \int_A^D dl$$

$$= -\frac{q_0\sigma a \cos \alpha}{2\epsilon_0}$$

since the path length $\overline{AD}$ is $a \cos \alpha$. It is interesting to note that this formula for the work is precisely the same as that found in (a). This confirms the fact that W_{AB} is independent of the path, at least in this case.

21-3 The electrostatic potential

Consider, in Figure 21-3, a region of space in which there exists an electrostatic field $\mathbf{E}$, and let A and B represent two fixed points in this region. Imagine calculating, by use of (21-2), the work W_{AB} required to take a particle of charge q_0 from point A to point B along each of the three paths C_1, C_2, and C_3. In general, since the electric field will assume different values along these paths, we might expect that W_{AB} would also vary depending on which of the paths C_1, C_2, or C_3 is chosen. However, this is found *not* to be

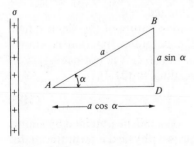

Figure 21-2

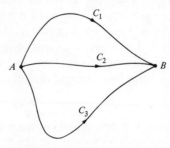

Figure 21-3

true. Instead, W_{AB} is found to be the *same* for *all* paths connecting points A and B. This property of the electrostatic field, that the integral in (21-2) is independent of the path, is described by saying that the electrostatic field is a *conservative field*. It is one of the two most important properties of this field; the other one is Gauss' law.

An equivalent way of describing the conservative nature of the electrostatic field is by the statement that no (zero) work is required to take a charged particle around any closed path. The equivalence of these two ways of characterizing a conservative field is established in the problems.

Now because of this fact that the integral

$$\int_A^B \mathbf{E} \cdot d\mathbf{l}$$

in (21-2) is independent of the path connecting the points A and B, it follows that the integrand $\mathbf{E} \cdot d\mathbf{l}$ must be a perfect differential. Hence, just as for the corresponding analysis of potential energy in Section 8-5, it follows that there must exist a scalar function of position V, called the *potential function*, with the property that its differential dV is related to the electric field $\mathbf{E}$ for an arbitrary infinitesimal displacement $d\mathbf{l}$ by

$$dV = -\mathbf{E} \cdot d\mathbf{l} \tag{21-3}$$

The minus sign here is conventional and has no other significance. Substituting this relation into the definition for the work W_{AB} in (21-2), we obtain

$$W_{AB} = -q_0 \int_A^B \mathbf{E} \cdot d\mathbf{l} = q_0 \int_A^B dV = q_0 V \Big|_A^B = q_0(V_B - V_A) \tag{21-4}$$

where the third equality follows since dV is a perfect differential, and where V_B and V_A are the values of the potential function at the points B and A, respectively. Note that W_{AB} is proportional to $(V_B - V_A)$ and thus has the desired property of depending *only* on the endpoints A and B and not on the path connecting them.

The minus sign, whose presence in (21-3) is only by convention, is important to keep in mind. Physically, it has the consequence that as a particle travels along the direction of a field line it goes to regions of lower potential. Thus the potential decreases as we recede from a positively charged particle or approach a negatively charged one.

An important consequence of the conservative nature of the electric field is that if for a given charge distribution the potential function is known everywhere, then to evaluate the work W_{AB} in (21-2) it is not necessary to evaluate the line integral directly. For, according to (21-4),

$$W_{AB} = q_0(V_B - V_A) \tag{21-5}$$

and since V is presumed known everywhere, W_{AB} can be obtained by simple substitution. This formula in (21-5) also suggests a physical interpretation for the *potential difference* $(V_B - V_A)$ as *the work per unit charge required to take a particle from A to B.*

The definition of the potential function in (21-3) is in terms of differentials and thus V is defined only up to an additive constant. This means in particular that only differences in potential, as in (21-5), for example, can be given a physical meaning. Following custom, we shall arbitrarily assign the value zero to the value for the potential function for the point at "infinity." With this convention, if point A in (21-5) is selected to be the point at infinity, so that $V_A = 0$, this relation may be interpreted by the statement that the *potential at a point is the work required to bring a unit charge from infinity to this point.*

The unit of potential, according to (21-5), is energy per unit charge, that is, the joule per coulomb. This unit potential is defined to be the *volt* (V), and thus, by definition,

$$1\,V = 1\,J/C$$

Thus it takes 1 joule of work to take a 1-coulomb particle through a potential difference of 1 volt. In terms of this unit, it follows from (21-3) that the volt per meter (V/m) is the unit of electric field strength. We shall express electric field strength in volts per meter from now on.

Note the distinction between the *italic* letter V for potential and the roman letter V for the unit of potential of the volt. As a general rule, the symbol for the unit V will usually be preceded by a numerical value.

Example 21-2 For the physical situation described in Example 21-1, calculate the potential difference between points:

(a) B and D.
(b) A and B.

Solution

(a) Along the path BD in Figure 21-2 the electric field is perpendicular to $d\mathbf{l}$ and hence $W_{BD} = 0$. It follows then from (21-5) that $V_B = V_D$ and thus $V_B - V_D = 0$.

(b) Comparing the result in Example 21-1 with (21-5), we find that

$$V_B - V_A = \frac{-\sigma a \cos \alpha}{2\epsilon_0}$$

21-4 Calculation of the electric field

A direct consequence of the defining relation in (21-3) is that if for a given charge distribution the potential function V is known, then the associated electric field $\mathbf{E}$ can be obtained by differentiation. The purpose of this section is to derive this important relation.

Suppose that for a given physical situation the potential function V is known everywhere, and that V_A represents its value at some given point A (Figure 21-4). If $\mathbf{E}$ is the electric field at this point, then, according to (21-3) the value of the potential at a point that is at an infinitesimal displacement $d\mathbf{l}$ from A is $(V_A + dV)$, with dV given by

$$dV = -\mathbf{E} \cdot d\mathbf{l} \tag{21-3}$$

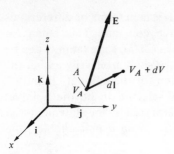

Figure 21-4

Note the important feature that for a fixed point A and a given value for $\mathbf{E}$ at that point, this formula gives the change in potential dV for *all* points at any infinitesimal displacement $d\mathbf{l}$ from A.

To derive the relation between $\mathbf{E}$ and V, let us assume a Cartesian coordinate system and make the particular choice $d\mathbf{l} = \mathbf{i}\, dx$ corresponding to a displacement along the x-axis. For this case the change in potential dV is

$$dV = -\mathbf{E} \cdot (\mathbf{i}\, dx) = -E_x dx \qquad (21\text{-}6)$$

with E_x the component of $\mathbf{E}$ along the x-axis of the given system. Noting that for this change the variables y and z are kept fixed, we find on making use of the definition of a partial derivative that (21-6) is equivalent to

$$E_x = -\frac{\partial V}{\partial x}$$

Analogous relations may be obtained by considering displacements $d\mathbf{l} = \mathbf{j}\, dy$ and $d\mathbf{l} = \mathbf{k}\, dz$ along the y- and z-axes, respectively. Hence

$$E_x = -\frac{\partial V}{\partial x} \qquad E_y = -\frac{\partial V}{\partial y} \qquad E_z = -\frac{\partial V}{\partial z} \qquad (21\text{-}7)$$

so that the *component of the electric field along any direction is the negative of the rate of change of the potential along that direction.*

From a practical point of view (21-7) is of considerable importance. By use of these formulas we can calculate the electric field for any charge distribution once its potential function V has been obtained. As will be seen below, the problem of calculating the potential function is, as a rule, appreciably simpler than that of calculating $\mathbf{E}$ directly.

Example 21-3 In Section 21-5 it is shown that the potential V associated with a dipole of moment p located at the origin of a certain coordinate system and oriented along the positive z-axis is

$$V = \frac{p}{4\pi\epsilon_0} \frac{z}{(x^2 + y^2 + z^2)^{3/2}}$$

Calculate the electric field everywhere for this dipole.

Solution To obtain **E** we need only evaluate the derivatives of V in accordance with (21-7). Thus

$$E_x = -\frac{\partial V}{\partial x} = -\frac{p}{4\pi\epsilon_0}\frac{\partial}{\partial x}\left[\frac{z}{(x^2+y^2+z^2)^{3/2}}\right]$$

$$= \frac{3p}{4\pi\epsilon_0}\frac{xz}{(x^2+y^2+z^2)^{5/2}}$$

and similarly for E_y

$$E_y = \frac{3p}{4\pi\epsilon_0}\frac{yz}{(x^2+y^2+z^2)^{5/2}}$$

For E_z it is necessary to apply the rule for differentiating a product. The result is

$$E_z = -\frac{p}{4\pi\epsilon_0}\frac{\partial}{\partial z}\frac{z}{(x^2+y^2+z^2)^{3/2}}$$

$$= -\frac{p}{4\pi\epsilon_0}\left[\frac{1}{(x^2+y^2+z^2)^{3/2}} - \frac{3z^2}{(x^2+y^2+z^2)^{5/2}}\right]$$

$$= -\frac{p}{4\pi\epsilon_0}\frac{1}{(x^2+y^2+z^2)^{5/2}}[x^2+y^2-2z^2]$$

Example 21-4 A solid conductor is placed into an electric field. Show that the potential associated with the resultant field must have the same value for every point inside and on the surface of the conductor.

Solution Let us first establish that every point at the surface of the conductor is at the same potential. To this end, consider in Figure 21-5 a small displacement $d\mathbf{l}_1$, which lies entirely *on* the surface of the conductor. If **E** is the electric field at this point, then the change dV in the potential associated with the displacement $d\mathbf{l}_1$ is given, according to (21-3), by

$$-\mathbf{E}\cdot d\mathbf{l}_1 = dV$$

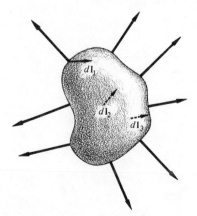

Figure 21-5

But the electric field on the surface of a conductor must be perpendicular to its surface. Hence, since dl_1 lies on the surface, it follows that $\mathbf{E} \cdot dl_1 = 0$, and thus $dV = 0$. In other words, as we go from one point on the surface of a conductor to a neighboring point, the change in the potential is zero. Since by repeated use of this argument we can establish that the change in potential is zero between any two points on the conducting surface, it follows that all points on the conducting surface must be at the same potential.

In a similar way, since for each point in the interior of the conductor the electric field vanishes, it follows that $\mathbf{E} \cdot dl_2 = 0$ for any displacement dl_2 lying entirely in the interior of the conductor. Thus, again, each point inside the conductor must be at the same potential. Finally, by considering a small displacement dl_3 from the interior of the conductor to its surface, we conclude that *all* elements of the conductor must be at the same potential.

21-5 The potential function for a system of charged particles

In order for (21-7) to be useful for calculating electric fields, it is necessary to have available formulas for the potential function of physically interesting systems. In this section we consider the problem of calculating the potential function for a collection of charged particles. The associated problem of continuous distributions will be considered in Section 21-6.

Let us consider first the case of a single particle of charge q, as in Figure 21-6. Since the potential associated with the point at infinity is assumed to vanish, it follows by integrating both sides of (21-3) along the radial path in the figure from infinity to point P that the potential V at a distance r from the particle is

$$V = -\int_{\infty}^{r} \mathbf{E} \cdot d\mathbf{l} \tag{21-8}$$

Now the electric field $\mathbf{E}$ at a distance s from the particle is

$$\mathbf{E} = \frac{q}{4\pi\epsilon_0} \frac{1}{s^2} \hat{\mathbf{r}}$$

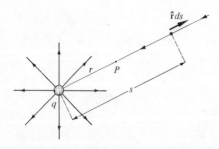

Figure 21-6

with $\hat{r}$ a unit vector along the radial direction. An infinitesimal displacement $d\mathbf{l}$ of length ds along this direction is $\hat{r}\,ds$. Substitution into (21-8) yields, for the potential at point P,

$$V = -\int_\infty^r \mathbf{E} \cdot d\mathbf{l} = -\frac{q}{4\pi\epsilon_0}\int_\infty^r \frac{ds}{s^2} = -\frac{q}{4\pi\epsilon_0}\left(-\frac{1}{s}\right)\Big|_\infty^r$$

where the last equality follows from the fact that $d(1/s)/ds = -1/s^2$. Thus we obtain the final formula

$$V = \frac{q}{4\pi\epsilon_0}\frac{1}{r} \tag{21-9}$$

for the potential due to a charged particle.

The compatibility of this formula with (20-3) is easily established by use of (21-7) and the fact that $r = [x^2 + y^2 + z^2]^{1/2}$. The details are left as an exercise. Equivalently, since V in (21-9) varies only along the radial direction, it follows from (21-3) that $\mathbf{E}$ must also. And its magnitude E may therefore be obtained with the choice $d\mathbf{l} = \hat{r}\,dr$ to be

$$E = -\frac{dV}{dr} = -\frac{q}{4\pi\epsilon_0}\frac{d}{dr}\left(\frac{1}{r}\right) = \frac{q}{4\pi\epsilon_0}\frac{1}{r^2}$$

The calculation of the potential function V associated with a collection of charged particles proceeds in the same way as that for (21-9). The result is

$$V = \frac{1}{4\pi\epsilon_0}\sum_{i=1}^{N}\frac{q_i}{r_i} \tag{21-10}$$

where q_i ($i = 1, 2, \ldots, N$) is the charge of the ith particle and r_i is its distance from the field point. Thus, whereas in computing the electric field due to a collection of charged particles it is necessary to evaluate the vector sum of the electric fields due to each one, in calculating the associated potential function we need only add together the scalar potentials associated with each particle. From a practical point of view, the latter procedure is considerably simpler.

Note that in (21-9) the symbol r, which represents the distance from the particle to the field point, is an inherently positive quantity. In particular, then, the sign of the potential function for a charged particle is the same as the sign of the particle's charge. An analogous statement applies to the more general formula in (21-10).

Example 21-5 How much work is required to bring a particle of charge 6.0 μC from infinity to a point 10^{-2} meter from a fixed particle of charge $-3.0\ \mu$C?

Solution According to (21-9), the potential V_0 at a distance of 10^{-2} meter from a particle of charge $-3.0\ \mu$C is

$$V_0 = \frac{q}{4\pi\epsilon_0}\frac{1}{r} = \left(9.0 \times 10^9\ \frac{\text{N-m}^2}{\text{C}^2}\right) \times \frac{(-3.0 \times 10^{-6}\ \text{C})}{10^{-2}\ \text{m}}$$

$$= -2.7 \times 10^6\ \text{V}$$

where the minus sign is due to the fact that the particle has a negative charge. The work required to bring a particle of charge q from infinity to a point where the potential has the value V_0 is qV_0, according to (21-5). Using the known values for q and V we thus obtain

$$W = (6.0 \times 10^{-6}\,\text{C}) \times (-2.7 \times 10^6\,\text{V}) = -16\,\text{J}$$

The negative sign here signifies that the *electric field* has carried out positive work on the particle.

Example 21-6   Four particles each of charge q are located at the vertices of a square of side a, as in Figure 21-7. Calculate the potential at the center of the square.

Solution   The diagonals of the square each have a length $a\sqrt{2}$, and hence the distance of each particle from the field point is $a\sqrt{2}/2$. Substituting this into (21-10), we obtain for V

$$V = \frac{1}{4\pi\epsilon_0} \sum \frac{q_i}{r_i} = \frac{1}{4\pi\epsilon_0} q \frac{4}{a\sqrt{2}/2}$$

$$= \frac{\sqrt{2}q}{\pi\epsilon_0 a}$$

What is the electric field at this field point?

Example 21-7   A charge q is placed on a conducting sphere of radius a. Show that (21-9) correctly gives the potential outside of the sphere $(r > a)$ and calculate the potential V_0 of the sphere.

Solution   By symmetry we know that the charge will distribute itself uniformly over the surface of the conducting sphere. It follows then by use of the result of Problem 38 of Chapter 20 that the electric field outside the sphere is the same as it would be if the entire charge q were concentrated at its center. Thus, for $r > a$, (21-9) correctly describes the potential due to the sphere.

   Now, by the definition in (21-3), the potential function is differentiable, and hence it must be continuous. It follows that the potential V_0 of the surface of the sphere is

$$V_0 = \frac{q}{4\pi\epsilon_0} \frac{1}{a}$$

The fact that all points of the conducting sphere must be at the same potential was established in Example 21-4.

   Figure 21-8 shows a plot of this potential. Note that even though the potential is continuous, its derivative—that is, the slope of this curve—is not continuous at $r = a$. This reflects the fact that the magnitude of the electric field vanishes inside the conductor and jumps discontinuously to the value $(q/4\pi\epsilon_0 a^2)$ just outside.

Example 21-8   Derive the formula

$$V = \frac{p}{4\pi\epsilon_0} \frac{z}{(x^2 + y^2 + z^2)^{3/2}}$$

for the potential of a dipole of moment p located at the origin and oriented along the positive z-axis.

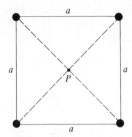

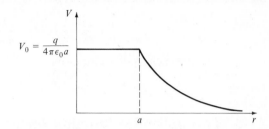

Figure 21-7   **Figure 21-8**

Solution   Consider, in Figure 21-9, two particles of charges $+q$ and $-q$ located at the respective points $(0, 0, a)$ and $(0, 0, -a)$ in the given coordinate system. According to (21-10), the potential V for this charge configuration is

$$V = \frac{q}{4\pi\epsilon_0}\left\{\frac{1}{[x^2 + y^2 + (z - a)^2]^{1/2}} - \frac{1}{[x^2 + y^2 + (z + a)^2]^{1/2}}\right\} \qquad \textbf{(21-11)}$$

since, for example, $[x^2 + y^2 + (z - a)^2]^{1/2}$ is the distance from the field point with coordinates (x, y, z) to the positive particle at $(0, 0, a)$.

Now, by definition, the dipole potential is obtained from the above formula by carrying out the limits $a \to 0$ and $q \to \infty$ in such a way that the product $p = 2aq$ remains finite. Making use of the formula

$$\frac{1}{(1 + \epsilon)^{1/2}} \cong 1 - \frac{1}{2}\epsilon \qquad [|\epsilon| \ll 1]$$

the first term in (21-11) becomes (in the limit as $a \to 0$)

$$\frac{1}{[x^2 + y^2 + (z - a)^2]^{1/2}} = \frac{1}{[r^2 - 2az + a^2]^{1/2}} = \frac{1}{r}\frac{1}{\left[1 + \dfrac{a^2 - 2az}{r^2}\right]^{1/2}} \cong \frac{1}{r}\left[1 + \frac{az}{r^2}\right]$$

$$\textbf{(21-12)}$$

where $r = (x^2 + y^2 + z^2)^{1/2}$. In a similar way the second term in (21-11) may be approximated by

$$-\frac{1}{[x^2 + y^2 + (z + a)^2]^{1/2}} \cong -\frac{1}{r}\left[1 - \frac{az}{r^2}\right] \qquad \textbf{(21-13)}$$

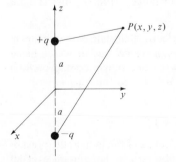

Figure 21-9

Substituting (21-12) and (21-13) into (21-11), we thus obtain

$$V = \frac{q}{4\pi\epsilon_0}\left\{\frac{1}{r} + \frac{az}{r^3} - \frac{1}{r} + \frac{az}{r^3}\right\} = \frac{2aqz}{4\pi\epsilon_0 r^3} = \frac{pz}{4\pi\epsilon_0 r^3}$$

where the final equality follows from the definition $p = 2aq$.

21-6 The potential function for continuous charge distributions

In this section we illustrate the application of (21-10) to continuous charge distributions by reference to several examples. For a discussion of the physical significance of these distributions, see Section 20-4.

Example 21-9 Calculate the potential on the axis of and at a distance z from the plane of a circular loop of radius a carrying a charge per unit length λ (Figure 21-10).

Solution According to (21-9), the potential dV at the given field point due to the element of charge $\lambda a\, d\theta$ located at the angular displacement θ from a fixed reference line is

$$dV = \frac{\lambda a\, d\theta}{4\pi\epsilon_0} \frac{1}{[a^2 + z^2]^{1/2}}$$

Integrating over all values of θ from 0 to 2π we obtain for the potential V due to the entire ring

$$V = \int dV = \int_0^{2\pi} \frac{\lambda a\, d\theta}{4\pi\epsilon_0} \frac{1}{[a^2 + z^2]^{1/2}} = \frac{\lambda a}{4\pi\epsilon_0(a^2 + z^2)^{1/2}} \int_0^{2\pi} d\theta$$

$$= \frac{\lambda a}{2\epsilon_0} \frac{1}{(a^2 + z^2)^{1/2}} \tag{21-14}$$

It is left as an exercise to confirm by substitution into (21-7) that this formula yields the electric field in (20-7), as it must.

Example 21-10 A circular disk of radius R has a uniform charge per unit area σ. Calculate the potential at a point on the axis and at a distance z from the disk (Figure 21-11).

Solution As for the corresponding problem of the electric field calculation, imagine the disk as consisting of a series of concentric rings, a typical one of which has a radius r and thickness dr. The area of this ring is $2\pi r\, dr$ and thus it contains a total charge $2\pi\sigma r\, dr$. According to (21-14), the potential dV due to this ring is

$$dV = \frac{1}{4\pi\epsilon_0} 2\pi\sigma r\, dr \frac{1}{(r^2 + z^2)^{1/2}}$$

where we have made the substitution $a \rightarrow r$ since the radius of the loop in the present case is r. The potential V due to the entire disk is obtained by integration over all

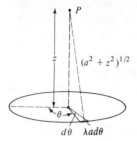

Figure 21-10

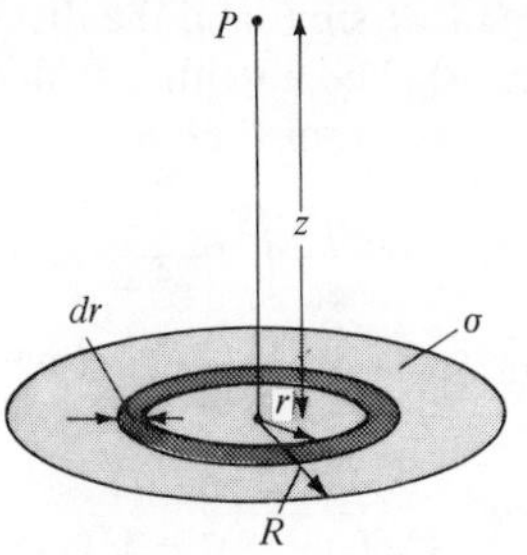

Figure 21-11

values of r from 0 to R. Thus we obtain

$$V = \int dV = \int_0^R \frac{2\pi\sigma r \, dr}{4\pi\epsilon_0} \frac{1}{(r^2 + z^2)^{1/2}}$$

$$= \frac{\sigma}{2\epsilon_0} \int_0^R \frac{r \, dr}{(r^2 + z^2)^{1/2}} = \frac{\sigma}{2\epsilon_0} (r^2 + z^2)^{1/2} \Big|_0^R$$

$$= \frac{\sigma}{2\epsilon_0} [(R^2 + z^2)^{1/2} - \sqrt{z^2}] \qquad \textbf{(21-15)}$$

Again, the consistency of this formula with (20-8) may be established by substitution into the third equation of (21-7).

Example 21-11   A sphere of radius a carries a uniform surface charge per unit area σ. Calculate the potential at a point P outside of the sphere by direct integration.

Solution   Consider, in Figure 21-12, a circular element of the sphere of radius $a \sin \theta$ and of thickness $a \, d\theta$. The area of this element is $(2\pi a \sin \theta)(a \, d\theta)$ and hence it has a total charge $2\pi\sigma a^2 \sin \theta \, d\theta$. According to (21-14), the potential dV at P due to this charge is

$$dV = \frac{2\pi\sigma a^2 \sin \theta \, d\theta}{4\pi\epsilon_0} \frac{1}{[(a \sin \theta)^2 + (r - a \cos \theta)^2]^{1/2}}$$

$$= \frac{\sigma a^2 \sin \theta \, d\theta}{2\epsilon_0} \frac{1}{[a^2 + r^2 - 2ar \cos \theta]^{1/2}}$$

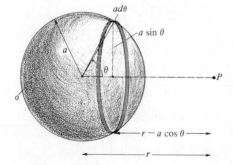

Figure 21-12

since the radius of the ring is $a \sin \theta$ and the distance from the center of the ring to the field point is $(r - a \cos \theta)$. The potential V due to the total sphere is obtained by integrating for all values of θ from 0 to π:

$$
\begin{aligned}
V = \int dV &= \frac{\sigma a^2}{2\epsilon_0} \int_0^\pi \frac{\sin \theta \, d\theta}{[r^2 + a^2 - 2ar \cos \theta]^{1/2}} \\
&= \frac{\sigma a^2}{2\epsilon_0} \frac{[a^2 + r^2 - 2ar \cos \theta]^{1/2}}{ar} \bigg|_0^\pi \\
&= \frac{\sigma a^2}{2\epsilon_0 ar} \{\sqrt{(a + r)^2} - \sqrt{(r - a)^2}\} = \frac{\sigma a^2}{2\epsilon_0 ar} 2a \\
&= \frac{Q}{4\pi\epsilon_0} \frac{1}{r}
\end{aligned}
\tag{21-16}
$$

where $Q = 4\pi a^2 \sigma$ is the total charge on the sphere. This result is consistent with that found less directly in Example 21-7.

21-7 Equipotential surfaces and conductors

Consider a region of space in which there exists an electric field $\mathbf{E}$, with its associated potential function V. A surface S (be it a purely mathematical or an actual physical surface) is said to be an *equipotential surface* if the value of the potential at each point of the surface is the same. Thus, for a given potential function $V \equiv V(x, y, z)$ a given surface S is an equipotential surface at the value V_0 provided that all points (x, y, z) that satisfy the relation

$$V(x, y, z) = V_0$$

lie on this surface S and conversely all points of S satisfy this relation.

Consider, for example, the potential function associated with two particles of charges $+q$ and $-q$ and located in a given coordinate system at the respective points $(0, 0, a)$ and $(0, 0, -a)$. The potential function V is

$$V(x, y, z) = \frac{q}{4\pi\epsilon_0} \left\{ \frac{1}{[x^2 + y^2 + (z - a)^2]^{1/2}} - \frac{1}{[x^2 + y^2 + (z + a)^2]^{1/2}} \right\}$$

Hence the points that lie on the equipotential surface characterized by the value V_0 satisfy the relation

$$V_0 = \frac{q}{4\pi\epsilon_0} \left\{ \frac{1}{[x^2 + y^2 + (z - a)^2]^{1/2}} - \frac{1}{[x^2 + y^2 + (z + a)^2]^{1/2}} \right\}$$

In particular, the entire x-y plane defined by the equation $z = 0$ is an equipotential surface corresponding to the value $V_0 = 0$.

Similarly, since for a particle of charge q the potential function is

$$V = \frac{q}{4\pi\epsilon_0} \frac{1}{r}$$

it follows that the associated equipotential surfaces are the concentric spheres, r = constant, centered at the given particle. Specifically, the sphere of radius b is equipotential at the value V_0 given by

$$V_0 = \frac{q}{4\pi\epsilon_0}\frac{1}{b}$$

Figure 21-13 shows some of the equipotential surfaces (the dashed lines) and the associated electric field lines for these two systems. The vertical dashed line in Figure 21-13a represents the zero potential surface.

Equipotential surfaces have various interesting properties. Two of these, which follow directly from the definition of an equipotential surface are:

1. The work required to take a charged particle from one point on an equipotential surface to another point on this same surface is zero.
2. The electric field lines are everywhere perpendicular to the equipotential surfaces.

The first of these properties follows directly from (21-5). For if the points A and B lie on the same equipotential surface, then by definition $V_A = V_B$ and thus $W_{AB} = 0$. The second property follows from (21-3):

$$\mathbf{E} \cdot d\mathbf{l} = -dV$$

For if $d\mathbf{l}$ is a small displacement that lies entirely on an equipotential surface, then by definition of such a surface $dV = 0$, and hence $\mathbf{E} \cdot d\mathbf{l} = 0$. In other words, $\mathbf{E}$ must be perpendicular to $d\mathbf{l}$. Since by hypothesis $d\mathbf{l}$ lies entirely on the equipotential surface, it follows then that $\mathbf{E}$ must be perpendicular to this surface. This property that the electric field lines and the equipotential surfaces are mutually orthogonal is exemplified in Figure 21-13.

One of the very important applications of this notion of equipotential

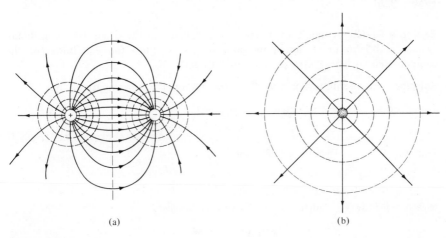

(a)   (b)

Figure 21-13

surfaces deals with conductors in an electric field. In Example 21-4 it was established that if a conductor is in an electric field, then the conducting surface must itself be an equipotential surface and have the same value as does the potential inside. Consider, in Figure 21-14a, a region of space where there exists an electric field and let the dashed surface S represent a closed equipotential surface with the value $V_0 = 0$. If, as in Figure 21-14b, we place a solid conductor of precisely the same shape as the surface S at its location and ground the conductor so that its potential is also zero, then the electric field outside of the conductor will be precisely the same as it was originally. In the interior of the conductor, on the other hand, the electric field must now vanish. Hence *if an equipotential surface is replaced by a conductor at the same potential as is the surface, then all external electric field lines remain unaltered.* However, those field lines that originally penetrated the equipotential surface will now terminate or originate on induced charge on the surface of the conductor.

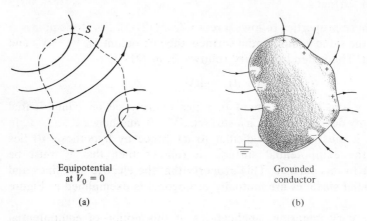

Equipotential
at $V_0 = 0$

(a)

Grounded
conductor

(b)

Figure 21-14

Example 21-12   A dipole of moment p_0 is located at the origin of a coordinate system and is oriented along its positive x-axis. Find the equations for the equipotential surfaces corresponding to $+V_0$ and $-V_0$.

Solution   According to Example 21-8, the potential V for this dipole is

$$V(x, y, z) = \frac{p_0}{4\pi\epsilon_0} \frac{x}{(x^2 + y^2 + z^2)^{3/2}}$$

Thus the $V = +V_0$ surface satisfies the equation

$$(x^2 + y^2 + z^2)^{3/2} = \alpha x$$

with $\alpha = p_0/4\pi\epsilon_0 V_0$, and the $V = -V_0$ surface satisfies

$$(x^2 + y^2 + z^2)^{3/2} = -\alpha x$$

Transforming to polar coordinates, $r = (x^2 + y^2)^{1/2}$ and $\cos\theta = x/r$, we obtain the projection of these two surfaces in the x-y plane:

$$r^2 = \pm\alpha \cos\theta$$

and these are plotted in Figure 21-15. The complete surface is obtained by rotating this figure about the x-axis.

Example 21-13 Two infinite, conducting planes are parallel, at a separation distance d and carry the respective uniform surface charge densities $+\sigma$ and $-\sigma$. Suppose the positively charged plate is grounded.

(a) Calculate the potential of the negatively charged plate.

(b) What is the potential V_0 of the equipotential surface at a distance x ($<d$) away from the positive plate?

Solution As shown in Figure 21-16, the field lines are uniform and perpendicular to the plates. They originate on the positively charged plate and terminate on the negative charge on the other plate. Since the plates are infinite in extent, the electric field is confined to the region between the plates. The equipotential surfaces must be perpendicular to the field lines and are thus planes parallel to the plates. These equipotential surfaces are designated by the dashed lines in the figure.

(a) According to (20-16), the magnitude of the electric field immediately to the right of the positive plate and to the left of the negative plate is

$$E = \frac{\sigma}{\epsilon_0}$$

and thus, according to the above argument, σ/ϵ_0 is the electric field strength everywhere between the plates. Hence the work required to take a unit charge from the positive to the negative plate is $-\sigma d/\epsilon_0$. Making use of (21-5) and the fact that the left-hand plate is grounded, we conclude then that the potential of the negatively charged plate is $-\sigma d/\epsilon_0$.

(b) By use of the same arguments as above, the potential $V(x)$ at a distance x ($<d$) from the positive plate is found to be

$$V(x) = -\frac{\sigma}{\epsilon_0}x$$

This time the work required to take a unit charge from the positive plate a distance x toward the negative one is $-\sigma x/\epsilon_0$.

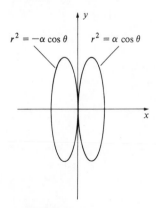

$r^2 = -\alpha \cos\theta$ $r^2 = \alpha \cos\theta$

Figure 21-15

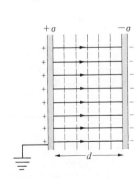

Figure 21-16

21-8 Energy in the electrostatic field

The notion of energy plays a very important and often central role in many branches of physical science. The purpose of this section is to examine this concept as it relates to the electrostatic field.

Consider first the case of two particles of charges q_1 and q_2 separated by a distance r_{12}. The electrostatic energy U_2 associated with this two-particle system is numerically equal to the work W_2 required to bring them from a state of infinite separation to the given configuration. According to (21-9), the potential function V associated with q_1 at a point a distance r_{12} away is

$$V = \frac{q_1}{4\pi\epsilon_0} \frac{1}{r_{12}}$$

But the work W_2 required to bring q_2 from a state of infinite separation to within a distance r_{12} of q_1 is $q_2 V$. Hence, equating this to the energy U_2 of this two-particle system, we find that

$$U_2 = \frac{q_1 q_2}{4\pi\epsilon_0} \frac{1}{r_{12}} \tag{21-17}$$

Consistent with the facts that charged particles of the same sign repel and those of unlike sign attract, U_2 is positive in the former case and negative in the latter. In other words, if the force between the particles is repulsive, then consistent with (21-17), positive work must be carried out on the system, and $U_2 > 0$. For the case $q_1 q_2 < 0$ the field does work on the agent since the electric force is attractive, and hence $U_2 < 0$ in this case.

The calculation of the electrostatic energy U_3 for three charged particles proceeds along similar lines. This time, U_3 is the sum of the work U_2 required to bring q_1 and q_2 to their final positions, plus that required to bring q_3 to its final position at the point P once q_1 and q_2 are in place; see Figure 21-17. The value of the potential V at the point P due to q_1 and q_2 is, according to (21-10),

$$V = \frac{1}{4\pi\epsilon_0} \left[\frac{q_1}{r_{13}} + \frac{q_2}{r_{23}} \right]$$

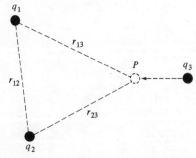

Figure 21-17

where r_{13} and r_{23} are the respective distances of P from q_1 and q_2. The work required to bring in q_3 is $q_3 V$, and adding this to U_2 we obtain

$$U_3 = \frac{1}{4\pi\epsilon_0}\left[\frac{q_1 q_2}{r_{12}} + \frac{q_1 q_3}{r_{13}} + \frac{q_2 q_3}{r_{23}}\right] \tag{21-18}$$

Note the symmetry of this formula under permutation of the indices. This is as it should be, since the order in which the particles are brought to the final configuration should have no physical significance. The corresponding formula for more than two particles should be evident by analogy to (21-18).

Example 21-14   Three particles, each of charge q, are at the vertices of an equilateral triangle of side a. Calculate:
(a) The electrostatic energy of this configuration.
(b) The change in this energy if the particles are pulled apart until the sides of the triangle have a length $2a$.

Solution
(a) In the notation of (21-18), the parameter values are $q_1 = q_2 = q_3 = q$ and $r_{12} = r_{23} = r_{13} = a$. Thus we find that

$$U = \frac{1}{4\pi\epsilon_0}\left(\frac{q_1 q_2}{r_{12}} + \frac{q_1 q_3}{r_{13}} + \frac{q_2 q_3}{r_{23}}\right) = \frac{3q^2}{4\pi\epsilon_0}\frac{1}{a}$$

(b) By use of this result, it follows that if the sides of the triangle were of length $2a$, then the energy would be

$$\frac{3q^2}{4\pi\epsilon_0}\frac{1}{2a}$$

Thus the energy change ΔU or the work required to go from the former to the latter configuration is

$$\Delta U = \frac{3q^2}{4\pi\epsilon_0}\left(\frac{1}{2a} - \frac{1}{a}\right) = -\frac{3q^2}{8\pi\epsilon_0 a}$$

The fact that $\Delta U < 0$ means that the energy of the electrostatic field has decreased. This is consistent with the fact that the external agent must carry out *negative* work to bring about this configuration change.

Example 21-15   A spherical conductor of radius a has a total charge Q. What is the energy U of this system?

Solution   To obtain U, let us calculate the work W required to bring the total charge Q from infinity to the surface of the sphere in successive, infinitesimal amounts. Suppose at some stage an amount of charge q has been brought up to the sphere. The work dW required to bring up an additional infinitesimal amount dq is

$$dW = V(q)\,dq = \frac{q}{4\pi\epsilon_0}\frac{1}{a}\,dq$$

where $V(q)$ is the potential of the sphere when it has charge q, and the second equality follows from (21-16) for $r = a$. The total work W and thus the electrostatic energy is obtained by integrating over all values of q from 0 to Q. The result is:

$$U = W = \int dW = \frac{1}{4\pi\epsilon_0 a} \int_0^Q q \, dq = \frac{Q^2}{8\pi\epsilon_0 a} \tag{21-19}$$

21-9   Spherically and cylindrically symmetric charge distributions

Although it is possible, in principle, to use the methods of Section 21-6 to derive formulas for the potential function associated with any continuous distribution of charge, the integrals involved are generally much too complex for the results to be practically useful. However, for the special case of charge distributions having spherical or cylindrical symmetry there is an alternate and practical method available. For these cases the electric field $\mathbf{E}$ points along the radial direction and varies only with the radial distance r. Hence the potential V is determined by

$$E = -\frac{dV}{dr} \tag{21-20}$$

the validity of which follows from (21-3) if we select for $d\mathbf{l}$ a small displacement dr along the radial direction. Note that the symbol r in this context is used in two senses. For situations having spherical symmetry it denotes the radial distance from the origin, whereas for cases of cylindrical symmetry it represents the radial distance from the axis.

To illustrate the usage of (21-20) let us consider two examples involving spherical symmetry. Applications to cases of cylindrical symmetry will be found in the problems.

Example 21-16   Find the potential function associated with a uniformly charged sphere of radius a and charge density ρ_0.

Solution   Substituting (20-20) and (20-21) into (21-20), and making use of the fact that the total charge Q_0 of the sphere is $4\pi\rho_0 a^3/3$, we find that

$$\frac{dV}{dr} = -E = -\frac{\rho_0}{3\epsilon_0} \begin{cases} a^3/r^2 & r \geq a \\ r & r \leq a \end{cases}$$

Hence since $r = d(r^2/2)/dr$ and $d(-r^{-1})/dr = 1/r^2$, this may be integrated to

$$V(r) = -\frac{\rho_0}{3\epsilon_0} \begin{cases} c_1 - a^3/r & r \geq a \\ c_2 + r^2/2 & r \leq a \end{cases}$$

with c_1 and c_2 constants of integration. The boundary condition that $V(\infty) = 0$ determines the value $c_1 = 0$, while the condition that $V(r)$ be continuous at $r = a$ determines the second constant c_2 to be $-3a^2/2$. Hence we obtain the final formula

$$V(r) = \frac{\rho_0}{3\epsilon_0} \begin{cases} a^3/r & r \geq a \\ \dfrac{3a^2}{2} - \dfrac{r^2}{2} & r \leq a \end{cases} \qquad (21\text{-}21)$$

Example 21-17 Two concentric, conducting spherical shells of radii a and b have charges $+q$ and $-q$, respectively (Figure 21-18). Calculate the potential $V(a)$ of the smaller sphere assuming that the outer one is grounded so that $V(b) = 0$.

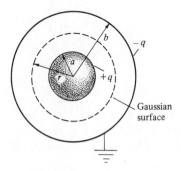

Figure 21-18

Solution Applying Gauss' law to a spherical surface of radius r $(a < r < b)$ we find, as in Example 20-13, that the electric field $E(r)$ $(a < r < b)$ is radial and is the same as that of a point charge q. Thus (21-20) becomes in this case

$$\frac{dV}{dr} = -E = -\frac{q}{4\pi\epsilon_0} \frac{1}{r^2} \qquad a < r < b$$

Integrating both sides, from $r = a$ to $r = b$, we find that

$$V(b) - V(a) = -V(a) = -\frac{q}{4\pi\epsilon_0} \int_a^b \frac{dr}{r^2} = -\frac{q}{4\pi\epsilon_0} \left[-\frac{1}{r} \right] \Big|_a^b$$

$$= \frac{q}{4\pi\epsilon_0} \left[\frac{1}{b} - \frac{1}{a} \right]$$

The fact that $V(a)$ is positive (for $q > 0$) indicates that the outer sphere with its negative charge is at the lower potential. As noted previously, the direction of the electric field is toward decreasing values for potential.

21-10 Summary of important formulas

The work W_{AB} carried out by an agent in the presence of an electric field **E** in slowly transporting a particle of charge q_0 from A to B is

$$W_{AB} = -q_0 \int_A^B \mathbf{E} \cdot d\mathbf{l} \qquad (21\text{-}2)$$

and this is the same for *all paths* connecting the two points A and B. The potential function V associated with $\mathbf{E}$ is defined so that the change dV between two points at a relative infinitesimal displacement $d\mathbf{l}$ is

$$dV = -\mathbf{E} \cdot d\mathbf{l} \tag{21-3}$$

Substitution into (21-2) yields the relation between work and potential difference

$$W_{AB} = q_0(V_B - V_A) \tag{21-5}$$

The potential function V at a distance r from an isolated particle of charge q is

$$V = \frac{q}{4\pi\epsilon_0} \frac{1}{r} \tag{21-9}$$

and for a collection of N particles of charges $q_1, q_2, \ldots, q_n$, the potential V is

$$V = \frac{1}{4\pi\epsilon_0} \sum_{i=1}^{N} \frac{q_i}{r_i} \tag{21-10}$$

where r_i is the distance of the ith particle from the field point.

The energy U_2 associated with a two-particle configuration is the work required to bring them from a state of infinite separation to the given state. It has the value

$$U_2 = \frac{q_1 q_2}{4\pi\epsilon_0} \frac{1}{r_{12}} \tag{21-17}$$

where r_{12} is the distance between the two particles.

QUESTIONS

1. Define or describe briefly what is meant by (a) conservative field; (b) potential difference; (c) equipotential surface; and (d) electrostatic energy.
2. Consider two potential functions V_1 and V_2, which characterize the *same* charge distribution. In what way or ways can they differ from each other? Assuming that $V_1 \neq V_2$, explain why they are physically indistinguishable.
3. The potential of the earth is customarily assigned the value of zero volts. What effect would a different assignment, say 10 volts, have on measured values of potentials? What effect would there be on measurements of potential differences?
4. If the force between two charged

particles varied as $r^{-2-\epsilon}$ ($\epsilon > 0$) instead of r^{-2}, as it actually does, would a potential function for the electrostatic field exist? Would Gauss' law be valid in this circumstance?
5. A positive charge q is transferred from conductor A to conductor B. Assuming A and B are originally electrically neutral, which of them will be at the higher potential? In what way would your answer be different if A and B originally carried a charge?
6. Explain in qualitative terms why the line integral of a conservative field around a *closed* path must vanish. Is this equivalent to the text's definition of such a field?

7. As you go along the direction of an electric field line, do you go to regions of higher or of lower potential? Justify your answer by use of (21-3).

8. A proton is initially at rest in an electric field. Explain why it will go to a region of lower potential. What would happen to an electron under the same circumstance?

9. Consider the physical situation in Example 21-1. If the particle goes from B to A in Figure 21-2, how much work is done by the external agent? (Make use of the result obtained in the example.) How much work is carried out by the electric field? Which of these two works must be positive? Assume that σ is positive.

10. How much work would be carried out for the system in Figure 21-2 if the particle were taken from A to B then to D and finally back to A? Is it necessary for your answer that this path be entirely along the linear paths in the figure?

11. Show that it is possible for the potential to be zero at a point where the electric field does *not* vanish. Is it possible for the electric field to vanish at a point where the potential function does not?

12. Assuming that the potential at infinity vanishes, in Example 21-6 we established that the potential at the center of the square in Figure 21-7 is nonzero. Nevertheless, the electric field vanishes at this point. Is this a contradiction? Explain.

13. Explain why the term $\sqrt{z^2}$ in (21-15) cannot be replaced by z. (*Hint*: Compare the direction in which the electric field would point on opposite sides of the disk if this replacement were made.)

14. Why must the potential be a continuous function of position? (*Hint*: Make use of (21-3) and show that this relation cannot be satisfied for finite $\mathbf{E}$ at a point where the potential is discontinuous.)

15. What can you say about the electric field in a region where the potential is constant? What can you say about the potential in a region of constant electric field?

16. Is it possible for surfaces, characterized by different values for potential, to intersect? Can two such surfaces be tangent to each other? Explain.

17. Must all equipotential surfaces be closed? If your answer is no, give an example of a physical situation for which an equipotential surface is open. If your answer is yes, explain.

18. What simplifications (if any) occur in using Gauss' law with a surface that is equipotential? Need the electric field have the same value everywhere on such a surface?

19. Consider an infinitely long cylinder that contains a uniform charge. Describe the equipotential surfaces.

20. Why must every point in the interior of a conducting shell be at the same potential as the shell is? Assume that there is no charge in the cavity, but that the conductor itself carries a charge.

21. A spherical conductor of radius a has a charge Q. What are the equipotential surfaces inside and outside the conductor? What is the potential of the conductor?

22. If two particles of the same sign of charge are brought closer together, does the electrostatic energy increase or decrease? What happens if they are of the opposite sign?

23. A particle of charge q is brought near an uncharged conductor. Does the electrostatic energy increase or decrease in this process? Is the sign of q relevant to your answer?

PROBLEMS

1. A particle of charge 6.0 μC is in a uniform field **E** of strength 200 V/m. How much work must be supplied by an external agent to move the particle a distance of 0.5 meter:
   (a) Parallel to the field?
   (b) Antiparallel to the field?
   (c) At right angles to the field?
   (d) Along a line inclined at 30° to the field?

2. By use of the fact that for a conservative field **E** the line integral

$$\int_A^B \mathbf{E} \cdot d\mathbf{l}$$

is independent of the path connecting the points A and B, show that the line integral of **E** around a closed path is zero. Derive the converse of this result.

3. An infinite nonconducting plane of charge has a density $\sigma = 10^{-6}$ C/m². (a) What is the electric field strength? (b) Describe the equipotential surfaces. (c) If the plane has zero potential, what is the potential of a point 0.2 meter from the plane?

4. A uniform electric field has a strength of 2.0×10^5 V/m. (a) How far apart are two equipotential surfaces whose potential difference is 2.0 volts? (b) What is the potential difference between two equipotential surfaces separated by a distance of 0.4 meter?

5. Two very large metal plates are separated by a distance of 1 cm and are maintained at a potential difference of 100 volts. (a) What is the strength of the uniform electric field in the region between the plates? (b) How much work is required to take a particle of charge $+6.0$ μC from the plate at the higher potential to the other?

6. What is the strength of the potential at a distance of 2.0×10^{-4} meter from a particle of charge 3.0 μC? At a distance of 2.0 km? Assume in each case that the potential at infinity vanishes.

7. Figure 21-19 shows a particle of charge $+3q$ at a distance d from a particle of charge $-q$. (a) What is the potential at the point P at a distance a from $3q$? (b) For what value of a will the potential at P vanish?

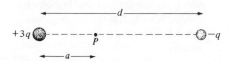

Figure 21-19

8. For the charge configuration in Figure 21-19 calculate:
   (a) The potential at the midpoint of the line joining the two particles.
   (b) The potential at a distance $d/2$ along the perpendicular bisector of the line joining the particles.
   (c) The work required to take a particle of charge q_0 from the point in (a) to the point in (b).

9. The electron and the proton in a hydrogen atom are separated by a distance of 0.53×10^{-10} meter. Assuming the charge values are $\mp 1.6 \times 10^{-19}$ coulomb, calculate the potential at:
   (a) The midpoint of the line joining them.
   (b) A distance of 0.1 mm along their perpendicular bisector.
   (c) How much work is required to bring a second electron from the point in (b) to that in (a)?

10. A particle of charge $+1.0$ μC is located at the point $(0, 0, 1)$ in a cer-

tain coordinate system. A second particle has charge $-2.0\ \mu C$ and is located at $(0, 0, -1)$. Assuming that all lengths are in meters, calculate the value of the potential at the following points: (a) $(0, 0, 0)$; (b) $(0, 0, 2)$; (c) $(1, 1, 0)$.

11. How much work is required to take a particle of charge $-3.0\ \mu C$ from the initial point $(2, 2, 2)$ to each of the three points where the potential has been evaluated in Problem 10?

12. Assume in Figure 21-7 that each particle has a charge q. (a) What is the potential at the midpoint of the upper side of the square? (b) How much work is required to take a particle of charge $-q$ from the midpoint of the upper side to the midpoint of the lower side? (c) Explain your answer to (b) in physical terms.

13. For the charge configuration in Figure 21-7 calculate: (a) The potential at a perpendicular distance y above the plane of the square and vertically above its center. (b) The electric field at the point in (a) by use of (21-7).

14. Three particles of charges q, q, and $-3q$ are at the vertices of an equilateral triangle of side a. Calculate the potential at (a) the midpoint of a side whose ends carry the same charge and (b) a point in the plane of the triangle and equidistant from the three vertices.

15. Calculate everywhere the electric field $\mathbf{E}$ associated with the potential function

$$V = -\alpha(x^2 + y^2 - 2z^2)$$

with α a positive constant.

16. The components of the electric field in a certain coordinate system are (in SI units) $E_x = -10$; $E_y = 3y$; and $E_z = 0$. Calculate a potential function for this field.

17. Show that the components of $\mathbf{E}$ must satisfy

$$\frac{\partial E_x}{\partial y} = \frac{\partial E_y}{\partial x}$$

and two similar relations obtained by cyclic permutation of the indices. (Hint: Express the components of the field in terms of the derivatives of the potential.)

18. A dipole of moment p is oriented along the positive x-axis in a certain coordinate system.
(a) Show that in the x-y plane the potential V may be written

$$V = \frac{p}{4\pi\epsilon_0}\frac{\cos\theta}{r^2}$$

where $r = (x^2 + y^2)^{1/2}$ and $\cos\theta = x/r$ are polar coordinates.
(b) Show by use of (21-3) that the components of $\mathbf{E}$ along the direction of increasing r and of increasing θ, namely E_r and E_θ, respectively, are given by

$$E_r = -\frac{\partial V}{\partial r} \qquad E_\theta = -\frac{1}{r}\frac{\partial V}{\partial\theta}$$

19. A circular disk of radius R_0 carries charge per unit area σ. How much work is required to take a particle of charge q_0 from a point on the axis of the disk and at a distance z from its plane to: (a) The point on the axis at a distance z on the other side of the disk? (b) The center of the disk?

20. By use of (21-15) calculate the electric field at any point on the axis of a disk of radius R and carrying a uniform charge per unit area σ.

*21. A line charge of length $2l$ has charge per unit length λ. Show by direct integration that the potential at a distance r from the line charge and on its perpendicular bisector is

$$V = \frac{\lambda}{4\pi\epsilon_0}\ln\frac{[l + (l^2 + r^2)^{1/2}]}{[-l + (l^2 + r^2)^{1/2}]}$$

22. For the physical system in Problem 21: **(a)** calculate the electric field along the perpendicular bisector and **(b)** show that, for an infinitely long line charge, that is, for $r/l \to 0$, your formula for E reduces to that in (20-6).

23. A nucleus of atomic weight A contains Z protons and may be thought of as a uniform sphere of charge of radius r_0 given by

$$r_0 = 1.1\ A^{1/3} \times 10^{-15}\ \text{m}$$

   (a) In terms of A, Z, and the proton charge e, what is the potential at any point outside of the nucleus?
   (b) Calculate the potential at any point inside the nucleus.
   (c) How much work is required to take an electron from the surface of the nucleus to a distance of 10^{-10} meter from its center? Assume that $A = 235$ and $Z = 92$.

24. A charge of 10^{-7} coulomb is placed on a conducting sphere having a radius of 0.5 meter. **(a)** What is the potential of the sphere? **(b)** What is the value of the potential at a distance of 2 meters from the center of the sphere? **(c)** What is the potential at any point inside the sphere?

25. Figure 21-20 shows two infinite conducting planes, which are parallel to each other and carry uniform surface charge densities $+\sigma$ and $-\sigma$, respectively. Assume that the positive plate is grounded and that an infinite planar conductor of thickness b is introduced at a distance a from the positively charged plate.
   (a) Show that the potential at a distance x ($<a$) from the positive plate is

$$V(x) = -\frac{\sigma}{\epsilon_0} x$$

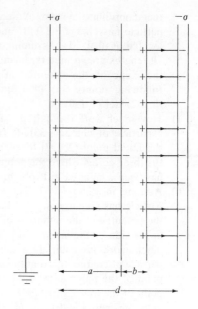

Figure 21-20

   (b) What is the potential of the conducting slab?
   (c) Calculate the potential of the negatively charged plate and contrast your answer with the corresponding one in Example 21-13.

26. Two particles of charges $-qa/b$ and q are located at the points with coordinates $(a^2/b, 0, 0)$ and $(b, 0, 0)$, respectively, with a and b ($>a$) certain positive distances. **(a)** Show that a sphere of radius a centered at the origin is an equipotential surface. **(b)** What value for the potential is associated with this surface?

27. Repeat both parts of Problem 26, but suppose the presence this time of a third particle of charge Q_0 located at the origin.

28. How much work is required to bring the three charges Q_0, $-qa/b$, and q in Problem 27 from a state of infinite separation to the given configuration?

29. What is the electrostatic energy as-

sociated with the charge configuration in Figure 21-7?

30. (a) Calculate the electrostatic energy associated with the two particles in Figure 21-19. (b) What would this energy be if a third particle of charge $-q$ were located at the point P in the figure?

31. Calculate the electrostatic energy associated with a hydrogen atom assuming that the electron and proton are at a separation of 0.53×10^{-10} meter.

32. Two particles, each of charge $2.0 \ \mu C$, are at a separation distance of 0.4 meter. (a) What is the energy associated with this configuration? (b) Where can a third particle of charge $-3.0 \ \mu C$ be placed so that the electrostatic energy of the resulting configuration will be zero? Is this point unique?

33. A regular tetrahedron of side a has a charged particle at each of its four vertices. If three of these have a charge $+q$ and the fourth a charge $-3q$, calculate the electrostatic energy of this configuration. How much work is required to double the sides of the tetrahedron?

34. Consider a nonconducting sphere of radius r, which carries a uniform charge per unit volume ρ_0.
(a) Show by use of (21-21) that the potential V at the surface of this sphere is

$$V = \frac{\rho_0}{3\epsilon_0} r^2$$

(b) Show that the change in electrostatic energy dU, on adding to this sphere a spherical shell of radius r and thickness dr and thus carrying a total charge $4\pi r^2 \ dr \ \rho_0$, is

$$dU = \frac{4\pi}{3\epsilon_0} \rho_0^2 r^4 \ dr$$

(c) By making use of (b), prove that the energy U associated with a sphere of radius a and carrying a uniform charge density ρ_0 is

$$U = \frac{3}{5} \frac{1}{4\pi\epsilon_0} \frac{Q^2}{a}$$

where

$$Q = \frac{4\pi}{3} a^3 \rho_0$$

is the total charge on the sphere.

35. (a) Show by use of the result of Problem 34 that the energy $U(A, Z)$ of a nucleus of atomic weight A and charge Ze, assuming that it is a sphere of uniform charge and of radius $a = 1.1 \times 10^{-15} A^{1/3}$ meter, is

$$U(A, Z) = \frac{3}{5} \frac{e^2}{4\pi\epsilon_0} \frac{Z^2}{1.1 \times 10^{-15} A^{1/3}}$$

Why do you suppose that this formula is normally written with the Z^2 factor replaced by $Z(Z-1)$?
(b) What is the "potential gradient"—that is, the electric field strength—at the surface of the nucleus $^{235}_{92}U$?

36. Using (21-19) calculate the electrostatic energy stored in a conducting sphere of radius 2.0 cm that contains a charge of $10 \ \mu C$.

37. Making use of the result of Problem 34 calculate the "electrostatic energy" of a proton assuming it to be a uniformly charged sphere of radius 3.0×10^{-16} meter (see Figure 19-19). Compare this with the proton's "rest energy," Mc^2 (c = speed of light $\cong 3 \times 10^8$ m/s).

38. A particle of charge q and mass m undergoes motion in an electrostatic field $\mathbf{E}$ with associated potential function V. Show that the energy $\mathscr{E}$ of the particle

$$\mathscr{E} = \frac{1}{2} mv^2 + qV$$

where **v** is the velocity of the particle, is constant in time. (*Hint*: Calculate $d\mathscr{E}/dt$ and show by use of
Newton's law and (21-7) that it vanishes.)

39. By use of the result of Problem 38,
calculate the potential difference
through which an electron travels if:
(a) Its velocity drops from 2.0×10^4 m/s to zero. (b) Its velocity
rises from 2.0×10^4 m/s to 3.0×10^6 m/s.

40. The unit of the "electron volt" (eV)
is an energy unit that represents the
gain or loss of kinetic energy of a
particle of charge $\pm 1.6 \times 10^{-19}$
coulomb when it goes through a
potential drop of 1 volt. Find the
relation between the electron volt
and the joule.

41. A proton starts from rest, travels in
a uniform field for a distance of
0.2 mm, and acquires a final velocity of 3.0×10^3 m/s. (a) Through
what potential difference has it
gone? (b) What is the strength of
the electric field?

42. In an electrostatic accelerator doubly ionized helium atoms or alpha
particles are accelerated from rest
through a potential difference of
2.0×10^6 volts. Calculate the final
velocity of the α particles.

43. A sphere of radius a contains a
uniform charge density ρ_0 (>0) and
is surrounded by a concentric
grounded conductor of radius b
(Figure 21-21). Calculate: (a) the

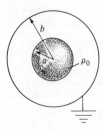

Figure 21-21

potential for $r \leqslant a$; (b) the potential
for $b \leqslant r \leqslant a$; and (c) the charge
density σ on the surface of the
conductor.

44. A particle of charge q is at the
center of a spherical conducting
shell of inner radius a and outer
radius b ($>a$), which in turn is
inside a concentric grounded spherical conductor of radius c ($>b$).
Calculate:
(a) the potential for $r \leqslant a$;
(b) the potential for $a \leqslant r \leqslant b$;
(c) the potential for $b \leqslant r \leqslant c$.

*45. A spherical hole of radius a is cut
out of the center of a sphere of
radius b ($>a$) and containing a uniform charge per unit volume ρ_0.
Show that the potential in the hole
is constant and determine the value
of this constant assuming that $V = 0$ at infinity.

46. A conducting sphere of radius a
carries a charge q.
(a) What is the potential of the
sphere?
(b) What is the electric field
strength at the surface of the
sphere?
(c) The air surrounding such a
sphere will break down (that is,
a spark will be generated) if the
electric field reaches a strength
$\cong 3 \times 10^6$ V/m. How much
charge can be placed on a
sphere with a radius of 0.2
meter without causing a break-
down?

47. An infinite cylinder of radius a has
a uniform charge per unit volume
ρ_0. Show that the potential at a
distance r from the axis of the
cylinder is

$$V(r) = \frac{\rho_0}{2\epsilon_0}$$

$$\times \left\{ \begin{array}{ll} -r^2/2 & r \leqslant a \\ \dfrac{a^2}{2} - a^2 \ln \dfrac{r}{a} & r \geqslant a \end{array} \right\}$$

provided $V(0) = 0$. Explain in physical terms why we *cannot* select $V(\infty) = 0$ in this case.

48. Consider two coaxial cylindrical conducting shells of radii a and b ($>a$). Suppose that the outer one is grounded and the inner one has a charge per unit area σ.

(a) Calculate the electric field everywhere.

(b) What is the potential of the inner cylinder?

(c) Calculate the charge density on the outer cylinder.

22 Dielectric materials and capacitors

The beauty of electricity... is not that the power is mysterious and unexpected... but that it is under law and that the taught intellect can now even govern it largely.

MICHAEL FARADAY (1791–1867)

22-1 Introduction

In our studies of the effects of electric charge on matter, up to this point we have been concerned mainly with conductors. The purpose of this section is to discuss the electric properties of a second class of materials, which are known as *insulators* or *dielectrics*.

A solid conductor, it will be recalled, may be visualized as a rigid lattice of positive ions, throughout which is interspersed a compensating negative charge of highly mobile electrons. When placed into an electric field, these electrons distribute themselves in such a way that the resultant electric field vanishes throughout the conductor.

By contrast, in a dielectric each electron is, as a rule, tightly bound to its parent ion, and thus undergoes very little translational motion even in the presence of an external electric field. However, any externally applied electric field will, in general, bring about some degree of charge separation between each electron and its associated ion, and in this way an originally neutral atom becomes a small electric dipole. As will be seen below, it is

667

because of the existence of these dipoles that a dielectric acquires its distinctive electric properties when placed into an electric field.

In connection with a study of dielectrics it is convenient to introduce first the notion of *capacitance*. Accordingly, the first several sections of this chapter are concerned with a discussion of capacitors and with the derivation of some of their important properties. The remainder of the chapter is then devoted to an analysis of the macroscopic, electric properties of dielectrics using capacitors as a tool.

22-2 Capacitors and capacitance

Consider, in Figure 22-1, two isolated conductors A and B and suppose that a charge q (>0) has been taken from one of these—originally neutral—conductors and placed on the other. The conductors will then have the charges q and $-q$, respectively, and as shown in the figure, electric-field lines will originate on the surface of the positively charged conductor and terminate on the surface of the other. A configuration of two conductors of this type is known as a *capacitor* and the conductors themselves are often referred to as its *plates*. It is customary to assume that $q > 0$, so the symbol q represents the net charge on the positive plate of the capacitor. The charge on the other plate will always be $-q$.

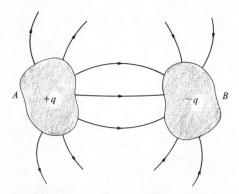

Figure 22-1

Figure 22-2 shows a simple way for charging a capacitor. Suppose that the two conductors A and B are originally electrically neutral and are connected by a conducting wire so that they are at the same potential. If a positive electric charge is brought near, say, conductor B, then some electrons will flow along the conducting wire from A to B. As shown in Figure 22-2b, at equilibrium A will have a certain positive charge q and its partner, B, will have a compensating charge $-q$ of the opposite sign. At this point they are still at the same potential. The remaining steps then involve disconnecting the wire joining A and B and finally removing the external

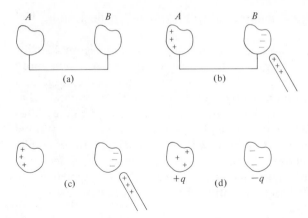

Figure 22-2

charge. As shown in Figure 22-2d, in this way we obtain two conductors carrying equal and opposite charges and thus have a capacitor at a certain charge q.

According to the analysis of Section 21-7, each of the two charged conductors in Figure 22-1 is an equipotential surface and, unless $q = 0$, they are at different potentials. Since the direction of the electric-field lines is from the positively charged conductor A to B, it follows that the potential V_A of A is larger than is the potential V_B of B. The potential V of a capacitor is defined to be the *potential drop* $(V_A - V_B)$ from the positive to the negatively charged conductor. Equivalently, V is the work per unit charge required to take a charged particle from B to A. Note that the potential V of a capacitor is an inherently positive quantity.

The capacitance C of a capacitor is defined by[1]

$$C = \frac{q}{V} \qquad\qquad (22\text{-}1)$$

where V is the potential of the capacitor when it has charge q. Since both V and q are positive quantities it follows that the same must be true for the capacitance C. Note that implicit in this definition is the fact that the capacitance C is a constant independent of V and q. In other words, the ratio q/V for a capacitor is a constant independent of its potential and charge. As will be exemplified for special cases in Section 22-3, C depends only on geometrical factors, such as the shape of the conductors, their relative separation, and so forth.

According to (22-1), the unit of capacitance is the coulomb per volt. A unit equivalent to this is the farad (abbreviated F), which is defined by

$$1\,\text{F} = 1\,\text{farad} = 1\,\text{C/V}$$

[1]The italic letter C, which stands for capacitance, should not be confused with the roman C, which is the abbreviation for the coulomb, the SI unit of charge. As a general rule, the symbol C for the coulomb will usually be preceded by a number.

Also in frequent usage is the microfarad (μF), which is defined to be 10^{-6} farad.

Although it is possible to do so, the procedure shown in Figure 22-2 for charging a capacitor is not always the most desirable. A much more convenient and more readily controllable procedure involves the usage of a device, such as a battery, which has two conducting terminals maintained at some potential difference V_0. If these terminals are connected to the plates of a capacitor, charge will flow from one plate to the other until the potential of the capacitor is the same as that between the terminals. If, for example, the potential difference between the terminals is V_0, then when connected to the plates of a capacitor of capacitance C, a charge $+CV_0$ will appear on the plate at the higher potential and a negative charge $-CV_0$ will appear on the other.

It is customary to represent a capacitor schematically by the symbol

where the two vertical lines represent the plates. In a similar way we shall represent any device with two terminals which are maintained at a fixed potential difference V_0 by the symbol

$$\text{————o } V_0 \text{ o————}$$
$$\qquad\;_+\qquad\,_-$$

where the $+$ ($-$) sign represents the terminal at the higher (lower) potential. In terms of these symbols, Figure 22-3 shows a capacitor of capacitance C whose plates are kept at a potential difference V_0. By our convention, the left-hand plate is at the higher potential and has the positive charge CV_0. The other plate has the compensating charge $-CV_0$.

Example 22-1 Suppose that the plates of a 5.0-μF capacitor are connected to a potential difference of 10 volts.

(a) What is the charge on the capacitor?

(b) If the 5.0-μF capacitor is disconnected from the battery and connected to an originally uncharged 3.0-μF capacitor, what is the final charge on each capacitor?

Solution

(a) According to (22-1), the charge is

$$q = CV = 5.0 \times 10^{-6}\,\text{F} \times 10\,\text{V} = 5.0 \times 10^{-5}\,\text{C}$$

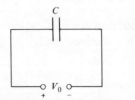

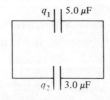

Figure 22-3   *Figure 22-4*

(b) If, as shown in Figure 22-4, q_1 represents the final charge on the 5.0-μF capacitor and q_2 that on the other, then $q_1 + q_2 = 5.0 \times 10^{-5}$ coulomb, since the original charge on the 5.0-μF capacitor is 5.0×10^{-5} coulomb. Moreover, since at equilibrium there is no flow of charge, the left-hand plates of the two capacitors must be at the same potential. And for the same reason the two negatively charged plates must be at the same potential. It follows that the potential across each capacitor must be the same, and therefore (since $V = q/C$ for any capacitor)

$$\frac{q_1}{5.0 \times 10^{-6} \text{ F}} = \frac{q_2}{3.0 \times 10^{-6} \text{ F}}$$

Solving for q_1 and q_2 we obtain

$$q_1 = 3.1 \times 10^{-5} \text{ C} \qquad q_2 = 1.9 \times 10^{-5} \text{ C}$$

22-3 The parallel-plate capacitor

There are a very small number of idealized, but very important, geometric configurations of two conductors for which an explicit formula for the capacitance can be obtained. The purpose of this section is to derive two of these.

Consider first the *parallel-plate* capacitor. This capacitor consists of two very large, parallel, conducting plates, each of area A and separated by certain distance d. If a potential difference V_0 is maintained between the plates, then as shown in Figure 22-5 a certain positive charge per unit area σ will appear on the plate at the higher potential while a compensating negative charge density $-\sigma$ will appear on the other. For regions of space not too close to the edges of the plates, the field lines, as shown in the figure, will originate on the positive charge on the left-hand plate, be perpendicular to the two conductors, and terminate on the negative charge on the other plate. Provided then that the linear dimensions of the plates are very large compared to their separation distance d—a feature we shall always assume

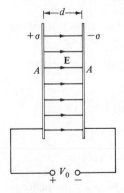

Figure 22-5

to be true—it follows that the electric field will be uniform and perpendicular to both plates and be confined to the region between them. Physically, this means that the nonuniform "fringing fields" that exist at the edges of the plates will be neglected.

Let us then calculate the capacitance C of the parallel-plate capacitor. According to (21-4), the potential V_0 of the capacitor in Figure 22-5 is the negative of the work required to take a unit charge from the left- to the right-hand plate. Since the electric field $\mathbf{E}$ is uniform and perpendicular to the plates it follows that

$$V_0 = Ed$$

Moreover, according to (20-16), E is related to the charge density σ on the positive plate by

$$E = \frac{\sigma}{\epsilon_0}$$

Combining these two relations and noting that the charge q of this capacitor is σA, we find on substitution into (22-1) that

$$C = \frac{q}{V_0} = \frac{\sigma A}{Ed} = \frac{\sigma A}{\sigma d/\epsilon_0}$$

and this leads to the final formula

$$C = \frac{\epsilon_0 A}{d} \tag{22-2}$$

As anticipated above, C is independent of V_0 and q and depends only on the area A of the plates and their separation distance d. Note that the closer together are the plates and the larger is their area, the greater is the capacitance.

The calculation of the capacitance of the spherical and the cylindrical capacitor proceeds, in principle, in the same way. The details will be found in Example 22-4 and in the problems. The results are given by (22-3) and (22-5), respectively.

Example 22-2 Calculate the capacitance of the parallel-plate capacitor whose plates have an area of 15 cm^2 and are separated by a distance of 3.0 mm.

Solution Substituting the given values for A and d into (22-2) and using $\epsilon_0 = 8.9 \times 10^{-12}$ F/m ($\equiv 8.9 \times 10^{-12}$ C^2/N-m^2), we find that

$$C = \frac{\epsilon_0 A}{d} = \frac{(8.9 \times 10^{-12} \text{ F/m}) \times (15 \times 10^{-4} \text{ m}^2)}{3.0 \times 10^{-3} \text{ m}}$$

$$= 4.5 \times 10^{-12} \text{ F} = 4.5 \times 10^{-6} \ \mu\text{F}$$

Example 22-3 It is desired to place a charge of 1.0 μC on a parallel-plate capacitor by use of a 10-volt source. If the separation distance between the plates is to be 1.0 mm, what must be the area of the plates?

Solution Solving (22-2) for A and substituting for C by use of (22-1), we find that

$$A = \frac{Cd}{\epsilon_0} = \frac{qd}{V\epsilon_0}$$

The substitution of the given values for q, V, d, and ϵ_0 then yields

$$A = \frac{qd}{V\epsilon_0} = \frac{(1.0 \times 10^{-6}\,\text{C}) \times (1.0 \times 10^{-3}\,\text{m})}{(10\,\text{V}) \times (8.9 \times 10^{-12}\,\text{F/m})}$$
$$= 11\,\text{m}^2$$

which is a very large area indeed. We see therefore, that in order to store a charge of the order of 1.0 μC on a capacitor of reasonable size, a substantially larger voltage source must be utilized.

Example 22-4 Calculate the capacitance of a spherical capacitor that consists of two concentric conducting spherical shells of radii a and $b (> a)$, respectively (Figure 22-6).

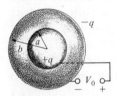

Figure 22-6

Solution As shown in the figure, suppose that there is a charge q on the inner sphere and a compensating charge $-q$ on the outer one. Because of the spherical symmetry, in the region between the conducting shells the potential $V(r)$ is the same as if the charge q were at the center of the inner sphere. Thus for $a \leqslant r \leqslant b$,

$$V(r) = \frac{q}{4\pi\epsilon_0} \frac{1}{r}$$

so that the potential $V[\equiv V(a) - V(b)]$ of the capacitor is

$$V = \frac{q}{4\pi\epsilon_0} \left(\frac{1}{a} - \frac{1}{b}\right)$$

Substitution into (22-1) then yields the desired formula

$$C = 4\pi\epsilon_0 \frac{ab}{(b - a)} \tag{22-3}$$

For the special case of a very large sphere, that is, $b \gg a$, this reduces to

$$C = 4\pi\epsilon_0 a \qquad \left(\frac{b}{a} \to \infty\right) \tag{22-4}$$

and when speaking of the capacitance of an "isolated" spherical conductor, it is this formula which we have in mind. According to this formula, for example, the capacitance of the earth is about 7×10^{-4} farad.

In the problems it is established that the capacitance of a cylindrical capacitor, consisting of two coaxial cylindrical conductors of radii a and b ($> a$) and of length L, is

$$C = \frac{2\pi\epsilon_0 L}{\ln(b/a)} \tag{22-5}$$

22-4 Capacitors in series and parallel

Two capacitors, C_1 and C_2, can be connected together across a potential difference V_0 in only one of two ways. If they are connected as shown in Figure 22-7a, then they are said to be in *parallel*. If, on the other hand, they are connected as in Figure 22-7b, then they are said to be in *series*. Note that for two capacitors in parallel, the potential across each one is the same, whereas the charges on each are, in general, different. As shown in Figure 22-7a, for the parallel connection the positively charged left-hand plates of *both* C_1 and C_2 are at the same potential as is the positive terminal of the external source, and similarly for the right-hand plates. On the other hand, as shown in Figure 22-7b, for two capacitors in series the positively charged plate of one is connected to and is thus at the same potential as the negatively charged plate of the other. Hence, in general, the potentials of two capacitors in series will be different. However, since the charges on the two plates of a capacitor are equal and opposite, and since the negative charge on the right-hand plate of C_1 can only come from electrons that originate on the left-hand plate of C_2, it follows that the charge on two capacitors in series must invariably be the same.

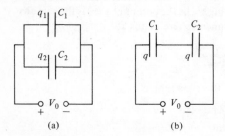

Figure 22-7

We shall now demonstrate that if two capacitors of capacitances C_1 and C_2 are connected in *parallel*, then they are equivalent to a single capacitor of capacitance C, given by

$$C = C_1 + C_2 \qquad \text{(parallel connection)} \tag{22-6}$$

whereas if they are connected in *series*, the corresponding formula is

$$\frac{1}{C} = \frac{1}{C_1} + \frac{1}{C_2} \qquad \text{(series connection)} \tag{22-7}$$

To establish (22-6) let us note that in Figure 22-7a the potential of each capacitor is the same as the potential difference V_0 of the external source. Hence

$$V_0 = \frac{q_1}{C_1} = \frac{q_2}{C_2} \tag{22-8}$$

where q_1 and q_2 are the charges on C_1 and C_2, respectively. Now, by definition, the capacitance C of the capacitor equivalent to these two has the property that when connected to the same source V_0, it will acquire the same *total* charge $(q_1 + q_2)$. Making use of (22-1) we then obtain

$$C = \frac{q_1 + q_2}{V_0} = \frac{C_1 V_0 + C_2 V_0}{V_0} = C_1 + C_2$$

where the second equality follows by use of (22-8). The validity of (22-6) is thereby established.

The derivation of (22-7) follows along similar lines. This time, as noted above and as shown in Figure 22-7b, the charge q is the same on each capacitor. Therefore the potentials of the two capacitors are q/C_1 and q/C_2, respectively. Moreover, from (21-5) and the additivity property of the work carried out in taking a charged particle between two points it follows that the sum of the potentials across the capacitors must be the same as that of the external source, V_0. Hence

$$V_0 = \frac{q}{C_1} + \frac{q}{C_2} = q\left(\frac{1}{C_1} + \frac{1}{C_2}\right)$$

and comparison with (22-1) shows that the capacitance C equivalent to C_1 and C_2 in series is given in (22-7).

Example 22-5 Find the capacitance C of a capacitor equivalent to two capacitors of capacitances $C_1 = 2\ \mu F$ and $C_2 = 4 \mu F$ if:
(a) They are connected in parallel.
(b) They are connected in series.

Solution
(a) According to (22-6), for the parallel connection we have

$$C = C_1 + C_2 = 2\ \mu F + 4\ \mu F = 6\ \mu F$$

(b) Expressing (22-7) in the form

$$C = \frac{C_1 C_2}{C_1 + C_2}$$

we find for the series connection that

$$C = \frac{(2\ \mu F) \times (4\ \mu F)}{2\ \mu F + 4\ \mu F} = \frac{4}{3}\ \mu F$$

Example 22-6 Calculate the equivalent capacitance C of the four capacitors connected as in Figure 22-8a. Assume the values $C_1 = C_2 = 2\ \mu F$ and $C_3 = C_4 = 4\ \mu F$.

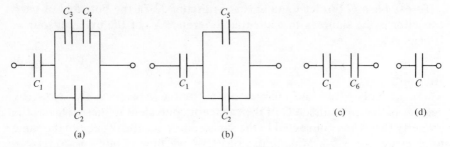

Figure 22-8

Solution Since only parallel and series connections are present, the equivalent capacitance can be found by repeated usage of (22-6) and (22-7).

First, since C_3 and C_4 are in series, they are equivalent to a capacitor of capacitance C_5:

$$C_5 = \frac{C_3 C_4}{C_3 + C_4} = \frac{(4\ \mu\text{F}) \times (4\ \mu\text{F})}{4\ \mu\text{F} + 4\ \mu\text{F}} = 2\ \mu\text{F}$$

and the original configuration is reduced to that in Figure 22-8b. Now C_2 and C_5 are in parallel, so by use of (22-6) the configuration in Figure 22-8c results, with C_6 given by

$$C_6 = C_2 + C_5 = 2\ \mu\text{F} + 2\ \mu\text{F} = 4\ \mu\text{F}$$

Finally, combining C_1 and C_6, we obtain the equivalent capacitance C for the original network

$$C = \frac{C_1 C_6}{C_1 + C_6} = \frac{(2\ \mu\text{F}) \times (4\ \mu\text{F})}{2\ \mu\text{F} + 4\ \mu\text{F}} = \frac{4}{3}\ \mu\text{F}$$

This example illustrates the method which can be used to reduce certain networks of capacitors to an equivalent single one. Note, however, that *not* all networks of capacitors are reducible by use only of (22-6) and (22-7). The network in Figure 22-9 is an example of this type. Why?

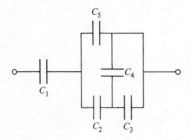

Figure 22-9

22-5   Energy of a charged capacitor

In Chapter 21 the energy stored in the electrostatic field was defined as the work required to assemble any given configuration of charged particles from an initial state of infinite separation. The purpose of this section is to show that the energy U of a capacitor C with charge Q is

$$U = \frac{1}{2}\frac{Q^2}{C} \tag{22-9}$$

Because of (22-1), equivalent forms for U are

$$U = \frac{QV}{2} = \frac{CV^2}{2} \tag{22-10}$$

where V is the potential of the capacitor when it has charge Q.

According to our definition of electrical energy, the energy U of a capacitor represents the work W required to transfer Q units of charge from one plate of the capacitor to the other. Imagine carrying out this charge transfer in a sequence of steps, each of which involves the transfer of an infinitesimal amount of charge from one plate to the other. If, at some intermediate state, q units of charge have been transferred, then the potential V of the capacitor at this stage is q/C. The work dW required to transport the next infinitesimal charge element dq is then

$$dW = V \, dq = \frac{q}{C} \, dq$$

since the charge dq must be carried through a potential difference $V = q/C$. Integrating both sides of this formula from the initial state, corresponding to $q = 0$, to the final one, $q = Q$, we find that the total work W is

$$W = \int dW = \int_0^Q \frac{q \, dq}{C} = \frac{1}{2C} q^2 \Big|_0^Q = \frac{Q^2}{2C}$$

Finally, equating this work W to the electrostatic energy U of the capacitor we obtain (22-9).

Example 22-7 A spherical "capacitor" has an inner radius of 0.50 meter and an outer radius of 0.51 meter. How much energy is stored in this capacitor if a potential difference of 100 volts is maintained across its plates?

Solution According to (22-3), the capacitance is

$$C = 4\pi\epsilon_0 \frac{ab}{b-a} = \frac{(0.5 \text{ m}) \times (0.51 \text{ m})}{(9.0 \times 10^9 \text{ N-m}^2/\text{C}^2) \times (0.51 \text{ m} - 0.50 \text{ m})}$$
$$= 2.8 \times 10^{-3} \, \mu\text{F}$$

Thus it follows by use of (22-10) that the energy is

$$U = \frac{CV^2}{2} = 0.5 \times (2.8 \times 10^{-9} \text{ F}) \times (100 \text{ V})^2$$
$$= 1.4 \times 10^{-5} \text{ J}$$

22-6 Force on a capacitor

As an application of the formula for the energy in a capacitor in (22-9) we shall now calculate the strength of the *external* force required to compensate for the attractive electric force that the plates of a charged parallel-plate capacitor exert on each other.

Consider the situation in Figure 22-10, in which the two plates of a capacitor, each of area A, are separated by a distance y and have the respective fixed charges $\pm Q$. We shall assume that the source which originally charged up the capacitor has been disconnected, so that the charge Q on the capacitor is fixed. If f represents the strength of the external force required to keep the plates apart, then the work dW required to increase the separation distance by an amount dy is

$$dW = f \, dy \qquad\qquad (22\text{-}11)$$

On the other hand, according to (22-2) and (22-9), the original energy U of this capacitor is

$$U = \frac{1}{2C} Q^2 = \frac{1}{2} \frac{1}{\epsilon_0 A / y} Q^2 = \frac{1}{2\epsilon_0 A} Q^2 y$$

Therefore, since A and Q^2 are constant, it follows that the change in energy dU as the distance between the plates is increased by dy is

$$dU = \frac{1}{2\epsilon_0 A} Q^2 \, dy$$

Finally, equating this to the work dW in (22-11) and canceling out the factor dy, we obtain

$$f = \frac{Q^2}{2\epsilon_0 A} \qquad\qquad (22\text{-}12)$$

The fact that this external force is positive reflects the fact that it must be directed, as shown in the figure, in a way to keep the plates from coming together.

Even though the formula for f in (22-12) has been derived on the

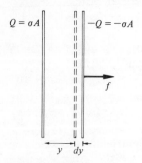

Figure 22-10

assumption that the charge on the capacitor is fixed—that is, that Q does not change as the separation distance between the plates is increased—it is also valid if a fixed potential V is maintained across the plates. This second possibility is more complex since the external source is also able to carry out work and thus it is not possible in this case simply to equate the external work $f\,dy$ to the energy change dU of the electrostatic field. However, making use of the fact that according to (21-5) the work carried out in transporting a charge dq is $V\,dq$, it is shown in the problems that (22-12) is valid in this more complex case nevertheless.

Example 22-8 Suppose a 100-volt potential difference is maintained between the plates of a 1.0-μF parallel-plate capacitor. What external force is required to keep the plates at the fixed separation distance of 10^{-3} meter?

Solution Making use of (22-12), (22-1), and (22-2) we have

$$f = \frac{Q^2}{2\epsilon_0 A} = \frac{Q^2}{2Cd} = \frac{1}{2d}\,CV^2$$

and the substitution of the given parameter values leads to

$$f = \frac{1}{2d}\,CV^2 = \frac{1}{2.0 \times 10^{-3}\,\text{m}} \times (1.0 \times 10^{-6}\,\text{F}) \times (100\,\text{V})^2$$
$$= 5.0\,\text{N}$$

22-7 Dielectrics—qualitative discussion

Having established some of the important properties of capacitors, in the remainder of this chapter we shall make use of this knowledge to study dielectrics.

By contrast to a solid conductor whose electrons are relatively free to wander through a fixed lattice of positive ions, the electrons in an insulator, as noted previously, are *not* able to move freely. Rather, we find that each such electron is strongly bound to its parent ion and is thereby permanently attached to it. In the absence of an external electric field, then, each electron in an insulator may be thought of as orbiting about its parent ion and in such a way that, overall, each atom is electrically neutral.

A typical value for the velocity of an electron as it moves about its parent ion is of the order of 10^6 m/s. Since the atomic radius is of the order of 10^{-10} to 10^{-9} meter, this means that an electron will make about 10^{15} to 10^{16} traversals about its associated ion in 1 second. On a macroscopic scale, only time averages of such high-speed microscopic motions can be observed. Hence the physical picture that emerges—which, incidentally, is very similar to that predicted by quantum mechanics—is that each atom of a dielectric consists of a positive core (the ion) surrounded by a "smear" of negative charge. Generally speaking, this smear of negative charge will be distributed symmetrically and its center of gravity will coincide with that of the positively charged ion. An application of Gauss' law then shows that the electric field outside of such an atom vanishes, and that the atom must therefore be electrically neutral. Figure 22-11 depicts such a macroscopic view of a hydrogen atom. Note that the center of gravity of the negative-charged cloud is located at the position of the nucleus.

Consider now what happens to an atom of a dielectric when it is in the presence of an external electric field E_0. The electron will again travel about the ion, but this time its orbit will not have the spherically symmetric form in Figure 22-11. Instead, because of the existence of the external field, the negative cloud will become distorted and will assume a shape such as that in Figure 22-12. The reason for this distortion is as follows. Since the electron has a negative charge, the downward electric force it experiences when it is at, say, point A is larger than is the upward force on it when it is at point B. For at A the force due to the external electric field E_0 and that due to the ion are in the same direction, whereas at B these forces are oriented in opposite directions. It follows that a time-averaged view of this electronic motion must have a structure similar to that in the figure.

As implied in Figure 22-12, in the presence of an electric field the center of gravity of the negative electron cloud does *not* coincide with that of the ion. This means that there is an effective charge separation between the electron and the ion, as a result of which the atom acquires a certain electric dipole moment **p**. We shall use the term *polarization* to characterize this property of an atom's acquiring an electric dipole moment when placed into an electric field. According to Figure 22-12, the induced dipole moment **p**, which is conventionally defined to be parallel to a vector from the negative to the positive charge, is parallel to the direction of the external field E_0. Moreover, experiment shows further that for electric fields of moderate strength this induced dipole moment **p** is proportional to the strength of the electric field. Thus

$$\mathbf{p} = \alpha \mathbf{E}_0 \qquad\qquad (22\text{-}13)$$

with α a proportionality constant known as the *atomic polarizability*. Table 22-1 lists the values[2] of α for typical atoms. Note that for the tightly bound

Table 22-1 Atomic polarizabilities in units of 10^{-40} (C-m^2)/V

Element	H	He	Li	C	Ne	Na	Ar	K
α	0.73	0.23	13	1.7	0.44	30	1.8	38

[2]Strictly speaking, the table lists the *static* values for α. If the external field varies in time, (22-13) is still valid, but with the value of α determined by the details of this variation.

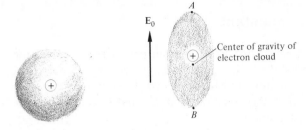

Figure 22-11   **Figure 22-12**

electrons in an inert gas, such as He, Ne, or Ar, the polarizability is very small compared with the values of α for the alkalis, such as Li, Na, and K, whose electrons are more loosely bound.

Consider, for example, the case of a hydrogen atom in an external field of strength 10^4 V/m. Using the value for α listed in the table, we find that the induced dipole moment p has the value

$$p = \alpha E = \left(7.3 \times 10^{-41} \frac{\text{C-m}^2}{\text{V}}\right) \times \left(10^4 \frac{\text{V}}{\text{m}}\right) = 7.3 \times 10^{-37} \text{ C-m}$$

Since the electric charge q of this dipole has the value 1.6×10^{-19} coulomb, the effective separation distance l between the proton and the center of gravity of the electron cloud is

$$l = \frac{p}{q} = \frac{7.3 \times 10^{-37} \text{ C-m}}{1.6 \times 10^{-19} \text{ C}} = 4.6 \times 10^{-18} \text{ m}$$

This corresponds to a very slight distortion of the originally spherically symmetric electronic orbit and is of the order of 10^{-7} that of the atomic radius.

Having made plausible the fact that in the presence of an electric field the individual atoms become polarized, let us consider what happens when an insulator is placed into an electric field. As for the isolated atom above, each atom in the dielectric will become polarized and set up its own dipole field. However, the individual dipoles in the dielectric will not, in general, line up along the direction of the external field! For the electric force acting on a given atom is due not only to the external field but to the dipole fields produced by its nearby neighbors as well. In general, these two fields are not parallel. In addition, there are effects associated with the thermal vibrations of the atoms that also tend to counteract the tendency for the dipoles to line up along $\mathbf{E}_0$. However, for isotropic and homogeneous dielectrics—the only substances with which we shall be concerned—it is valid to think of the dipoles in the dielectric as being lined up with the external field, and we shall do so in all of the following.

22-8 The dielectric constant

The purpose of this section is to define a certain parameter κ, called the *dielectric constant*, in terms of which the macroscopic, electric properties of isotropic, homogeneous insulators may be described. The relation between κ and the dipole moments **p** of the constituent atoms and molecules will be discussed in the following sections.

Consider a parallel-plate capacitor of area A and separation distance between the plates d. According to (22-2), its capacitance, for which we now use the symbol C_0, is

$$C_0 = \frac{\epsilon_0 A}{d} \tag{22-14}$$

Imagine now carrying out various measurements on this capacitor to determine the effect of placing between its plates a slab of dielectric of area A and of thickness d. First, as shown in Figure 22-13, let us insert the slab

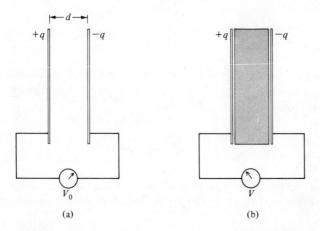

Figure 22-13

after the capacitor has first been given a charge q and has then been disconnected from the line. A measurement of the potential difference across the plates shows that the potential V of the capacitor with the dielectric in is less than the original value $V_0 = q/C_0$. Second, as shown in Figure 22-14, let us insert the dielectric slab while a fixed potential difference V is maintained across the plates. This time, the final charge q on the capacitor is found to exceed the original charge $q_0 = C_0 V$. Moreover, comparison of the results of these two experiments—one with a fixed charge on the capacitor and the second with a fixed potential across its plates—shows that, as the dielectric is inserted, the factor by which the potential decreases in the first case is precisely the same as the factor by which the charge increases in the second. In other words, the ratio V_0/V of the

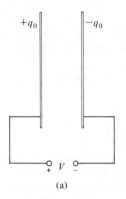
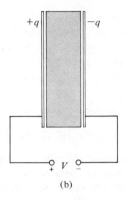

Figure 22-14

potential difference measurements in Figure 22-13 has precisely the same value as does the ratio q/q_0 of the charges in Figure 22-14.

To interpret these results, it is useful to turn to the defining relation for capacitance in (22-1). Evidently the insertion of the dielectric slab between the plates of the capacitor has increased its capacitance by the equivalent factors V_0/V or q/q_0 in the two cases above. Thus, we shall write

$$C = \kappa C_0 \qquad (22\text{-}15)$$

where $C_0 = \epsilon_0 A/d$ is the capacitance of the capacitor in the absence of the dielectric slab and C is the corresponding capacitance after it has been inserted. The coefficient κ is called the dielectric constant for the given material. Note that κ is a dimensionless number and satisfies the inequality $\kappa \geq 1$.

If the above experiments are repeated with the same (homogeneous and isopropic) dielectric material but with capacitors of various sizes and shapes, (22-15) is found to be valid for these cases also. This implies that the dielectric constant κ, as defined above, is a property of the given material only. In particular, it does not depend on the charge or the potential of the capacitor, nor does it depend on its geometrical structure.

Table 22-2 lists some experimental values for the dielectric constants of a number of materials. Note that $\kappa > 1$ for all of them.

Table 22-2 Dielectric constants at 20°C

Substance	Air	Beeswax	Paraffin	Pyrex	Water
κ	1.0006	2.9	2.3	4.0	80

Example 22-9 Suppose that a 10-volt source is connected across a 2.0-μF parallel-plate capacitor.

(a) What is the charge on the capacitor?

(b) Suppose that the source is disconnected and a dielectric slab is introduced between the plates. If in this process the potential across the capacitor drops to 6.0 volts, what is the dielectric constant κ?

(c) What is the capacitance of the capacitor with the dielectric slab in?

Solution

(a) According to (22-1), the original charge q is

$$q = C_0 V_0 = (2.0 \times 10^{-6}\,\text{F}) \times (10\,\text{V}) = 2.0 \times 10^{-5}\,\text{C}$$

(b) Since the charge is fixed, it follows from (22-1) that the product CV is independent of the presence or absence of the slab. According to (22-15), the capacitance increases by the factor κ and thus the potential across the capacitor must decrease by the same factor. Using the given parameter values, we thus obtain

$$\kappa = \frac{10\,\text{V}}{6\,\text{V}} = 1.7$$

(c) Substituting this value for κ into (22-15), we find that

$$C = \kappa C_0 = 1.7 \times 2.0\,\mu\text{F} = 3.4\,\mu\text{F}$$

22-9 Polarization charge

In order to relate the dielectric constant of an insulator to an induced charge, consider again the physical situation in Figure 22-13 of a capacitor with a fixed charge q. According to (22-15), the insertion of the dielectric slab between the plates increases its capacitance by the factor κ or, equivalently, decreases the potential difference between the plates by this factor. But for a uniform electric field **E**, the potential difference between any two equipotential planes a distance d apart is Ed. Hence, from the fact that the potential difference between the plates in Figure 22-13 decreases by the factor κ when the slab is inserted, it follows that the electric field in this region must also decrease by this factor. Thus if E_0 represents the magnitude of the uniform electric field between the plates *before* the dielectric slab is inserted and E_i is the corresponding value afterward, then

$$E_i = \frac{E_0}{\kappa} \tag{22-16}$$

Moreover, if the steps leading to (22-2) are repeated but with E replaced by this value E_i, then since $E_0 = \sigma/\epsilon_0 = q/\epsilon_0 A$, the defining relation in (22-15) results, as it should. We shall make extensive use of (22-16) in all of the following.

Further confirmation of the validity of (22-16) may be obtained in the following way. In Example 22-10, it is established that the capacitance C of the capacitor in Figure 22-15, whose plates have an area A and are separated by a distance d, and which contains a dielectric slab of constant κ and thickness t $(< d)$, is

$$C = \frac{\epsilon_0 A}{(d - t) + t/\kappa} \tag{22-17}$$

In deriving this relation we shall find it necessary to make essential use of (22-16). Hence the experimental verification in the laboratory of (22-17) for all thicknesses t $(< d)$ is in effect a confirmation of the validity of (22-16).

The physical significance of the internal field E_i is illustrated in Figure 22-16. As shown in Example 22-12, on the face of the dielectric slab nearest the positive plate of the capacitor there is induced a negative polarization charge density, $-\sigma_p$, where

$$\sigma_p = \epsilon_0 \frac{\kappa - 1}{\kappa} E_0 \tag{22-18}$$

and a compensating charge density $+\sigma_p$ is induced on the opposite face. For the isotropic and homogeneous dielectrics of interest to us, no polarization charge comes into existence in the interior of the dielectric. As shown in the figure, some of the field lines that originate on the positive plate of the capacitor terminate on the polarization charge on the left face of the dielectric while others continue on through to the charge on the negative plate. The field lines that originate on the positive charge on the right face of the dielectric terminate on the negative plate of the capacitor. Both of these features are consistent with (22-16) and its implication that $E_i < E_0$.

Example 22-10 Derive (22-17).

Solution Suppose that the capacitor in Figure 22-15 has a charge q. According to (20-16), the electric field E_0 in the region between the plates but *outside* of the dielectric is

$$E_0 = \frac{\sigma}{\epsilon_0} = \frac{q}{\epsilon_0 A}$$

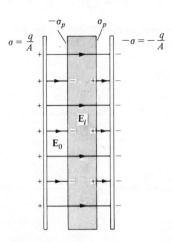

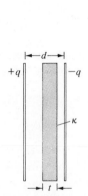

Figure 22-15   *Figure 22-16*

which is directed to the right in the figure if the plate on the left has the positive charge. The field E_i inside the dielectric lies along the same direction and, according to (22-16), has the magnitude

$$E_i = \frac{E_0}{\kappa} = \frac{q}{\epsilon_0 \kappa A}$$

Making use of these formulas and the result in Example 21-13, we find that the potential V across the capacitor is

$$V = E_0(d - t) + E_i t = E_0 \left[(d - t) + \frac{t}{\kappa} \right]$$

$$= \frac{q}{\epsilon_0 A} \left[(d - t) + \frac{t}{\kappa} \right]$$

Finally, solving for the ratio q/V and making use of the definition in (22-1), we obtain (22-17).

Example 22-11   For the dielectric slab in Figure 22-16, suppose that $\kappa = 2$ and that the capacitor is charged so that $E_0 = 100$ V/m. Calculate the field inside the dielectric and the surface polarization charge on the dielectric.

Solution   Substituting the given data into (22-16), we find that

$$E_i = \frac{E_0}{\kappa} = \frac{100 \text{ V}}{2} = 50 \text{ V}$$

The surface polarization charge $\sigma_p{}^l$ on the left side of the dielectric is negative and, according to (22-18), has the value

$$\sigma_p{}^l = -\sigma_p = -\frac{\epsilon_0(\kappa - 1)}{\kappa} E_0$$

$$= -\left(8.9 \times 10^{-12} \frac{\text{C}^2}{\text{N-m}^2} \right) \times \left(\frac{2-1}{2} \right) \times \left(\frac{100 \text{ V}}{\text{m}} \right)$$

$$= -4.5 \times 10^{-10} \text{ C/m}^2$$

A similar calculation shows that an equal charge density, but of the opposite sign, is induced on the right-hand face of the slab.

Example 22-12   Derive (22-18).

Solution   Consider, in Figure 22-17, one face of a homogeneous and isotropic dielectric of constant κ in a uniform external field E_0 perpendicular to this face.

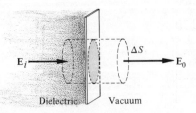

Figure 22-17

Inside the dielectric there will be an electric field $\mathbf{E}_i$ also perpendicular to the surface and with magnitude given in (22-16). To calculate the polarization charge density σ_p induced on the dielectric surface, let us construct a Gaussian surface in the shape of a cylinder of cross-sectional area ΔS with its axis parallel to $\mathbf{E}_0$. Reference to Figure 22-17 shows that since one end face of the cylinder is inside the dielectric and the other outside, the total flux out of the cylinder is $(E_0 \Delta S - E_i \Delta S)$. Therefore, since the total charge inside is $\sigma_p \Delta S$, it follows from Gauss' law that

$$(E_0 - E_i) \Delta S = \frac{1}{\epsilon_0} \sigma_p \Delta S$$

The result in (22-18) then follows by canceling the factor ΔS and substituting for E_i by use of (22-16).

22-10 Energy considerations

The purpose of this section is to extend the energy considerations of Section 22-5 to the case of a parallel-plate capacitor that contains a dielectric slab.

Consider first the case of a capacitor without a dielectric but with a fixed charge Q. According to (22-9), its energy U_0 is

$$U_0 = \frac{1}{2C_0} Q^2$$

with C_0 given in (22-14). If a dielectric slab of constant κ is introduced between the plates, then, according to (22-15), the capacitance is increased by the factor κ. Since the charge on the capacitor is assumed to be fixed and is thus unaltered in this process, it follows that the final energy U is

$$U = \frac{1}{2C} Q^2 = \frac{1}{2\kappa C_0} Q^2 = \frac{1}{\kappa} U_0 \qquad (22\text{-}19)$$

with U_0 the energy of the capacitor without the dielectric. Since the work W required to insert the dielectric between the plates is equal to the change in energy $(U - U_0)$, it follows by use of (22-19) that

$$W = U - U_0 = \frac{1}{\kappa} U_0 - U_0 = -U_0 \left(\frac{\kappa - 1}{\kappa} \right) \qquad (22\text{-}20)$$

The fact that this work is negative, since $\kappa > 1$, means that the electrostatic field carries out positive work on the dielectric and thus pulls it into the region between the plates. The physical mechanism that underlies this effect is the fringing field at the edge of the capacitor. Figure 22-18 shows schematically some of the field lines at the edges of a capacitor and thus illustrates how it is possible for the dielectric to be pulled into the region between the plates.

Let us now reexamine this problem, but suppose this time that the potential across the plates is kept at a fixed value V. In the absence of a dielectric, the energy U_0 stored in the capacitor is $C_0 V^2/2$, with C_0 defined in

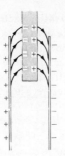

Figure 22-18

(22-14). With the insertion of the dielectric, the capacitance C is increased by the factor κ, and thus the final energy U is

$$U = \frac{CV^2}{2} = \frac{\kappa C_0 V^2}{2} = \kappa U_0$$

This corresponds to an energy change

$$U - U_0 = \kappa U_0 - U_0 = (\kappa - 1)U_0 \qquad (22\text{-}21)$$

which, by contrast to the fixed-charge case in (22-20), is positive. However, this increase in energy does *not* necessarily mean that the external agent must carry out positive work to insert the dielectric. For the external source that keeps the plates at the potential V will also, in general, carry out work while the dielectric is being inserted.

To calculate the work W_s carried out by the source, let us note that the charge Q of the capacitor after the dielectric has been inserted is

$$Q = CV = \kappa C_0 V = \kappa Q_0 \qquad (22\text{-}22)$$

where Q_0 is the charge without the dielectric. This means that a net charge $(Q - Q_0) = (\kappa - 1)Q_0$ flows as a result of the insertion of the dielectric slab. Hence the work W_s carried out by the source at the fixed potential V is

$$\begin{aligned} W_s &= V(Q - Q_0) = VQ_0(\kappa - 1) \\ &= C_0 V^2(\kappa - 1) \\ &= 2U_0(\kappa - 1) \end{aligned} \qquad (22\text{-}23)$$

where the last equality follows since the original energy U_0 is $C_0 V^2/2$. Now if W is the work required of an external agent to insert the dielectric into the capacitor, then according to the energy-conservation principle

$$W_s + W = U - U_0$$

which in words states that the change in electrostatic energy $(U - U_0)$ must come from the work produced by *all* external sources. Substituting (22-21) and (22-23) into this formula we find that the external work W is

$$W = U - U_0 - W_s = U_0(\kappa - 1) - 2U_0(\kappa - 1) = -U_0(\kappa - 1) \quad (22\text{-}24)$$

The fact that this is negative implies that the dielectric will be pulled into the region between the plates, as in the other case.

To summarize, then:

If a dielectric of constant κ is inserted between the plates of a capacitor which is kept at a fixed potential V, then the battery carries out a positive amount of work $2(\kappa - 1)U_0$, *where* U_0 *is the original energy. Half of this energy shows up as an increase of the electrostatic field energy, and the other half is expended in the form of work required to draw the dielectric into the field.*

Example 22-13   A 3.0-μF parallel-plate capacitor is connected across a 100-volt line. If a dielectric of constant $\kappa = 2.2$ is inserted between the plates, calculate:
(a) The original electrostatic energy.
(b) The work carried out by the battery as the dielectric is inserted.
(c) The increase in field energy.
(d) The final charge on the capacitor.

Solution
(a) The original energy U_0 is

$$U_0 = \frac{1}{2} C_0 V^2 = \frac{1}{2} \times (3.0 \times 10^{-6} \text{ F}) \times (100 \text{ V})^2$$
$$= 1.5 \times 10^{-2} \text{ J}$$

(b) The work W_s carried out by the source is, according to (22-23),

$$W_s = 2U_0(\kappa - 1) = 2 \times 1.5 \times 10^{-2} \text{ J} \times (2.2 - 1) = 3.6 \times 10^{-2} \text{ J}$$

(c) Of the 3.6×10^{-2} joule expended by the source, one half, or 1.8×10^{-2} joule, goes to increase the field energy from its original value of 1.5×10^{-2} joule to its final value of 3.3×10^{-2} joule.
(d) The original charge on the capacitor Q_0 is

$$Q_0 = C_0 V = (3.0 \times 10^{-6} \text{ F}) \times 100 \text{ V} = 3.0 \times 10^{-4} \text{ C}$$

and thus, by use of (22-22), we obtain, for the final charge,

$$Q = \kappa Q_0 = 2.2 \times 3.0 \times 10^{-4} \text{ C} = 6.6 \times 10^{-4} \text{ C}$$

†22-11   The displacement vector

In the above discussion of dielectrics we considered only those substances which could be characterized as being homogeneous and isotropic dielectrics, and treated only those cases for which the external field was uniform and directed perpendicularly to the dielectric surface. The purpose of this section is to describe briefly how to deal with situations for which these conditions are not satisfied.

If a dielectric is introduced into an electric field, then, in general, its constituent molecules become polarized and tend to line up with the resultant field. It is convenient to describe this electrical behavior of the dielectric in terms of a certain quantity known as the *dipole moment per unit volume* **P**. This quantity **P** represents the macroscopic average of the individual molecular dipole moments **p**, and is defined so that $\mathbf{P}\Delta V$ represents the dipole moment associated with any given small volume element ΔV. In general, **P** varies from point to point in space and is nonvanishing only inside and on the surface of dielectric materials.

On physical grounds we expect that if a dielectric is inserted into an electric field, a certain polarization charge will be induced within, and on the surface of, the sample. Let $q_p(S)$ represent this polarization charge contained in an arbitrary, but fixed, closed surface S (Figure 22-19). The relation between $q_p(S)$ and the dipole moment per unit volume **P** is

$$q_p(S) = -\oint_S \mathbf{P} \cdot d\mathbf{S} \tag{22-25}$$

where the integral is over the closed surface S. The validity of this relation for the special case for which **P** is constant inside the dielectric will be confirmed in the problems.

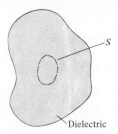

Figure 22-19

Consider now a dielectric characterized by a dipole moment per unit volume **P** and let **E** represent the electric field produced by the associated polarization charge plus any other charge that may be present. We define the *displacement vector* **D** associated with this system by

$$\mathbf{D} = \epsilon_0 \mathbf{E} + \mathbf{P} \tag{22-26}$$

so that outside of the dielectric, where $\mathbf{P} = 0$, the displacement vector **D** is—up to a factor ϵ_0—precisely the same as is the electric field **E**. However, inside the dielectric where, in general, $\mathbf{P} \neq 0$, the relation between **D** and **E** is given by the more complex form in (22-26). We shall now give a physical meaning to **D** by showing that it represents that part of the electric field produced by all electric charges except those induced in the dielectric.

To this end, consider an arbitrary closed surface S in a region of space that may contain both ordinary charge as well as polarization charge which it induces in dielectric materials. Let $q(S)$ represent the ordinary or "true" charge inside S and $q_p(S)$ the polarization charge contained within this same surface. The total charge contained within S is the sum of these; that is, $[q(S) + q_p(S)]$. Applying Gauss' law to the surface S, we have

$$\oint_S \mathbf{E} \cdot d\mathbf{S} = \frac{1}{\epsilon_0}[q(S) + q_p(S)] \tag{22-27}$$

where $\mathbf{E}$ is the total electric field and is produced by all charges. Let us now take the dot product of both sides of (22-26) with an infinitesimal area element $d\mathbf{S}$ and integrate over the closed surface S. The result is

$$\oint \mathbf{D} \cdot d\mathbf{S} = \epsilon_0 \oint \mathbf{E} \cdot d\mathbf{S} + \oint \mathbf{P} \cdot d\mathbf{S}$$

and, substituting for the surface integrals on the right-hand side by use of (22-25) and (22-27), we find that

$$\oint_S \mathbf{D} \cdot d\mathbf{S} = \epsilon_0 \left\{ \frac{1}{\epsilon_0}[q(S) + q_p(S)] \right\} - q_p(S)$$

which simplifies to

$$\oint_S \mathbf{D} \cdot d\mathbf{S} = q(S) \tag{22-28}$$

Comparison with Gauss' law shows then that up to a factor ϵ_0 the *flux of the displacement vector* $\mathbf{D}$ *out of the closed surface S is equal to the "true" charge inside S.* This leads to the physical interpretation for $\mathbf{D}$ as that part of the electric field produced by the "true" charge, that is, the nonpolarization charge $q(S)$. Since in general we have much more knowledge about $q(S)$ than we do about $q_p(S)$, the calculation of $\mathbf{D}$ is often much simpler than is that of $\mathbf{E}$. This is one of the main reasons for introducing this notion of the displacement vector.

For many cases of physical interest the induced dipole moment per unit volume $\mathbf{P}$ is proportional to the electric field $\mathbf{E}$. Materials for which this relation is valid are known as *linear materials*. For a linear material, then,

$$\mathbf{P} = \epsilon_0 \chi_e \mathbf{E} \tag{22-29}$$

where the dimensionless coefficient of proportionality χ_e is known as the *electric susceptibility.* Outside of the dielectric χ_e vanishes, of course. Experiment shows that the quantity χ_e is related to the dielectric constant of the material by

$$\chi_e = \kappa - 1 \tag{22-30}$$

Indeed, this relation is often taken as the definition of the electric susceptibility.

Additional confirmation of the validity (22-30) may be obtained in the following way. Take the dot product of (22-29) with $d\mathbf{S}$ and integrate over the closed dashed surface in Figure 22-17. According to (22-25), we obtain on the left-hand side $(-\sigma_p \, \Delta S)$. On the right-hand side, since $\chi_e = 0$ outside of the dielectric we obtain, similarly, $-\epsilon_0 \chi_e E_i \, \Delta S$. Equating these two, we obtain $\sigma_p = \epsilon_0 \chi_e E_i$, and this leads directly to (22-30) after comparison with (22-16) and (22-18).

For linear materials, for which (22-29) is applicable, there is a very simple relation between $\mathbf{D}$ and $\mathbf{E}$. Eliminating $\mathbf{P}$ and χ_e between (22-26), (22-29), and (22-30), we obtain

$$\mathbf{D} = \kappa \epsilon_0 \mathbf{E} \qquad (22\text{-}31)$$

22-12 Summary of important formulas

A capacitor is a device consisting of two isolated conductors having equal and opposite charges q (>0) and $-q$, respectively. If V is the potential of the positive plate relative to the negative one, then the capacitance C is an inherently positive quantity, defined by

$$C = \frac{q}{V} \qquad (22\text{-}1)$$

For a parallel-plate capacitor of area A and plate separation distance d, C is

$$C = \frac{\epsilon_0 A}{d} \qquad (22\text{-}2)$$

The electrostatic energy U stored in a capacitor of capacitance C is

$$U = \frac{Q^2}{2C} \qquad (22\text{-}9)$$

where Q is the charge on the capacitor. By use of (22-1), various equivalent forms for U can be obtained.

If a dielectric slab of constant κ is inserted between the plates of a capacitor so that it occupies all available space, then its capacitance C is

$$C = \kappa C_0 \qquad (22\text{-}15)$$

QUESTIONS

1. Define or describe briefly the following: (a) parallel-plate capacitor; (b) spherical capacitor; (c) dielectric; (d) dielectric constant; and (e) fringing field.
2. According to (22-2), the capacitance of a parallel-plate capacitor can be made arbitrarily large by decreasing

the separation distance d between the plates. What practical considerations set an upper limit to the capacitance of such a capacitor?
3. Why are the two conductors A and B in Figure 22-2b at the same potential even though they have equal and opposite charges? Why are they still

at the same potential in Figure 22-2c?

4. If C_0 is the capacitance of the two conductors in Figure 22-1 in a vacuum, present a qualitative argument showing that if the intervening space is filled with a dielectric of constant κ, then the capacitance is increased to the value κC_0.

5. Is it necessary that the inner conducting sphere in Figure 22-6 be hollow? What would the capacitance be if it were solid?

6. Why is it necessary to supply an external force to keep the plates of a parallel-plate capacitor from coming into contact?

7. Show explicitly that the unit of ϵ_0 is the farad per meter (F/m).

8. If two charged capacitors are connected together so that the positive plate of one is in contact with the positive one of the other, and similarly for the negative plates, are the capacitors in series or in parallel?

9. Repeat Question 8, but suppose this time that the positive plate of each one is connected to the negative plate of the other.

10. Consider a parallel-plate capacitor of area A. Is it legitimate to think of it as consisting of two parallel-plate capacitors each of area $A/2$ and connected together? Are they connected in series or in parallel?

11. A flat, metal plate is inserted between the plates of a parallel-plate capacitor, as in Figure 22-20. Is the

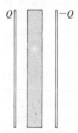

Figure 22-20

capacitance increased or decreased thereby? What is the sign of the work that must be carried out to insert the metal plate?

12. Would the capacitance of the capacitor in Figure 22-20 increase or decrease if the central conductor were connected to one of the plates by a conducting wire?

13. Describe in what sense the configuration in Figure 22-20 may be thought of as two capacitors connected in series.

14. List some practical difficulties associated with the calculation of the capacitance of two spherical conductors whose centers are separated by a distance greater than the sum of their radii.

15. Explain why the capacitance C of two conductors separated by a distance L, which is very large compared to their linear dimensions, is

$$C = 4\pi\epsilon_0 L$$

(*Hint*: What would be the potential difference between these conductors if they had charges q and $-q$?)

16. Two identical capacitors are separately charged up to the same potential, and then disconnected from the voltage source and reconnected together so they discharge. What happens to the energy stored in the capacitors? Explain.

17. A parallel-plate capacitor is charged to a potential V and lies on a horizontal surface. Neglecting friction, describe qualitatively the subsequent motion of a dielectric slab that is introduced at one end of the capacitor.

18. A parallel-plate capacitor has a fixed charge. If a dielectric slab is introduced between the plates, describe what happens to the following: (a) the potential across the plates; (b) the capacitance; (c) the electric field

strength; (d) the energy; and (e) the charge on the plates.

19. Repeat Question 18, but this time assume that a fixed potential difference is maintained across the plates.

20. Explain why in the interior of a dielectric slab in which there exists a uniform electric field there can be no electric charge.

21. Two capacitors C_1 and C_2 have the respective charges q_1 and q_2 and are connected together in parallel. If a dielectric slab is inserted into C_2, what happens to (a) q_1 and q_2? (b) C_1 and C_2? (c) The potential across the capacitors?

22. Show that the total energy stored in two capacitors connected in series across a certain source is the same as the energy stored in the equivalent capacitor when connected across the same source.

23. Repeat Question 22, but this time assume that the two capacitors are connected in parallel across the source.

24. Show, by use of (22-18), that under some circumstances a conductor may be thought of as a dielectric of infinite dielectric constant.

PROBLEMS

1. If a 6.0-volt source is connected across a 2.0-μF capacitor, how much charge will be transferred between the conductors?

2. If a charge of 0.3 μC appears on the positive plate of a capacitor when a 100-volt source is connected across the plates, what is its capacitance?

3. The area of each plate of a parallel-plate capacitor is 20 cm^2 and the plates are separated by a distance of 1.5 mm. (a) What is the capacitance? (b) If a 20-volt potential is connected across the capacitor, what charge density appears on each plate?

4. The distance between the plates of a parallel-plate capacitor is doubled. (a) By what factor must the area be increased to keep the capacitance fixed? (b) What must be the dielectric constant of an insulator that if placed between the plates keeps the capacitance unaltered?

5. A spherical capacitor consists of two concentric spheres of radii 10 cm and 9.9 cm, respectively. (a) What is the capacitance?

(b) If this capacitor is connected to a 10-volt source, what is the total charge on the positive sphere?

(c) Assuming that the larger sphere is at the higher potential, calculate the surface charge density on the two spheres. Explain why these two densities are *not* equal and opposite.

6. A parallel-plate capacitor of area 12 cm^2 and separation distance 2.0 mm has a charge of 0.1 μC. (a) What is the potential difference between the plates? (b) What is the electric field strength?

7. Suppose a dielectric slab of constant $\kappa = 1.5$ is inserted between the plates of the capacitor in Problem 6. (a) What is the potential of the capacitor now? (b) What is the electric field strength in the dielectric?

8. Calculate the electrostatic energy stored in the capacitor in Problem 1 when it is fully charged? Explain why this is *not* the same as the work carried out by the 6.0-volt source that was used to charge it up.

9. Assuming the sun to be a conducting sphere of diameter 1.4×10^6 km, calculate its capacitance.

10. What must be the capacitance of a capacitor that will store 10 joules of energy if connected to a 100-volt potential difference? What would the area of its plates have to be if it were a parallel-plate capacitor with a separation distance of 5.0×10^{-4} meter?

11. What is the capacitance C of a capacitor equivalent to two capacitors of respective capacitances 3.0 μF and 5.0 μF if:
    (a) They are connected in parallel?
    (b) They are connected in series?

12. If C is the capacitance equivalent to two capacitors C_1 and C_2, show that:
    (a) $C < C_1$ and $C < C_2$ if they are connected in series.
    (b) $C > C_1$ and $C > C_2$ if they are connected in parallel.

13. In Figure 22-21 find the capacitance equivalent to C_1, C_2, and C_3 connected as shown. Assume that $C_1 = C_2 = 2.0$ μF and $C_3 = 6.0$ μF.

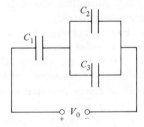

Figure 22-21

14. Suppose a potential difference of 10 volts is applied across the capacitors of Problem 13.
    (a) What is the charge on C_1?
    (b) What is the potential difference across C_1?
    (c) What is the charge on C_2 and C_3?

15. Consider the five capacitors connected as shown in Figure 22-22.

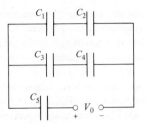

Figure 22-22

Assuming that $C_1 = 2$ μF, $C_2 = 4$ μF, $C_3 = 6$ μF, $C_4 = 8$ μF, and $C_5 = 10$ μF, calculate the capacitance C of the equivalent capacitor.

16. Suppose that for the network described in Problem 15, $V_0 = 100$ volts. (a) How much charge passes through the positive terminal of the source? (b) What is the charge on each of the five capacitors? (c) What is the potential across each capacitor?

17. Consider the five capacitors as arranged in Figure 22-23. Assume that $C_1 = C_3 = 2$ μF, $C_2 = C_4 = 6$ μF, $C_5 = 5$ μF, and $V_0 = 10$ volts. (a) Calculate the charge on each capacitor. (b) Find the capacitance of a single capacitor equivalent to the given arrangement.

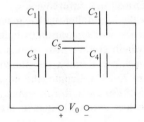

Figure 22-23

*18. Repeat Problem 17, but assume this time that $C_i = i$ μF ($i = 1, 2, 3, 4, 5$).

19. You are given a set of four 3-μF capacitors.
    (a) What equivalent capacitances can be obtained by connecting together two of these?
    (b) What equivalent capacitances

can be obtained by connecting together three of them?

(c) What equivalent capacitances can be obtained by using all four?

20. A 2-μF and a 6-μF capacitor are connected in parallel across a 5-volt line. (a) What energy is stored in the electrostatic field of the capacitors? (b) If the capacitors are disconnected from the line and reconnected with the positive plate of each connected to the negative plate of the other, what is the final value for the energy stored in the capacitors?

21. A 2-μF and a 6-μF capacitor are connected in series across a potential difference of 10 volts.

(a) How much energy is stored in each capacitor?

(b) Suppose that they are now disconnected from the source and reconnected in parallel. What is the final charge on each capacitor?

(c) How much energy is now stored in the capacitors?

(d) Compare your answers to (a) and (c) and explain, in physical terms, any differences.

22. Consider a parallel-plate capacitor of area A and separation distance d and which contains two dielectrics of constants κ_1 and κ_2, as shown in Figure 22-24. Calculate the capacitance of this capacitor. Assume that the dielectrics are identical in size.

Figure 22-24

23. A parallel-plate capacitor of area A and separation distance d contains two dielectric slabs, of thicknesses a and b ($\equiv d - a$) of area A and of dielectric constants κ_1 and κ_2, respectively; see Figure 22-25. Calculate its capacitance. (*Hint:* Use (22-16) and the method of Example 22-10.)

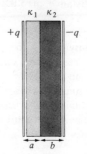

Figure 22-25

24. Consider the capacitor in Figure 22-25. Making use of (22-18) calculate (a) the charge density on the surfaces of the dielectrics nearest the plates and (b) the charge density at the interface between the dielectrics, and show that the total induced charge vanishes.

25. Show that if a dielectric slab of constant κ is introduced into the region between the plates of a parallel-plate capacitor whose plates have charge per unit area σ, then

$$\frac{\sigma_p}{\sigma} = 1 - \frac{1}{\kappa}$$

where σ_p is the induced charge on a face of the slab.

26. Show that the energy per unit volume u stored in a parallel-plate capacitor of area A and separation distance d and containing a dielectric slab of constant κ is

$$u = \frac{\kappa \epsilon_0}{2} E^2$$

where **E** is the electric field between the plates.

27. Show that the formula in Problem 26 is also applicable to a spherical capacitor. (*Note*: $|\mathbf{E}|$ is not constant in this case but varies with radius. Indeed, it can be shown that the result of Problem 26 is very generally valid for any electric field.)

28. Suppose that the plates of the capacitor in Figure 22-15 have area A, and the slab of thickness t is a conductor.
    (a) Show that the capacitance is

$$C = \frac{\epsilon_0 A}{(d - t)}$$

   (b) What charge is induced on each face of the slab if the capacitor has a charge Q?
   (c) Is a force required to move the slab closer to one of the plates? Explain.

29. Making use of (22-17), repeat (b) and (c) of Problem 28. Assume this time, however, that the slab is a dielectric of constant κ.

30. A spherical capacitor of inner radius a and outer radius b contains a concentric spherical shell of matter of inner radius c and outer radius d ($a < c < d < b$). (a) Calculate its capacitance if the shell is made of copper. (b) What would the capacitance be if the copper shell were connected to the outer sphere?

31. Repeat (a) of Problem 30 if the spherical shell is a dielectric of constant κ instead of copper. Does your answer reduce to the expected value for the special case $(d - c) = (b - a)$?

32. Assume that the entire region between the spheres in Figure 22-6 contains a dielectric of constant κ. (a) What is its capacitance? (b) If the total charge on the inner conductor is Q, what is the electric field everywhere?

33. Consider two very long, coaxial, conducting cylinders of radii a and b ($>a$), respectively, and of length L. Assuming that $L \gg a$ and b, show that the capacitance is given by (22-5). (*Hint*: Determine the electric field in the region between the cylinders by the use of Gauss' law, assuming that the capacitor has a charge Q.)

*34. A particle of charge q is at the center of a spherical dielectric shell of constant κ and radii a and b ($>a$). By applying Gauss' law and (22-16), show that the electric field E at a distance r from the particle is

$$E = \frac{q}{4\pi\epsilon_0} \frac{1}{r^2} \begin{cases} 1 & r < a \\ \dfrac{1}{\kappa} & a \leqslant r \leqslant b \\ 1 & r > b \end{cases}$$

*35. For the dielectric sphere in Problem 34, calculate (a) the charge density induced on the inner surface of the sphere and (b) the induced charge density on the outer surface, and show that the total induced charge vanishes. .

36. Show that the external force required to keep the plates of a parallel-plate capacitor at the separation distance d is U/d, where U is the energy stored in the capacitor.

*37. A parallel-plate capacitor of area A and separation distance y is connected to a fixed potential difference V_0. Suppose the separation distance is increased by an external force f from y to $(y + dy)$.
    (a) Show that the change in the electrostatic energy dU is

$$dU = -\frac{1}{2}\frac{\epsilon_0 A}{y^2} V_0^2 \, dy$$

   (b) Show that the charge dQ transported through the positive terminal of the source is

$$dQ = -\frac{\epsilon_0 A}{y^2} V_0 dy$$

(c) What work dW_s is carried out by the source in this process?

(d) Write down the conservation of energy relation and thus show that $f = U/y$.

*38. A parallel-plate capacitor of separation distance d and area A has a length b and a width a, so that $A = ab$. Suppose that a slab of dielectric constant κ is placed a distance y into the region between the plates; see Figure 22-18.

(a) Show that the capacitance C is

$$C = \frac{\epsilon_0}{d}[A + ay(\kappa - 1)]$$

(b) Calculate the energy $U(y)$, assuming a fixed charge Q on the capacitor.

(c) Show that the force f required to push the dielectric a distance dy is

$$f = -\frac{Q^2}{2\epsilon_0} \frac{(\kappa - 1) ad}{[A + ay(\kappa - 1)]^2}$$

*39. Repeat all parts of Problem 38, but assume this time that the potential V_0, across the capacitor is kept fixed.

40. In Problem 38 calculate the work W

$$W = \int_0^b F(y) \, dy$$

required to insert the dielectric completely. Compare this with the energy change in the electrostatic field during this process.

*41. Consider four parallel metal plates, each of area A, and spaced a distance d apart so that the distance between the first and last plate is $3d$. Show that if alternate plates are connected by conducting wires the

capacitance C is

$$C = \frac{3\epsilon_0 A}{d}$$

(*Hint:* Why, if q is the charge on the first plate, must the charge on the next be $-2q$?)

†42. For the physical situation described in Problem 34, show that the displacement vector **D** is everywhere radial and has the magnitude

$$D = \frac{q}{4\pi r^2}$$

†43. Consider, in Figure 22-26, a rectangular slab of a dielectric in which there is a *uniform* dipole moment per unit volume **P** directed perpendicular to a face.

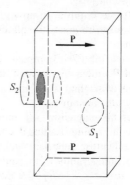

Figure 22-26

(a) Show that if S_1 is a closed surface contained entirely within the dielectric, then the total induced charge inside, $q_p(S_1)$, vanishes. (*Hint:* Use (22-25) and the fact that $\oint d\mathbf{S} = 0$ for any closed surface.)

(b) By making use of the closed surface S_2 in the figure, show that the polarization charge σ_p on the dielectric surface is

$$\sigma_p = -P.$$

23 The electric current

Nothing is too wonderful to be true if it be consistent with the laws of nature and in such things as this experiment is the best test of such consistency.

MICHAEL FARADAY (1791–1867)

23-1 Introduction

The preceding four chapters have dealt mainly with physical situations involving *static* distributions of charge. We now turn from these considerations to study a completely different set of phenomena, which are associated with electric charge in motion. In this connection an *electric current* or a *current* will be said to exist in a certain region of space if macroscopic amounts of charge undergo ordered motion in that region. As an introduction to a study of phenomena associated with electric currents, in this and the following chapter we first consider the problem of their generation, measurement, and control.

Figure 23-1 shows how a current may be generated very simply by use of a charged capacitor. Part (a) depicts the initial situation of a capacitor of charge Q and some of the electric-field lines associated with it. Suppose that these conductors are suddenly connected by a metallic wire. Because of the existence of a potential difference between the conductors, at least initially, the electric field in the interior of the connecting wire is *not* zero

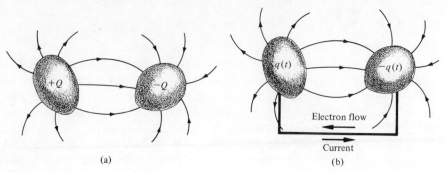

(a)   (b)

Figure 23-1

and thus electrons will start to flow from the negative conductor to its positively charged partner. As a result of this flow, the charge on the capacitor will start to decrease and will continue to do so as long as any charge remains on the conductors. According to the above definition, this flow of electrons along the connecting wire in Figure 23-1b constitutes an electric current. Because of its generally short-lived nature, this particular current is called a *transient* current.

Mainly for historical reasons, it is customary to define electric current, not in terms of the actual flow of negatively charged electrons, but rather in terms of an equivalent flow of positive charge in the opposite direction. Thus, even though the electrons travel along the connecting wire from right to left in Figure 23-1b, the electric current is by definition said to be directed in the opposite sense from left to right. This choice is arbitrary but the one conventionally made.

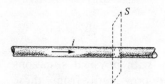

Figure 23-2

Consider, in Figure 23-2, a short segment of a wire through which flows a certain electric current. Let us assign a sense of direction to this current by arbitrarily drawing an arrowhead directed, say, to the right. Also, define S to be a surface whose plane is perpendicular to the current and thus to the wire at some point. Let Δq represent the charge that during an infinitesimal time interval Δt crosses S in the assumed direction of the current flow, that is, to the right in the figure. The current i passing through S is then defined by the limit

$$i = \lim_{\Delta t \to 0} \frac{\Delta q}{\Delta t}$$

which, according to the definition of a derivative, is the same as

$$i = \frac{dq}{dt} \qquad (23\text{-}1)$$

Note that i can be positive or negative depending on the sign of Δq. A negative value for i means that the current flows in a direction opposite to that assumed in the figure. Thus, $i < 0$ means that during the time interval Δt either a certain amount of positive charge moves through S directed to the left in the figure or, equivalently, negative charge moves to the right. Corresponding statements apply for positive values for i.

It follows from the definition in (23-1) that the coulomb per second is the unit of current. The SI unit of current is called the *ampere* (abbreviated A), and is given by

$$1 \text{ A} \equiv 1 \text{ ampere} = 1 \text{ C/s} \qquad (23\text{-}2)$$

Two units related to this are the milliampere (mA) and the microampere (μA), which are defined to be 10^{-3} ampere and 10^{-6} ampere, respectively.

Example 23-1 Suppose that the charge $q(t)$ on the capacitor in Figure 23-1b varies in time as

$$q(t) = Qe^{-t/\tau}$$

with $Q = 3.0\ \mu$C and $\tau = 1.0\ \mu$s. Calculate the current i in the connecting wire at any time t.

Solution Let us assume that, contrary to the actual situation, the direction of the current in the connecting wire is to the *left* in the figure. Then the amount of charge Δq that crosses a surface perpendicular to the connecting wire in a time interval Δt is the same as the *increase* of the charge on the capacitor. Hence, if $q(t)$ is the charge on the capacitor at time t, then, according to (23-1),

$$i = \frac{dq}{dt} = \frac{d}{dt}(Qe^{-t/\tau}) = -\frac{Q}{\tau}e^{-t/\tau}$$
$$= -3.0e^{-t/\tau} \quad \text{(A)}$$

where we have used the relation $d(e^{\alpha t})/dt = \alpha e^{\alpha t}$ and in the last equality the given values for Q and τ. The minus sign here reflects the fact that contrary to the stated assumption, but consistent with our physical expectations, the charge flow along the wire is directed to the right in the figure; that is, from the positive to the negative plate of the capacitor.

23-2 Electromotive force

The current associated with the capacitor discharge in Figure 23-1 is a *transient* since it exists only for the short period of time that the capacitor has a charge. To produce a *steady current*—that is, one which persists for long periods of time—it is necessary to use a different approach. The purpose of this section is to describe, in general terms, a class of devices that can be used to produce steady currents.

In the following, by the term *battery* we shall mean any one of a number of devices that are capable of maintaining a fixed potential difference between two points, and thus of producing a steady current between them. The terms a *source* or a *seat of electromotive force* are also in common usage. A complete listing of batteries would include chemical batteries, solar cells, thermoelectric cells, and electromagnetic generators, where in each case the preceding adjective describes the source of the energy involved. Thus the energy output of a chemical battery results from the chemical processes that take place in its interior, and the output of a solar cell is the result of radiant energy being incident on—and thereby interacting with— the material of the cell. Of these, we shall examine in detail only the electromagnetic generator. This device, as will be discussed in Chapter 27, makes possible the direct conversion of mechanical energy of motion into electrical energy.

An essential and visible feature of any battery, regardless of its type, is a pair of conducting terminals, which are outside the battery and are often labeled plus (+) and minus (−). As a general rule, the plus terminal has a positive charge and is therefore at a higher potential than is the other terminal with its negative charge. These terminals are always connected *inside the battery* by a conducting path of some type, such as a metallic strip or an electrolytic bath. If no conductor connects the terminals *outside* the battery, then the battery is said to be on *open circuit*. In general, no current flows inside or outside of a battery when it is on open circuit.

Figure 23-3 shows the external essentials of a battery on open circuit, with its plus and minus terminals labeled A and B, respectively. As is implied by the electric-field lines in the figure, there is a potential difference V_{AB} between these terminals. According to (21-2) and (21-5), V_{AB} may be expressed in the form

$$V_{AB} \equiv V_A - V_B = -\int_B^A \mathbf{E} \cdot d\mathbf{l} = \int_A^B \mathbf{E} \cdot d\mathbf{l} \qquad (23\text{-}3)$$

where, for example, V_A is the potential of the plus terminal and $\mathbf{E}$ is the electric field evaluated at points along the path connecting the terminals. As we saw in Chapter 21, the electrostatic field is conservative, and thus this

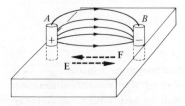

Figure 23-3

integral is independent of the path followed in going from A to B. This path may be entirely outside the battery, entirely inside, or partially inside and partially outside.

Consider now the situation *inside* a battery on open circuit. Because of the potential difference between the terminals, and the existence inside the battery of a conducting path between them, it follows that inside the battery there must exist a force field $\mathbf{F}$ that is distinct from the electric field $\mathbf{E}$ produced by the charge at the terminals. Otherwise, in response to the electric field, current would flow inside the battery and neutralize the charge on the two terminals. In particular, then, since on open circuit equilibrium conditions prevail, this force $\mathbf{F}$ on an electron or ion of charge q_0 must be related to the electric field $\mathbf{E}$ *inside* the battery by

$$\mathbf{F} + q_0\mathbf{E} = 0 \qquad (23\text{-}4)$$

It is convenient to characterize this force field $\mathbf{F}$ inside a battery in terms of a certain quantity called the *electromotive force* or *emf* $\mathscr{E}$ of the battery. The emf $\mathscr{E}$ is defined to be the work carried out by the force field $\mathbf{F}$ in taking a particle of unit positive charge from the minus to the plus terminal *inside* the battery. Thus

$$\mathscr{E} = \frac{1}{q_0}\int_B^A \mathbf{F}\cdot d\mathbf{l}$$

and therefore the product $q\mathscr{E}$ represents the work carried out by a battery in transporting a charge q between its terminals. On substituting into this relation, by use of (23-3) and (23-4), we find that

$$\mathscr{E} = V_{AB} \qquad \text{(open circuit)} \qquad (23\text{-}5)$$

so the emf $\mathscr{E}$ of a battery is precisely the potential difference between its terminals on *open circuit*.

It is important, despite the equality in (23-5), to keep the physical difference between $\mathscr{E}$ and V_{AB} in mind. The former represents the work per unit charge carried out by the internal energy source of the battery, whereas the latter is the potential difference between the terminals of the battery due to charge separation. The validity of (23-5) depends on (23-4), and therefore holds only under conditions of open circuit.

According to the above definition, the unit of electromotive force is the same as that of potential. Thus the volt is the unit of emf and when speaking of a "10-volt battery" what is meant is one that, on open circuit, will sustain a potential difference of 10 volts between its terminals.

Let us now consider briefly what happens to a battery not on open circuit. Imagine connecting a conducting wire between the terminals in Figure 23-3. Because of the initial potential difference V_{AB} between these terminals, current will flow along this wire from A to B. Associated with this current there will be a momentary decrease of charge on the battery terminals. Thus the electric field $\mathbf{E}$ decreases inside the battery and (23-4) ceases to be valid.

Instead, the strength of the force field **F** now exceeds that of the electric force, so that effectively positive charge is forced to go in the direction from B to A *inside* the battery. In other words, a current flows inside the battery from the negative to the positive terminals. Thus there is a current from A to B along the external conducting path and one from B to A inside the battery. Under steady-state conditions these two currents will be identical, so a continuous stream of charge will be going around a closed path. Unfortunately, the relation in (23-5) between the emf of the battery and the potential difference between its terminals is not generally valid under these conditions. However, assuming that there are no other batteries in the circuit, we find that the difference $(\mathscr{E} - V_{AB})$ is generally very small. Hence it will usually be neglected. A battery for which the difference $(\mathscr{E} - V_{AB})$ is significant is said to have an *internal resistance*.

23-3   Drift velocity and current density

Under conditions of static equilibrium there can be no electric field in the interior of a conductor. The free electrons inside behave as a dilute gas and wander through the positive lattice in such a way that from a macroscopic viewpoint the conductor is electrically neutral. The density of these electrons is very high, and in ordinary metals at room temperature it is of the order of 10^{29} per cubic meter. Just as for an ordinary gas, we may think of these electrons as being in a state of constant motion, with a velocity that typically is of the order of 10^6 m/s. The velocity distribution of these electrons is isotropic, so at any point in the lattice the average flow of electrons in any given direction is compensated for by an equal flow in the opposite direction. This means that under normal circumstances when the electric field inside the metal vanishes there is no electric current there either.

Let us now contrast this picture with the corresponding one when the electric field inside the conductor is not zero. Figure 23-4 depicts the motion of a typical electron in a metal when there is an electric field present. Assuming this field acts to the left, this electron will accelerate to the right. However, before it can pick up very much speed, it will collide with an ion,[1] and as a result it will, in general, lose energy and change its direction of motion. After this collision, it will again accelerate to the right and continue to do so until it suffers a second inelastic collision with an ion at some point b. As shown in the figure, the motion of the electron thus consists of a zigzag pattern of short, segmented trajectories *abcdefgh*. However, this pattern is not random. Because of the electric field, the electrons tend to drift

[1]According to the ideas of quantum mechanics, an electron traveling through a "perfect" lattice will not collide with a lattice ion. However, such electrons may undergo collisions with lattice imperfections, such as impurities or vacancies, and all text references to collision with ions should be interpreted in this sense.

gradually to the right. In other words, because of the ions, the electrons do not accelerate to arbitrarily high velocities, but rather when viewed over long periods of time they seem to move gradually in a direction opposite to that of the electric field. The velocity associated with this motion of the electrons is called the *drift velocity* and will be represented by the symbol $\mathbf{v}_d$. We shall see below that even though the velocity distribution of the electrons is isotropic and peaks at a value of the order of 10^6 m/s, the drift velocity $\mathbf{v}_d$ is directed opposite to the direction of the field and is, in general, very much smaller.

Let us estimate the drift velocity in a typical case. To this end, consider, in Figure 23-5, a segment of a wire of cross-sectional area A in which there flows a uniform current i. In a small time interval Δt, the charge ΔQ that flows past a given surface S is that contained in a cylinder of cross-sectional area A and length $v_d \Delta t$. That is, since we view the electrons as moving at the drift velocity $\mathbf{v}_d$, all electrons in this region of volume $Av_d \Delta t$ will cross the surface S and thus contribute to the current. If n represents the number of electrons per unit volume, it follows that the net charge ΔQ, which flows through S along the direction of the current, is

$$\Delta Q = en(Av_d \Delta t)$$

where e is the magnitude of the elementary quantum of charge and has the value 1.6×10^{-19} coulomb. Dividing both sides by Δt and equating the current i to the ratio $\Delta Q/\Delta t$, we thus obtain

$$i = neAv_d \tag{23-6}$$

Equivalently, this may be reexpressed in the form

$$\mathbf{j} = -ne\mathbf{v}_d \tag{23-7}$$

where $\mathbf{j}$ is a vector *along* the direction of the current and is known as the *current density*. It has the magnitude

$$j = \frac{i}{A} \tag{23-8}$$

The minus sign in (23-7) is necessary since, by our convention, the direction of the current is opposite to the drift velocity $\mathbf{v}_d$ of the electrons.

As a typical case, consider a wire of cross-sectional area $1 \text{ mm}^2 = 10^{-6} \text{ m}^2$

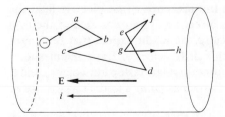

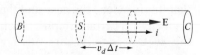

Figure 23-4   *Figure 23-5*

through which flows a current of 1.0 ampere. Assuming that $n \cong 10^{29}/m^3$, on solving (23-6) for v_d we obtain

$$v_d = \frac{i}{neA} = \frac{1.0 \text{ A}}{(10^{29}/m^3) \times (1.6 \times 10^{-19} \text{ C}) \times 10^{-6} \text{ m}^2}$$
$$= 6.3 \times 10^{-5} \text{ m/s}$$

As anticipated above, this is a very small velocity compared to the value 10^6 m/s, which characterizes the velocity of the electrons in a metal.

Example 23-2   An aluminum wire has a density of 2.7 g/cm^3 and a cross-sectional area of 1.0×10^{-5} m^2, and carries a current of 2.0 ampere.

(a) What is the current density in the wire?

(b) Assuming three free electrons per atom, calculate the density of electrons in the wire.

(c) What is the drift velocity of these electrons?

Solution

(a) Making use of the definition for current density in (23-8) and the given values for i and A, we find

$$j = \frac{i}{A} = \frac{2.0 \text{ A}}{10^{-5} \text{ m}^2} = 2.0 \times 10^5 \text{ A/m}^2$$

(b) Since the atomic mass of aluminum is 27, 1 mole, or 27 grams of this material, contains 6.0×10^{23} atoms. Further, since the density of the aluminum wire is 2.7 g/cm^3, it follows that 10 cm^3 of aluminum contains 1 mole. In other words, 1 cm^3 of aluminum contains 6.0×10^{22} atoms so there are altogether 6.0×10^{28} atoms/m^3. With three free electrons per atom, it follows then that the electron density is

$$n = 1.8 \times 10^{29}/m^3$$

(c) On solving (23-7) for v_d and substituting the values for the various parameters we obtain

$$v_d = \frac{j}{ne} = \frac{2.0 \times 10^5 \text{ A/m}^2}{(1.8 \times 10^{29}/m^3) \times (1.6 \times 10^{-19} \text{ C})}$$
$$= 6.9 \times 10^{-6} \text{ m/s}$$

23-4   Ohm's law

We have seen in Section 23-3 that in the presence of an electric field the electrons in a metal gradually drift in a direction opposite to that of the field with a certain drift velocity v_d. Physically, we might expect that the larger the electric field, the greater will be this drift velocity. Experiment shows that this is indeed the case and that the precise relationship between v_d and E is given by a certain empirical relation known as *Ohm's law*. The purpose of this section is to discuss this law.

Consider a region inside a conductor in which there is an electric field E. Following convention, we shall continue to interpret the associated electron flow opposite to the field in terms of an equivalent flow of positive charge along **E**. As for the case of the uniform current in Section 23-3, it is convenient to characterize this flow by a vectorial current density **j** defined so that the quantity **j** · *d***S** represents the flow of charge per unit time across any area element *d***S** inside the conductor. Since the quantity **j** represents a flow of charge and since the cause of this flow is the electric field **E** itself, we might expect a relation of some type to exist between them. Indeed, experiment shows that, for many conductors, **j** and **E** are simply proportional to each other. For these substances we may write

$$\mathbf{j} = \sigma \mathbf{E} \qquad (23\text{-}9)$$

a relation known as *Ohm's law*. The coefficient of proportionality, σ, is a constant independent of **E** and is called the *conductivity*.

It should be noted that since it is not feasible to measure **j** and **E** separately inside a conductor, Ohm's law in the form of (23-9) cannot be verified directly. We shall derive below an alternate form of this law, which involves quantities more amenable to experimental measurement. It is this latter form that is used almost exclusively in most applications.

It is convenient to define a unit of electrical resistance called the *ohm* (abbreviated Ω) by the relation

$$1 \text{ ohm} \equiv 1 \ \Omega = 1 \ \text{V/A}$$

Using the fact that the unit of electric field is the volt per meter (V/m) and that of current density is the ampere per square meter (A/m²), we find by use of (23-9) that the unit of conductivity is $(\Omega\text{-m})^{-1}$. A quantity related to the conductivity of a material is its *resistivity* ρ. This is defined to be the reciprocal of the conductivity σ:

$$\rho = \frac{1}{\sigma} \qquad (23\text{-}10)$$

It follows that the Ω-m is the appropriate unit of resistivity.

Table 23-1 lists the conductivities and the associated resistivities of typical metals. Note that Ag and Cu have relatively high conductivities and thus are better conductors than are the other elements listed. Physically, this means that for a given electric field strength, more current will flow in a silver or copper wire than in an iron wire of the same cross-sectional area. As will be

Table 23-1 Conductivities and resistivities at 293 K

Substance	$\sigma \ (\Omega\text{-m})^{-1}$	$\rho \ (\Omega\text{-m})$
Ag	6.3×10^7	1.6×10^{-8}
Al	3.6×10^7	2.8×10^{-8}
Cu	5.9×10^7	1.7×10^{-8}
Fe	1.0×10^7	1.0×10^{-7}
Ni	1.4×10^7	7.1×10^{-8}
W	1.8×10^7	5.6×10^{-8}

described in Section 23-8, σ also depends on temperature, and for most substances the higher the temperature the lower is the conductivity.

To cast Ohm's law into a more useful form, consider the wire in Figure 23-5 and suppose that it has a conductivity σ, a length l, and a cross-sectional area A. Assuming that the current i is uniform across the wire, we shall establish below that the potential difference $V \equiv V_B - V_C$ between its end faces is

$$V = Ri \tag{23-11}$$

where R is a parameter known as the *resistance* of the wire and is defined by

$$R = \frac{l}{A\sigma} \tag{23-12}$$

In words, the relation in (23-11), which is also known as Ohm's law, states that the ratio of the potential drop V between the ends of a wire to the current i flowing in it, is constant, independent of V and i. According to (23-12), this ratio R varies directly as the length l of the wire and inversely with the cross-sectional area A.

To establish (23-11), note first that the potential difference $V \equiv (V_B - V_C)$ between the end faces in Figure 23-5 may be expressed as a line integral of the electric field $\mathbf{E}$ along a straight line parallel to the current flow. Thus

$$V = V_B - V_C = \int_B^C \mathbf{E} \cdot d\mathbf{l} = \int_B^C \frac{1}{\sigma} \mathbf{j} \cdot d\mathbf{l}$$

$$= \frac{j}{\sigma} \int_B^C dl = \frac{jl}{\sigma}$$

where the third equality follows from (23-9) and the fourth from the fact that the current flow has been assumed to be uniform. The validity of (23-11), with R defined in (23-12), now follows by use of (23-8).

Just as for a capacitor, it is convenient to have available a symbol to denote a resistor in a circuit. Following convention we shall use the symbols

the first of which represents a resistor of resistance R and the second a variable resistor. Similarly, we shall use either of the symbols

to represent a battery of emf $\mathscr{E}$. The positive terminal is in each case associated with the left vertical line and the negative terminal with the shorter vertical line to the right.

Example 23-3 An aluminum wire has a cross-sectional area of $2.0\,\text{mm}^2$ and a length of 30 meters. Using the data in Table 23-1, calculate its resistance.

Solution Substituting into (23-12) and using the value $\sigma = 3.6 \times 10^7 \, (\Omega\text{-m})^{-1}$, we find that

$$R = \frac{l}{A\sigma} = \frac{30 \text{ m}}{(2.0 \times 10^{-6} \text{ m}^2) \times [3.6 \times 10^7 \, (\Omega\text{-m})^{-1}]}$$

$$= 0.42 \, \Omega$$

Example 23-4 A battery of emf $\mathcal{E} = 10$ volts is connected across a 15-Ω resistor, as in Figure 23-6. What current flows?

Solution According to (23-5), the emf of the battery is equal to the potential drop across the resistor. But the latter is Ri, according to (23-11). Thus

$$\mathcal{E} = Ri$$

or

$$i = \frac{\mathcal{E}}{R} = \frac{10 \text{ V}}{15 \, \Omega} = 0.67 \text{ A}$$

The direction of this current through the resistor is, as shown in the figure, from the plus to the minus terminal of the battery.

Example 23-5 Suppose that a current of 2.0 amperes flows clockwise in the circuit in Figure 23-7. If V_f, the potential at point f, is zero, calculate the following potentials: (a) V_a; (b) V_b; (c) V_c; (d) V_d; and (e) V_e.

Solution

(a) Since the emf of the lower battery is 50 volts and since $V_f = 0$, it follows that

$$V_a - V_f = V_a = 50 \text{ V}$$

(b) Since there is no resistance from a to b (straight lines on circuit diagrams always represent resistance-free paths) it follows that

$$V_b = V_a = 50 \text{ V}$$

(c) A 2.0-ampere current flowing through a 5-Ω resistor produces, according to (23-11), a potential drop of 10 volts. Thus, $V_b - V_c = 10$ volts or, in other words,

$$V_c = V_b - 10 \text{ V} = 50 \text{ V} - 10 \text{ V} = 40 \text{ V}$$

(d) Because of the 20-volt battery,

$$V_c - V_d = 20 \text{ V}$$

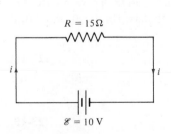

Figure 23-6

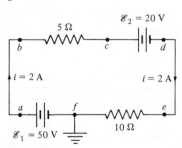

Figure 23-7

Thus

$$V_d = V_c - 20 \text{ V} = 40 \text{ V} - 20 \text{ V} = 20 \text{ V}$$

(e) As in (b), $V_d = V_e$ and thus $V_e = 20$ volts.

It is interesting to note that the current through the upper battery ($\mathscr{E}_2$ in the figure) is directed the "wrong" way; that is, the current flows from the positive to the negative terminal *inside* the battery. Under these circumstances we say that the battery is *being charged*. Indeed, in this case, part of the energy output of $\mathscr{E}_1$ goes to increase that of $\mathscr{E}_2$.

23-5   Discharge of a capacitor

The purpose of this section is to reexamine the physical situation described in Section 23-1 involving the discharge of a capacitor through a resistor. This time, however, we shall analyze the charge flow quantitatively.

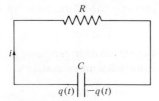

Figure 23-8

Consider, in Figure 23-8, a capacitor of capacitance C, which has an original charge Q_0 and across which a resistor of resistance R is connected. If $q \equiv q(t)$ is the charge on the capacitor at time t after the discharge has begun, then by definition of capacitance the potential across the capacitor at this instant is q/C. Assuming, as shown in Figure 23-8, that the left-hand plate has the positive charge and is thus at the higher potential, it follows that the current i in the circuit will be directed clockwise, as shown. According to (23-11), the potential difference between the ends of the resistor is Ri. Since this must be the same as the potential drop across the capacitor, it follows that

$$\frac{q}{C} = Ri \tag{23-13}$$

Finally, since a positive value for the current i means in the present case that the charge $q(t)$ on the capacitor is decreasing, that is, for $i > 0$, $dq < 0$, it follows that

$$i = -\frac{dq}{dt} \tag{23-14}$$

Substitution into (23-13) thus leads to

$$R\frac{dq}{dt} + \frac{1}{C} q = 0 \tag{23-15}$$

This relation is known as the *circuit equation* for the electric circuit in Figure 23-8. Its solution yields the charge q on the capacitor, as well as the current i in the circuit at any time t.

To solve (23-15), it is convenient first to cast it into the form

$$\frac{dq}{q} = -\frac{1}{RC}\,dt$$

which may be integrated to

$$\ln q = -\frac{t}{RC} + \alpha$$

with α an integration constant. Since at $t = 0$ the charge on the capacitor is Q_0, that is, $q(0) = Q_0$, it follows that $\alpha = \ln Q_0$, and thus

$$\ln q = \ln Q_0 - \frac{t}{RC}$$

Finally, taking the exponential of both sides and recalling that $e^{\ln q} = q$, we obtain, as the solution for (23-15),

$$q(t) = Q_0 e^{-t/RC} \tag{23-16}$$

Thus the charge on the capacitor drops exponentially to zero from its initial value Q_0. The current i in the circuit is found by substitution into (23-14) to be

$$i = \frac{Q_0}{RC}\, e^{-t/RC} \tag{23-17}$$

and thus i also drops off exponentially from its initial value of Q_0/RC. Figure 23-9 shows a plot of q and i as functions of time.

In connection with a study of the charging and the discharging of capacitors, it is convenient to define a *time constant* τ, which for the case just considered represents the time it takes the charge on the capacitor to drop to $1/e$ ($\cong 0.368$) of its initial value. Thus τ satisfies

$$q(\tau) = \frac{1}{e}\, Q_0$$

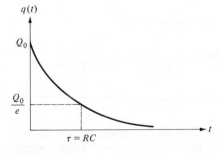
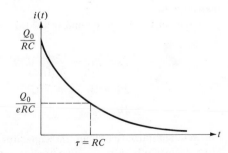

Figure 23-9

and, substituting for $q(\tau)$ from (23-16), we obtain

$$\tau = RC \tag{23-18}$$

Example 23-6   What is the time constant τ associated with the discharge of a 3.0-μF capacitor through a 100-Ω resistor?

Solution   Substituting the given values for R and C into (23-18), we find that

$$\tau = RC = 100 \ \Omega \times 3.0 \times 10^{-6} \ F = 3.0 \times 10^{-4} \ s$$

where we have used the fact that the unit of the (Ω-F) is the same as the second. Thus in a time interval of 3.0×10^{-4} second the charge on the capacitor drops from its original value, whatever it may have been, to $1/e \cong 37$ percent of this value.

Example 23-7   A 2.0-μF capacitor is charged by a battery of emf $\mathcal{E} = 10$ volts. It is then disconnected from the battery and shorted out by a $1.0 \times 10^3 \ \Omega$ resistor.
(a) What is the maximum charge on the capacitor?
(b) What is the time constant?
(c) What is the maximum current flowing in the resistor?
(d) How much charge is left on the capacitor 1.0 second after the discharge starts?

Solution
(a) According to (23-5), the maximum charge Q_0 on the capacitor is

$$Q_0 = C\mathcal{E} = (2.0 \times 10^{-6} \ F) \times (10 \ V) = 2.0 \times 10^{-5} \ C$$

(b) The substitution of the given values for R and C into (23-18) yields

$$\tau = RC = (1.0 \times 10^3 \ \Omega) \times (2.0 \times 10^{-6} \ F) = 2.0 \times 10^{-3} \ s$$

(c) The maximum current, according to (23-17), flows at $t = 0$, and thus

$$i_{max} = i(0) = \frac{Q_0}{RC} = \frac{Q_0}{\tau} = \frac{2.0 \times 10^{-5} \ C}{2.0 \times 10^{-3} \ s}$$
$$= 1.0 \times 10^{-2} \ A$$

where in the fourth equality we have used the value for Q_0 and τ from (a) and (b), respectively.
(d) On setting $t = 1$ second in (23-16), we obtain for $q(1 \ s)$

$$q(1 \ s) = Q_0 \exp\left\{-\frac{1 \ s}{\tau}\right\} = (2.0 \times 10^{-5} \ C) \times \exp\left\{-\frac{1 \ s}{2.0 \times 10^{-3} \ s}\right\}$$
$$= 2.0 \times 10^{-5} \times e^{-500} \ C$$
$$\approx 10^{-200} \ C$$

For practical purposes, then, the capacitor is completely discharged after 1 second. More realistically, after a capacitor has been discharging for a time interval of the order, say, of 10τ, it is essentially discharged.

23-6 The charging of a capacitor

Having seen how to describe quantitatively the discharge of a capacitor, in this section we examine the corresponding problem of the charging of a capacitor by use of a battery.

Consider, in Figure 23-10, a capacitor of capacitance C in series with a resistor R, both connected across a battery of emf $\mathscr{E}$. For convenience we have also included in the circuit a switch S and we shall assume that at $t = 0$ the switch is closed.

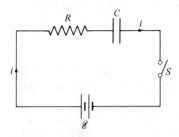

Figure 23-10

To obtain the circuit equation, assume, as shown in the figure, that the current flows in a clockwise sense. The potential difference across the terminals of the battery is the sum of the potential drops Ri and q/C across the resistor and the capacitor, respectively. According to (23-5), we have then

$$\mathscr{E} = \frac{q}{C} + Ri$$

This time, since the capacitor is charging up, a positive value for i means that $dq > 0$. Hence the current i is related to the capacitor charge q by (23-1), and substitution into the above formula leads to the circuit equation

$$\mathscr{E} = \frac{q}{C} + R\frac{dq}{dt} \tag{23-19}$$

Note that for $\mathscr{E} = 0$ this reduces to (23-15), as it should.

To solve (23-19), let us reexpress it as

$$\frac{dq}{\mathscr{E}C - q} = \frac{dt}{RC}$$

and then integrate both sides. The result is

$$\ln(\mathscr{E}C - q) = -\frac{t}{RC} + \alpha$$

and since at $t = 0$, $q = 0$, the integration constant α is found to be $\ln(\mathscr{E}C)$. Solving the resulting formula for q, we obtain

$$q = \mathscr{E}C(1 - e^{-t/RC}) \tag{23-20}$$

which by use of (23-1) yields the associated formula for the current

$$i(t) = \frac{\mathscr{E}}{R} e^{-t/RC} \tag{23-21}$$

Note that at $t = 0$, when the switch is closed, the current $i(0)$ in the circuit is the same as the current that would flow if the capacitor were *not* present.

Figure 23-11 shows a plot of q and i as functions of time. Initially, the current has the value $\mathscr{E}/R$ and from there it decreases exponentially with the same time constant $\tau = RC$ as found in Section 23-5. Thus, at $t = \tau$, the current has dropped to $1/e$ of its initial value $\mathscr{E}/R$. In a similar way, the charge on the capacitor rises from its initial value zero, to within the factor

$$1 - \frac{1}{e} \approx 1 - 0.37 = 0.63$$

of its saturated value $\mathscr{E}C$ in the same time interval, of length $\tau = RC$. Thus, in a period of the order of 10τ the charge on the capacitor assumes its final value to a high degree of precision.

To summarize, if a resistor and capacitor are connected in series across a battery, as in Figure 23-10, then initially a current $i = \mathscr{E}/R$ will flow in the circuit as if the capacitor were not present. In a time interval of the order of $\tau = RC$, however, this current decreases as the charge on the capacitor increases. This buildup of charge will continue until finally enough charge has accumulated on the capacitor that its potential is the same as the emf of the battery. At this point the potential drop across the resistor must be zero, and thus current can no longer flow.

Example 23-8 A battery of emf $\mathscr{E} = 10$ volts is connected across a 3.0-μF capacitor and a 100-Ω resistor in series. Assuming that at $t = 0$ the switch is closed, calculate:
   (a) The initial current in the circuit.
   (b) The final charge on the capacitor.
   (c) The charge on the capacitor at $t = 3.0 \times 10^{-4}$ second.
   (d) The current at $t = 9.0 \times 10^{-4}$ second.

Solution
   (a) The initial current i_0 has the same value as that of the current which would flow

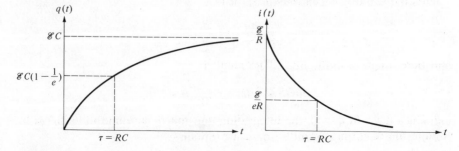

Figure 23-11

if the capacitor were not present. Thus

$$i_0 = \frac{\mathscr{E}}{R} = \frac{10 \text{ V}}{100 \, \Omega} = 0.1 \text{ A}$$

(b) The maximum charge q_f on the capacitor is that value for which the potential across it is the same as the emf of the battery. Thus

$$q_f = C\mathscr{E} = (3.0 \times 10^{-6} \text{ F}) \times (10 \text{ V})$$
$$= 3.0 \times 10^{-5} \text{ C}$$

(c) Since $R = 100 \, \Omega$ and $C = 3.0 \, \mu\text{F}$, the time constant τ has the value

$$\tau = RC = 100 \, \Omega \times 3.0 \times 10^{-6} \text{ F} = 3.0 \times 10^{-4} \text{ s}$$

Thus, according to (23-20), at time t

$$q(t) = C\mathscr{E}(1 - e^{-t/RC})$$
$$= 3.0 \times 10^{-5}(1 - e^{-t/3 \times 10^{-4} \text{s}}) \text{ C}$$

and, in particular, at $t = 3.0 \times 10^{-4}$ second

$$q(3 \times 10^{-4} \text{ s}) = 3.0 \times 10^{-5} (1 - e^{-1}) \text{ C}$$
$$= 1.9 \times 10^{-5} \text{ C}$$

(d) Substituting the given values for $\mathscr{E}$, R, and C we find in the same way that at $t = 9.0 \times 10^{-4}$ second $\equiv 3\tau$ the current in the circuit is

$$\frac{\mathscr{E}}{R} e^{-3\tau/\tau} = 0.1 e^{-3} \text{ A} = 5.0 \times 10^{-3} \text{ A}$$

23-7 Electric power—Joule heating

In Section 23-3 we noted that in the presence of an electric field the electrons in a conductor accelerate in a direction opposite to that of the field, and in the resulting collisions with the lattice ions they tend to lose energy. This energy loss by the electrons is gained by the lattice and manifests itself macroscopically by the heating up of any substance through which a current flows. This phenomenon is called *Joule heating*. The purpose of this section is to obtain a measure of this irreversible energy loss associated with all resistive electric circuits.

Consider again the circuit containing a resistor R and a capacitor C connected in series across a battery of emf $\mathscr{E}$ (Figure 23-10). To obtain a measure of the Joule heating in this circuit, let us multiply the circuit equation in (23-19) throughout by the instantaneous value of the current i. The result is

$$i\mathscr{E} = \frac{qi}{C} + Ri^2 \tag{23-22}$$

where use has been made of (23-1). We shall now establish that the term Ri^2 represents the rate at which energy in the form of heat is dissipated in the resistor by relating the other two terms $\mathscr{E}i$ and qi/C to the energy associated with the battery and the capacitor, respectively.

According to the definition in Section 23-2, the emf $\mathscr{E}$ of a battery represents the work per unit charge performed by the battery in carrying a charged particle between its terminals. The quantity $q\mathscr{E}$ thus represents the work that the battery carries out to transport charge q so that its rate of doing work, or power, P_B, is

$$P_B = \frac{d}{dt}(\mathscr{E}q) = \mathscr{E}i \tag{23-23}$$

Comparison with (23-22) shows that the left-hand side of that relation represents, therefore, the rate at which the battery carries out work.

In a similar way we can show that the term qi/C in (23-22) is the rate at which the energy stored in the capacitor is increasing. According to (22-9), the energy U stored in a capacitor is $q^2/2C$, and hence the rate of increase of U is

$$\frac{dU}{dt} = \frac{d}{dt}\left(\frac{1}{2C}q^2\right) = \frac{q}{C}\frac{dq}{dt} = \frac{qi}{C} \tag{23-24}$$

where the last equality follows by use of (23-1). Thus the first term on the right-hand side of (23-22) is the rate at which energy is being stored in the capacitor.

Now from a physical point of view the work expended by the battery must show up either as energy stored in the capacitor or else as energy lost in the resistor by electron collisions with the lattice. But, according to (23-24), the term qi/C on the right-hand side of (23-22) represents the rate at which the external agent (the battery in the present case) carries out work on the capacitor. Therefore since $\mathscr{E}i$ represents the power output of the battery we conclude that the residual term in (23-22), namely Ri^2, *must* represent the rate at which energy is dissipated in the resistor. Thus we have established that:

The rate P_R at which energy is dissipated in the form of heat in a resistor of resistance R through which flows a current i is given by

$$P_R = Ri^2 \tag{23-25}$$

Note that the power Ri^2 dissipated in the resistor is invariably positive, regardless of the direction of the current. Thus it represents an irreversible energy loss in the thermodynamic sense.

Example 23-9   Show that if a capacitor is charged by a battery of emf $\mathscr{E}$ through a resistor R, as in Figure 23-10, the total energy lost as Joule heat is numerically equal to that stored in the capacitor.

Solution   One way of establishing this property is by making use of the explicit formulas in (23-20) and (23-21) and integrating (23-23) through (23-25).

An alternate and simpler way is by integrating both sides of (23-22) for all positive times:

$$\int_0^\infty \mathscr{E}i \, dt = \int_0^\infty \frac{qi}{C} \, dt + \int_0^\infty Ri^2 \, dt \tag{23-26}$$

Now the term on the left is the total work carried out by the battery: $\mathscr{E}q(\infty) = \mathscr{E}^2 C$, where $q(\infty) = C\mathscr{E}$ is the final charge on the capacitor. Correspondingly, the first term on the right is, according to (23-24), the final energy stored in the capacitor. Since this has the value $C\mathscr{E}^2/2$, we find by substitution into (23-26) that

$$\int_0^\infty Ri^2 \, dt = C\mathscr{E}^2 - \frac{1}{2}C\mathscr{E}^2 = \frac{1}{2}C\mathscr{E}^2$$

which is the desired result. Note that this fact, that one half of the total energy output of the battery is dissipated in the resistor, is independent of the resistance R. It is a price that must be paid each time a capacitor is charged.

23-8   Temperature variation of conductivity

In connection with the discussion of Ohm's law in (23-9) we have noted that the conductivity σ of a conductor is a constant independent of the strength of the electric field **E** and of the current density **j**. In terms of (23-11) this means that the resistance R of a resistor is independent of the current i and the potential difference V developed across it. A graph of the current flowing through a resistor as a function of the potential difference V must therefore have the form of a straight line, as in Figure 23-12. Materials for which the plot of current versus voltage is a straight line are said to be *linear materials*.

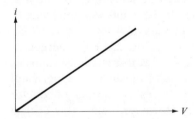

Figure 23-12

Now even though for many conductors the conductivity σ is a constant as far as current and electric field are concerned, in general σ varies with temperature. For most conductors, the conductivity rises as the temperature falls, or equivalently, the resistance R rises with temperature. Experiment shows further that if a material, which at some reference temperature T_0 has a resistivity ρ_0, undergoes a small temperature change ΔT, then the associated fractional change in resistivity, $\Delta\rho/\rho_0$, is directly proportional to ΔT. That is,

$$\Delta\rho = \alpha\rho_0\,\Delta T \qquad (23\text{-}27)$$

The coefficient of proportionality α is known as the *coefficient of resistivity*. Table 23-2 lists values for α at 293 K for a number of metals. The fact that α is positive for each of the materials listed implies, according to (23-27), that the resistivity rises with temperature. For some materials (for example, amorphous carbon with $\alpha \cong -5 \times 10^{-4}/\text{K}$) the coefficient of resistivity is found to be negative. This implies that it is the conductivity σ that rises with temperature for these cases. In this connection it will be established in the problems that in terms of σ (23-27) may be expressed in the equivalent form

$$\Delta\sigma = -\alpha\sigma_0\,\Delta T \qquad (23\text{-}28)$$

where $\sigma_0 = 1/\rho_0$ is the value for the conductivity at the reference temperature.

Table 23-2   Coefficients of resistivity at 293 K

Substance	α (1/K)
Al	4.0×10^{-3}
Cu	4.0×10^{-3}
Ag	4.0×10^{-4}
Au	3.0×10^{-4}
Fe	5.0×10^{-3}
Pb	4.2×10^{-3}
Hg	8.9×10^{-4}
W	4.7×10^{-3}

Of considerable interest in connection with a study of the temperature variation of resistivity is a phenomenon known as *superconductivity*. In 1908 Heike Kamerlingh Onnes (1853–1926) succeeded in achieving, for the first time, liquid helium temperatures (approximately 1 K) and was thus able to measure resistivities of various substances at very low temperatures. He found that, at 4.2 K, mercury (which is a solid below 234 K) undergoes a phase transition, in that for temperatures below 4.2 K its resistivity vanishes exactly. In Figure 23-13 we plot what the results of such an experiment would look like. Above 4.2 K, solid mercury behaves as a normal conductor in that its resistance rises with temperature. However, below this critical temperature its resistance drops abruptly to zero. Subsequent experiments by Kamerlingh Onnes and others have established the fact that this property of zero resistivity, or *superconductivity* as it is called today, is shared by other substances, such as Sn, Pb, and In, which become superconductors at 3.7 K, 7.2 K, and 3.4 K, respectively. Today we know that many materials including various alloys become superconductors at sufficiently low temperatures.

It has been found that superconductors have various interesting properties.

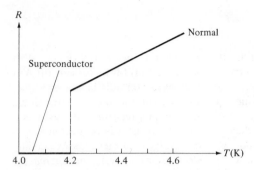

Figure 23-13

One of these is that if a current is started in a superconducting loop it will, in principle, persist forever. That is, since for a superconductor $R = 0$, it follows that the Joule heating Ri^2 must vanish exactly. Thus there is no power dissipation, and no mechanism is available to cause the current to decrease. Physically, this means that the electrons comprising the current in the superconductor undergo no energy-losing collisions with the lattice, a phenomenon that can be understood only within the framework of quantum mechanics.

23-9 Summary of important formulas

The current i flowing past a fixed point in a wire is defined by

$$i = \frac{dq}{dt} \tag{23-1}$$

and is the rate at which charge flows past this point, along a fixed and previously assigned direction. Negative values for i correspond to charge flow in the opposite sense.

The emf $\mathscr{E}$ of a battery on open circuit is related to the potential difference across its terminals by

$$\mathscr{E} = V_{AB} \tag{23-5}$$

If n represents the number of charge carriers per unit volume, each of charge $-e$, in a metal, then the current density $\mathbf{j}$ associated with their motion at the *drift velocity* $\mathbf{v}_d$ is

$$\mathbf{j} = -ne\,\mathbf{v}_d \tag{23-7}$$

In terms of the resistance R of a thin wire, Ohm's law is given by

$$V = Ri \tag{23-11}$$

where V is the potential difference between the ends of the wire when a current i flows through it. The power P_R dissipated in the resistor is given by

$$P_R = Ri^2 \tag{23-25}$$

QUESTIONS

1. Define or describe briefly what is meant by the following: (a) current; (b) electromotive force; (c) conductivity; (d) Ohm's law; and (e) Joule heating.

2. Why can the electric field inside the connecting wire in Figure 23-1b not vanish as long as the capacitor has a charge? Explain.

3. A wire is connected to one terminal of a battery, with the other end of the wire remaining free. Will a current flow in the wire? Explain.

4. Why must a circuit be closed in order for a steady current to flow in it? If a circuit is broken at some point, electric charge collects at the break. Why?

5. Explain why, in general, the electric field is *nonvanishing* in the region of space *outside* the circuit elements of a circuit in which a current flows?

6. What, in physical terms, is the distinction between the emf of a battery and the electric potential between its terminals? To be specific, consider a chemical battery.

7. Explain in physical terms why you would expect the conductivity of a conductor to be a decreasing function of temperature. Can you account for the fact that for some conductors σ is an increasing function of temperature?

8. A battery of emf $\mathscr{E}$ is connected across a wire. What happens to the current if (a) the length of the wire is doubled? (b) the diameter of the wire is doubled?

9. What is the distinction between the drift velocity of the electrons in a conductor and the velocity associated with their random motions?

10. A battery of emf $\mathscr{E}$ is connected across a wire of length l, of cross-sectional area A, and of conductivity σ. What is the effect on the drift velocity if (a) l is increased? (b) A is increased? (c) σ is increased?

11. In view of the fact that the Joule heating in a resistor is Ri^2, why is it that if a resistor is connected across a battery, then these losses actually *decrease* as the resistance *increases*?

12. Explain in physical terms why the initial current flow in the circuit in Figure 23-10 is the same as if the capacitor were not present. Similarly explain why the saturated value of the current is the same as if the resistor were not in the circuit.

13. Consider the circuit in Figure 23-14 and suppose that at $t = 0$ the switch is closed. Explain in physical terms why you would expect the initial value of the current to be $\mathscr{E}/R_2$.

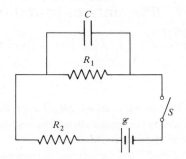

Figure 23-14

14. Explain why a battery must supply $2U$ units of energy to charge up a capacitor so that it has a final energy U. What is the maximum fraction of the output energy $2U$ of the battery that is convertible to useful work?

15. Consider the circuit in Figure 23-14. Explain, in physical terms, why you would expect after a long time that the current through the battery is $\mathscr{E}/(R_1 + R_2)$. What would be the charge on the capacitor at this time?

16. A 75-watt and a 100-watt light bulb

are designed to be powered by a 110-volt line. Through which will the higher current flow?

17. Consider the electron path $\overline{ab}$ in Figure 23-4. Explain why this path cannot be straight! Which way would its curvature be?

18. Consider an electron whose motion contributes to the current in a wire. Will this electron collide with electrons as well as with ions? Do both types of collisions contribute to the Joule heat? Explain.

19. Explain the negative sign in (23-14) and explain why this relation is *not* in contradiction with the defining relation (23-1). (*Hint:* Do the symbols q in these two relations represent the same physical quantity?)

20. If the resistance R in Figure 23-10 were doubled, explain what would happen to (a) the initial current through the battery; (b) the time constant τ; (c) the final charge on the capacitor; and (d) the energy output of the battery.

PROBLEMS

1. A battery having an emf of 100 volts is connected across a resistor of resistance 500 Ω. Calculate (a) the current in the wire and (b) the current density in the wire, assuming it has a cross-sectional area of 5.0 mm^2.

2. A current of 0.5 mA flows in a wire. (a) How much charge passes a given point in 10 minutes? (b) How many electrons pass a given point in the wire in 1 second?

3. A wire of cross-sectional area 2.0×10^{-5} m^2 is welded end-to-end to a second wire of cross-sectional area 4.0×10^{-5} m^2. If a steady current of 5.0 amperes flows through this composite wire, what is the current density in each segment? Assume uniform flow.

4. A wire is made of copper and has a cross-sectional area of 0.2 cm^2. Use the fact that the atomic mass of copper is 64 and assume that it has a density of 9.0 g/cm^3.
   (a) What is the number of copper atoms per unit volume in the sample?
   (b) If a current of 10 amperes flows along this wire, what is the current density?
   (c) Assuming one free electron per

atom in copper, calculate the drift velocity of the electrons.

5. A 10-volt battery is connected across a 3-Ω resistor.
   (a) How much current flows through the resistor?
   (b) How much charge flows through the battery during a 1-minute time interval?
   (c) How many electrons flow through the resistor in 15 minutes?

6. A uniform current of 5.0 amperes flows along an aluminum wire of cross-sectional area 2.0×10^{-5} m^2.
   (a) What is the current density in the wire?
   (b) What is the electric field strength in the wire? Use the data in Table 23-1.
   (c) If the wire has a length of 10 meters, what is the potential drop across it?

7. A silver wire and an aluminum wire have the same length and the same cross-sectional areas. What is the ratio of the currents that flow through them if they are separately connected to two batteries each with an emf of 5 volts?

8. Show that the unit of the farad-ohm (F-Ω) is the same as the second.

9. A copper wire has a resistance of 10 Ω and a length of 3 meters. It is then stretched so that its length becomes 12 meters. Assuming that the volume of the wire is *not* altered in the stretching process, calculate the final resistance of the wire. Assume that the temperature of the copper remains constant.

10. A wire has a length of 100 meters and a cross-sectional area of $10^{-6}\,m^2$. It is found that a 50-volt battery will produce a 10-ampere current when connected across the wire. (a) Calculate the conductivity of the wire. (b) What is its resistivity?

11. A certain light bulb dissipates 100 watts when connected across a 100-volt line. (a) What is the filament resistance? (b) What current flows through the filament?

12. Calculate the energy dissipated as Joule heat in 1 hr in the resistor in Problem 1.

13. A 15-volt battery is connected across two resistors, as shown in Figure 23-15.

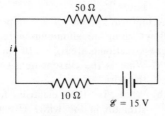

Figure 23-15

(a) In terms of the current i flowing in the circuit, what is the potential drop across the 10-Ω resistor?

(b) Repeat (a) for the other resistor.

(c) Making use of (23-5) show that a current of 0.25 ampere flows in the circuit.

14. Consider again the circuit in Figure 23-15.

(a) What is the power expended in the 10-Ω resistor?

(b) How much energy in the form of heat is dissipated in 1 hr in the 50-Ω resistor?

(c) How much energy will be expended by the battery if it operates for 10 hr?

15. What must be the resistance of a resistor that, when connected in series with a 2.5-μF capacitor across a battery, will yield a time constant for the circuit of (a) 1 μs? (b) 1 second?

16. A 5.0-μF capacitor is charged up by a 100-volt battery and is then disconnected from the battery and reconnected across a 200-Ω resistor.

(a) What is the original charge on the capacitor?

(b) What is the time constant τ for the circuit?

(c) What is the current through the resistor at $t = \tau$?

17. A charge of 10 μC is placed on a 0.05-μF capacitor and then allowed to discharge through a 50-Ω resistor.

(a) What is the time constant for this circuit?

(b) What is the initial value for the current?

(c) How much energy is originally stored in the capacitor?

(d) What is the rate at which energy is dissipated in the resistor at any time t?

18. By use of (23-17) show, by explicit calculation, that all the energy originally stored in the capacitor is dissipated as Joule heat in the resistor.

19. For the circuit in Figure 23-10, assuming $\mathscr{E} = 100$ volts, $C = 5.0\,\mu F$, and $R = 200\,\Omega$, calculate:

(a) The initial current in the circuit.

(b) The charge on the capacitor at the times 2.0×10^{-4} second, and 2.5×10^{-3} second.

(c) The current in the circuit at the times in (b).

20. Consider again the circuit of Problem 19.

(a) What is the final value of the potential across the capacitor?

(b) What is the final charge on the capacitor?

(c) What fraction of the energy put out by the battery is stored in the capacitor? What happens to the remainder of this energy?

21. Consider again the circuit in Problem 19.

(a) What is the charge on the capacitor at time t?

(b) What is the current at time t?

(c) Calculate the total energy lost in the resistor.

22. Suppose that the capacitor in Figure 23-10 has an original charge Q_0. Assume that at $t = 0$ the switch is closed.

(a) Show that (23-19) is still the circuit equation.

(a) Show that

$$q(t) = C\mathscr{E} + (Q_0 - C\mathscr{E})e^{-t/RC}$$

(c) Calculate the current at any time t and explain in physical terms why if $Q_0 = C\mathscr{E}$, there is no current at any time.

23. Consider again the situation in Problem 22 for the special case $Q_0 = 2C\mathscr{E}$.

(a) Show that the current is

$$i(t) = -\frac{\mathscr{E}}{R} e^{-t/RC}$$

and discuss the physical significance of the minus sign.

(b) Calculate the total energy expended in the resistor.

(c) Show that the amount of work carried out *on* the battery is $C\mathscr{E}^2$.

(d) Compare your results of (b) and (c) to the difference between the original energy stored in the capacitor and its final energy. Account for any differences.

24. Suppose that the switch S in Figure 23-10 is closed at time $t = 0$, and then reopened at time $t = T$.

(a) Show that for $t < T$ the current through the resistor is

$$i(t) = \frac{\mathscr{E}}{R} e^{-t/RC} \qquad (t < T)$$

(b) Show that at $t = T$, the ratio of the energy stored in the capacitor to that dissipated in the resistor is $\tanh(T/2RC)$.

25. Assuming that the capacitor in Figure 23-10 is a parallel-plate capacitor of area A and plate separation d, calculate the electric field in the region between the plates at any time t. Assume that the switch is closed at $t = 0$.

26. A spherical capacitor of inner radius a and outer radius b is connected through a resistor R to a battery of emf $\mathscr{E}$, as shown in Figure 23-16.

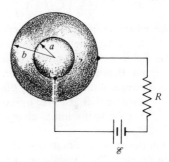

Figure 23-16

(a) Show that the time constant τ for this circuit is

$$\tau = 4\pi\epsilon_0 R \frac{ab}{b - a}$$

(b) Calculate the electric field in the region between the conducting spheres at any time t. Assume that the current first begins to flow at $t = 0$.

*27. Consider two capacitors C_1 and C_2 connected to a resistor R as in Figure 23-17. If i is the current directed as shown, and q_1 and q_2 are the charges on C_1 and C_2 at time t:

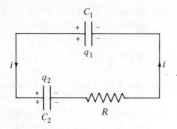

Figure 23-17

(a) Show that the circuit equation is

$$\frac{q_1}{C_1} = Ri + \frac{q_2}{C_2}$$

where if q_1, $q_2 > 0$, then the left-hand plates of C_2 and C_1 have the positive charge.

(b) Show that this may be reexpressed as

$$R\frac{dq_2}{dt} + q_2\left(\frac{1}{C_2} + \frac{1}{C_1}\right) - \frac{Q_0}{C_1} = 0$$

where Q_0 is the total charge on the two capacitors originally.

(c) Solve the circuit equation in (b) and thus find a formula for the current through R at any time t. Assume that at $t = 0$, $q_2 = 0$, so that Q_0 is the initial charge on C_1.

*28. For the physical situation described in Problem 27, calculate in terms of Q_0, C_1, C_2, and R:

(a) The initial current in the circuit.
(b) The final charge on each capacitor.
(c) The total energy dissipated in the resistor.

29. The resistivity of aluminum at 20°C is 2.8×10^{-8} Ω-m, and its coefficient of resistivity at the same temperature is 4.0×10^{-3}/°C. Calculate the resistivity of this substance at (a) 0°C and (b) 100°C.

30. If α is the coefficient of resistivity of a substance, derive (23-28) by use of (23-27) and (23-10).

31. Show that if α is the coefficient of

resistivity and β is the coefficient of linear expansion of a given wire, and if the temperature is raised by a small amount ΔT, then the fractional change in the wire's resistance, $\Delta R/R_0$, is

$$\frac{\Delta R}{R_0} = (\alpha - \beta)\,\Delta T$$

(*Note*: For typical metals the coefficient of linear expansion β has a value of about 10^{-5}/K. On the other hand, we see from Table 23-2 that $\alpha \approx 10^{-3}$/K, and thus $\alpha \gg \beta$.)

32. A copper wire at room temperature carries a current of 2.0 amperes if a 50-volt battery is connected across it.

(a) What is the resistance of the wire at room temperature?
(b) If the temperature of the wire is raised by 30 K, what is its resistance now?
(c) What current would flow (assuming again a 50-volt battery) through this wire at the elevated temperature?

33. A tungsten wire has a resistance of 100 Ω at 273 K. Calculate its resistance at 1500 K, assuming that (23-27) is valid for such a large temperature increment.

*34. The motion of an electron in a conductor in the presence of an electric field $\mathbf{E}_0$ is often described in terms of a model originally proposed by Drude. In his model Drude assumed that a frictional force

$$\mathbf{F} = -m\omega_c\,\mathbf{v}_d$$

acts on the electron. In this expression m is the electron's mass, $\mathbf{v}_d$ is its drift velocity, and ω_c is a certain parameter called the *collision frequency*, which represents the number of times per second that an electron collides with a lattice ion.

(a) Using the fact that an electron moves at the uniform velocity v_d under the combined action of this force and that due to the electric field $\mathbf{E}_0$, show that

$$\omega_c = \frac{eE_0}{mv_d}$$

(b) Show that the current density $\mathbf{j}$ in a conductor may be written in the form

$$\mathbf{j} = \frac{ne^2}{m\omega_c} \mathbf{E}_0$$

and that therefore we may express the conductivity by the formula

$$\sigma = \frac{ne^2}{m\omega_c}$$

(c) Make use of this formula for σ to calculate ω_c for a copper conductor for which $\sigma = 5.9 \times 10^7 \ (\Omega\text{-m})^{-1}$. Use the data for copper in Problem 4.

24 Elementary circuit theory

24-1 Introduction

In Chapter 23 we studied the flow of electric currents in circuits consisting of only a single loop. These circuits, as exemplified in Figures 23-7 and 23-10, are characterized by the fact that at any given instant the same current i flows through each element of the circuit. The purpose of this chapter is to generalize these results to the case of multiloop circuits, such as that in Figure 24-1, for which the currents in the various elements are, in general, different.

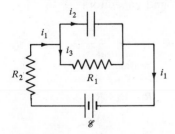

Figure 24-1

24-2 Resistors in series and parallel

To open the discussion of multiloop circuits, let us consider first the problem of replacing certain combinations of resistors by an equivalent single resistor.

Two resistors R_1 and R_2 can be connected to each other in either of two ways. If they are connected as in Figure 24-2a they are said to be in *series*, whereas if they are connected as in 24-2b they are said to be connected in *parallel*. The current through each of two resistors in series must be the same, whereas the current in each of the two resistors connected in parallel will, in general, be different. On the other hand, the potential drops across two resistors in parallel must be the same since each end of each resistor is at the same potential as is that end of the other with which it is in electrical contact. By contrast, the potential drops across two resistor in series are, in general, different.

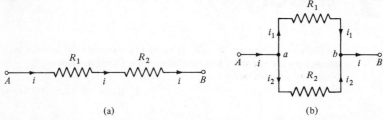

(a) (b)

Figure 24-2

We shall now establish that if two resistors R_1 and R_2 are connected in series, then they can be replaced by a single equivalent resistor with resistance

$$R = R_1 + R_2 \qquad \text{(series connection)} \qquad \textbf{(24-1)}$$

whereas if they are in parallel, the equivalent resistance R is

$$\frac{1}{R} = \frac{1}{R_1} + \frac{1}{R_2} \qquad \text{(parallel connection)} \qquad \textbf{(24-2)}$$

Note that the rules for combining resistors in series and parallel are the reverse of those governing capacitors, as given in (22-6) and (22-7). Table 24-1 summarizes these rules for both resistors and capacitors.

Table 24-1

Connection	*Equivalent capacitance C*	*Equivalent resistance R*
Series	$\dfrac{1}{C} = \dfrac{1}{C_1} + \dfrac{1}{C_2}$	$R = R_1 + R_2$
Parallel	$C = C_1 + C_2$	$\dfrac{1}{R} = \dfrac{1}{R_1} + \dfrac{1}{R_2}$

To derive (24-1) suppose that the endpoints A and B of the resistors in series in Figure 24-2a are kept at a fixed potential difference V_{AB}. The current flow i must then have a magnitude such that the *sum* of the potential drops iR_1 and iR_2 across the two resistors has the value V_{AB}. Hence,

$$i = \frac{V_{AB}}{R_1 + R_2}$$

If now R_1 and R_2 are replaced by a single resistor R, the current will adjust to some new value V_{AB}/R. For the equivalent resistor R, by definition, the current must assume the above value $V_{AB}/(R_1 + R_2)$. Hence (24-1) follows.

The derivation of (24-2) proceeds along similar lines. Suppose that points A and B in Figure 24-2b are kept at some fixed potential difference V_{AB}. The current i that leaves point A will travel to point a, where it branches into two currents: one of strength i_1 going through R_1 and the second of strength i_2 going through R_2. At point b these currents merge again and continue on to point B. If steady-state conditions are assumed to prevail, then there will be no accumulation of charge at any point in the circuit, including points a and b. Hence it follows that

$$i = i_1 + i_2 \tag{24-3}$$

which is another way of saying that all of the charge that arrives at the point a via the current i leaves this point by means of either the current i_1 or the current i_2. Since the potential drops across a resistor R is iR, it follows by reference to the figure that

$$V_{AB} = i_1 R_1 = i_2 R_2 = iR \tag{24-4}$$

where R represents the resistance of the single resistor which, when connected across AB, would draw the current i. Finally, combining (24-3) and (24-4), we conclude that the resistance R equivalent to R_1 and R_2 in parallel is given by (24-2).

By repeated application of the results in (24-1) and (24-2), it is also possible to calculate the equivalent resistance associated with combinations of resistors connected in certain ways. Thus, the equivalent resistance R_a for R_1, R_2, and R_3 connected as in Figure 24-3a is

$$R_a = R_1 + R_2 + R_3$$

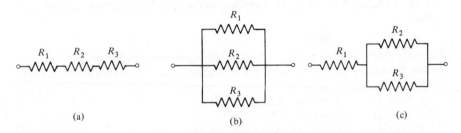

(a)

(b)

(c)

Figure 24-3

and that for the connection in Figure 24-3b satisfies

$$\frac{1}{R_b} = \frac{1}{R_1} + \frac{1}{R_2} + \frac{1}{R_3}$$

Similarly, the equivalent resistance R_c for the resistors connected as in Figure 24-3c is

$$R_c = R_1 + \left(\frac{1}{R_2} + \frac{1}{R_3}\right)^{-1} = R_1 + \frac{R_2 R_3}{R_2 + R_3}$$

where the second term is the equivalent resistance of R_2 and R_3 in parallel.

Example 24-1 What current i is drawn from a 10-volt battery if 10 resistors, each of resistance 20 Ω, are connected across it in: (a) Series? (b) Parallel?

Solution

(a) For resistors in series, the equivalent resistance, according to (24-1), is the sum of the individual resistances. Thus the equivalent resistance R is

$$R = 20\ \Omega + 20\ \Omega + \cdots + 20\ \Omega$$
$$= 200\ \Omega$$

According to Ohm's law, the current drawn from the battery is

$$i = \frac{\mathscr{E}}{R} = \frac{10\ \text{V}}{200\ \Omega} = 0.05\ \text{A}$$

(b) For the parallel connection, the reciprocal of the equivalent resistance is given by the sum of the reciprocals of the individual resistances. Hence

$$\frac{1}{R} = \frac{1}{20\ \Omega} + \frac{1}{20\ \Omega} + \cdots + \frac{1}{20\ \Omega} = \frac{10}{20\ \Omega} = \frac{1}{2\ \Omega}$$

or, equivalently,

$$R = 2\ \Omega$$

Thus, the current i drawn by the battery is

$$i = \frac{\mathscr{E}}{R} = \frac{10\ \text{V}}{2\ \Omega} = 5\ \text{A}$$

and this is 100 times as large as the current that flows if the resistors are connected in series.

Example 24-2 A 100-Ω and a 300-Ω resistor are connected in parallel across a battery having an emf of 50 volts and a negligible internal resistance (see Figure 24-4).

(a) What current flows through the battery?
(b) What current flows in each resistor?
(c) Calculate the ohmic losses in each resistor.
(d) Calculate the power output of the battery.

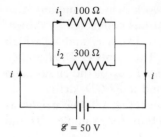

Figure 24-4

Solution

(a) According to (24-2), the resistance R equivalent to the two resistors is

$$R = \frac{R_1 R_2}{R_1 + R_2} = \frac{100\ \Omega \times 300\ \Omega}{100\ \Omega + 300\ \Omega} = 75\ \Omega$$

and hence the current i through the battery is

$$i = \frac{\mathscr{E}}{R} = \frac{50\ \text{V}}{75\ \Omega} = \frac{2}{3}\ \text{A}$$

(b) Since the potential drop across the two resistors is the same as the emf of the battery, it follows that

$$50\ \text{V} = (100\ \Omega) \times i_1 = (300\ \Omega) \times i_2$$

and this leads to

$$i_1 = \frac{1}{2}\ \text{A} \qquad i_2 = \frac{1}{6}\ \text{A}$$

(c) The Joule heat is Ri^2, so the power P_1 dissipated in the 100-Ω resistor is

$$P_1 = (100\ \Omega) \times i_1{}^2 = (100\ \Omega) \times \left(\frac{1}{2}\ \text{A}\right)^2 = 25\ \text{W}$$

while for the other it is

$$P_2 = (300\ \Omega) \times i_2{}^2 = (300\ \Omega) \times \left(\frac{1}{6}\ \text{A}\right)^2 = 8.3\ \text{W}$$

(d) The power output P_B of the battery is

$$P_B = \mathscr{E}i = (50\ \text{V}) \times \left(\frac{2}{3}\ \text{A}\right) = 33.3\ \text{W}$$

and this is the same as the sum $(P_1 + P_2)$ of the rates at which energy is dissipated in the resistors, as it must be.

24-3   Kirchhoff's rules

The purpose of this section is to introduce two rules, known as Kirchhoff's rules, which are indispensable for calculating the current flows in multiloop circuits.

Consider, in Figure 24-5, a three-loop network which contains two batteries of respective emf's $\mathscr{E}_1$ and $\mathscr{E}_2$ with polarities connected as shown, six resistors $R_1, R_2, \ldots, R_6$, and two additional resistors r_1 and r_2, which represent the internal resistances of the batteries. The points C, D, H, and G, where three or more wires meet, are called the *junctions* of the circuit, and the closed circuit paths *ABCHA*, *CDGHC*, and *DEFGD* define three possible *loops* in this circuit. The circuit path connecting any two neighboring junctions is called a *branch* of the circuit, and thus, for example, *CH* and *DEFG* are two branches.

Since the current that flows in a given branch will in general differ from that flowing in any other, in order to analyze this circuit it is necessary first to assign a mathematical symbol and a sense of direction for the current in each branch. (Should the assignment of this direction for any current be incorrect, the subsequent calculations would show that current to have a negative value.) Since there are altogether six branches in the network in Figure 24-5, there will be six currents; these have been labeled i_1, i_2, i_3, i_4, i_5, and i_6 and have been arbitrarily assigned the directions shown. The problem now is to evaluate these six currents in terms of the various resistances and battery emf's.

The procedure that is most commonly used to obtain these currents involves the application, to the given network, of two principles which are known as *Kirchhoff's rules*. Generally speaking, these rules, when properly applied, lead to just the right number of linearly independent relations to determine these currents. In particular, for the network in Figure 24-5, this means, as will be seen below, that Kirchhoff's rules yield six linearly independent relations for the six unknown currents $i_1, i_2, \ldots, i_6$.

The first of these two rules is based on the physical fact that under steady conditions, electric charge does not accumulate at any point in a circuit except possibly on the plates of a capacitor. To satisfy this condition at the junctions of the circuit, it is necessary that the net current flowing into each junction must vanish. Hence follows Kirchhoff's rule 1:

At any junction in a network, the sum of the currents approaching this junction must be equal to the corresponding sum of the currents leaving the same junction.

By way of illustration, consider the circuit in Figure 24-5. Applying this rule successively to the junctions C, D, G, and H, we find that the six unknown currents must satisfy the four relations:

$$i_1 = i_2 + i_3$$
$$i_2 + i_5 + i_6 = 0$$
$$i_4 = i_5 + i_6 \qquad \text{(24-5)}$$
$$i_3 = i_1 + i_4$$

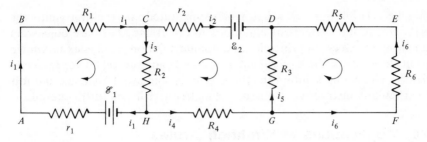

Figure 24-5

In order to obtain additional relations among these currents it is necessary to utilize the second rule. Imagine an external agent taking a unit (positive) charge around a given loop of a circuit in (say) a clockwise sense. Since the electrostatic field is conservative, it follows that the work he must carry out against the electrostatic field in traversing the loop vanishes. But on going through a resistor R carrying a current i, the work that he carries out will be $-Ri$ for traversals along the direction of the current and $+Ri$ for traversals in the opposite direction. Similarly, according to the analysis of Section 23-2, on going *through* a battery of emf $\mathscr{E}$ the work he must carry out will be $+\mathscr{E}$ if the battery is traversed from the negative $(-)$ to the positive $(+)$ terminals and $-\mathscr{E}$ if it is traversed in the opposite direction. Hence follows Kirchhoff's rule 2:

As a loop is traversed in (say) a clockwise sense the algebraic sum of the potential drops across all batteries in the loop plus the algebraic sum of the potential drops across all resistors in the loop will vanish provided the signs of the potential drops across the resistors and the seats of emf are as defined immediately above.

As an illustration, if the single-loop circuit in Figure 23-6 is traversed in a clockwise sense, this rule leads directly to the known result $-Ri + \mathscr{E} = 0$. If the loop is traversed counterclockwise against the direction of the current, we obtain the equivalent result: $Ri - \mathscr{E} = 0$.

Let us return now to the circuit in Figure 24-5. The application of rule 2 successively to the loops *ABCHA*, *CDGHC*, and *DEFGD* (assuming a clockwise traversal in each case) leads to

$$\mathscr{E}_1 - i_1 R_1 - i_3 R_2 - i_1 r_1 = 0$$
$$-\mathscr{E}_2 + i_5 R_3 + i_4 R_4 + i_3 R_2 - i_2 r_2 = 0 \qquad \textbf{(24-6)}$$
$$i_6 R_5 + i_6 R_6 - i_5 R_3 = 0$$

The fact that these relations, when combined with (24-5), comprise seven relations to determine six currents is no essential problem, since they are *not* linearly independent. For the addition of the second of (24-5) to the third yields the relation $i_2 = -i_4$, which when substituted into the first yields the fourth. Hence the systems of equations in (24-5) and (24-6) are consistent and lend themselves, in principle, to a unique solution for the six unknown currents in terms of the battery emf's $\mathscr{E}_1$ and $\mathscr{E}_2$ and the values for the eight resistors. The actual solution for this circuit is somewhat complex and it is not particularly instructive to pursue the matter here in any further detail.

24-4 Applications of Kirchhoff's rules

In this section the general utility of Kirchhoff's rules will be illustrated by reference to a number of simple multiloop networks.

Example 24-3 Two resistors R_1 and R_2 are connected in parallel across a battery of emf $\mathscr{E}$ and internal resistance r. Calculate the current in each resistor.

Solution Rather than solve this problem as we did in Section 24-2 by calculating the resistance of the equivalent resistor, let us apply Kirchhoff's laws directly.

As shown in Figure 24-6, the circuit has three branches, two junctions, and two independent loops. If we assign currents i_1, i_2, and i_3 directed as shown, then an application of rule 1 to the left junction leads to

$$i_1 = i_2 + i_3$$

and an identical relation is obtained at the right-hand junction. Applying rule 2 to the upper loop we obtain

$$-R_1 i_3 + R_2 i_2 = 0$$

and an application of this rule to the lower loop yields

$$-i_1 r - i_2 R_2 + \mathscr{E} = 0$$

Solving these relations we find that

$$i_1 = \frac{\mathscr{E}}{r + \dfrac{R_1 R_2}{R_1 + R_2}}$$

$$i_2 = \frac{\mathscr{E} R_1}{r(R_1 + R_2) + R_1 R_2}$$

$$i_3 = \frac{\mathscr{E} R_2}{r(R_1 + R_2) + R_1 R_2}$$

Note that, as expected, the current i_1 through the battery is that which would flow through the resistor R equivalent to r connected in series with the combination of R_1 and R_2 in parallel. Note also that the ratio i_2/i_3 of the currents through R_2 and R_1 is equal to the ratio, R_1/R_2.

Example 24-4 A 5-volt battery and a 20-volt battery are connected as in Figure 24-7. Calculate the current through each battery and the power output of the smaller battery.

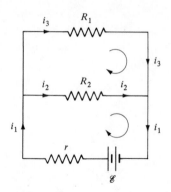

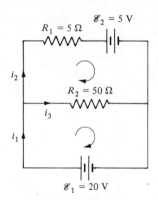

Figure 24-6   **Figure 24-7**

Solution On assigning the currents i_1, i_2, and i_3 as shown, we find by use of Kirchhoff's first rule that

$$i_1 = i_2 + i_3$$

The application of the second rule to the upper and lower loops, respectively, leads to

$$-\mathscr{E}_2 + R_2 i_3 - R_1 i_2 = 0$$

$$\mathscr{E}_1 - R_2 i_3 = 0$$

Solving these three relations for i_1, i_2, and i_3, and inserting the given values for R_1, R_2, $\mathscr{E}_1$, and $\mathscr{E}_2$, we obtain

$$i_1 = 3.4 \text{ A} \qquad i_2 = 3.0 \text{ A} \qquad i_3 = 0.40 \text{ A}$$

Since the current i_2 is positive, it follows by reference to the figure that this current is directed (through the smaller, or 5-volt, battery) from the positive to the negative terminal. This means that its power output P_2 is negative and has the value

$$P_2 = -i_2 \mathscr{E}_2 = -(3.0 \text{ A}) \times (5.0 \text{ V}) = -15 \text{ W}$$

In other words, work at the rate of 15 watts is being carried out *on* this battery. The power for this charging up comes from the other battery, whose power output P_1 is

$$P_1 = i_1 \mathscr{E}_1 = (3.4 \text{ A}) \times (20 \text{ V}) = 68 \text{ W}$$

Of the residual 53 watts ($= 68 \text{ W} - 15 \text{ W}$), the amount

$$R_1 i_2^{\,2} = (5 \text{ }\Omega) \times (3.0 \text{ A})^2 = 45 \text{ W}$$

is dissipated as heat in R_1 and the amount

$$R_2 i_3^{\,2} = (50 \text{ }\Omega) \times (0.4 \text{ A})^2 = 8.0 \text{ W}$$

is dissipated in R_2. Overall, energy is conserved, as it must be.

Example 24-5 Consider the "Wheatstone bridge" circuit in Figure 24-8, with the resistor R being variable. Calculate the value of R so that the current through the 100-Ω resistor is zero. Also calculate the power output of the battery.

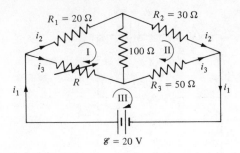

Figure 24-8

Solution Since, by hypothesis, there is no current through the 100-Ω resistor, it follows that this three-loop circuit may be described uniquely in terms of only three currents: i_1, i_2, and i_3. According to Kirchhoff's first rule, these are related by

$$i_1 = i_2 + i_3 \tag{24-7}$$

Applying Kirchhoff's second rule to loop I we find that

$$-R_1 i_2 + i_3 R = 0 \tag{24-8}$$

and, similarly, for loop II we obtain

$$-R_2 i_2 + R_3 i_3 = 0 \tag{24-9}$$

In order for (24-8) and (24-9) to be consistent, it is necessary that

$$\frac{i_2}{i_3} = \frac{R}{R_1} = \frac{R_3}{R_2} \tag{24-10}$$

or, in other words, R must satisfy

$$R = \frac{R_1 R_3}{R_2} = \frac{20\,\Omega \times 50\,\Omega}{30\,\Omega} = 33\,\Omega$$

To calculate the current i_1 let us apply Kirchhoff's second rule to loop III. The result is

$$-i_3(R + R_3) + \mathscr{E} = 0$$

Using the above value, $R = 33\,\Omega$, we thus obtain for i_3

$$i_3 = \frac{\mathscr{E}}{(R + R_3)} = \frac{20\,\text{V}}{83\,\Omega} = 0.24\,\text{A}$$

which, when substituted into (24-9), leads to

$$i_2 = i_3 \frac{R_3}{R_2} = (0.24\,\text{A}) \times \frac{50\,\Omega}{30\,\Omega} = 0.40\,\text{A}$$

Finally, substituting these values for i_2 and i_3 into (24-7), we obtain for the battery current i_1

$$i_1 = i_2 + i_3 = 0.40\,\text{A} + 0.24\,\text{A} = 0.64\,\text{A}$$

Hence the power output P_B of the battery is

$$P_B = \mathscr{E}i_1 = (20 \text{ V}) \times 0.64 \text{ A} = 13 \text{ W}$$

When used in practice, the 100-Ω resistor in the Wheatstone bridge circuit in Figure 24-8 is usually replaced by a current-measuring device, such as a galvanometer (see Section 24-7). For given resistors R_1 and R_2 and an unknown resistor R_3, the variable resistor R is adjusted until no current flows through the galvanometer. The resistor R_3 is then determined by (24-10) to be the ratio RR_2/R_1.

24-5   Circuits with capacitors

To extend the ideas of Section 24-3 to circuits involving capacitors, let us consider the particular circuit in Figure 24-9. It has three branches; let the currents in these be represented by i_1, i_2, and i_3, directed as shown in the figure. Applying Kirchhoff's first rule to the junctions, we obtain

$$i_1 = i_2 + i_3 \tag{24-11}$$

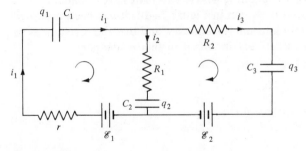

Figure 24-9

In order to apply rule 2 let us adopt the convention that the charges on the left-hand plate of C_1 and on the top plates of C_2 and C_3 are the positive charges q_1, q_2, and q_3 on these three capacitors, respectively. Then according to the definition of current,

$$i_1 = \frac{dq_1}{dt} \qquad i_2 = \frac{dq_2}{dt} \qquad i_3 = \frac{dq_3}{dt} \tag{24-12}$$

since in each case positive current corresponds to an increase of charge on the capacitor. (If, by contrast, the charge on the bottom plate of C_2 had been assumed to be the "charge" on this capacitor, then the second of these relations would have been $i_2 = -dq_2/dt$.)

Let us now generalize rule 2 to include the potential drops across capacitors and apply it to each of the two loops in Figure 24-9. Traversing the loops in a clockwise sense, for the left-hand loop we obtain

$$-i_1r - \frac{q_1}{C_1} - i_2R_1 - \frac{q_2}{C_2} + \mathscr{E}_1 = 0 \tag{24-13}$$

and, correspondingly, for the other loop we obtain

$$-i_3R_2 - \frac{q_3}{C_3} + \frac{q_2}{C_2} + i_2R_1 - \mathscr{E}_2 = 0 \qquad (24\text{-}14)$$

In writing down these relations we have used the fact that the work required of an external agent to take a unit (positive) charge from the positive to the negative plate of a capacitor is $-q/C$ and the above convention that the top plates of C_2 and C_3 and the left plate of C_1 carry the positive charge.

The six relations in (24-11) through (24-14) constitute a set of six coupled differential equations for the three capacitor charges q_1, q_2, and q_3, and for the three currents i_1, i_2, and i_3. Their solution constitutes complete information of the charges and currents in the circuit at any time t.

24-6 The charging of a shunted capacitor

As an application of the method outlined in Section 24-5 let us consider the problem of the charging of a capacitor, whose two plates are connected by a resistor. The circuit diagram is shown in Figure 24-10. It has been assumed that the battery has a certain internal resistance r and that the resistances of the various leads can be effectively included in this resistance.

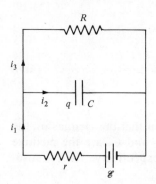

Figure 24-10

Examination of this circuit diagram shows that it has three branches, and that the currents defined there must satisfy

$$i_1 = i_2 + i_3 \qquad (24\text{-}15)$$

Further, the current i_2 will be related to the charge q on the capacitor by

$$i_2 = \frac{dq}{dt} \qquad (24\text{-}16)$$

provided that the left-hand plate of the capacitor is assumed to carry the positive charge. Applying rule 2 in a clockwise sense to the upper loop we find that

$$-i_3 R + \frac{q}{C} = 0 \tag{24-17}$$

and, correspondingly, for the lower loop

$$-i_1 r - \frac{q}{C} + \mathscr{E} = 0 \tag{24-18}$$

The four relations, (24-15) through (24-18), constitute a complete description for this system.

Fortunately, the circuit in Figure 24-10 is sufficiently simple that it can be solved in its entirety. To this end, let us substitute the forms for i_2 and i_3 given by (24-16) and (24-17), respectively, into (24-15). Combining the resultant formula for i_1 with (24-18), we find that

$$\frac{dq}{dt} + \frac{q}{C}\left[\frac{1}{r} + \frac{1}{R}\right] = \frac{\mathscr{E}}{r}$$

To simplify matters, let us make use of the fact that in most cases of physical interest the internal resistance r is small compared to R so that the term $1/R$ is negligible in comparison with $1/r$. With this approximation this relation takes on the simpler form

$$\frac{dq}{dt} + \frac{q}{rC} = \frac{\mathscr{E}}{r} \tag{24-19}$$

Now a comparison of (24-19) with (23-19) shows that mathematically they are essentially the same. Therefore, assuming that the charge on the capacitor is initially zero—that is, $q(0) = 0$—we find the solution of (24-19) to be

$$q(t) = C\mathscr{E}(1 - e^{-t/rC}) \tag{24-20}$$

In words, this states that the charge on the capacitor approaches its asymptotic value $C\mathscr{E}$ exponentially in a time interval measured by the time constant $\tau = rC$. The substitution of (24-20) into (24-15) through (24-18) yields, for the various currents, the explicit formulas:

$$i_1 = \frac{\mathscr{E}}{R} + \mathscr{E}\left[\frac{1}{r} - \frac{1}{R}\right] e^{-t/rC}$$

$$\approx \frac{\mathscr{E}}{R} + \frac{\mathscr{E}}{r} e^{-t/rC}$$

$$\tag{24-21}$$

$$i_2 = \frac{\mathscr{E}}{r} e^{-t/rC}$$

$$i_3 = \frac{\mathscr{E}}{R}(1 - e^{-t/rC})$$

where the approximate formula for i_1 follows since $r \ll R$. Figure 24-11 shows a plot of the complete solution in (24-20) and (24-21). For the sake of

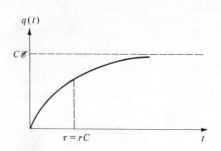

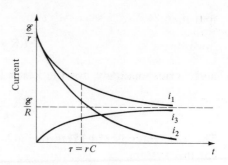

Figure 24-11

pictorial clarity only, the value $R/r = 3$ has been assumed in these graphs. Consistent with what we might anticipate on physical grounds, these graphs show that for times $t \gg rC$ the charge on the capacitor achieves its final charge $C\mathscr{E}$ and that the current i_1 through the battery becomes identical to that flowing through the resistor R. In other words, for times much greater than a time constant $\tau = rC$ the charge on the capacitor becomes the same as if the resistor R were not in the circuit, and the current through the battery assumes the value it would have if the capacitor were shorted out.

Example 24-6 For the circuit in Figure 24-10, if $\mathscr{E} = 50$ volts, $r = 1.0\,\Omega$, $R = 100\,\Omega$, and $C = 3.0\,\mu$F, calculate:
(a) The time constant of the circuit.
(b) The initial current through the battery and the resistor R.
(c) The final charge on the capacitor.

Solution
(a) The time constant τ of this circuit is rC. Hence

$$\tau = rC = (1.0\,\Omega) \times (3.0 \times 10^{-6}\,\text{F}) = 3.0\,\mu\text{s}$$

(b) The initial current through the battery may be obtained from the first of (24-21) by setting $t = 0$. Hence

$$i_1(0) = \frac{\mathscr{E}}{R} + \mathscr{E}\left(\frac{1}{r} - \frac{1}{R}\right)e^{-0} = \frac{\mathscr{E}}{r}$$

$$= \frac{50\,\text{V}}{1.0\,\Omega} = 50\,\text{A}$$

and this very large initial surge of current is the same as if neither R nor the capacitor were in the circuit. The initial current i_3 through R vanishes since the capacitor effectively shorts it out.
(c) The final charge q_f on the capacitor is

$$q_f = C\mathscr{E} = (3.0 \times 10^{-6}\,\text{F}) \times (50\,\text{V})$$
$$= 1.5 \times 10^{-4}\,\text{C}$$

Example 24-7 Consider the circuit in Figure 24-12 and suppose that at $t = 0$ the switch S is closed. Without writing down the circuit equations, calculate the following:
(a) The initial current through the battery.
(b) The steady current through the battery after a long time.
(c) The final charge on the capacitor.

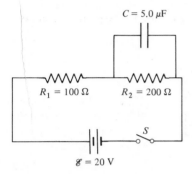

Figure 24-12

Solution

(a) Immediately after the switch is closed, the current through the battery is the same as if the capacitor were replaced by a *short*; that is, by a conductor of negligible resistance. Thus the initial current assumes the value it would have if only the 100-Ω resistor were connected across the battery. Since the total resistance across the battery is then 100 Ω, it follows that the initial current i_0 through it is

$$i_0 = \frac{\mathscr{E}}{R_1} = \frac{20 \text{ V}}{100 \text{ } \Omega} = 0.2 \text{ A}$$

(b) After a sufficiently long time has elapsed, the capacitor will be fully charged, and current will cease to flow in that branch of the circuit which has the capacitor. It follows that the sum of the potential drops across the 100-Ω and the 200-Ω resistors will approach the emf of the battery. Hence the steady-state current i_s through the battery is

$$i_s = \frac{\mathscr{E}}{R_1 + R_2} = \frac{20 \text{ V}}{100 \text{ } \Omega + 200 \text{ } \Omega} = 0.067 \text{ A}$$

(c) Reference to the figure shows that the final potential across the capacitor is the same as the final value for the iR drop across the 200-Ω resistor. Since the latter has the value

$$R_2 i_s = 200 \text{ } \Omega \times 0.067 \text{ A} = 13 \text{ V}$$

it follows that the final charge q_f on the capacitor is

$$q_f = C(R_2 i_s) = 5.0 \times 10^{-6} \text{ F} \times 13 \text{ V}$$
$$= 6.5 \times 10^{-5} \text{ C}$$

It is interesting to note that in order to obtain the initial and final values of charges and currents in a circuit, it is *not* always necessary to write down the full set of circuit equations. As illustrated in this example, by using the fact that a capacitor acts as a short immediately after the switch is closed and as an infinite resistor after a sufficiently long time has elapsed, we can obtain a considerable amount of information about the various currents and charges in a circuit.

Example 24-8 Consider the three-loop network in Figure 24-13. Assuming that at $t = 0$ the switch S is closed, calculate:

(a) The initial current i_0 through the battery.

(b) The final charges q_1 and q_2 on the capacitors C_1 and C_2, respectively, and the final current i_f through the battery.

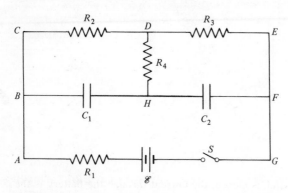

Figure 24-13

Solution

(a) Since initially there is no charge on the capacitors, it follows that the current path BHF is effectively a short. Hence, just as the switch is closed, the current will be nonzero only along the path $ABHFGA$. The total resistance along this path is R_1 and thus

$$i_0 = \frac{\mathscr{E}}{R_1}$$

must be the initial current in the battery.

(b) After the capacitors are both fully charged, no current can flow through the two branches BH and HF. Thus, after a long time a certain current i_f will flow along the path $ABCDEFGA$. Since the total resistance along this path is $(R_1 + R_2 + R_3)$, it follows that

$$i_f = \frac{\mathscr{E}}{R_1 + R_2 + R_3}$$

Since the potential drops across R_2 and R_3 must be in the same, respectively, as those across C_1 and C_2, it follows that

$$q_1 = C_1(i_f R_2) = \frac{C_1 R_2 \mathscr{E}}{R_1 + R_2 + R_3}$$

$$q_2 = C_2(i_f R_3) = \frac{C_2 R_3 \mathscr{E}}{R_1 + R_2 + R_3}$$

where in both cases the second equality is obtained by substituting the above value for i_f.

24-7 Galvanometers, ammeters, and voltmeters

In carrying out experiments on electric circuits in the laboratory, it is frequently useful to have available a means for measuring the current flow through—and the potential drop across—a given circuit element. The two instruments that have been developed for these purposes are the *ammeter* and the *voltmeter*.

An *ammeter*, as the name implies, is a device used to measure current. Normally, it is connected in series with the other circuit elements in that branch of the circuit whose current is to be measured. Correspondingly, a *voltmeter* measures the potential difference between any two points in a circuit. It is always connected across the two points whose potential difference is being measured. In a circuit diagram ammeters and voltmeters are customarily represented by the respective symbols

Figure 24-14 shows a branch of a circuit on which measurements are being made. Note that the ammeter is connected in series with the other elements in that branch through which the current flow is being measured. Its position relative to the other elements in that branch is of no significance since the same current flows through each. By contrast, the voltmeter is connected in parallel with (or across) the two points whose potential difference it is desired to measure. Thus the voltmeter in Figure 24-14 measures the potential drop across R_2. The ratio of the voltmeter reading to the ammeter reading in this circuit is numerically equal to the resistance R_2.

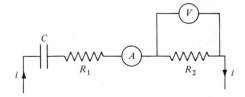

Figure 24-14

In connecting an ammeter or a voltmeter into a circuit, care must be taken so that the direction of the current is appropriate for the given instrument. To assist the experimenter, the two terminals of ammeters and voltmeters are often distinguished by a plus (+) sign on one terminal and a minus (−) sign on the other. When properly connected, current flows from the plus to the minus terminal *inside* the instrument. Also, care must be exercised in using voltmeters and ammeters so that the strength of the current does not exceed the limiting value for which the given instruments are designed.

Underlying the operation of both ammeters and voltmeters is an instrument known as a *galvanometer*. The conventional symbol for a galvanometer in a circuit diagram is

Like the ammeter and the voltmeter, the galvanometer also has two terminals, which, as a rule, are also marked plus (+) and minus (−). When a current passes through a galvanometer (from the plus to the minus terminals), the pointer on the galvanometer is deflected by an amount proportional to the strength of this current. Thus, once a galvanometer has been appropriately calibrated, the strength of the current flowing through it is given by the position of the pointer on its scale. The physics underlying the operation of a galvanometer is based on the experimental fact that if a current flows through a coil of wire that is in a magnetic field, a torque proportional to the strength of this current is exerted on the coil. A detailed consideration of the principles underlying the operation of the galvanometer will be found in Chapter 26. For the moment let us view the galvanometer pragmatically as simply a device that is capable of giving a pointer reading proportional to the strength of the current flowing through it.

Consider a resistor R connected across a battery of emf $\mathscr{E}$ and of negligible internal resistance. According to Ohm's law, its resistance is the ratio of $\mathscr{E}$ to the battery current i. This suggests that to measure R we connect two galvanometers into this circuit, as shown in Figure 24-15. Applying Kirchhoff's second rule to the upper loop in the figure, we find that

$$R = \left(\frac{i_G}{i_R} - 1\right)R_G \qquad (24\text{-}22)$$

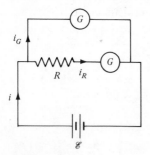

Figure 24-15

where i_G is the measured value of the current flowing through the upper galvanometer, i_R is the measured current flow through the resistor R, and R_G is the presumed known internal resistance of each of the two galvanometers. In effect, then, we can measure in this way the ratio of $\mathscr{E}/i$ for the circuit in Figure 24-15. However, this method does not yield the value of $\mathscr{E}$ or i separately. We shall now describe how by judicious use of galvanometers and resistors it is possible to construct ammeters and voltmeters and thereby to measure currents and voltages directly.

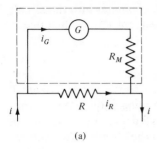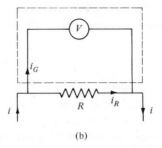

(a)   (b)

Figure 24-16

As shown in Figure 24-16a a voltmeter consists of a galvanometer connected in series with a very high resistance R_M, called the *multiplier*. Figure 24-16b shows how the circuit diagram would normally be drawn. In order not to perturb the original current that flows through R, it is necessary that R_M be selected to be very large in comparison with all other resistors in the circuit. With this choice, the potential drop across the resistance R will be essentially the same whether or not the voltmeter is present in the circuit. Applying Kirchhoff's second rule to the upper loop in Figure 24-16a we conclude that

$$i_G(R_M + R_G) = i_R R = V_R \qquad (24\text{-}23)$$

with V_R the potential drop across the resistor R. Since the coefficient of proportionality $(R_M + R_G)^{-1}$ depends only on the properties of the voltmeter, it follows that the current i_G through the galvanometer is proportional to V_R. We see therefore that by suitably calibrating the voltmeter, the device in Figure 24-16 may be used to measure potential differences.

Let us now consider the operation of an ammeter. As shown in Figure 24-17a, this time a galvanometer is connected in series with the line at the same time that a very low resistor R_s, called a *shunt*, is connected in parallel with the galvanometer. Figure 24-17b shows the relevant part of the circuit diagram as it normally would be drawn. To be effective, the shunt resistance R_s must be selected to be very small compared to the resistance of the galvanometer R_G. If, in addition, $R_s \ll R$, then the original current i will not

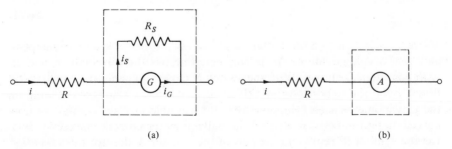

(a)   (b)

Figure 24-17

be appreciably affected by the insertion of the galvanometer into the circuit. Applying Kirchhoff's second rule to the loop in Figure 24-17a, we find that

$$i_G = \left(\frac{R_s}{R_G}\right) i_s$$

Hence the galvanometer current i_G is proportional to the shunt current i_s, with the coefficient of proportionality, R_s/R_G, depending only on the parameters characterizing the ammeter. Since i_s is, to a high degree of precision, the same as the current i flowing in the given branch of the circuit before the ammeter is inserted, it follows that a properly calibrated ammeter does indeed measure the current through that branch of the circuit into which it is inserted.

24-8   The potentiometer

It has been noted in the above discussion that the introduction of a voltmeter or an ammeter into a circuit modifies the original current flow to some extent. Even if the associated galvanometer were precisely calibrated, there would inevitably remain small experimental errors in any measurements made by their usage. The essential difficulty is that the measuring instruments themselves must draw some current from the circuit in order to operate.

By contrast, a *potentiometer* is a device that may be used to compare any two voltages without having to draw current from the original circuit in any way. If, for example, the emf of a battery is measured by use of a voltmeter, an error would arise because of the ir drop due to the internal resistance r of the battery. The corresponding measurement by use of a potentiometer would give a true reading of this emf since no current would be drawn from the battery. Evidently, the potentiometer is a very useful laboratory instrument, and the purpose of this section is to describe the principles underlying its operation.

Figure 24-18 shows the rudiments of a potentiometer. First, with the switch S open, the current i that flows around the lower loop is given, according to Ohm's law, by

$$i = \frac{\mathscr{E}}{r + R} \tag{24-24}$$

where $\mathscr{E}$ is the emf of the battery and r is its internal resistance. Suppose now that a voltage source V_1 is connected across the terminals A and B, with polarities as shown. With the switch S closed, in general, a current will flow around the upper loop and this will give rise to a nonzero reading on the galvanometer scale. Suppose that the variable contact point C is now moved along the resistor R until the galvanometer current is exactly zero. Let the symbol R_1 represent the part of the resistor R that for this condition comprises part of the upper loop. Since the current around the upper loop

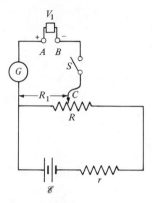

Figure 24-18

vanishes, it follows that the current i in the lower loop is precisely that in (24-24). Hence the unknown potential V_1 must be

$$V_1 = iR_1 = \frac{\mathcal{E}R_1}{R + r} \qquad (24\text{-}25)$$

Note that in this process no current whatsoever is drawn from the source of the V_1 potential.

Let us now disconnect V_1 from the potentiometer and reconnect across it a second voltage source V_2. Repeating the same procedure as above, we find that in general the contact point C will be at a different position corresponding to a portion R_2 of the resistor R. The analogue of (24-25) for this second voltage source is

$$V_2 = iR_2 = \frac{R_2 \mathcal{E}}{R + r}$$

Dividing this relation into (24-25) we find that

$$\frac{V_1}{V_2} = \frac{R_1}{R_2} \qquad (24\text{-}26)$$

and thus the ratio of the two voltages V_1 and V_2 is independent of the battery emf $\mathcal{E}$, its internal resistance r, and the magnitude of the resistance R. It depends only on the ratio of the two intercepted resistances R_1 and R_2. In practice, a fairly accurate determination of the ratio R_1/R_2 can be obtained by making use of fixed resistors connected in series with slide wires, for which the resistance can be simply related to length. (Recall (23-12).) The ratio R_1/R_2, and thus that of V_1/V_2, can therefore be obtained with considerable precision. In this way, then, the potentiometer can be used to relate *any* potential difference to that of any convenient standard.

Figure 24-19 shows how the potentiometer can be used to calibrate accurately both an ammeter and a voltmeter. In both cases the terminals A' and B' are to be connected to the correspondingly labeled points A and B of

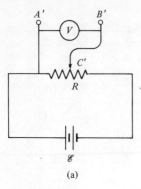
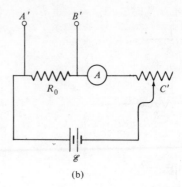

(a)   (b)

Figure 24-19

the potentiometer in Figure 24-18. As the contact point C' in Figure 24-19a is moved, the potential across the voltmeter will vary, and the voltmeter settings can be adjusted by connecting the potentiometer, for each such setting, to the points A' and B'. Similarly, in Figure 24-19b, if R_0 represents a precisely known resistor, then by moving the contact point C' both the current through the ammeter and the potential across R_0 will change. If this potential drop across R_0 is measured by use of a potentiometer, the value of the current through R_0 can thus be inferred and the ammeter calibrated.

QUESTIONS

1. Define or describe briefly the terms (a) junction; (b) resistors in series; (c) ammeter; (d) potentiometer; and (e) shunt resistance.

2. Two resistors of respective resistances $5\,\Omega$ and $10\,\Omega$ are connected in series. If a 2-ampere current flows through one of them, what is the current in the other?

3. If the current through each of two resistors in parallel is the same, what can be said about the two resistors? If they were connected in series, what conclusions could be drawn?

4. Twelve resistors, each of resistance R_0, are connected so that they comprise the edges of a cube. What is meant by a resistance R equivalent to this combination? How many equivalent resistances are there for this cube?

5. How much current would flow through a 10-Ω resistor that is connected in parallel with a 5-Ω resistor if a 1.0-ampere current flows through the latter?

6. Show by constructing an example that not all combinations of resistors can be decomposed into subunits of resistors in series and parallel.

7. Explain in physical terms why the current must be everywhere (except in the region between two capacitor plates) the same in a given branch of a circuit.

8. Explain in physical terms why the algebraic sum of the currents at a junction must vanish. At what other points in a circuit must this sum also vanish?

9. At what point(s), if any, in a circuit need the algebraic sum of the currents approaching this point not vanish?

10. Explain Kirchhoff's second rule in physical terms by reference to a particular two-loop circuit.

11. Would it be valid in Figure 24-5 to consider the loop *ABCDGHA* as one for which Kirchhoff's second rule would be applicable? Do you suppose if we did apply the rule to this loop that we would find a relation linearly independent of (24-6)?

12. How many loops are there altogether in the circuit in Figure 24-5? How many are there in the circuit in Figure 24-6? How many linearly independent relations do we find for each of these circuits by applying Kirchhoff's second rule?

13. A circuit contains capacitors, resistors, and batteries. Explain in physical terms why the *initial* values (after all switches are closed) for all currents are those that would be obtained if all capacitors (assumed to be originally uncharged) are shorted out—that is, if they are all replaced by resistors of zero resistance.

14. A circuit contains capacitors, resistors, and batteries. Explain in physical terms why, after the currents reach the steady state, we may calculate their values by replacing each capacitor in the original network by a resistor of infinite resistance and applying Kirchhoff's rules to the resulting circuit.

15. In a circuit a voltmeter is connected across a resistor. Assuming that the voltmeter is precisely calibrated, explain why you would expect its reading to be slightly *larger* than the potential across the resistor before the voltmeter was connected.

16. Explain in physical terms why the introduction of an ammeter into a branch of a circuit will, in general, cause a slight drop in the current in this branch. How can this error be minimized?

17. What are the comparative advantages and disadvantages of using a voltmeter or a potentiometer in measuring potential differences?

18. Explain how a potentiometer, which measures potential differences, can be used to calibrate an ammeter which measures current.

19. A voltmeter is connected across a fully charged but isolated capacitor. Describe the subsequent behavior of the voltmeter pointer. Assume that the multiplier resistance is so large that the time constant for the circuit is 2 seconds.

20. A voltmeter is connected across a resistor, and the pointer goes to maximum deflection. If the multiplier resistance is doubled, what is the new position of the pointer? What would happen if the multiplier resistance were halved?

21. If the shunt resistance in an ammeter is doubled, what happens to the pointer, assuming that it is originally at maximum deflection? What would happen if the shunt resistance were halved?

22. Why need the galvanometer in the potentiometer in Figure 24-18 *not* be calibrated precisely? Explain.

PROBLEMS

1. Find the resistance equivalent to a 3-Ω, a 5-Ω, and a 15-Ω resistor if they are connected in (a) series; (b) parallel.

2. A 10-volt battery is connected across a 10-Ω and a 40-Ω resistor connected in parallel. (a) What current flows through the battery? (b) Calculate the ratio of the currents through each resistor.

3. Three resistors, each of resistance R, are connected so that they comprise the three sides of a triangle. Calculate the equivalent resistance between any two vertices. Calculate also the ratio of the currents that would flow in each resistor if a battery were connected across any two vertices.

4. (a) Show that if R_1 and R_2 are connected in parallel, then the equivalent resistance R satisfies: $R < R_1$ and $R < R_2$.

   (b) State and derive the analogous result if R_1 and R_2 are connected in series.

5. If R_1 and R_2 are connected across a battery of emf $\mathcal{E}$, show that the ratio of the power output of the battery for the series connection to that for the parallel connection is

$$\frac{R_1 R_2}{(R_1 + R_2)^2}$$

6. Three 60-watt bulbs are connected in series across a 120-volt line. (a) What is the resistance of each bulb? (b) What is the current through each bulb? (c) How much power is dissipated in each bulb?

7. Repeat Problem 6, assuming that this time the bulbs are connected in parallel across the 120-volt line.

8. Consider the circuit in Figure 24-20. Calculate:

   (a) The equivalent resistance across the 12-volt battery.

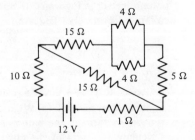

Figure 24-20

(b) The current through the battery.

(c) The total power dissipated in all the resistors.

9. In Figure 24-20 what is the ratio of the respective currents through the upper and the lower 15-Ω resistors?

10. Calculate the equivalent resistance between points A and B of the network in Figure 24-21. If a 50-volt battery were connected across these terminals, what current would flow?

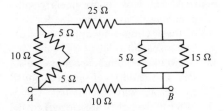

Figure 24-21

11. Twelve identical resistors, each of resistance R, are connected together as shown in Figure 24-22 to form the edges of a cube. Show that the equivalent resistance across any diagonally opposite vertices such as A and B is $5R/6$. (*Hint:* Show, by use of symmetry, that if I is the current through a battery that is connected across A and B, then the current in any of the 12 branches can only be $I/3$ or $I/6$.)

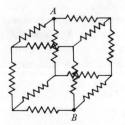

Figure 24-22

12. Suppose a thirteenth resistor also of resistance R is connected directly across the diagonal from A

to B in Figure 24-22. What is the equivalent resistance between these points now?

*13. Consider the infinite network of resistors in Figure 24-23. Show that the equivalent resistance between points A and B is

$$R = \frac{1}{2}(R_1 + R_2)$$

$$+ \frac{1}{2}[(R_1 + R_2)(R_1 + R_2 + 4R_3)]^{1/2}$$

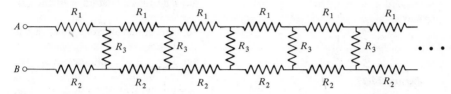

Figure 24-23

14. Consider the circuit in Figure 24-6.
   (a) How many branches and junctions are there in this circuit?
   (b) Assign a symbol and direction for the current in each branch and write down the implications of Kirchhoff's rules.
   (c) Find the value of the current in each branch, assuming that $R_1 = 20\ \Omega$, $R_2 = 10\ \Omega$, $r = 0.5\ \Omega$, and $\mathscr{E} = 15$ volts.
15. Consider the circuit in Figure 24-24.
   (a) Assign a symbol for the current in each branch and write down the implications of Kirchhoff's rules to this network.
   (b) Calculate the rate at which energy is being put out by each of the batteries.

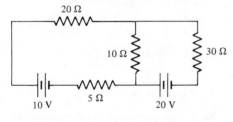

Figure 24-24

16. For the circuit in Figure 24-25 calculate the direction and magnitude of the current in the 20-Ω and 30-Ω resistors.

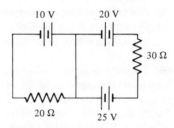

Figure 24-25

17. For the circuit depicted in Figure 24-26:
   (a) Calculate the current in the 100-Ω resistor.
   (b) Set up and solve Kirchhoff's rules to find the current through each of the batteries.
   (c) Is work being carried out *on* any of the batteries?

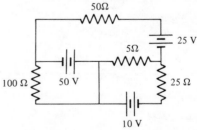

Figure 24-26

18. Without setting up Kirchhoff's rules in their entirety, for the circuit in Figure 24-27 calculate:

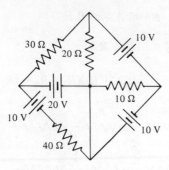

Figure 24-27

(a) The current (magnitude and direction) in the 10-Ω resistor.
(b) The current in the 40-Ω resistor.
(c) The current in the 20-Ω resistor by making use of your result to (a).
(d) The current in the 30-Ω resistor.

*19. Consider the special Wheatstone bridge circuit depicted in Figure 24-28. Set up the implications of Kirchhoff's rules for this circuit and show that the current i through the galvanometer is

$$i = \frac{\mathscr{E}(R - R_1)}{(R_0 + 2R_G)(R + R_1) + 2RR_1}$$

where R_G is the resistance of the galvanometer.

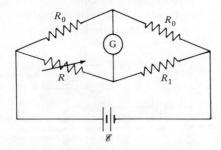

Figure 24-28

20. Consider the circuit in Figure 24-10. Carry out the analysis of this circuit, but this time assume that the resistor r and R are comparable. Assuming that $r = R = 100 \ \Omega$ and $C = 10 \ \mu F$, what is the time

constant associated with this circuit?

21. For the circuit in Figure 24-10, assuming that $r = 2 \ \Omega$, $R = 10^3 \ \Omega$, $\mathscr{E} = 60$ volts, and $C = 5.0 \ \mu F$, calculate:
    (a) The time constant τ of the circuit.
    (b) The initial current through the battery.
    (c) The final charge on the capacitor.

22. Assuming that $R = 10^3 \ \Omega$, $r = 1 \ \Omega$, $C = 10 \ \mu F$, and $\mathscr{E} = 0.5$ volt, and that just before the currents are allowed to flow the capacitor had a charge $q_0 = 2 \ \mu C$, calculate the three currents, as functions of time, flowing in the circuit in Figure 24-10.

23. In Figure 24-29 if the switch S is closed at $t = 0$ and the capacitors are originally uncharged, calculate:
    (a) The initial current through the battery.
    (b) The final charge on each capacitor and the final value for the current through the battery.

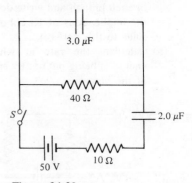

Figure 24-29

24. If the capacitors in the circuit in Figure 24-30 are originally uncharged and the switch S is closed at $t = 0$, calculate:
    (a) The initial value of the current i_0 through the battery.
    (b) The value of this current after a long time.

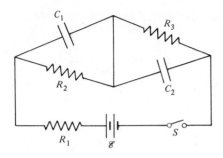

Figure 24-30

(c) The ultimate value of the charge on each capacitor.

25. Repeat both parts of Problem 23 if the 2.0-μF capacitor in Figure 24-29 has a charge of 25 μC initially. Assume that the upper plate has the positive charge initially.

26. Consider the circuit in Figure 24-12, but suppose that initially the capacitor has a charge of 50 μC. If the *left-hand* plate in the figure has the positive charge initially, calculate:
   (a) The initial current in the 200-Ω resistor.
   (b) The initial current through the battery.
   (c) The initial current to the capacitor.

27. Repeat all parts of Problem 26, but assume that initially the right-hand plate has the positive charge.

28. For the circuit in Figure 24-31, assuming that initially the capacitors

are uncharged, calculate:
   (a) The initial currents through R_1, R_2, and R_3.
   (b) The final charge on each capacitor.
   (c) The final current through the battery.

29. Repeat (a) of Problem 28, but assume this time that initially C_1 has a charge $C_1\mathscr{E}/2$ (with the left plate positive).

30. Repeat Problem 29, but assume that initially C_1 and C_2 have the respective charges Q_1 and Q_2, with the left-hand plate positive in each case.

31. Consider the circuit in Figure 24-29 and let $q_1\,(>0)$ and $q_2\,(>0)$ be the respective charges at time t on the left-hand plate of the 3.0-μF capacitor and the upper plate of the 2.0-μF capacitor.
   (a) Write down Kirchhoff's rules in terms of the currents i_1, i_2, and i_3 through the battery, the 40-Ω resistor, and the 3.0-μF capacitor, respectively.
   (b) What must be the initial values for i_1, i_2, and i_3, assuming that the capacitors are originally uncharged?
   (c) Solve the system of equations in (a) for the currents i_1, i_2, and i_3 at any time t, and compare with your result to (b) at $t = 0$.

32. Apply Kirchhoff's rules to the circuit in Figure 24-31. Are the initial values of the capacitor charges relevant in writing down these circuit equations? At what point in the solution do the values for these charges play a role?

33. In an attempt to measure a resistance R, an ammeter and a voltmeter are connected with a battery and the resistor as shown in Figure 24-32. Show that

$$\frac{1}{R} = \frac{i}{V} - \frac{1}{R_M}$$

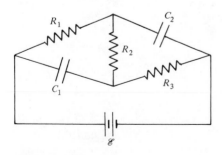

Figure 24-31

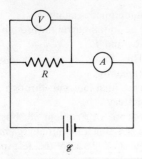

Figure 24-32

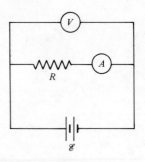

Figure 24-33

where i is the ammeter reading, V is the voltmeter reading, and R_M is the multiplier resistance. In general, $R_M \gg R$, and thus the ratio i/V is a good measure of R.

34. Consider the circuit in Figure 24-33. Show that if the shunt resistance R_s of the ammeter is negligible com-

pared to the galvanometer resistance, then

$$R = \frac{V}{i} - R_s$$

where V/i is the ratio of the voltmeter reading to the ammeter reading.

25 The magnetic field

In amber there is a flammeous and spiritous nature and this by rubbing on the surface is emitted by hidden passages and does the same that lodestone does.

PLUTARCH

25-1 Introduction

The very existence of the word *electromagnetism* implies that there must be a connection of some type between electric and magnetic phenomena. In order to explore the nature of this relationship, in this and the next two chapters we turn from our studies of purely electrical effects to a consideration of physical systems that exhibit magnetic behavior.

As noted in Chapter 19, the phenomenon of magnetism was discovered in antiquity. The word "magnetism" itself is derived from the name for a certain region in Asia Minor, *Magnesia*, where the iron ore $FeO-Fe_2O_3$, or lodestone (leading stone), occurred in abundance. These stones were found to be able to exert attractive and repulsive forces on each other—depending on their relative orientation. Further, they could also impart their magnetic property to other nearby iron and steel objects by *magnetizing* them. Today we know that the earth itself is a huge magnet and is capable of magnetizing iron and steel rods. Also well known is the related fact that a thin, magnetized needle, which is free to rotate about a vertical axis through its center, will invariably orient itself along the north-south direction. This distinctive behavior of a magnetized needle was discovered in the twelfth century and is basic to the operation of the mariner's compass.

25-2 Magnetic poles

Since early in the nineteenth century it has been known that magnetic effects can be produced by electric currents. Our discussion of magnetism will reflect this fact throughout. By way of an introduction to these ideas, in this section we review briefly some of the earlier known and thus more qualitative aspects of magnetism.

Consider a thin, magnetized iron needle, which is suspended so that it is free to rotate in a horizontal plane. As in the mariner's compass and as shown in Figure 25-1, the needle experiences a torque about its center that causes it to rotate until it is lined up along the north-south direction. That end of the needle pointing north is said to be its *north* (or north-seeking) *pole* and the other end is known as its *south* (or south-seeking) *pole*. Once the north pole of a magnetized needle has been identified and suitably marked, if suspended, the needle will invariably point in a northerly direction from any point on the earth. Two notable exceptions to this are the regions near the two magnetic poles of the earth itself.

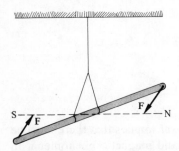

Figure 25-1

Figure 25-2a shows two thin, magnetized needles, suspended so that they are free to rotate in a plane (the plane of the page in the figure) about their respective centers. Experiment shows that, just as for the case of electric charge, a north and south pole attract each other and that two north poles or two south poles repel each other. Further, just as for electric charge, the strength of these forces also decreases as the separation distance between the poles increases and, as shown in the figure, the force between any two poles lies along the line joining them. As a consequence of these forces, each of the needles experiences a torque, which tends to line it up as in Figure 25-2b. Only in this latter configuration is there no torque acting on either needle.

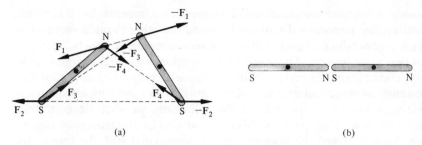

Figure 25-2

By analogy to our studies of ːlectrostatics, the above properties of magnetic poles suggest that we study magnetism by analyzing these forces between magnetic poles in quantitative terms. The main reason this is not usually done has to do with the fact that it is *impossible to isolate a magnetic pole completely.* This is often described by saying that *magnetic monopoles* do not exist. To obtain an experimental verification of this feature consider, in Figure 25-3a, a thin, magnetized needle with its two poles. On cutting this needle in two we find, as in Figure 25-3b, that instead of physically separating the original two poles, a new north and south pole appear at the cut surfaces. Thus two magnetized needles, each with its own north and south pole, result. Even if this subdivision process is continued down to a microscopic level, an isolated pole never appears.

Figure 25-3

It is worth emphasizing that magnetic forces are very much distinct from electrostatic forces. If, for example, a charged body is brought near a magnetized rod, no force arises between them except possibly for that due to any charge that may be induced on the rod. However, as will be seen in the next section, there is an unambiguous force between a charged body and a magnetized rod *if they are in relative motion.*

25-3 The discoveries of Oersted and Ampere

After the invention of the voltaic cell by Alessandro Volta (1745–1827), it became possible for the first time to produce steady electric currents at will, and to study phenomena associated with them. In 1820 Hans Christian Oersted (1777–1851) discovered that a wire in which there flows such a current has properties similar to that of a permanent magnet. That is, he found that a wire carrying a current behaves as if it had been magnetized and that this magnetic property ceases when the current does. Not unexpectedly, the study of magnetism received a tremendous impetus by this work. Thereafter, one important discovery followed the other in rapid succession. Today it is generally recognized that all observed magnetic effects are due to one of two basic sources. These are: (1) the motion of electric charge as in an electric current; and (2) certain intrinsic magnetic properties of the microscopic constituents of matter, particularly those associated with a property of the electron known as its *spin*. In the following, we shall be concerned mainly with describing magnetic effects produced by the motion of electric charge. To understand the magnetic effects associated with the spin of the electron, some knowledge of quantum mechanics is required; therefore further consideration of this matter will be postponed at this point.

Consider, in Figure 25-4, a tightly wound coil of wire, or solenoid, carrying a current i. Experiment shows that this coil behaves in many respects like the magnetized iron rods or needles considered in the previous section, and that its magnetic behavior becomes more pronounced the greater is the current. For the assumed direction of the current, as shown in the figure, the ends of the coil take on the attributes of a north and a south pole. If, for example, a magnetized needle is brought near the coil, it will rotate in the direction shown. The situation here is very much analogous to that in Figure 25-2.

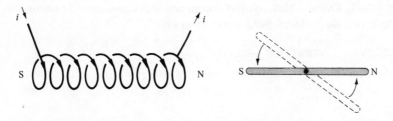

Figure 25-4

Since currents appear to have magnetic properties, we might expect that two currents will exert forces on each other. The fact that this is indeed the case was first established, shortly after Oersted's discovery, by André Marie Ampère (1775–1836). Figure 25-5 shows schematically how such an experiment may be carried out. Two stiff wires are connected across a seat of electromotive force and allowed to hang freely in a pool of liquid mercury. They are thus in electrical contact and can at the same time move relative to each other. Figure 25-5a shows the situation when the switch S is open, so that there is no current. The wires hang in a vertical position along the force of gravity in this case. If the switch S is closed, then, as in Figure 25-5b, certain currents i will flow in opposite directions through the two wires. As shown, the two wires exert repulsive forces on each other in this circumstance. Similar experiments can be used to establish that parallel currents attract each other.

It is also interesting to confirm that the force between currents is due to

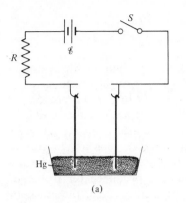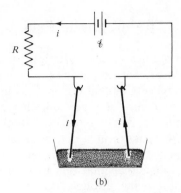

Figure 25-5

the fact that charges are in relative motion and not to any of the material properties of the wires. For this purpose, it is convenient to carry out an experiment by use of an *oscilloscope*, or *cathode-ray tube*. The main elements of such a tube are an electron gun, and directional controlling electric plates by means of which an electron stream can be sent down the tube to make a spot on the screen at its lower end. Figure 25-6a shows schematically such a tube, in which an electron beam makes a spot on the center of the screen. If now, as in Figure 25-6b, an upward current *i* opposite to the original direction of the electron stream is generated, the spot is deflected to the right. Similarly, if as in Figure 25-6c the external current *i* flows downward, the electron stream will be deviated to the left. Recalling that electrons have a negative charge, so that a downward flow of electrons corresponds to an upward flow of current, these deviations of the electron beam are easily accounted for. For example, in Figure 25-6c the two currents flow in opposite directions and thus repel each other, just as in Figure 25-5b.

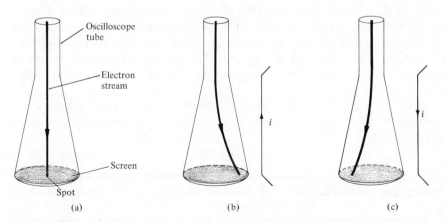

Figure 25-6

25-4   The magnetic field

In our studies of electrostatics it was convenient to think of the region of space near charged particles to be modified by virtue of the existence of an electric field **E** in that region. This field **E** is a vector field and represents at each point the force per unit charge on a particle placed at that point.

Similarly, it is convenient to think of a *magnetic field* as existing at each point of space near a collection of magnetic sources such as currents or magnetized bodies. We say that a magnetic field exists at such a point if a force, other than that due to an electric field, is produced on a charged particle *moving* through this point. As for the electric field, the magnetic field is a vector field and associates with each point in space both a magnitude and a direction. For the present, this field will be specified by a certain vector function **B**, which is known as the *magnetic induction field*[1] or the **B**-*field*. The purpose of this section is to define this field and to discuss some of its properties.

Imagine a region of space containing various magnetic sources. Experiments involving the observation of the trajectories of charged particles traveling through this region show that the force **F** acting on these particles has the following properties:

1. **F** is directly proportional to the charge q of the particle.
2. **F** is directly proportional to the magnitude v of the velocity **v** of the particle.
3. **F** is perpendicular to **v** throughout the trajectory of the particle.

Thus, doubling the charge or the velocity of a particle results in the doubling of the strength of the magnetic force. The third property of **F**, that it is invariably at right angles to the particle's velocity, is perhaps its most unusual property. It rests on the fact that the kinetic energy of a particle moving in a **B**-field does not vary throughout its motion. Hence **F** must always be at right angles to **v**. For, if **F** had a component along **v**, then work would be carried out on the particle and its kinetic energy would change according to the work-energy theorem. The fact that the kinetic energy of the particle does not vary thus confirms the fact that **F** is invariably perpendicular to **v**.

Because of the above experimental properties of the magnetic force **F**, we can define the magnetic induction field **B** associated with the given sources by the relation

$$F = q \mathbf{v} \times \mathbf{B} \tag{25-1}$$

where the symbol **v** × **B** stands for the *cross product* of the vectors **v** and **B**. According to the definition in Section 10-3, the cross product **v** × **B** is

---

[1] For historical reasons, the term *magnetic field* is reserved for a different quantity, which is, however, directly related to the **B**-field. It will be defined in Chapter 28.

perpendicular to both **v** and **B** and thus, as required by experiment, **F** is perpendicular to **v**. Further, since experiment shows **F** to be directly proportional to both **q** and |**v**|, it follows from (25-1) that the **B**-field as defined here is independent of both **q** and **v**; *it is determined exclusively by the magnetic sources.*

Figure 25-7 summarizes the relations between the directions of the three vectors **v**, **B**, and **F**. Assuming that **B** and **v** lie in the x-y plane, **F** will be parallel to the z-axis. As shown, **F** points along the positive sense of the z-axis for $q > 0$ and in the opposite direction for $q < 0$. The magnitude F of this force is $|qvB \sin \theta|$ with θ ($< 180°$) the angle between **v** and **B**.

It follows from (25-1) that the unit of **B** is the newton second per coulomb meter (N-s/C-m), or the newton per ampere meter (N/A-m). Let us define the unit of the tesla (abbreviated T) by

$$1 \text{ T} = 1 \text{ N/A-m}$$

so that the unit of the **B**-field is the tesla. A related unit is the weber (Wb). It is defined by

$$1 \text{ Wb} = 1 \text{ T-m}^2 \tag{25-2}$$

and thus 1 Wb/m^2 is an equivalent unit for **B**. Parenthetically, let us note the unit of the gauss (G), which is defined by

$$1 \text{ G} = 10^{-4} \text{ T} = 10^{-4} \text{ Wb/m}^2$$

but which will not be used in this book. The **B**-field of the earth, for example, has at its surface a magnitude of the order of 5×10^{-5} tesla or, equivalently, 0·5 gauss.

Just as for the electric field, it is convenient to imagine **B**-field lines to exist in the space around magnetic sources. Each of these field lines is defined so that the tangent to a line at a given point is parallel to **B** at that point. Also, the density of field lines in any region is proportional to the magnitude of **B** in that region. By way of illustration, Figure 25-8 shows the field lines about

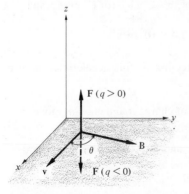

Figure 25-7

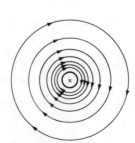

Figure 25-8

a long, straight wire with current directed perpendicularly down into the plane of the diagram. Note that the field lines are circles, concentric with the axis of the wire, and that their density decreases with increasing distance from the current. Figure 25-9 shows some of the field lines produced by the current in a circular loop of wire.

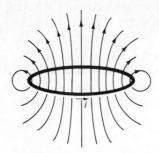

Figure 25-9

Example 25-1 Consider a proton traveling due east in the upper reaches of the atmosphere at a velocity of 5.0×10^5 m/s. Assuming the earth's field at this point has a strength of 0.5×10^{-4} tesla and is directed due south, calculate:

(a) The magnitude of the force on the proton.
(b) The direction òf this force.
(c) The acceleration of the proton.

Solution

(a) Since **v** and **B** are at right angles, according to (25-1) the magnitude of **F** is

$$F = qvB = (1.6 \times 10^{-19}\,\text{C}) \times (5.0 \times 10^5\,\text{m/s}) \times (0.5 \times 10^{-4}\,\text{T})$$
$$= 4.0 \times 10^{-18}\,\text{N}$$

(b) From the facts that the charge on the proton is positive, and **v** is directed east while **B** is south, it follows by reference to Figure 25-7 or to the definition of a cross product and (25-1) that the force on the proton is directed vertically downward.

(c) According to Newton's law, the acceleration of the proton is the ratio F/m. Using the known value $m = 1.67 \times 10^{-27}$ kg, and the above value for F, there results

$$a = \frac{F}{m} = \frac{4.0 \times 10^{-18}\,\text{N}}{1.67 \times 10^{-27}\,\text{kg}} = 2.4 \times 10^9\,\text{m/s}^2$$

Since the proton has a positive charge, the direction of this acceleration is also directed downward.

Example 25-2 Show explicitly that if a particle of charge q and mass m moves in a static magnetic induction **B**, then its kinetic energy is constant in time.

Solution Making use of the given data, (25-1), and Newton's law, we have

$$m\frac{d\mathbf{v}}{dt} = q\mathbf{v} \times \mathbf{B}$$

where $d\mathbf{v}/dt$ is the acceleration of the particle at the instant when its velocity is **v**.

Taking the dot product of this relation with $\mathbf{v}$, we find that

$$m\mathbf{v} \cdot \frac{d\mathbf{v}}{dt} = q\mathbf{v} \cdot (\mathbf{v} \times \mathbf{B})$$

But by definition of a cross product, the vector $\mathbf{v} \times \mathbf{B}$ is at right angles to $\mathbf{v}$, and thus according to the definition of the dot product the right-hand side vanishes. Hence

$$m\mathbf{v} \cdot \frac{d\mathbf{v}}{dt} = 0$$

and since this can be expressed in the equivalent form

$$\frac{d}{dt}\left(\frac{m}{2}v^2\right) = 0$$

the desired result follows.

25-5 The Biot-Savart law

In this section we consider the problem of calculating the **B**-field associated with current flows in *thin* wires by use of the experimentally based law of *Biot-Savart*. The problem of calculating **B** for other current distributions will be discussed in Sections 25-8 and 25-9.

Consider, in Figure 25-10, an infinitesimal current element of length $d\mathbf{l}$ directed parallel to the current flow i in a thin wire. The law of Biot-Savart states that the magnetic induction $d\mathbf{B}$ at a point P at the vectorial displacement $\mathbf{r}$ from this current element is

$$d\mathbf{B} = \frac{\mu_0}{4\pi} i \frac{d\mathbf{l} \times \mathbf{r}}{r^3} \tag{25-3}$$

where μ_0 is a constant defined by

$$\mu_0 = 4\pi \times 10^{-7}\,\text{Wb/A-m}$$

Recalling the definition of a cross product, we see from (25-3) that the magnitude of $d\mathbf{B}$ is

$$|d\mathbf{B}| = \frac{\mu_0}{4\pi} i \frac{|d\mathbf{l}|\sin\theta}{r^2} \tag{25-4}$$

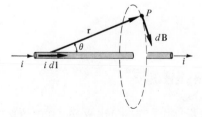

Figure 25-10

with θ the angle between $d\mathbf{l}$ and $\mathbf{r}$. The direction of $d\mathbf{B}$ may be obtained, as in the figure, by drawing a circle through the point P so that $d\mathbf{l}$ lies on its axis. The vector $d\mathbf{B}$ is invariably tangent to this circle. The *sense* of $d\mathbf{B}$ may be obtained by use of the right-hand rule: If the current element $d\mathbf{l}$ is grasped in the right hand, with the outstretched thumb pointing along the direction of the current, then the fingers will be oriented along the direction of $d\mathbf{B}$.

The implications of (25-3), as far as the direction of $d\mathbf{B}$ is concerned, are illustrated in Figure 25-11. Since $d\mathbf{B}$ is always at right angles to both $d\mathbf{l}$ and $\mathbf{r}$, the results in the figure follow directly either by use of the definition of the cross product or from the right-hand rule. The fact that $d\mathbf{B}$ vanishes at all points along the *y*-axis is a consequence of (25-3), since for these points $\mathbf{r}$ and $d\mathbf{l}$ are parallel vectors, and for these the cross product always vanishes.

Just as for the electrostatic field, experiment shows that the *superposition principle* is also valid for the **B**-field. That is, the magnetic induction due to two or more current elements is the vector sum of the magnetic inductions produced by the individual elements separately. Hence the magnetic induction **B** produced by the current in a closed circuit is

$$\mathbf{B} = \frac{\mu_0}{4\pi} \oint \frac{i \, d\mathbf{l} \times \mathbf{r}}{r^3} \tag{25-5}$$

where the integral is to be carried out over every element $d\mathbf{l}$ of the closed loop in which the current i flows. Since no truly isolated current element $i \, d\mathbf{l}$ exists, (25-5) is really more useful, and it is for this latter form that the term the *Biot-Savart law* is generally reserved.

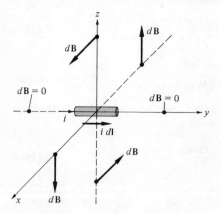

Figure 25-11

25-6 Applications of the Biot-Savart law

The purpose of this section is to derive the **B**-field associated with a number of simple current distributions. Specifically, we shall establish the following:

1. The **B**-field at a perpendicular distance r from a very long, straight wire carrying a current i has the magnitude

$$B = \frac{\mu_0 i}{2\pi r} \tag{25-6}$$

and its direction is tangent to the circle of radius r with center on the wire. The sense of **B** is shown in Figure 25-12 and is given by the same "right-hand" rule as above for a small current element $i\,d\mathbf{l}$:

*Grasp the wire in the right hand with the thumb pointing along the direction of the current. The fingers will then circle the wire in the sense of the direction of the **B**-field.*

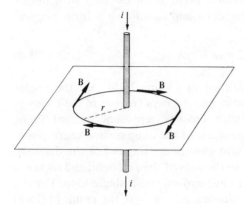

Figure 25-12

2. The magnitude of the **B**-field at a field point on the axis of a circular loop of wire of radius a and carrying a current i is

$$B = \frac{\mu_0 i a^2}{2(a^2 + b^2)^{3/2}} \tag{25-7}$$

where b is the distance from the center of the loop to the field point. See Figure 25-13. The direction of **B** is parallel to the axis of the loop and its sense is given by a rule also known as a right-hand rule. It may be obtained from the preceding one associated with the current in a long wire by interchanging the roles of i and **B**:

*Grasp the loop in the right hand with the fingers pointing along the direction of the current. The thumb then points along **B**.*

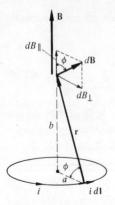

Figure 25-13

3. The magnitude of the **B**-field at a point *P on the axis* of a tightly wound solenoid carrying a current *i* and containing *n* turns per unit length is

$$B = \frac{\mu_0 ni}{2} [\cos \alpha_1 + \cos \alpha_2] \tag{25-8}$$

The angles α_1 and α_2, as defined in Figure 25-14, are the angles between the axis of the solenoid and the rays from *P* to the edges of the solenoid. The circles with crosses represent current going perpendicularly down into the plane of the diagram and correspondingly circles with dots represent current coming out of the diagram. The direction of **B** is parallel to the axis of the solenoid, and its sense is determined by the above right-hand rule of a single loop. For the special case of a very long solenoid, $\alpha_1 = \alpha_2 \approx 0$, the result in (25-8) reduces to

$$B = \mu_0 ni \tag{25-9}$$

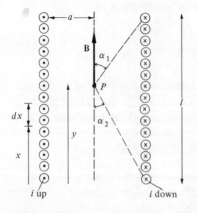

Figure 25-14

The validity of (25-6) through (25-8) is established below in Examples 25-3 through 25-5, respectively.

Example 25-3   Establish (25-6) by use of the Biot-Savart law.

Solution   As shown in Figure 25-15, let us set up a coordinate system with the current along the z-axis and with the field point P on the x-axis a distance x_0 from the origin. The magnetic induction $d\mathbf{B}$ due to a current element $i\,dz$ at a distance z from the origin is perpendicular to the x-z plane and points along the positive y-axis according to the Biot-Savart law. (See also Figure 25-10 in this connection.) According to (25-4), its magnitude, dB, is

$$dB = \frac{\mu_0 i}{4\pi}\frac{dz}{r^2}\sin\theta = \frac{\mu_0 i}{4\pi}\frac{dz}{(z^2+x_0^2)}\frac{x_0}{(z^2+x_0^2)^{1/2}}$$

$$= \frac{\mu_0 i}{4\pi}x_0\frac{dz}{[z^2+x_0^2]^{3/2}}$$

where the second equality follows since $r = (x_0^2 + z^2)^{1/2}$ and $\sin\theta = \sin(\pi-\theta) = x_0/(x_0^2+z^2)^{1/2}$. The total field is obtained by integrating over all values of z from $-\infty$ to $+\infty$. The result is

$$B = \int dB = \frac{\mu_0 i x_0}{4\pi}\int_{-\infty}^{\infty}\frac{dz}{[z^2+x_0^2]^{3/2}} = \frac{\mu_0 i x_0}{4\pi}\left[\frac{1}{x_0^2}\frac{z}{[z^2+x_0^2]^{1/2}}\right]\Big|_{-\infty}^{\infty}$$

$$= \frac{\mu_0 i}{4\pi x_0}[1-(-1)] = \frac{\mu_0 i}{2\pi x_0}$$

where the third equality follows by consulting a table of integrals. Finally, identifying the distance x_0 with the radial distance r, the validity of (25-6) is established.

Example 25-4   Establish the validity of (25-7) by use of the Biot-Savart law.

Solution   Consider, in Figure 25-13, the field $d\mathbf{B}$ due to a current element $i\,d\mathbf{l}$ of the loop. As shown, $d\mathbf{B}$ is perpendicular to both $d\mathbf{l}$ and $\mathbf{r}$ in accordance with (25-3). Let its components along and perpendicular to the axis be $dB_\parallel$ and $dB_\perp$, respectively. By symmetry, the sum of the contributions to $dB_\perp$, as we integrate around the loop, will

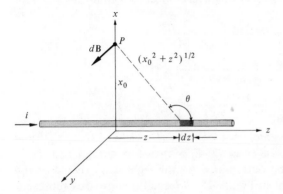

Figure 25-15

cancel. Hence the total field **B** will be directed along the axis of the loop and can be obtained by integrating $dB_\parallel$ around the loop.

To determine B let us first note that

$$dB_\parallel = |d\mathbf{B}| \cos \phi = |d\mathbf{B}| \frac{a}{(a^2 + b^2)^{1/2}}$$

where ϕ is as defined in the figure. Since the angle θ between dl and $\mathbf{r}$ is here $\pi/2$, and since $r = (a^2 + b^2)^{1/2}$, it follows by use of (25-4) that

$$dB_\parallel = \frac{\mu_0 i \, dl}{4\pi} \frac{1}{(a^2 + b^2)} \frac{a}{(a^2 + b^2)^{1/2}} = \frac{\mu_0 i a \, dl}{4\pi [a^2 + b^2]^{3/2}}$$

The total field B is obtained by integrating $dB_\parallel$ over the entire loop:

$$B = \int dB_\parallel = \int \frac{\mu_0 i a}{4\pi [a^2 + b^2]^{3/2}} \, dl = \frac{\mu_0 i a}{4\pi [a^2 + b^2]^{3/2}} \int dl$$

$$= \frac{\mu_0 i a (2\pi a)}{4\pi [a^2 + b^2]^{3/2}}$$

$$= \frac{\mu_0 i a^2}{2[a^2 + b^2]^{3/2}}$$

where the fourth equality follows since the length of the loop is $2\pi a$. The validity of (25-7) is thereby established.

Example 25-5 Derive (25-8) for the **B**-field on the axis of a solenoid. As shown in Figure 25-14, assume that it consists of N turns, has a radius a, length l, and $n \equiv N/l$ turns per unit length.

Solution To calculate the field at a point P on the axis and at a distance y from one end of the solenoid, let dB be the field at P due to $(n \, dx)$ turns at a distance x from the same end. See Figure 25-14. According to (25-7), dB is

$$dB = \frac{\mu_0 i a^2}{2[a^2 + (y - x)^2]^{3/2}} n \, dx$$

since in the present case $b = y - x$. Integrating over all values of x from 0 to l yields for the total field

$$B = \int dB = \frac{\mu_0 i a^2 n}{2} \int_0^l \frac{dx}{[a^2 + (y - x)^2]^{3/2}}$$

$$= \frac{\mu_0 i a^2 n}{2} \frac{1}{a^2} \left[\frac{x - y}{[a^2 + (y - x)^2]^{1/2}} \right]\Big|_0^l$$

$$= \frac{\mu_0 i n}{2} \left[\frac{l - y}{[a^2 + (l - y)^2]^{1/2}} + \frac{y}{[a^2 + y^2]^{1/2}} \right]$$

Reference to Figure 25-14 shows that $\cos \alpha_1 = (l - y)/[a^2 + (l - y)^2]^{1/2}$ and $\cos \alpha_2 = y/[y^2 + a^2]^{1/2}$. The validity of (25-8) is therefore established.

The direction of **B** may be determined by the right-hand rule. For the assumed sense of the current in Figure 25-14 (recall that crosses represent current flow perpendicularly down and dots represent current flow up), **B** is directed as shown.

25-7 Further applications

To obtain a feeling for orders of magnitude in this section we shall apply some of the above results to particular cases.

Example 25-6 Two very long, parallel wires are 0.5 meter apart and carry currents of 3.0 amperes in opposite directions, as shown in Figure 25-16. Calculate the **B**-field at a point between the wires and at a perpendicular distance of 0.4 meter from one of them.

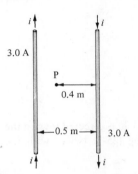

Figure 25-16

Solution The field $\mathbf{B}_L$ at P due to the wire on the left is directed perpendicularly down into the plane of Figure 25-16. Its magnitude is

$$B_L = \frac{\mu_0 i}{2\pi r} = \frac{(4\pi \times 10^{-7}\,\text{Wb/A-m}) \times (3.0\,\text{A})}{2\pi \times 0.1\,\text{m}}$$

$$= 6.0 \times 10^{-6}\,\text{T}$$

Correspondingly, the field $\mathbf{B}_R$ at P due to the other current is also directed perpendicularly down and it has the magnitude

$$B_R = \frac{\mu_0 i}{2\pi r} = \frac{(4\pi \times 10^{-7}\,\text{Wb/A-m}) \times (3.0\,\text{A})}{2\pi \times 0.4\,\text{m}}$$

$$= 1.5 \times 10^{-6}\,\text{T}$$

The total field $(\mathbf{B}_L + \mathbf{B}_R)$ is therefore also directed perpendicularly down and its magnitude B is

$$B = B_L + B_R = 6.0 \times 10^{-6}\,\text{T} + 1.5 \times 10^{-6}\,\text{T}$$

$$= 7.5 \times 10^{-6}\,\text{T}$$

Example 25-7 Calculate the **B**-field at the center of a loop of radius 10 cm. Assume a current of 5.0 amperes.

Solution The direction of the field is perpendicular to the plane of the loop and its sense is given by the right-hand rule. Its magnitude is obtained by substituting into

(25-7) the parameter values $i = 5.0$ amperes, $a = 0.1$ meters, and $b = 0$. The result is:

$$B = \frac{\mu_0 i a^2}{2(a^2 + b^2)^{3/2}} = \frac{(4\pi \times 10^{-7}\ \text{Wb/A-m}) \times (5.0\ \text{A})}{2 \times 0.1\ \text{m}}$$
$$= 3.1 \times 10^{-5}\ \text{T}$$

Example 25-8 Two circular loops, each of radius a, carry parallel currents each of strength i. Assuming they are at a separation distance $2c$, calculate the magnetic induction on the axis of the loops and at a point a distance z from one of the loops. See Figure 25-17.

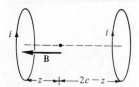

Figure 25-17

Solution Applying the right-hand rule, we see that the **B**-field is directed toward the left in the figure. Its magnitude is obtained by applying (25-7) twice, with the following result:

$$B = \frac{\mu_0 i a^2}{2} \left\{ \frac{1}{(a^2 + z^2)^{3/2}} + \frac{1}{[a^2 + (2c - z)^2]^{3/2}} \right\}$$

It is established in the problems that for the choice $z = c = a/2$, the field at this point on the axis and midway between the two loops is relatively uniform. When used with these parameter values this system defines an important laboratory tool known as a *Helmholtz coil*.

Example 25-9 A solenoid has a radius of 2 cm, a length of 10 cm, and 500 turns. Assuming a current of 5.0 amperes, calculate the magnitude of **B** along the axis at:
(a) The center of the coil.
(b) One end of the coil.

Solution
(a) Since there are 500 turns in 10 cm, it follows that the number of turns per unit length is

$$n = \frac{N}{l} = \frac{500}{0.1\ \text{m}} = 5 \times 10^3/\text{m}$$

Reference to Figure 25-14 shows that at the center

$$\cos \alpha_1 = \cos \alpha_2 = \frac{5}{\sqrt{5^2 + 2^2}} = 0.93$$

Substitution into (25-8) yields

$$B = \frac{\mu_0 n i}{2} [\cos \alpha_1 + \cos \alpha_2]$$
$$= \frac{(4\pi \times 10^{-7}\ \text{Wb/A-m}) \times (5 \times 10^3/\text{m}) \times 5\ \text{A}}{2} \times [0.93 + 0.93]$$
$$= 2.9 \times 10^{-2}\ \text{T}$$

(b) At an end of the solenoid,

$$\cos \alpha_1 = 0 \qquad \cos \alpha_2 = \frac{10}{\sqrt{10^2 + 2^2}} = 0.98$$

since $\alpha_1 = 90°$ in this case. Proceeding as above with these values, we find that

$$B = 1.5 \times 10^{-2}\,\text{T}$$

25-8 Gauss' law for magnetism

As we saw in our studies of electrostatics, Gauss' law and the existence of a potential function determine, to a large extent, all of the essential features of the electrostatic field. There are two analogous laws, called *Gauss' law for magnetism* and *Ampère's law*, which play the same role for the **B**-field. The purpose of this and the next section is to discuss these two very important characterizations of the magnetic induction field.

As will be seen in the following chapters, the importance of these two laws is due, in the main, to the fact that they are basic to Maxwell's equations. Indeed, Gauss' law for magnetism is one of these four basic relations. Further, for any given distribution of currents, the laws of Ampère and of Gauss when taken together comprise a complete specification of the **B**-field everywhere. Hence they constitute the necessary generalization of the more restricted Biot-Savart law, which applies only to current flows in thin wires.

By analogy to the definition of electric flux in (20-13), we define the magnetic flux Φ_m through a surface S by

$$\Phi_m = \int_S \mathbf{B} \cdot d\mathbf{S} \tag{25-10}$$

where $d\mathbf{S}$ is an area element normal to S, **B** is the value of the **B**-field at that point and the integral is over the surface S. In terms of this quantity, Gauss' law for magnetism states that the *magnetic flux out of every closed surface vanishes*. Thus

$$\oint_S \mathbf{B} \cdot d\mathbf{S} = 0 \tag{25-11}$$

with $d\mathbf{S}$ a vectorial area element directed *outward* from the *closed* surface S. Comparison with Gauss' law for the **E**-field in (20-15) leads to the conclusion that there is no magnetic analogue of electric charge. We often describe this by saying that there are no *magnetic monopoles*. The validity of (25-11) has been established by a vast number of experiments and despite continuing research no one has as yet ever detected the presence of a magnetic monopole.

One of the important consequences of (25-11) is that all **B**-field lines must be continuous. The reasoning here is precisely the same as that referred to in Chapter 20 to establish the continuity property of the E-field lines in a charge-free region. Since there seem to be no magnetic monopoles, the **B**-field lines must be continuous everywhere. Hence, the property of **B**-field lines, as exemplified in Figure 25-8, of always closing on themselves is true in general.

25-9 Ampère's law

Consider, in Figure 25-18, an open surface S bounded by a closed curve l. Let us assign a positive sense of direction along this curve by constructing an infinitesimal tangent vector $d\mathbf{l}$ at an arbitrary point of l. As viewed from the top, then, the positive sense of the curve in Figure 25-18 is counterclockwise. Associated with this sense for l let us also assign an *outward sense* to every area element $d\mathbf{S}$ of the surface S bounded by l, in accordance with the following rule:

---

*Grasp the curve l in the right hand with the fingers pointing along the predefined sense of l; the outstretched thumb will then point in the direction of the outward sense of the area elements d**S** of S.*

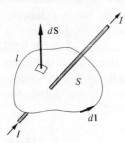

Figure 25-18

According to this rule, for example, the infinitesimal vectorial area element $d\mathbf{S}$ in Figure 25-18 points along the outward sense of S. Note that the surface S in this context is always *open*, by virtue of the fact that it is bounded by a closed curve. The assignment of a sense to this bounding curve will always be assumed in the following to associate the above outward sense for this surface.

Consider now an arbitrary open surface S with its bounding curve l in a region of space through which flow certain electric currents. *Ampère's law* states that the components of the **B**-field along the curve l are related to the net current I, which flows through S along its outward sense, by

$$\oint_l \mathbf{B} \cdot d\mathbf{l} = \mu_0 I \tag{25-12}$$

where the integral is a *line integral* (Section 21-2) and represents the sum of the products $\{\mathbf{B} \cdot d\mathbf{l}\}$ for all elements $d\mathbf{l}$ of the closed curve l. It is important to note that (25-12) applies to *all* closed curves l, and that for a given l it is applicable to all open surfaces S bounded by that curve. Also implicit in (25-12) is the fact that the current I through S will be positive or negative, depending on the assumed positive sense for l. The reversal of the positive sense for l, for example, will reverse that for the outward sense for S and thus the sign of the current I. However, Ampère's law in (25-12) must be, and is, independent of this choice; it is valid for either one.

To help fix these ideas, consider in Figure 25-19a a current i, say in a thin wire, going through a surface S in the direction shown. For the given sense of the bounding curve l, the current through S is positive. Hence Ampère's law in this case becomes

$$\oint_l \mathbf{B} \cdot d\mathbf{l} = \mu_0 i$$

On the other hand, if the sense of the curve relative to the direction of the current flow is as shown in Figure 25-19b, the corresponding relation is

$$\int_l \mathbf{B} \cdot d\mathbf{l} = -\mu_0 i$$

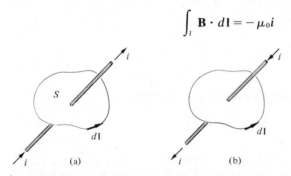

Figure 25-19

In a similar way, since the total current passing through the surface S in Figure 25-20a is $(i_1 + i_2)$, Ampère's law here assumes the form

$$\oint_l \mathbf{B} \cdot d\mathbf{l} = \mu_0 (i_1 + i_2)$$

Correspondingly, for the currents shown in Figure 25-20b the analogous

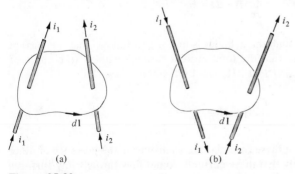

Figure 25-20

formula is

$$\oint_l \mathbf{B} \cdot d\mathbf{l} = \mu_0(i_2 - i_1)$$

Example 25-10 A uniform current i flows in a cylinder of radius a, and an identical current i flows in a nearby thin wire (see Figure 25-21). Evaluate, by use of Ampère's law, the line integral $\oint \mathbf{B} \cdot d\mathbf{l}$ about each of the paths A, B, C, and D, with the sense of direction specified in the figure.

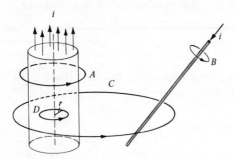

Figure 25-21

Solution The total current through path A is i, and hence

$$\oint_A \mathbf{B} \cdot d\mathbf{l} = \mu_0 i$$

Similarly, through path B the total current is i, but this time the current is directed opposite to that associated with the assumed sense of the path. Thus we have

$$\oint_B \mathbf{B} \cdot d\mathbf{l} = -\mu_0 i$$

The total current through path C consists of an upward flow i through the cylinder plus a downward flow i through the wire. Accordingly the net current through path C vanishes, and thus

$$\oint_C \mathbf{B} \cdot d\mathbf{l} = 0$$

Finally, since the current through the cylinder is uniform, it follows that the current density is $i/\pi a^2$. This implies that the total current through path D (the circle of radius $r < a$ in the figure) is $i\pi r^2/\pi a^2$. Hence, by Ampère's law,

$$\oint_D \mathbf{B} \cdot d\mathbf{l} = \frac{\mu_0 i r^2}{a^2}$$

It should be emphasized that these formulas are valid even in the presence of other nearby currents, provided only that these currents do not flow through any surfaces bounded by these four integration paths.

25-10 Applications of Ampère's law

We have seen previously that by use of the Biot-Savart law it is possible to calculate, in principle, the magnetic induction associated with *any* given distribution of currents. The purpose of this section is to show how these calculations may be simplified in certain cases by use of Ampère's law. Just as for the analogous calculations of the electrostatic field by use of Gauss' law, this method is applicable only for highly symmetric current distributions.

As a first application consider again, in Figure 25-12, the **B**-field at a perpendicular distance r from a very long wire carrying a current i. Let us apply Ampère's law to this case by selecting a circular path of radius r and with the center on the wire. Assuming that the **B**-field lines are circles concentric with the wire, we find by use of (25-12) that

$$\mu_0 i = \oint_l \mathbf{B} \cdot d\mathbf{l} = B \oint dl = 2\pi r B$$

where the second equality follows since **B** and $d\mathbf{l}$ are parallel, so that $\mathbf{B} \cdot d\mathbf{l} = B\,dl$, and from the fact that B is constant along the path of integration and may thus be taken out from under the integral. Solving the last equality for B we regain (25-6). The validity of Ampère's law is thus confirmed for this case.

As a second application of Ampère's law, consider the magnetic induction in the *interior* of an infinitely long cylinder of radius a, through which flows a uniform axial current density **j** corresponding to a current $i = j\pi a^2$. Just as for the external field, the lines of magnetic induction in the interior of the wire must be circles concentric with the axis of the wire. For if there were, say, a radial component to **B** at any point inside, then either the symmetry or the continuity property of the **B** lines would be violated. Further, the strength of **B** can vary only with the distance r from the axis, so **B** must have the same magnitude at all points of any circle concentric with the axis.

To calculate B let us apply Ampère's law to a circle of radius r ($<a$) centered on the axis of the wire, as shown in Figure 25-22. The total current

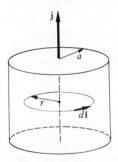

Figure 25-22

through this circular path is

$$\int_S \mathbf{j} \cdot d\mathbf{S} = j \int_S dS = \pi r^2 j = \frac{r^2}{a^2} i$$

where S has been taken to be the disk that is bounded by the circular path. Substitution into Ampère's law in (25-12) leads to

$$\frac{\mu_0 i r^2}{a^2} = \oint \mathbf{B} \cdot d\mathbf{l} = B \oint dl = 2\pi r B$$

since the field lines are assumed to be circles concentric with the axis. Solving for B we obtain

$$B = \frac{\mu_0 i r}{2\pi a^2} \qquad (r \leqslant a) \tag{25-13}$$

so B vanishes along the axis of the cylinder and rises linearly with radius to assume its maximum value $\mu_0 i / 2\pi a$ on the surface. Outside the wire, B decreases for increasing values for r in accordance with (25-6). The field lines both inside and outside are circles concentric with the cylinder axis, and with their sense given by the right-hand rule. Figure 25-23 shows a plot of B as a function of r both inside and outside the wire.

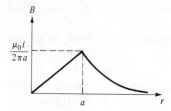

Figure 25-23

As a third and final application of Ampère's law we now show that the relation

$$B = \mu_0 n i \tag{25-9}$$

which characterizes the **B**-field on the axis of an infinitely long solenoid is valid *throughout the interior* of what is known as an *ideal solenoid*. A solenoid is said to be ideal if it is very long and is sufficiently tightly wound that the current flow on its surface may be thought of as a continuous sheet of current. Strictly speaking, no solenoid is truly ideal. However, just as for the parallel-plate capacitor—for which we neglected the fringing field at the edges of the plates—if we agree to neglect the leakage of **B**-lines out of the coil, then we can consider it to be ideal.

Figure 25-24 shows the field lines associated with the current in an ordinary solenoid. Note that there is an outward "leakage" of field lines between the wires, and that in the region near the center the field lines tend to be parallel to the axis. It is plausible to expect that the longer is the

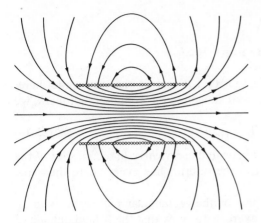

Figure 25-24

solenoid and the closer together are the turns, the more extended will be this region in which the field lines are parallel to the axis. The *ideal solenoid* is then the limiting form for which the field outside vanishes, while inside the field lines are everywhere parallel to the axis.

To demonstrate the validity of (25-9) for all points inside an ideal solenoid, let us apply Ampère's law to the rectangular path *abcd* in Figure 25-25 with the segment $\overline{ab}$ on the axis of the coil. Since the entire path lies inside the coil, no current passes through it. Hence, by Ampère's law,

$$0 = \oint \mathbf{B} \cdot d\mathbf{l} = \int_a^b \mathbf{B} \cdot d\mathbf{l} + \int_b^c \mathbf{B} \cdot d\mathbf{l} + \int_c^d \mathbf{B} \cdot d\mathbf{l} + \int_d^a \mathbf{B} \cdot d\mathbf{l}$$

$$= \int_a^b \mathbf{B} \cdot d\mathbf{l} + \int_c^d \mathbf{B} \cdot d\mathbf{l}$$

where the third equality follows since by hypothesis the field lines are parallel to the axis and hence perpendicular to the paths $\overline{bc}$ and $\overline{da}$. Now since the path $\overline{ab}$ lies along the axis of the coil, and since there the field has according to (25-9) the value $\mu_0 ni$, it follows that

$$\int_a^b \mathbf{B} \cdot d\mathbf{l} = \mu_0 ni \, \Delta h$$

where Δh is the distance from *a* to *b*. In a similar way, if B is the value of

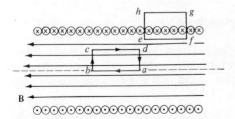

Figure 25-25

the field along the path $\overline{cd}$, then

$$\int_c^d \mathbf{B} \cdot d\mathbf{l} = -B \, \Delta h$$

where the minus sign reflects the fact that $\mathbf{B}$ and $d\mathbf{l}$ are antiparallel here. Combining the last three relations and canceling the common factor Δh, we find that

$$B = \mu_0 n i \qquad \text{(25-14)}$$

Finally, since this argument can be repeated *everywhere* in the interior of the solenoid, it follows that the field in the interior has everywhere the same value as on the axis.

In the same way, but making use of the rectangular path *efgh*, with $\overline{gh}$ *outside* of the coil, it is left as an exercise to show that the **B**-field outside of an ideal solenoid vanishes.

Example 25-11 A uniform current of density j flows along (down into the plane of Figure 25-26) an infinitely long and conducting cylinder of inner radius a and outer radius b. Calculate the **B**-field everywhere.

Solution Because of the fact that the cylinder is infinitely long, it follows, just as above, that the field lines are circles concentric with the axis. Accordingly, let us apply Ampère's law successively to the three regions: $r > b$; $b \geqslant r \geqslant a$; and $r < a$, by use in each case of a circle of appropriate radius. In the region $r > b$ there results

$$\oint \mathbf{B} \cdot d\mathbf{l} = 2\pi r B = \mu_0 j \pi (b^2 - a^2)$$

where the second equality follows since the total current through the circle of radius $r \ (>b)$ is $j\pi(b^2 - a^2)$. Hence

$$B = \frac{\mu_0 j (b^2 - a^2)}{2r} \qquad (r > b)$$

In a similar way, for the region $b \geqslant r \geqslant a$,

$$B = \frac{\mu_0 j (r^2 - a^2)}{2r} \qquad (b \geqslant r \geqslant a)$$

Finally, for the region $r \leqslant a$,

$$B = 0 \qquad (r < a)$$

since no current flows through a circle of radius $r < a$.

Example 25-12 A *toroid* is a solenoid that has been bent into the form of a *torus*. See Figure 25-27. Assuming that the N turns of the toroid are very close together, so that the field lines are confined to its interior, calculate **B** at a distance r $(a < r < b)$ from the center of the toroid. See Figure 25-27b.

Solution Using symmetry arguments, it follows that the **B**-field lines inside the torus must be circles concentric with the center of the torus. Their sense is

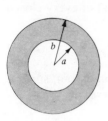 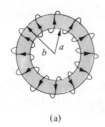 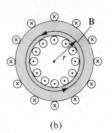

(a)   (b)

Figure 25-26   **Figure 25-27**

determined by the right-hand rule, and for the current flow in the figure they are directed as shown.

To calculate B, let us apply Ampère's law to a circle of radius r inside the torus. Since there are N turns about the torus, it follows that the total current through this path is Ni. (Note that for values of r for which this path is *outside* the torus, the associated current flow vanishes.) Assuming, as is implied in the figure, that **B** depends only on r and is tangent to this circular path, we find that

$$\mu_0 Ni = \oint \mathbf{B} \cdot d\mathbf{l} = B \oint dl = 2\pi r B$$

and this leads to

$$B = \frac{\mu_0 Ni}{2\pi r} \qquad (25\text{-}15)$$

Unlike the ideal solenoid, inside of which the field is constant, in the interior of a toroid, **B** varies with radius in accordance with this formula. In both cases, **B** vanishes outside.

25-11 Summary of important formulas

The force **F** on a particle of charge q traveling at the velocity **v** through a magnetic induction field **B** is

$$\mathbf{F} = q\mathbf{v} \times \mathbf{B} \qquad (25\text{-}1)$$

Experiment shows that the field **B** is independent of q and **v** and thus is determined exclusively by the magnetic sources. The properties of the particle enter *only* through the factor $q\mathbf{v}$.

The magnetic induction $d\mathbf{B}$ due to an infinitesimal current element $i\, d\mathbf{l}$ in a thin wire is given by the Biot-Savart law

$$d\mathbf{B} = \frac{\mu_0 i}{4\pi} \frac{d\mathbf{l} \times \mathbf{r}}{r^3} \qquad (25\text{-}3)$$

with **r** the position vector from $d\mathbf{l}$ to the field point. The total field produced by a collection of currents is found by integrating (25-3) around all closed current paths.

Gauss' law for magnetism states that the magnetic flux out of every closed surface vanishes:

$$\oint_S \mathbf{B} \cdot d\mathbf{S} = 0 \tag{25-11}$$

where the integral is to be carried out over the arbitrary, but closed, surface S. Ampère's law states that for l an arbitrary, closed curve,

$$\oint_l \mathbf{B} \cdot d\mathbf{l} = \mu_0 I \tag{25-12}$$

where I is the net current flowing along the outward sense of an arbitrary *open* surface bounded by l. See Figure 25-18.

QUESTIONS

1. Define or describe briefly what is meant by the following terms: (a) north pole; (b) magnetic monopole; (c) Gauss' law for magnetism; (d) magnetic flux; and (e) Ampère's law.

2. Consider the earth as a huge magnet. In view of the fact that the "north pole" of a magnetized needle points north, is the northernmost magnetic pole of the earth a north or a south magnetic pole? Explain.

3. Suppose that the magnetic induction field of the earth were produced by circular currents flowing in its equatorial plane. Would these currents flow eastward or westward around the earth?

4. If a magnetized needle were free to rotate simultaneously in both a vertical and a horizontal plane, which way would the needle be pointing near the north pole of the earth? Near the south pole?

5. An electron moving at some fixed velocity enters a region of space where there is a uniform **B**-field. Explain how it is possible for the electron to experience *no* force under these circumstances.

6. Consider the three vectors **F**, **v**, and **B** in (25-1). Which two pairs of these three are always perpendicular to each other?

7. A proton traveling at a velocity of 5.0×10^5 m/s enters a region in which there is a time-independent magnetic induction **B**. What must be its speed when it leaves this region? Can anything be said in general about its final direction of motion? Explain.

8. Explain why (25-1) suffices to define the **B**-field for a given magnetic source even though the component of **B** along **v** does not enter this formula.

9. A particle of charge q is at rest in a region of space where there exists a **B**-field produced by nearby currents. What is the force on the particle, assuming that there is no electric field present?

10. Consider again the situation of Question 9, but this time from the viewpoint of a moving observer with respect to whom the particle has the velocity **v**. What, according to this observer, is the force on the particle? Explain your answer in light of (25-1) and the fact that there were no electric fields present originally.

11. Contrast and compare the two right-hand rules, which give the directions of **B** associated, respectively, with the current in a straight wire and that in a solenoid. Review the arguments used for deriving these two rules.

12. An observer sees current flowing counterclockwise about a loop of

wire shaped in the form of a triangle. Are the field lines in the plane of the loop directed toward or away from this observer? Justify your answer.

13. An explorer is looking at his compass when suddenly a horizontal bolt of lightning flashes overhead. Assuming that the associated current is directed due north, will the compass needle be deflected eastward or westward? Explain.

14. Devise a circuit analogous to that in Figure 25-5 so that the currents in the two hanging wires are in the same direction.

15. In view of result in (25-13) is it correct to say that the magnetic induction at a distance r from the axis of a uniform current is due only to those currents flowing inside the cylinder of radius r? Explain.

16. Consider the flow of a uniform current along the axis of a cylinder that has an elliptical cross section. Why can we *not* calculate the magnetic induction by use only of Ampère's law? Explain.

17. In view of the fact that parallel currents attract, do you expect the neighboring turns of wire in a solenoid to attract or to repel each other?

18. A particle is suspended from the lower end of a spring, whose axis is vertical. If a steady current is somehow sent through the spring, will the new equilibrium position of the particle be above or below its original one? Does it matter in which direction current flows?

19. A proton is traveling parallel to a straight wire along which flows a current i. If the proton is moving in the direction of the current, what is the direction of the force on the proton?

20. Suppose that a current of $(20/4\pi)$ amperes flows around the rectangular loop in Figure 25-28. Evaluate $\oint \mathbf{B} \cdot d\mathbf{l}$ for each of the four paths A, B, C, and D.

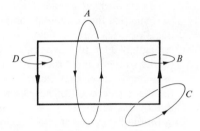

Figure 25-28

21. Show by use of Ampère's law that there does not exist a scalar potential function whose derivatives are the components of the **B**-field. (*Hint*: Assume the existence of such a function and show that it leads to a contradiction.)

22. Devise an experiment to confirm *Gauss' law* for magnetism.

23. Suppose you have available an apparatus to measure the magnetic induction at any point on earth. Devise a *thought experiment* by means of which you could ascertain whether the magnetic induction of the earth is due to currents flowing in its interior or to currents in the space outside. (*Hint*: Use the apparatus in conjunction with Ampère's law.)

PROBLEMS

1. A particle having a charge of 2.0 μC is traveling at a velocity of 3.0×10^5 m/s in a region of space where there exists a magnetic induction of strength 2.5×10^{-3} tesla directed due north. Calculate the magnitude and direction of the force on this particle at an instant when it is traveling: **(a)** vertically upward; **(b)** due south; **(c)** in a northeasterly direction.

2. Calculate the magnitude of the

force on a proton traveling through the upper atmosphere, where the earth's field has the value 4.0×10^{-5} tesla. Assume the proton to be traveling in a direction making an angle $45°$ with respect to **B** and at a speed 6.0×10^5 m/s.

3. A cathode-ray tube is oriented so that the electron steam is directed vertically upward. Assuming that the electrons have a velocity of 10^7 m/s, and that the horizontal component of the earth's field has a strength of 7.0×10^{-5} tesla, calculate (a) the strength and direction of the magnetic force on each electron and (b) the acceleration of an electron due to this force.

4. A proton beam is sent into a region of space where there exists a uniform magnetic induction. In terms of a Cartesian coordinate system set up in this region, it is observed that the beam is undeflected if it travels along the positive z-axis, but if it is sent along the direction of the positive x-axis, then it is deflected along the direction of the positive y-axis. Based on these data, what is the direction of the magnetic induction?

5. Repeat Problem 4, but assume that the observed results are for an electron beam.

6. A proton makes circular orbits in a cyclotron with a speed of 3.0×10^7 m/s. If its acceleration is observed to be 1.2×10^{15} m/s², calculate (a) the force on the proton and (b) the strength of the **B**-field perpendicular to the plane of the orbit.

7. In a certain Cartesian coordinate system a current element $i\,d\mathbf{l}$ is located at the origin and oriented along the positive z-axis. Calculate the direction and magnitude of **B** produced by this element at the following points (assume that the parameter a is in each case a posi-

tive distance): (a) $(0, 0, a)$; (b) $(0, a, 0)$; (c) $(-a, 0, 0)$; and (d) $(a, a, 0)$.

8. A current of 5.0 amperes flows along a very long and straight wire. Calculate the strength of the **B**-field at two points, which are at perpendicular distances of 1.0×10^{-3} meter and 2.0 meters from the wire, respectively. Under what circumstance will the fields at these two points be parallel? Under what circumstance will they be antiparallel?

*9. Show, by use of the Biot-Savart law, that the **B**-field at a distance **r** from a particle of charge q that is traveling at a velcoity **v** is

$$\mathbf{B} = \frac{\mu_0}{4\pi} q \frac{\mathbf{v} \times \mathbf{r}}{r^3}$$

10. Consider an electron orbiting about a proton in a hydrogen atom in a circle of radius 0.53×10^{-10} meter with a speed 2.2×10^6 m/s. Making use of the result of Problem 9, calculate the strength of the **B**-field at the position of the proton due to the electron's motion.

11. If the wire in Figure 25-15 had a total length $2l$ and the point P is on the perpendicular bisector with coordinates $(x, 0, 0)$, show that **B** is directed along the y-axis in the figure and has the magnitude

$$B = \frac{\mu_0 i}{2\pi x} \frac{l}{[x^2 + l^2]^{1/2}}$$

(*Hint:* Use the methods of Example 25-3, but restrict the region of integration to the interval $-l \le z \le l$.)

12. A current i flows about a square loop of side a. By use of the result of Problem 11 show that B at the center of the loop is given by

$$B = \frac{2\sqrt{2}\mu_0 i}{\pi a}$$

and that its direction is given by the right-hand rule for a circular loop.

13. A 10-ampere current flows around a

loop of wire in the shape of an equilateral triangle 50 cm on a side. By use of the result of Problem 11, calculate B at the center of the loop.

14. Repeat Problem 12, but this time calculate B at a point P a distance b above the plane of the square loop and along the perpendicular through its center. (*Hint*: Recall that **B** is a vector, and vector addition must be used to add together the contributions from each side of the square.)

*15. A wire of length $2l$ carries a current i and lies along the z-axis of a certain coordinate system with its center at the origin. Show that the **B**-field at a point with the coordinates $(x, 0, z)$ has the magnitude

$$B = \frac{\mu_0 i}{4\pi x}\left\{\frac{l-z}{[(l-z)^2 + x^2]^{1/2}} + \frac{l+z}{[(l+z)^2 + x^2]^{1/2}}\right\}$$

and determine its direction. Show also that this formula reduces to the result in Problem 11 for an appropriate choice of variables.

16. Two infinite parallel wires are separated by a distance $2a$ and carry currents i in opposite directions, as shown in Figure 25-29. Calculate **B** at a point P, which lies at a distance b along the perpendicular bisector, so that in terms of the coordinate system shown, the coordinates of P are (b, a). What should your answer reduce to when $b = 0$?

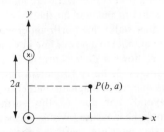

Figure 25-29

*17. Repeat Problem 16, but calculate the field at a general point with coordinates (x, y). Is it an essential restriction that we do not introduce the z-coordinate of the field point?

18. Repeat Problem 17, but assume this time that the currents in the two wires are *both* directed perpendicularly down into the plane of Figure 25-29.

19. What current must flow around a circular loop of wire of radius 50 cm so that the field at the center is 5×10^{-5} tesla? What current must flow if the field is to have a strength of 2 teslas?

20. Show, by use of the result of Example 25-8, that if $\partial B/\partial z = 0$ and $\partial^2 B/\partial z^2 = 0$, then $z = c = a/2$. Explain why you might expect the resultant *Helmholtz coil* to have a relatively constant magnetic induction along the axis of, and midway between, the two loops.

21. A current i flows in a segment of a circular loop of radius a and angle α, as shown in Figure 25-30. Calculate **B** at the center C of the loop, disregarding the wires that feed the current.

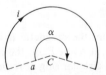

Figure 25-30

22. Suppose that a circular current circled the earth at the equator. What would the strength of this current have to be to produce the observed field at the poles of about 7.5×10^{-5} tesla? Would the current flow from east to west, or in the opposite direction?

23. Calculate at the point P the strength of the magnetic induction due to the current i through the wire in Figure 25-31. (*Hint*: Use the result of Problem 21.)

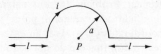

Figure 25-31

24. A circular *line charge* of radius a and charge per unit length λ rotates at an angular velocity ω about its axis.

(a) Show that this motion corresponds to a current i flowing in a circular loop of radius a and given by

$$i = \lambda a \omega$$

(b) Calculate **B** at a point on the axis of the line charge and at a distance b from its center.

*25. A disk of radius a carries a uniform charge per unit area σ and rotates with an angular velocity ω about its axis. See Figure 25-32.

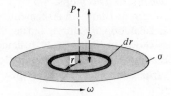

Figure 25-32

(a) Show that the magnetic induction dB at point P on the axis due to a circular ring of radius r and thickness dr is

$$dB = \frac{\mu_0 \sigma \omega r^3 \, dr}{2[r^2 + b^2]^{3/2}}$$

(*Hint*: Make use of the results of Problem 24.)

(b) Show that the total field B at the point P is

$$B = \frac{\mu_0 \sigma \omega}{2}\left[\frac{a^2 + 2b^2}{(a^2 + b^2)^{1/2}} - 2b\right]$$

26. A solenoid of radius 0.5 cm and length 20 cm carries a current of 10 amperes and has 1000 turns. Calculate the strength of the magnetic

induction on the axis of the coil at the following points: (a) the center of the coil; (b) the edge of the coil; (c) a distance of 100 cm from the center of the coil.

27. For the solenoid described in Problem 26 make a plot of B as a function of position on the axis of the coil.

28. How many turns per unit length are required so that a very long solenoid which carries a current of 2.0 amperes will have a magnetic induction on its axis of 1.5×10^{-2} tesla.

29. A current of 0.5 ampere flows around a solenoid of radius 1.0 cm and of length 40 cm. If the strength of the uniform magnetic induction near the center of the solenoid is 10^{-3} tesla, how many turns per unit length are there in the solenoid?

30. Suppose that the field on the axis of a very long solenoid having 1000 turns per meter is 5.0×10^{-3} tesla. (a) What is the current? (b) If now a wire that carries a current of 10 amperes is wrapped with n turns per unit length around the original solenoid in such a way that the field on the axis is reduced to 2.5×10^{-3} tesla, calculate n. Are the current flows parallel in these two coils?

31. A proton is traveling at a speed of 5.0×10^5 m/s on the axis of a solenoid, which has 1000 turns per meter and carries a current of 2.0 amperes. Calculate the acceleration of the proton. In what way would your answer differ if it were moving parallel to but *not* on the axis.

32. A man walks due north underneath and parallel to a power line in which there flows a direct current of 100 amperes. If he is 10 meters below the line, what magnetic induction— beyond that due to the earth— would he measure? Would this seriously interfere with a compass reading at this point?

*33. Figure 25-33 shows a cut through an infinite array of n (per unit length) parallel wires, each carrying a current i directed as shown.

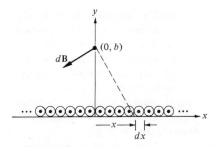

Figure 25-33

(a) Show that the field $d\mathbf{B}$ at the point with coordinates $(0, b)$ due to the $n\,dx$ wires located between x and $(x + dx)$ is

$$d\mathbf{B} = -\frac{\mu_0 ni}{2\pi(b^2 + x^2)}(\mathbf{i}b + \mathbf{j}x)\,dx$$

where $\mathbf{i}$ and $\mathbf{j}$ are unit vectors along the x- and y-axes, respectively.

(b) By integration show that the field for $y > 0$ due to all of the currents is uniform—that is, independent of b—and is given by

$$\mathbf{B} = -\frac{1}{2}\mu_0 ni\,\mathbf{i} \qquad y > 0$$

(c) Show in a similar way that for points below the currents, that is, for $y < 0$,

$$\mathbf{B} = \frac{1}{2}\mu_0 ni\,\mathbf{i} \qquad y < 0$$

The field lines for this current distribution are shown in Figure 25-34.

34. Confirm the validity of the results for (b) and (c) of Problem 33 by applying Ampère's law to the rectangular path $abcd$ in Figure 25-34.

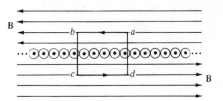

Figure 25-34

35. Suppose that the cylinder in Figure 25-35 represents a solenoid of N turns and length l, around which flows a current i so that the field on the axis of the solenoid is directed upward. Evaluate $\oint \mathbf{B} \cdot d\mathbf{l}$ for the paths **(a)** $ABCDA$; **(b)** $abcda$; **(c)** l_1; and **(d)** l_2.

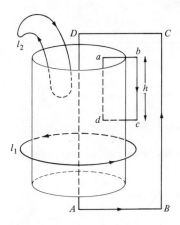

Figure 25-35

*36. Show that it is impossible for a *uniform* magnetic induction to drop abruptly to zero in a current-free region. (*Hint:* Suppose, as in Figure 25-36, that there is a uniform field in the half-space $x < 0$, and that B is zero for $x > 0$. By applying Ampère's law to the path $abcd$ show that this leads to a contradiction. What conclusions can you draw about the magnetic induction based on this result?)

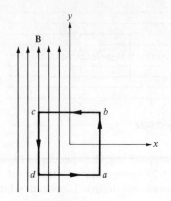

Figure 25-36

*37. Consider, in Figure 25-37, a circular coil of wire of radius a and carrying a current i.

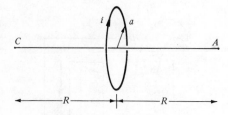

Figure 25-37

(a) Evaluate the line integral

$$\mathscr{J} = \int_A^C \mathbf{B} \cdot d\mathbf{l}$$

along the axis of the loop a symmetric distance of $2R$.

(b) Show that in the limit as $R \to \infty$,

$$\mathscr{J} = \mu_0 i$$

(c) Justify your result to (b) by use of Ampère's law. (*Hint:* Close the integration path AC by a large semicircle and show that $\int \mathbf{B} \cdot d\mathbf{l}$ along this path vanishes as $R \to \infty$.)

*38. Repeat Problem 37, but this time along the axis of a solenoid of length l and N turns.

*39. A very large coaxial cable consists, as in Figure 25-38, of an inner solid conductor of radius a and an outer one of inner radius b and outer radius c. Assuming equal and opposite uniform currents i in these conductors, calculate the magnetic induction for (a) $r \le a$; (b) $a \le r \le b$; (c) $b \le r \le c$; and (d) $r \ge c$.

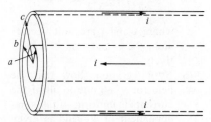

Figure 25-38

26 Magnetic forces

The only justification for our concepts and systems of concepts is that they serve to represent the complex of our experiences; beyond this, they have no legitimacy.

ALBERT EINSTEIN (1879–1955)

26-1   Introduction

In Chapter 25 the magnetic induction field was defined in terms of the force acting on a moving charged particle. The focus there, however, was not on this force itself, but rather on the problem of calculating the fields associated with certain idealized current distributions. The purpose of this chapter is to present a more detailed analysis of this force and to consider it as it affects both the motions of individual charged particles as well as macroscopic current flows. The principles underlying the operation of a number of devices including cyclotrons, mass spectrometers, and galvanometers are, as we shall see, readily explained in terms of this force.

26-2   The motion of a charged particle in a uniform B-field

Consider a particle of charge q and mass m moving in a uniform magnetic induction **B**. According to (25-1) the force on it is $q\mathbf{v} \times \mathbf{B}$ and thus it follows, from Newton's second law, that the equation of motion for the particle is

$$m\frac{d\mathbf{v}}{dt} = q\mathbf{v} \times \mathbf{B} \tag{26-1}$$

with **v** its instantaneous velocity. All the details of the particle's trajectory can be deduced from this relation by solving it subject to an appropriate set of initial conditions. However, rather than proceeding in this way it is instructive to solve it indirectly by use of a much simpler heuristic approach which is available in this particular case.

Suppose, first, that initially the velocity of the particle is perpendicular to the uniform **B**-field. Specifically, let us assume that the **B**-field lines are directed perpendicularly down into the plane of Figure 26-1 and, following convention, let us represent these lines by an array of crosses. (Correspondingly, an array of dots is normally used to represent **B**-field lines emerging from the plane of a figure toward the reader.) The force **F** on the particle is directed as shown in Figure 26-1a for $q > 0$ and in Figure 26-1b for $q < 0$. In both cases **F** is perpendicular to **v** throughout the motion. Hence, only the direction, but not the magnitude, of the particle's velocity can change during its subsequent motion (see Example 25-2). It follows that the orbit of the particle must be a circle whose plane is perpendicular to **B**. Further, the magnitude of the velocity of the particle along this orbit must be constant and equal to its initial value $v = |\mathbf{v}|$.

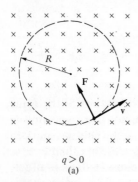

$q > 0$
(a)

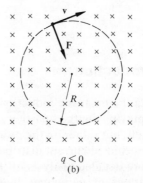

$q < 0$
(b)

Figure 26-1

Now according to the analysis of Section 3-10, a particle such as this which travels with a uniform speed v around a circle of radius R undergoes a *centripetal* acceleration $\mathbf{a}_c$, which is directed radially inward and has the magnitude v^2/R. Reference to Figure 26-1 shows, then, that for a uniform field we may express (26-1) in the form

$$qvB = m\frac{v^2}{R} \tag{26-2}$$

and this is the fundamental relation from which all orbital parameters can be determined. For example, solving (26-2) for the radius we obtain

$$R = \frac{mv}{qB} \qquad (26\text{-}3)$$

so that R is directly proportional to the speed, v, of the particle and with the coefficient of proportionality the known ratio m/qB. The angular velocity of a particle going with a uniform speed, v, around a circle of radius R is v/R. Hence, solving (26-3) for this ratio, we obtain

$$\omega_c = \frac{v}{R} = \frac{qB}{m} \qquad (26\text{-}4)$$

where the quantity ω_c is the *cyclotron frequency* associated with this orbit. A most striking feature of this formula for ω_c is that it depends *only* on B and on the ratio q/m and is thus independent of the velocity of the particle. This means that a collection of *identical* charged particles will orbit, in a given **B**-field, with the same angular velocity ω_c regardless of their initial velocities. According to (26-3), however, the radii of these orbits will, in general, be different. The period P of this motion is

$$P = \frac{2\pi}{\omega_c} = \frac{2\pi m}{qB} \qquad (26\text{-}5)$$

If the initial velocity **v** of a particle is *not* perpendicular to **B**, then the orbit of the particle is not confined to a plane perpendicular to **B** as in Figure 26-1. To treat this more complex case, suppose that $v_\perp$ is the component of **v** perpendicular to **B** and $v_\parallel$ the corresponding component along **B**. As far as the motion at right angles to **B** is concerned, all of the relations in (26-2) through (26-5) are still valid provided that v is everywhere replaced by $v_\perp$. With regard to $v_\parallel$, note from (26-1) and the definition of a cross product that there is no component of the force **F** along the direction of **B**. Hence the particle will continue to move with its initial velocity $v_\parallel$ along this direction. The motion of the particle thus consists of two parts: (1) a motion at the initial velocity $v_\parallel$ along the direction of **B**; and (2) a motion in a circular orbit at the uniform speed $v_\perp$ at right angles to **B**. When viewed by an external observer, the particle appears to move in a *helix*, as shown in Figure 26-2.

Figure 26-3 illustrates the circular paths followed by charged particles in a uniform **B**-field. Shown are the actual tracks produced in the collision of a 300 GeV (3.0×10^{11} eV) proton with a stationary one in the 30-in. hydrogen bubble chamber at the Fermi National Accelerator laboratory. Twenty-six charged particles are produced. Assuming that the inward spiraling tracks are due to electrons, is the **B**-field directed into or out of the plane of the photograph?

Example 26-1 A proton is moving at right angles to a uniform magnetic induction having a strength of 2.0 tesla.

(a) Calculate the cyclotron frequency.

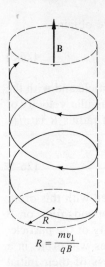

$$R = \frac{mv_\perp}{qB}$$

Figure 26-2

(b) How long does it take the proton to make one full turn around its orbit?

(c) If the radius of the orbit is 5.0 cm, what is the velocity of the proton?

(d) What is the proton's energy?

Solution

(a) Using the fact that for a proton $m = 1.67 \times 10^{-27}$ kg and $q = 1.6 \times 10^{-19}$ coulomb, we find, by use of (26-4), that

$$\omega_c = \frac{qB}{m} = \frac{(1.6 \times 10^{-19} \text{ C}) \times (2.0 \text{ T})}{1.67 \times 10^{-27} \text{ kg}}$$

$$= 1.9 \times 10^8 \text{ rad/s}$$

(b) Making use of this result we find for the period P of the motion

$$P = \frac{2\pi}{\omega_c} = \frac{2\pi}{1.9 \times 10^8 \text{ rad/s}} = 3.3 \times 10^{-8} \text{ s}$$

In other words, the proton makes one complete revolution about its circular orbit in 3.3×10^{-8} second.

(c) Solving (26-3) for v we find

$$v = R\frac{qB}{m} = R\omega_c = (0.05 \text{ m}) \times (1.9 \times 10^8 \text{ rad/s})$$

$$= 9.5 \times 10^6 \text{ m/s}$$

(d) The energy E of the proton is all kinetic. Hence

$$E = \tfrac{1}{2}mv^2 = \tfrac{1}{2} \times 1.67 \times 10^{-27} \text{ kg} \times (9.5 \times 10^6 \text{ m/s})^2$$

$$= 7.5 \times 10^{-14} \text{ J}$$

In connection with elementary particles it is customary to express energies in units of the "MeV." This is defined to be the energy gained by a proton in dropping through a

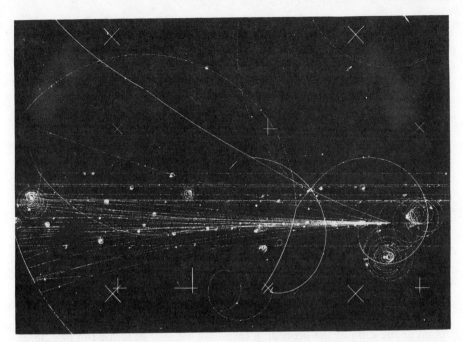

Figure 26-3 *A proton with an energy of 300 GeV producing 26 charged particles in a 30-in. hydrogen bubble chamber. (Courtesy Fermi National Accelerator Laboratory.)*

potential difference of 10^6 volts. It has the approximate value

$$1 \text{ MeV} = 1.6 \times 10^{-13} \text{ J}$$

In terms of this unit, E has the value

$$E = 7.5 \times 10^{-14} \text{ J} \times \left(\frac{1 \text{ MeV}}{1.6 \times 10^{-13} \text{ J}} \right)$$

$$= 0.47 \text{ MeV}$$

Example 26-2 Consider in Figure 26-4 a collimated beam of particles of various charges, masses, and velocities entering via slit A a region of space in which there is a uniform electric field **E** directed vertically downward and a uniform magnetic

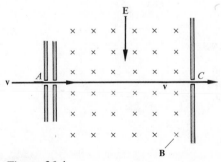

Figure 26-4

induction perpendicular to **E**. What can be concluded about the velocity, the charge, and the mass of the particles that eventually emerge at slit C?

Solution   The subset of particles that arrive at C consists of those that continue to travel in a straight line and thus experience no force on traversing this region. All others will be accelerated at right angles to this direction. Hence if **v** is the velocity of any particle which reaches C, then

$$\mathbf{F} = q(\mathbf{E} + \mathbf{v} \times \mathbf{B}) = 0$$

where q is its charge. Since the vectors **E**, **B**, and **v** are, as shown in Figure 26-4, mutually orthogonal it follows that the speed v of all particles emerging at C must be $v = E/B$. Since this is independent of mass and charge, it follows that nothing can be concluded about the values for q and m for the particles that arrive at slit C. By varying the ratio E/B we can, in effect, control the velocities of the particles which arrive at C; thus the apparatus in Figure 26-4 is known as a *velocity selector* for charged particles. Neutral particles would go from A to C regardless of their velocities.

26-3   The mass spectrometer

As an application of the preceding formulas we now describe the principles underlying the instrument known as a *mass spectrometer*, which can be used to measure the mass of a charged particle such as a heavy ion. This device was invented and first constructed in 1919 by Francis W. Aston (1877–1945), who was interested in establishing the masses of the isotopes associated with the various chemical elements. Since its invention, many refinements of the mass spectrometer have been made, and today it continues to be one of the important tools of chemists and physicists.

Figure 26-5 shows the key elements of a mass spectrometer. A collimated beam of charged particles of various masses and velocities enters via slit system A a region of space in which there exist an electric and a magnetic

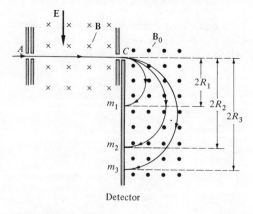

Detector

Figure 26-5

field, **E** and **B**, at right angles to each other. According to the results of Example 26-2, this configuration of fields acts as a *velocity selector*, and the magnitude of the common velocity **v** of the particles that get through to slit *C* is

$$v = \frac{E}{B}$$

Upon emerging from slit *C*, these particles of velocity **v** now enter a region where, as indicated by the dots, there is only a magnetic induction B_0 perpendicular to and out of the plane of the diagram. The particles here traverse a semicircular path until they strike a screen or some other detecting instrument. According to (26-3), the distance $2R$ below the slit *C* at which a particle will be detected is

$$2R = \frac{2mv}{qB_0}$$

and substituting the known velocity $v = E/B$ this distance may be expressed in the equivalent form

$$2R = 2m \left(\frac{E}{qBB_0} \right) \tag{26-6}$$

Assuming that the charge *q* is known, since *E*, *B*, and B_0 are each separately measurable, we conclude that a measurement of the distance $2R$ enables us, in effect, to deduce the mass *m* of the particle. Alternatively, for purposes of measuring relative masses, as in the determination of isotopes, we may express (26-6) in the form

$$\frac{2R_1}{2R_2} = \frac{m_1}{m_2} \tag{26-7}$$

where $2R_1$ and $2R_2$ are the distances defined in Figure 26-5 and correspond, respectively, to the isotopes of masses m_1 and m_2. In words, (26-7) states that the distance below slit *C* that a particle strikes the detector in Figure 26-5 is directly proportional to its mass. The coefficient of proportionality, according to (26-6), is $(2E/qBB_0)$.

26-4 The cyclotron

As a second application of the formulas in Section 26-2 let us discuss briefly the principles underlying the operation of a cyclotron.

The cyclotron is an instrument that was invented and made operational in 1931 by E. O. Lawrence (1901–1958) and by means of which charged particles, such as protons or ionized helium atoms, can be accelerated to very high speeds. Although the cyclotron is a rather complex instrument, the basic principles underlying its operation are simple, and thus let us focus only on these.

The essential elements of a cyclotron are an electromagnet, which can produce a **B**-field of the order of a few tesla, and two metal chambers, called "dees," which are shaped like the halves of a large pillbox cut in two. Figure 26-6a shows a side view of these elements, and the same arrangement is viewed from the top in Figure 26-6b. As shown, the two dees are connected to an oscillator so that the potential difference between them can be reversed at some fixed frequency of the order of 10^6 Hz.

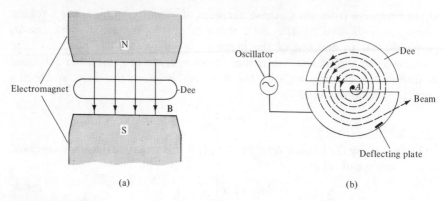

(a)   (b)

Figure 26-6

Suppose a positively charged ion is introduced at point A in Figure 26-6b. Assuming that it has a small velocity v_0 perpendicular to **B**, the ion will traverse a semicircular orbit in the upper dee with a radius R_0 given, according to (26-3), by

$$R_0 = \frac{mv_0}{qB} \tag{26-8}$$

where m and q are, respectively, the mass and charge of the particle. Further, according to (26-5), the time τ required to traverse this path is independent of the velocity and has the constant value

$$\tau = \frac{1}{2} P = \frac{\pi m}{qB} \tag{26-9}$$

There is essentially no electric force acting on the particle while it is within the dee, since we know from studies in electrostatics that the electric field in the interior of a conducting shell vanishes.

After the particle has traversed this first semicircular orbit, it leaves the upper dee, and while in the gap between the dees it experiences an electric force due to the potential difference between the dees. As a result, assuming that the lower dee is at this instant at the lower potential, it follows that this ion will be accelerated and subsequently enter the lower dee with a velocity v_1 larger than was its initial velocity v_0. In the lower dee, because of the magnetic induction, it will again traverse a semicircular orbit but with a somewhat larger radius corresponding, in accordance with (26-3), to its new velocity v_1. However, and this is a most important point, the time τ required to traverse this orbit around the lower dee is the *same* as it was in the upper one; namely, the time τ in (26-9). If the frequency of the oscillator is set so that the potential difference between the dees is periodically reversed in this time interval τ, then each time the particle leaves one dee to go to the other it will accelerate. In other words, if the angular frequency ω_{os} of the oscillator is adjusted to be the same as that of the cyclotron frequency in (26-4)

$$\omega_{os} = \omega_c = \frac{qB}{m} \qquad (26\text{-}10)$$

then each time the particle crosses between the dees it will be speeded up. The radius of its orbit will also increase each time because of the associated increases in velocity. Finally, when the orbital radius is sufficiently large, the particle is ejected from the cyclotron through an exit port by means of a charged deflecting plate.

To sum up, if the angular frequency of the oscillator connected to the dees is adjusted to be the cyclotron frequency, then the orbit of the particle consists of a series of semicircular paths of ever-increasing velocities and corresponding radii. The time τ required to traverse each semicircular orbit is given in (26-9) and is a constant independent of the velocity. Thus each time the ion traverses the gap its energy increases, until finally it is ejected by a deflecting plate at the exit port of the cyclotron.

To calculate the energy E of the particles accelerated in a cyclotron let R represent the "radius" of the cyclotron; that is, the radius of the largest orbit of an ion just prior to ejection. According to (26-3), the maximum velocity v_{max} of an accelerated ion is

$$v_{max} = \frac{qB}{m} R \qquad (26\text{-}11)$$

Hence its energy, and that of the cyclotron, is

$$E = \frac{1}{2} mv_{max}^2 = \frac{1}{2m} q^2 B^2 R^2 \qquad (26\text{-}12)$$

where the second equality follows by substitution from (26-11). Note that the energy of a cyclotron is directly proportional to B^2 and to R^2 and varies inversely with the mass m of the ion being accelerated.

Example 26-3 A cyclotron used to accelerate protons has a field of strength 0.8 tesla and a maximum orbital radius of 0.3 meter.
   (a) What must be the angular frequency of the oscillator connected to the dees?
   (b) What is the velocity of the protons after they are ejected?
   (c) Calculate the final energy of the protons.
   (d) What would the final energy be if deuterons were accelerated in this cyclotron?

Solution

(a) Making use of (26-10) we find that

$$\omega_{os} = \frac{qB}{m} = \frac{1.6 \times 10^{-19}\,\text{C} \times 0.8\,\text{T}}{1.67 \times 10^{-27}\,\text{kg}}$$

$$= 7.7 \times 10^{7}\,\text{rad/s}$$

where use has been made of the known values of q and m for a proton.

(b) On substituting the given values for q, B, m, and R into (26-11) we obtain

$$v_{max} = \frac{qBR}{m} = \frac{1.6 \times 10^{-19}\,\text{C} \times 0.8\,\text{T} \times 0.3\,\text{m}}{1.67 \times 10^{-27}\,\text{kg}}$$

$$= 2.3 \times 10^{7}\,\text{m/s}$$

(c) Substitution into (26-12) yields for the energy E

$$E = \frac{1}{2m}(qBR)^2$$

$$= \frac{1}{2}\frac{[1.6 \times 10^{-19}\,\text{C} \times 0.8\,\text{T} \times 0.3\,\text{m}]^2}{1.67 \times 10^{-27}\,\text{kg}}$$

$$= 4.4 \times 10^{-13}\,\text{J}$$

$$= 2.8\,\text{MeV}$$

where the final equality follows since by definition

$$1.6 \times 10^{-13}\,\text{J} = 1\,\text{MeV}$$

The final energy of 2.8 MeV means that the protons have the same energy as if they had been accelerated through a potential difference of 2.8 million volts.

(d) Since the mass of the deuteron is twice that of the proton, while their charges are the same, (26-12) shows that the final energy of an accelerated deuteron will be only half that acquired by a proton. Accordingly, making use of the result of (c) we find that

$$E = 1.4\,\text{MeV} \qquad \text{(deuterons)}$$

Example 26-4   A particular cyclotron produces 18-MeV alpha particles. If the frequency of the oscillator is $1.2 \times 10^{7}\,\text{Hz}$, calculate:

(a) The strength of the magnetic induction.

(b) The number of times that an α particle orbits inside the cyclotron if its energy is increased by 0.15 MeV during each orbit.

Solution

(a) Solving (26-10) for B we obtain

$$B = \frac{m\omega_{os}}{q}$$

Now a frequency of $1.2 \times 10^{7}\,\text{Hz}$ corresponds to an angular frequency ω_{os}:

$$\omega_{os} = 2\pi \times 1.2 \times 10^{7}\,\text{rad/s}$$

$$= 7.5 \times 10^{7}\,\text{rad/s}$$

Hence

$$B = \frac{m\omega_{os}}{q} = \frac{6.67 \times 10^{-27} \text{ kg} \times 7.5 \times 10^7 \text{ rad/s}}{3.2 \times 10^{-19} \text{ C}}$$

$$= 1.6 \text{ T}$$

where we have used the fact that the charge of an α particle is twice that of a proton, while its mass is four times greater.

(b) Since the particle picks up 0.15 MeV during each full orbit and its final energy is 18 MeV, it follows that the number n of such traversals is

$$n = \frac{18 \text{ MeV}}{0.15 \text{ MeV}} = 120$$

Hence, since the frequency of the oscillator is 1.2×10^7 Hz, the total acceleration time is

$$\frac{120}{1.2 \times 10^7 \text{ Hz}} = 1.0 \times 10^{-5} \text{ s}$$

and this is a very short time indeed.

Example 26-5 According to the theory of relativity, for speeds close to that of light ($c = 3 \times 10^8$ m/s), (26-2) must be replaced by

$$qvB = \frac{mv^2/R}{[1 - v^2/c^2]^{1/2}} \qquad (26\text{-}13)$$

(a) What is the cyclotron frequency in this case?
(b) How long does it take such a particle to traverse a dee?
(c) Explain why a cyclotron, as described above, does *not* work for relativistic particles.

Solution

(a) A comparison of (26-2) and (26-13) shows that if we make the replacement

$$m \rightarrow \frac{m}{[1 - v^2/c^2]^{1/2}}$$

in the former we obtain the latter. Hence, the relativistic generalization of (26-4) is

$$\omega_c = \frac{qB}{m} \left[1 - \frac{v^2}{c^2}\right]^{1/2}$$

and thus for high speeds the cyclotron frequency decreases as v increases.

(b) Making the same replacement in (26–9) we find that

$$\tau = \frac{\pi m}{qB} \frac{1}{[1 - v^2/c^2]^{1/2}}$$

(c) According to this formula for τ, the time a particle spends in a dee varies with its speed. Hence if an oscillator of angular frequency ω_{os} given in (26-10) is used, then, as the velocities of the accelerating particles approach the speed of light, they will not consistently cross the gap between the dees in phase with the oscillating field.

This means that they will not be consistently accelerated when crossing the gap; indeed, during some crossings they will be slowed down. Hence the effectiveness of a cyclotron is restricted to particle speeds small compared to the speed of light.

Because of this restriction, during recent decades much effort has gone into the design and construction of a variety of accelerators that are not subject to this limitation. Figure 26-7 shows one of these. This accelerator is 2 km in diameter and accelerates protons to a peak energy of about $300 \text{ GeV} = 3 \times 10^5 \text{ MeV}$! That is, it can accelerate protons to energies about 1000 times greater than is possible for a cyclotron. As of this writing (1973) these are the most energetic particles produced, in a controlled way, by man.

Figure 26-7 *Aerial view of the main accelerator at the Fermi National Accelerator Laboratory (FNAL) in Batavia, Illinois. The circumference of the accelerator is about 4 miles. (Courtesy FNAL.)*

26-5 The force on current

Consider, in Figure 26-8, an infinitesimal element of length dl of a wire of cross-sectional area A and through which flows a uniform current i. Suppose also the existence of an external **B**-field—that is, one produced by magnetic sources other than that due to this current i—and let $d\mathbf{F}$ represent the magnetic force on the current due to this **B**-field. According to the physical picture of macroscopic currents developed in Chapter 23, the current flow through the element in Figure 26-8 can be thought of as being due to a certain number, n, of charge-carriers per unit volume, each of which travels at the drift velocity v_d along the direction of the current. The total charge ΔQ involved is therefore

$$\Delta Q = qnA \, dl$$

where q is the charge on each of these particles. Substituting this formula into (26-1), we may express $d\mathbf{F}$ as

$$d\mathbf{F} = qnA \, dl \, (ev_d) \times \mathbf{B} \qquad (26\text{-}14)$$

where $\mathbf{e}$ is a unit vector along the direction of the current. Moreover, according to (23-6) the current i is related to the above parameters by

$$i = qnAv_d$$

and therefore (26-14) may be reexpressed in the form

$$d\mathbf{F} = i \, dl \, \mathbf{e} \times \mathbf{B}$$

Finally, defining the vector $d\mathbf{l}$ to have the magnitude dl and to be directed along the current, that is, along the unit vector $\mathbf{e}$, we may write

$$d\mathbf{F} = i \, d\mathbf{l} \times \mathbf{B} \qquad (26\text{-}15)$$

a formula which is basic for calculating the magnetic force on a wire that carries a current.

For the more general case of a closed current loop, it is necessary to add together contributions of the form in (26-15) for each current element $i \, d\mathbf{l}$ of the loop. Expressing this sum in the form of an integral, we find

$$\mathbf{F} = i \oint d\mathbf{l} \times \mathbf{B} \qquad (26\text{-}16)$$

where $\mathbf{F}$ is the total force on the loop, and the circle superimposed on the integral means that the integral is to be carried out around the entire loop.

Example 26-6 A wire 2 meters long carries a current of 5.0 amperes and is in a uniform $\mathbf{B}$-field of strength 0.03 tesla, which makes an angle of 30° with the wire. See Figure 26-9. Calculate the force on the wire.

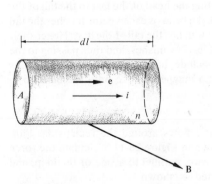

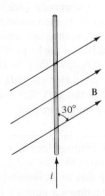

Figure 26-8　　　　　　　　　　　*Figure 26-9*

Solution According to the definition of a cross product and (26-15), the force on each element of the wire is directed perpendicular to and down into the plane of Figure 26-9. The magnitude dF of this force on an arbitrary element of length dl is

$$dF = |i\,d\mathbf{l} \times \mathbf{B}| = i\,dl\,B\,\sin 30°$$

Since i and B are constant, the magnitude of the *total* force on the wire is therefore

$$F = ilB\,\sin 30°$$
$$= (5.0\ \text{A}) \times (2.0\ \text{m}) \times (0.03\ \text{T}) \times 0.5$$
$$= 0.15\ \text{N}$$

where in the second equality we have used $\sin 30° = 0.5$.

Example 26-7 Show that a closed loop of current in a uniform magnetic induction field experiences no force.

Solution Since the **B**-field is constant in this case, the factor **B** can be taken out from under the integral sign in (26-16), and it assumes the form

$$\mathbf{F} = i\left[\oint d\mathbf{l}\right] \times \mathbf{B}$$

To show that **F** vanishes, we shall now establish that the integral

$$\mathbf{I} = \oint d\mathbf{l}$$

vanishes over any closed path.

Consider, in Figure 26-10, the replacement of a closed loop by a sequence of N infinitesimal vectorial elements $d\mathbf{l}_i (i = 1, 2, \ldots, N)$. In the figure $N = 9$. Recalling that an integral is the limit of a sum, it follows that if the elements $\{d\mathbf{l}_i\}$ are sufficiently small, then

$$\mathbf{I} \cong \sum_{i=1}^{N} d\mathbf{l}_i \tag{26-17}$$

where the approximate equality becomes equality in the limit as each of the $\{d\mathbf{l}_i\}$ tends to zero. But the sum of a collection of vectors may be obtained graphically by placing the vectors in sequence with the head of one touching the tail of the next, with the sum then given by the vector connecting the head of the last to the tail of the first. For a closed loop, such as in Figure 26-10, the head of each vector touches the tail of another and in particular the head of the last touches the tail of the first. Hence their vector sum is zero. It follows that the sum in (26-17) vanishes, and thus passing to the limit of vanishingly small elements $\{d\mathbf{l}_i\}$ we conclude that the integral **I** in (26-17) does also. This establishes the fact that there is no magnetic force on a closed loop in a *uniform* magnetic induction field.

Example 26-8 A counterclockwise current i flows around the rectangular loop $ACDEA$ of dimensions $(b - a)$ and c, as shown in Figure 26-11. Calculate the force on this current loop due to an infinitely long wire, parallel to a side of the loop, and through which there flows a current I directed as shown.

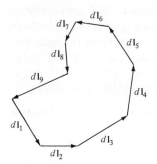

Figure 26-10

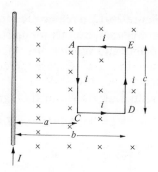

Figure 26-11

Solution   The magnitude of the magnetic induction **B** produced by an infinitely long wire at a perpendicular distance r from the wire is, according to (25-6),

$$B = \frac{\mu_0 I}{2\pi r} \tag{26-18}$$

and as indicated by the crosses its direction is perpendicular and down into the plane of the figure.

Making use of the definition of a cross product and (26-15), we find that the force F_{AC} on the wire segment AC is directed toward the right and has the strength

$$F_{AC} = icB = \frac{ic\mu_0 I}{2\pi a}$$

where the second equality follows from (26-18). In a similar way the force F_{ED} on the segment ED has the magnitude

$$F_{ED} = \frac{ic\mu_0 I}{2\pi b}$$

but this is directed toward the left. Since the currents in the upper and lower segments AE and CD flow in opposite directions, the forces on these segments are equal and opposite; thus there is no contribution to the force due to them. Combining these results we obtain for the magnitude of the total force **F** on the entire loop

$$F = F_{AC} - F_{ED} = \frac{ic\mu_0 I}{2\pi}\left(\frac{1}{a} - \frac{1}{b}\right)$$

and this is directed to the right in the figure.

Note that since **B** is not uniform in this case, the force on the closed loop need not vanish as for the uniform field in Example 26-7.

26-6   Torque on a current loop in a uniform B-field

In Example 26-7 we saw that there is no net force on a current loop in a uniform **B**-field. Nevertheless, depending on the orientation of the current loop relative to **B**, there may be a torque on the loop, and the purpose of this section is to examine this possibility.

Consider, in Figure 26-12, a rectangular loop of sides a and b and carrying a current i. For convenience let us set up a coordinate system, with the loop lying in the y-z plane, and suppose first the existence of a uniform magnetic induction $\mathbf{B}$ along the x-axis, perpendicular to the plane of the loop. According to (26-15), the forces $\mathbf{F}_1$, $\mathbf{F}_2$, $\mathbf{F}_3$, and $\mathbf{F}_4$ on each of the straight segments are directed as shown in the figure and have the magnitudes

$$|\mathbf{F}_1| = |\mathbf{F}_3| = ibB \qquad |\mathbf{F}_2| = |\mathbf{F}_4| = iaB \tag{26-19}$$

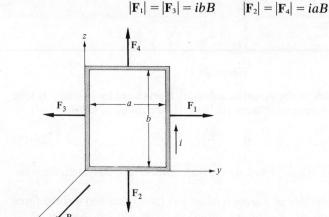

Figure 26-12

Consistent with the result of Example 26-7, these equations imply that there is no net force on the loop. Furthermore, since all four forces lie in the y-z plane and are pairwise equal and opposite, it follows that the net torque on the loop produced by these forces also vanishes.

Consider now the same physical situation as above, but with the $\mathbf{B}$-field no longer perpendicular to the plane of the loop. By analogy to the discussion in Section 25-9, let us define a unit vector $\mathbf{n}$ normal to the plane of the loop and with sense such that:

If the loop is grasped in the right hand with fingers pointing along the current, then the outstretched thumb poinst along $\mathbf{n}$.

With this convention, suppose in Figure 26-13 that $\mathbf{B}$ is directed along the x-axis, but now the normal $\mathbf{n}$ to the plane of the loop makes an angle α with the positive x-axis. Applying (26-15) to this system, we find that the forces $\mathbf{F}_1$ $\mathbf{F}_2$, $\mathbf{F}_3$, and $\mathbf{F}_4$ are directed as shown in the figure; that is, $\mathbf{F}_4$ and $\mathbf{F}_2$ are along the positive and negative z-axis, respectively, and correspondingly for $\mathbf{F}_1$ and $\mathbf{F}_3$ along the y-axis. Their magnitudes are precisely the same as in (26-19) and, again, there is no net force acting on the loop. However, this time, the forces $\mathbf{F}_1$ and $\mathbf{F}_3$ no longer lie in the plane of the loop. Hence $\mathbf{F}_1$ and $\mathbf{F}_3$ produce a

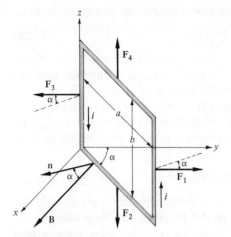

Figure 26-13

torque τ, about the center of the loop, which tends to line up the normal **n** to the plane of the loop with the direction of **B**. That is, the torque on the loop in Figure 26-13 is directed so that the loop tends to turn to the equilibrium arrangement in Figure 26-12.

Having established the direction of the torque on the loop, let us calculate its magnitude. To this end, let us view the loop in Figure 26-13, but this time along the direction of the positive z-axis, as in Figure 26-14. According to (26-19), the forces $\mathbf{F}_1$ and $\mathbf{F}_3$ both have the magnitude ibB, and thus they contribute equally to the torque about the axis through the center of the loop and parallel to the z-axis. Because of the fact that the torque τ produced by a force **F** is

$$\tau = \mathbf{r} \times \mathbf{F}$$

where **r** is a vector from the origin to the point where the force acts, it follows that the magnitude of the torque produced by each of $\mathbf{F}_1$ and $\mathbf{F}_3$ is

$$|\mathbf{r} \times \mathbf{F}_3| = |\mathbf{r} \times \mathbf{F}_1| = \frac{a}{2}|\mathbf{F}_1| \sin \alpha = \frac{iab \sin \alpha}{2} B$$

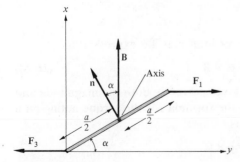

Figure 26-14

Hence the magnitude of the total torque τ on the loop in Figure 26-13 is

$$\tau = iabB \sin \alpha \qquad (26\text{-}20)$$

and its direction is along the positive z-axis. That is, the direction of this torque acts in such a way that the normal $\mathbf{n}$ to the plane of the loop tends to line up with the direction of the magnetic induction.

Although (26-20) has been derived only for the special case of a rectangular loop, in a certain modified form, as established in the problems, it is true in general. Since the area A of a rectangular loop is ab, (26-20) may be expressed in the equivalent form

$$\tau = iAB \sin \alpha \qquad (26\text{-}21)$$

a formula which is valid also for loops of arbitrary shape. As above, this torque tends to line up the normal $\mathbf{n}$ to the plane of the loop along $\mathbf{B}$.

If there are N turns of wire in the loop, then effectively it carries a current Ni. The generalization of (26-21) to this more general case is

$$\tau = iNAB \sin \alpha \qquad (26\text{-}22)$$

To obtain a vector form for (26-22), consider, in Figure 26-15, a planar loop of N turns, area A, and current i, and let $\mathbf{n}$ be a unit vector normal to the loop with its sense determined as above. If we define the *magnetic dipole moment* $\boldsymbol{\mu}$ of this loop to be a vector directed along the normal $\mathbf{n}$ and to have the magnitude iNA, then

$$\boldsymbol{\mu} = iNA\mathbf{n} \qquad (26\text{-}23)$$

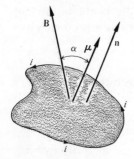

Figure 26-15

and the torque τ acting on the current loop may be expressed as

$$\tau = \boldsymbol{\mu} \times \mathbf{B} \qquad (26\text{-}24)$$

It is easily verified that this formula gives correctly both the magnitude and the direction of τ; it is analogous to the formula for the torque acting on a dipole $\mathbf{p}$ in an electric field $\mathbf{E}$; that is,

$$\tau = \mathbf{p} \times \mathbf{E}$$

as derived in Problem 6, Chapter 20.

Example 26-9 A thin, circular loop of wire of radius 10 cm carries a current of 5.0 amperes. Calculate:

(a) The magnetic dipole moment of the loop.

(b) The torque on it if it is in a uniform field of strength 0.2 tesla and making an angle of $\pi/4$ with the normal to the loop.

Solution

(a) Since the area A of the loop is

$$A = \pi r^2 = \pi (0.1 \text{ m})^2 = 3.14 \times 10^{-2} \text{ m}^2$$

we find, by use of (26-23), that

$$\mu = niA$$
$$= \mathbf{n}(5.0 \text{ A}) \times (3.14 \times 10^{-2} \text{ m}^2)$$
$$= \mathbf{n}0.16 \text{ A-m}^2$$

where **n** is the unit vector normal to the plane of the loop with its sense defined above.

(b) The magnitude of τ may be obtained, by use of (26-24) and (26-23), as

$$\tau = |\boldsymbol{\mu} \times \mathbf{B}| = \mu B \sin \alpha$$
$$= (0.16 \text{ A-m}^2) \times (0.2 \text{ T}) \times 0.71$$
$$= 2.3 \times 10^{-2} \text{ N-m}$$

since $\sin(\pi/4) = 0.71$.

26-7 The galvanometer

In Chapter 24 we introduced the *galvanometer* as a device that can be used to measure current. The purpose of this section is to discuss briefly the principles underlying its operation.

First, however, let us consider an early form of the galvanometer developed by Oersted. Suppose, in Figure 26-16, a compass is placed in a horizontal position at a distance d above a very long wire oriented along the

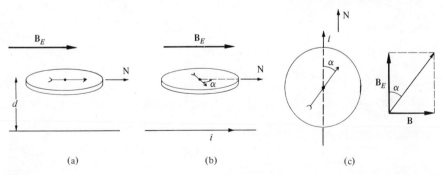

(a)   (b)   (c)

Figure 26-16

north-south direction. In the absence of any current in the wire, the compass needle will line up with the horizontal component $\mathbf{B}_E$ of the earth's magnetic induction. As shown, in this circumstance it will point north.

Suppose now a current i flows north along the wire. Then in addition to the northerly earth-field $\mathbf{B}_E$, there will also be a magnetic induction $\mathbf{B}$ directed eastward and of strength

$$B = \frac{\mu_0 i}{2\pi d}$$

The compass needle will point along the direction of the total field and thus, as shown in Figure 26-16b, it will now point in a direction which makes a certain angle α with respect to north. Reference to the top view of this situation in Figure 26-16c shows that

$$\tan \alpha = \frac{B}{B_E} = \frac{\mu_0 i}{2\pi d B_E}$$

and solving for i yields

$$i = \frac{2\pi d B_E}{\mu_0} \tan \alpha$$

Finally, since all quantities on the right-hand side are known or independently measurable, this instrument effectively measures current.

There are several practical difficulties associated with the above galvanometer and it is not used in practice. Most galvanometers in use today are of the *d'Arsonval* moving-coil or pivoted-coil type. These galvanometers contain a movable coil of wire, in a fixed $\mathbf{B}$-field, and the current to be measured is allowed to pass through the coil. A measurement of the associated torque on the coil thus enables us, by use of (26-22), to measure the desired current.

Figure 26-17 shows the main elements of a galvanometer of the pivoted-coil type. A flat, rectangular coil of wire, which will carry the current to be measured, encircles a fixed, soft-iron core, and is suspended in a way so that it is free to rotate about an axis perpendicular to the plane of the figure. The coil

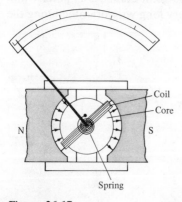

Figure 26-17

itself is maintained in a fixed equilibrium position by a spring, and a pointer is attached so that rotations about the axis are easily observed. As shown, the core is fixed between the cylindrically concave pole tips of a permanent magnet and helps to produce the required radial **B**-field. If a current of strength i flows through the galvanometer, then the coil will experience a torque, which, according to (26-22), is proportional to i. For the direction of current shown, the coil and its attached pointer will move clockwise until this motion is counteracted by the torque produced by the stretched spring. Assuming the latter torque to be proportional to the angle of rotation of the coil, it follows that this angle of rotation is also proportional to the current. In other words, the angle through which the pointer moves is directly proportional to the current, so that once the galvanometer has been calibrated at some known value of the current it may be used to measure the strength of other currents.

The roles played by the soft-iron core and the shapes of the poles of the permanent magnet should be carefully noted. As will be clarified in Chapter 28, the iron-core concentrates the field lines and thereby helps to produce a nearly radial and uniform **B**-field at the position of the coil windings. The angle α between **B** and the normal to the plane of the coil is then always 90° and the factor $\sin \alpha$ ($= 1$) in (26-22) is the same for all currents. Hence the torque on the coil, (and the associated angular displacement of the pointer) is proportional to the current, with the same coefficient of proportionality for all angles.

26-8 The Hall effect

A striking illustration of the fact that a magnetic induction field exerts a force on current is provided by a phenomenon discovered by E. H. Hall in 1879 and now known as the *Hall effect*. This effect is very important in the experimental studies of the electrical properties of matter and can be readily understood at an elementary level.

Consider, in Figure 26-18a, a thin strip of material of cross-sectional area A, height h, and length l. If a voltage source (not shown in the figure) is connected between the end faces of the strip, an electric field **E** directed as shown will appear inside and associated with it will be a certain current i. If further, a voltmeter is connected transversely across the strip, it will give a zero reading since the electric field has no component along the vertical direction. The electrons in the sample will simply move with their characteristic drift velocity $\mathbf{v}_d$ in a direction opposite to the field.

Suppose now, as shown in Figure 26-18b, the existence of a magnetic induction field **B** directed perpendicular to the slab. In response to this field, the electrons will no longer move along the direction of $-\mathbf{E}$, but instead they will move downward in a curved path. As a consequence of this motion, a piling up of negative charge takes place along the bottom of the strip and a compensating positive charge appears at the top. As shown in Figure 26-18c,

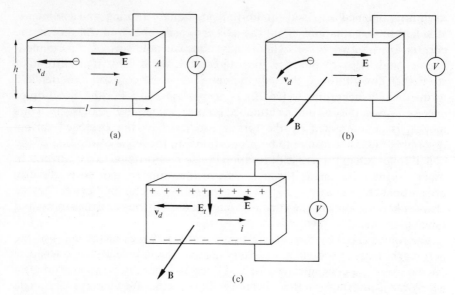

Figure 26-18

this piling up of charge will continue until its associated transverse electric field $\mathbf{E}_t$ cancels out the effect of the magnetic force. In other words, if q is the charge on an electron, then this transverse electric field $\mathbf{E}_t$ must satisfy

$$q\mathbf{E}_t + q\mathbf{v}_d \times \mathbf{B} = 0$$

for only then will this accumulation of charge cease. Taking into account the directions of $\mathbf{E}_t$, $\mathbf{v}_d$, and $\mathbf{B}$ in the figure, we find on canceling the common factor of q that

$$E_t = v_d B \qquad \textbf{(26-25)}$$

The existence of this transverse field $\mathbf{E}_t$ is readily confirmed by virtue of the fact that for $B \neq 0$ the voltmeter in the figure will give a nonzero reading. The value of E_t itself is given by

$$E_t = \frac{V}{h}$$

where V is the reading on the voltmeter and h is the height of the slab. Since B can be measured independently, it follows from (26-25) that the Hall effect can be used to measure the drift velocity of the charge carriers in any material.

Additional information on the properties of the charge carriers can also be ascertained by measuring the strength of the current i flowing in the slab. Let us recall in this connection the relation (23-6)

$$i = nqAv_d$$

which relates the current i in a conductor to the number of charge carriers per unit volume n, to the cross-sectional area A, and to the drift velocity. Solving this relation for v_d and substituting into (26-25), we obtain

$$E_t = \left(\frac{1}{nq}\right)\frac{iB}{A} \tag{26-26}$$

The coefficient $1/nq$ in this formula is called the *Hall coefficient* for the given material. Since i, E_t, B, and A are independently measurable, it follows that the Hall coefficient for a given substance can also be readily ascertained.

By use of (26-26), Hall coefficients for many metals have been measured. The fact that the charge carriers have a negative sign, and are thus presumably electrons, can also be determined, as in Figure 26-18c, by observing the direction of the transverse electric field $\mathbf{E}_t$. If, as shown there, the charge carriers are electrons, then a negative charge accumulates at the bottom of the slab, and $\mathbf{E}_t$ points downward. A most interesting fact that was discovered in this connection was that for some materials the direction of $\mathbf{E}_t$ is opposite to that in the figure. This implies that the charge carriers are positive for these cases! This remained something of a paradox until it was ultimately explained by the quantum theory of matter.

Example 26-10 A strip of copper is placed into a magnetic induction of strength 5.0 T (see Figure 26-18c), and a measurement of E_t yields the value 3.0×10^{-3} V/m.
   (a) Calculate v_d.
   (b) Assuming that $n \cong 8.0 \times 10^{28}$ electrons per cubic meter for copper and that the cross-sectional area $A = 5.0 \times 10^{-6}$ m², calculate the current that flows along the strip.
   (c) What is the Hall coefficient, $1/nq$?

Solution
   (a) Substitution into (26-25) yields

$$v_d = \frac{E_t}{B} = \frac{3.0 \times 10^{-3}\text{ V/m}}{5.0\text{ T}}$$

$$= 6.0 \times 10^{-4}\text{ m/s}$$

   (b) Solving (26-26) for i,

$$i = nqA\frac{E_t}{B} = nqAv_d$$

and substituting the given data we find that

$$i = (8.0 \times 10^{28}/\text{m}^3) \times (1.6 \times 10^{-19}\text{ C}) \times (5.0 \times 10^{-6}\text{ m}^2) \times (6.0 \times 10^{-4}\text{ m/s})$$

$$= 38\text{ A}$$

   (c) Substituting the given values for q and n we obtain

$$\frac{1}{nq} = \frac{1}{(8.0 \times 10^{28}/\text{m}^3) \times (-1.6 \times 10^{-19}\text{ C})} = -7.8 \times 10^{-11}\text{ m}^3/\text{C}$$

26-9   The ampere and the coulomb

In our discussion of Coulomb's law, the unit of charge, the coulomb, was defined as the amount of charge transported in 1 second through a wire in which there flows a current of 1.0 ampere. In order to complete this definition of the coulomb we now give an operational definition of the ampere.

To this end consider, in Figure 26-19, two very long and thin parallel wires, A and C, through which flow the respective currents i_A and i_C. If d is the separation distance between them, the magnetic induction $\mathbf{B}_A$ at wire C due to the current i_A has the magnitude

$$B_A = \frac{\mu_0 i_A}{2\pi d}$$

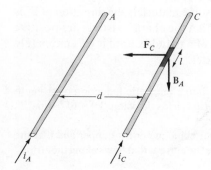

Figure 26-19

and, in accordance with the right-hand rule, its direction is as shown in the figure. Substituting this value for B_A into (26-15), we find that the force $\mathbf{F}_C$ on a length l of the wire has the magnitude

$$F_C = i_C l B_A = \frac{\mu_0 i_A i_C}{2\pi d} l \tag{26-27}$$

and its direction is perpendicular to wire C and directed toward wire A. Correspondingly, the force on wire A due to i_C is equal and opposite to this force, and thus has the same magnitude as in (26-27).

In terms of (26-27) we now define the ampere as follows:

An ampere is the amount of current that, if flowing in each of two very long, parallel wires separated by a distance of 1 meter, will cause each wire to experience a force of precisely 2.0×10^{-7} N for each meter of wire.

Using the values $i_A = i_C = 1$ ampere, $d = 1$ meter, and $l = 1$ meter, we find by substitution into (26-27) that

$$F_C = \frac{\mu_0 i_A i_C l}{2\pi d} = \frac{(4\pi \times 10^{-7}\, \text{Wb/A-m}) \times (1.0\, \text{A}) \times (1.0\, \text{A}) \times (1.0\, \text{m})}{2\pi \times 1.0\, \text{m}}$$
$$= 2.0 \times 10^{-7}\, \text{N}$$

thus confirming the consistency of this definition with the preceding formulas.

In practice, it is not convenient to measure the ampere by use of two very long, straight wires, as above, but rather to make use of two parallel coils of wire separated by a few centimeters. The force of attraction, or repulsion—depending upon whether the currents are parallel or antiparallel—is then measured by a spring balance. At the National Bureau of Standards, precise measurements of currents are carried out in this way by a device known as a *current balance.*

To sum up, then, the ampere can be defined in terms only of measurements of force and of length by means of an apparatus housed at the Bureau and known as a current balance. Once the value of the ampere has been standardized in this way, the coulomb is defined as the flow of charge in 1 second through a wire carrying a current of 1 ampere.

26-10 Summary of important formulas

A particle of charge q and mass m traveling at a velocity $\mathbf{v}$ perpendicular to a uniform $\mathbf{B}$-field will orbit about the field with a constant speed v in a circle of radius R:

$$R = \frac{mv}{qB} \tag{26-3}$$

The plane of the orbit is perpendicular to $\mathbf{B}$.

The force $d\mathbf{F}$ on a current element $i\, d\mathbf{l}$ of length dl in a magnetic induction $\mathbf{B}$ is

$$d\mathbf{F} = i\, d\mathbf{l} \times \mathbf{B} \tag{26-15}$$

The torque τ on a loop of N turns, area A, and carrying a current i is

$$\tau = \mu \times \mathbf{B} \tag{26-24}$$

where $\mathbf{B}$ is the external field and where μ is its magnetic dipole moment

$$\mu = \mathbf{n} N i A \tag{26-23}$$

with $\mathbf{n}$ a unit vector defined in Figure 26-15.

QUESTIONS

1. Define or describe briefly what is meant by the following: (a) cyclotron; (b) mass spectrometer; (c) MeV; and (d) Hall effect.

2. A proton in the upper reaches of the atmosphere is traveling in a direction parallel to the lines of the earth's magnetic induction. Will it experience a force? If it were traveling due east at some instant, what would be the direction of the force on it?

3. Suppose that you look down on a cyclotron and find that the lines of magnetic induction are directed toward you. Would you observe the protons to be traveling clockwise or counterclockwise?

4. Could a cyclotron be used to accelerate electrons? Why is this *not* done in practice? (*Hint:* See (26-13).)

5. Before entering a mass spectrometer, the particles are usually accelerated by a few kilovolts after they emerge from the ion source. Why?

6. A charged particle is at some instant traveling parallel to a long, straight wire in which there flows a current. Does it experience a force?

7. A charged particle is traveling at right angles to a straight wire carrying a current. Does it experience a force? Explain.

8. How could you prove experimentally that the charged particles in a cathode-ray-tube beam are negatively charged?

9. Describe a procedure by means of which it is possible to determine the sign of the charge carriers in any substance in which a current can be caused to flow.

10. Can you think of a way that it might be possible physically for the charge carriers in certain conductors to have a positive charge?

11. It is desired to change the direction of motion of a proton beam by an angle of 60°. Assuming that the protons are *monoenergetic*, that is, they all have the same velocity, describe an experimental setup by means of which this beam deflection can be accomplished.

12. A straight wire carrying a current i lies on the axis of a solenoid which carries a current of the same strength. Is there a force on the solenoid? Is there a torque?

13. A long solenoid that carries a current i is placed into a uniform magnetic induction. Does it experience a force? Will the solenoid exert a torque on the currents that produce the original uniform field?

14. According to (26-22) and (26-24), the torque on a current loop exerted by a uniform induction will vanish if the normal to the loop n is either parallel or antiparallel to B. Describe in physical terms why it is that nevertheless the loop tends to line up preferentially with n parallel to B.

15. To what extent do you think (26-26) is applicable if the magnetic induction is not uniform?

16. Need the force on a closed current loop in a *nonuniform* magnetic induction vanish? Explain.

17. Why is it *not* possible to design a galvanometer as in Figure 26-17 for which the pointer can swing through 360°? (*Hint:* Consider the B-field lines in the figure and the implications of this if the pointer had this degree of freedom.)

18. What do you suppose is the justification for our calling the quantity μ in (26-23) the magnetic dipole moment of a current loop?

19. In our studies of electrostatics we saw that it is not possible for an electric field to penetrate into the

interior of a charge-free conductor. Does the same feature hold for the magnetic induction? Would a cyclotron be able to operate if the **B** lines could not penetrate conducting shells?

20. In the discussion of magnetic forces on wires we actually calculated the force on the *current* in the wire and *not* on the wire itself. Explain why it is that our formulas for the magnetic force can be applied directly to the wire itself.

21. Suppose that in the apparatus in Figure 26-18 a slab of circular cross section instead of the rectangular one were used. Would there still be a Hall effect? Explain why a slab of rectangular cross section is to be preferred.

PROBLEMS

1. A proton in the upper reaches of the atmosphere is traveling at a velocity of 10^6 m/s at right angles to the earth's field, which has a strength of 5.0×10^{-5} tesla at that point.
   (a) What is the radius of the proton's orbit?
   (b) How long does it take the proton to complete one orbit?
   (c) What is its cyclotron frequency?

2. An electron is traveling at right angles to a uniform magnetic induction of strength 3.0×10^{-2} tesla. If its energy is 50 keV (1 keV = 10^{-3} MeV), calculate (a) its velocity; (b) the radius of its orbit; and (c) the cyclotron frequency.

3. A proton and an alpha particle are traveling in parallel directions at the same speed when they enter a region of space where there is a uniform magnetic induction **B**. Assume that they travel at right angles to the field.
   (a) What is the ratio of the radii of their orbits?
   (b) What is the ratio of their cyclotron frequencies?
   (c) What is the ratio of their energies?

4. A 2-MeV proton is traveling in a region of space where there is a uniform electric field of strength 10^5 V/m and a uniform magnetic induction **B** at right angles to it. If the direction of motion of the proton is perpendicular to the direction of both the electric field and the magnetic induction, and the proton is *not* accelerated, calculate the strength and the sense of direction of the **B**-field.

5. A beam containing a mixture of the isotopes ^{6}Li and ^{7}Li enters the region of the uniform magnetic induction B_0 via slit C in the magnetic spectrometer in Figure 26-5. If the ^{6}Li ions are detected at a distance of 10 cm below slit C, where will the ^{7}Li ions appear? What is the ratio of the kinetic energies of these two isotopes?

6. A cyclotron that is used to accelerate protons has a radius of 0.5 meter and a magnetic introduction of 0.75 tesla.
   (a) What is the energy of the emerging protons? Express your answer in joules and in MeV.
   (b) What is the final velocity of the ejected protons?
   (c) What would be the energies of α particles if they were accelerated by this cyclotron?

7. Consider the cyclotron of Problem 6.
   (a) What is the oscillator frequency

ω_{os} for this cyclotron when it accelerates protons?

(b) If the protons pick up 100 keV each time they cross the space between the dees, how many semicircular orbits will the protons complete before they are ejected?

(c) Calculate the time required to accelerate the protons up to their final velocities.

8. A cyclotron with a magnetic induction of 2.0 teslas is used to accelerate protons. (a) What must the frequency (in Hz) of the oscillating field between the dees be? (b) If this cyclotron is used to accelerate deuterons, to what must the frequency of this oscillating field be set?

9. If the cyclotron of Problem 6 were used to accelerate electrons, then, neglecting relativistic effects, what would be the final energies of the electrons? Are we justified in assuming that relativistic effects are negligible?

10. If E_0 is the energy of a cyclotron when it accelerates protons, show that it can accelerate ions, of atomic mass A and with Z units of charge, to the energy

$$E = \left(\frac{Z^2}{A}\right) E_0$$

11. If a magnetic monopole of strength ϵ existed, then the magnetic induction associated with it would be

$$\mathbf{B} = \left(\frac{\epsilon}{r^3}\right)\mathbf{r}$$

where $\mathbf{r}$ is the position in space measured from the monopole.

(a) Write down the equation of motion for a particle of charge q and mass m moving in the field of a monopole.

(b) Show that the kinetic energy of the particle is a constant of the motion.

*12. For the physical system in Problem 11 show that:

(a) The angular momentum $m\mathbf{r} \times \mathbf{v}$ (relative to the monopole) of the particle is *not* constant in general.

(b) The quantity $d(m\mathbf{r} \cdot \mathbf{v})/dt$ is constant, and determine its value in terms of the initial velocity $\mathbf{v}_0$.

13. It has been estimated that at the surface of a *neutron star* the B-field may be as high as 10^9 teslas. For a 10-MeV proton moving in such a field find (a) its cyclotron frequency and (b) the radius of its orbit.

14. A rectangular coil of wire carries a current i and has a width a. If the lower end of the coil is held between the poles of an electromagnet of strength B and directed as shown in Figure 26-20, calculate the magnitude of the force **F**, beyond that of gravity, required to support the coil. (*Note*: By suspending a coil from the arm of an analytical balance, the principle illustrated here has been used at the National Bureau of Standards for making precise measurements of magnetic fields.)

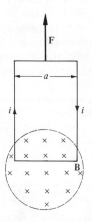

Figure 26-20

15. By use of the method of Example 26-7 show that the force on a portion of a wire carrying a current in a uniform magnetic induction is the same for all wires having the same endpoints. That is, show that the force on the wires 1 and 2 in Figure 26-21 are the same if the same currents flow in each.

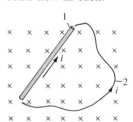

Figure 26-21

16. A current of 5.0 amperes flows in a square loop of side 10 cm. Calculate the total force on two of its adjacent sides produced by an external magnetic induction perpendicular to the plane of the loop and of strength 0.1 tesla.

17. A current of strength i flows in a coil in the shape of a regular pentagon of side a. Assume that there is a uniform induction of strength B perpendicular to the plane of the coil. **(a)** Calculate the force produced by the external field on each segment. **(b)** By making use of the results of (a), show explicitly that the total force on the loop is zero.

*18. Consider two parallel wires each of length l, carrying the same currents i and separated by a distance a, as in Figure 26-22.

   (a) Show that the magnetic induction $\mathbf{B}$ at the point (x, a) of the upper wire due to the current in the lower one has the magnitude

$$B = \frac{\mu_0 i}{4\pi} \frac{1}{a} \left\{ \frac{(l/2) - x}{[(x - l/2)^2 + a^2]^{1/2}} + \frac{(l/2) + x}{[(x + l/2)^2 + a^2]^{1/2}} \right\}$$

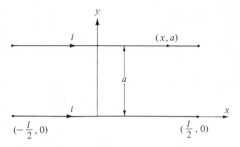

Figure 26-22

What is the direction of $\mathbf{B}$?

   (b) Calculate the force $d\mathbf{F}$ on the element of length dx located at the point (x, a) of the upper wire.

   (c) Show by integration that the total force $\mathbf{F}$ on the upper wire is downward and has the strength

$$F = \frac{\mu_0 i^2}{2\pi} \left(\frac{1}{a} \right) [(l^2 + a^2)^{1/2} - a]$$

19. Show that for appropriate values of the various parameters, the result in (c) of Problem 18 is consistent with (26-27).

20. In Figure 26-11, if $a = 3.0$ cm, $b = 5.0$ cm, $c = 3.0$ cm, $i = 5.0$ amperes, and $I = 2.0$ amperes, calculate **(a)** the force on segment AC due to the current in the long wire and **(b)** the force on segment AE due to the current in the long wire.

21. A rectangular loop of wire of sides 10 cm and 30 cm carries a current of 15 amperes. Make use of the result of Problem 18 to calculate the mutual forces of repulsion between the two pairs of opposite wires.

*22. Consider the two wires of length l and L at right angles to each other in Figure 26-23. Assume that currents i and I are directed as shown.

   (a) Show that the magnetic induction $\mathbf{B}$ at the position of the element of length dy located at

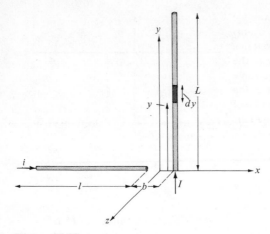

Figure 26-23

the point $(0, y, 0)$ is

$$\mathbf{B} = \mathbf{k}\,\frac{\mu_0 i}{4\pi y}\left\{\frac{-b}{(b^2 + y^2)^{1/2}}\right.$$

$$\left.+ \frac{b + l}{[(b + l)^2 + y^2]^{1/2}}\right\}$$

where $\mathbf{k}$ is a unit vector along the z-axis.

(b) Show that the magnetic force $d\mathbf{F}$ on the element dy is

$$d\mathbf{F} = \mathbf{i}\,\frac{\mu_0 i I}{4\pi y}\,dy\left\{\frac{-b}{(b^2 + y^2)^{1/2}}\right.$$

$$\left.+ \frac{b + l}{[(b + l)^2 + y^2]^{1/2}}\right\}$$

where $\mathbf{i}$ is a unit vector along the x-axis.

(c) Calculate the total force $\mathbf{F}$ on the wire of length L.

23. Consider again the system in Figure 26-11. This time calculate the force on the long, straight wire due to the magnetic induction produced by the rectangular loop. Should your answer be the same as that given in Example 26-8?

24. Two current elements $i_1\,d\mathbf{l}_1$ and $i_2\,d\mathbf{l}_2$ are positioned relative to each other as shown in Figure 26-24.

(a) Calculate the force that the

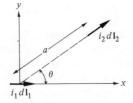

Figure 26-24

element $i_1\,d\mathbf{l}_1$ exerts on the other.

(b) Calculate the force which the element $i_2\,d\mathbf{l}_2$ exerts on the other element.

(c) Explain why, even though your results to (a) and (b) are not equal and opposite, there is no essential contradiction with Newton's law of action and reaction.

*25. In a certain coordinate system a current I flows along an infinitely long wire that lies along the x-axis. Show that the magnetic force on a second wire of length l and carrying a current i whose endpoints are at the points $(0, 0, a)$ and $(0, l, a)$ is

$$\mathbf{F} = -\mathbf{i}\,\frac{\mu_0 i I}{4\pi}\ln\left[1 + \frac{l^2}{a^2}\right]$$

26. Consider the same situation as in Problem 25, but this time suppose the endpoints of the shorter segment are at the points $(0, -l/2, a)$ and $(0, l/2, a)$. (a) Show that this time there is no net force on the shorter segment. (b) Calculate the torque about the point $(0, 0, a)$ on the shorter segment.

27. Calculate the magnetic dipole moments associated with each of the following planar loops (assume that in each case the current is 2.0 amperes and that there are 10 turns in each loop):
    (a) A circular loop of radius 10 cm.
    (b) A rectangular loop of sides 2 cm and 10 cm.
    (c) An elliptical-shaped loop of semimajor axis 10 cm and semiminor axis 5 cm.

28. By calculating the work required to rotate a current loop of dipole moment μ in a magnetic induction $\mathbf{B}$, show that the energy U associated with it is

$$U = -\mu \cdot \mathbf{B}$$

(Hint: The work dW required to rotate a dipole by a small angle $d\alpha$ is $\tau \, d\alpha$, where τ is the torque that must be applied.)

29. A circular coil of wire of radius 10 cm and 150 turns carries a current of 10^{-2} ampere. What is the maximum torque that can be exerted on this coil by a uniform magnetic induction of strength 0.2 tesla?

30. A rectangular loop of sides a and b is suspended so that it is free to rotate about the horizontal axis AB (see Figure 26-25). If it has a mass

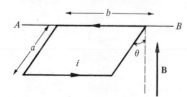

Figure 26-25

m and if the current around it is i, calculate the angle θ at which it will be in equilibrium in the presence of a uniform vertical magnetic induction $\mathbf{B}$.

*31. Consider, in Figure 26-26, a planar loop of wire carrying a current i in the presence of a uniform magnetic induction $\mathbf{B}$.

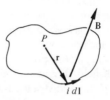

Figure 26-26

(a) Show that the torque τ about a point P inside the loop may be expressed, by use of (26-15), in the form

$$\tau = i \int \mathbf{r} \times (d\mathbf{l} \times \mathbf{B})$$

where $\mathbf{r}$ is the vector from the point P to the element $i \, d\mathbf{l}$.

(b) By making use of the fact that the integral is to be carried out around a closed loop show that an equivalent formula for τ is

$$\tau = \frac{1}{2} i \left[\oint \mathbf{r} \times d\mathbf{l} \right] \times \mathbf{B}$$

(c) Finally, show that the integral

$$\frac{1}{2} \oint \mathbf{r} \times d\mathbf{l}$$

is the vectorial area of the loop, and thus establish the general validity of (26-24).

*32. Show that (26-24) is valid for a loop of arbitrary shape by replacing the given current loop by a collection of contiguous, very thin rectangular loops each carrying the same current i.

*33. Consider a wire of fixed length l

and carrying a current *i*. This wire can be formed into various loops, such as a square of side $l/4$, N circular loops each of radius $l/2\pi N$, and so on. Show that the maximum torque on any of these in a uniform magnetic induction **B** is achieved when the wire forms a circle of radius $l/2\pi$, and calculate the torque in this case.

34. A circular loop of wire of mass m and radius a carries a current i and is free to rotate about a horizontal diameter AC in the presence of a uniform **B**-field directed vertically upward. If it is distributed by a slight amount from its equilibrium position, show that it will oscillate about AC with simple harmonic motion of period

$$P = \left[\frac{2\pi m}{iB}\right]^{1/2}$$

(*Note*: The moment of inertia I of this loop about a diameter is $ma^2/2$.)

35. In an experiment to measure the Hall effect in sodium, suppose that a magnetic induction of strength 0.8 tesla is used and a current of 10 amperes is measured. Assuming that the cross-sectional area of the metallic strip is $2\,\text{cm}^2$ and $E_t = 10^{-5}\,\text{V/m}$, calculate:

(a) The drift velocity of the electrons.

(b) The value for the Hall coefficient.

(c) The number of charge carriers per unit volume and compare this with the number of sodium atoms per unit volume. The atomic mass of sodium is 23 and its density $\cong 10^3\,\text{kg/m}^3$.

*36. Suppose that the pivoted-coil galvanometer in Figure 26-17 has a radial **B**-field of strength 0.3 tesla, and the coil itself has 200 turns and an area of $3.0\,\text{cm}^2$. Calculate the spring constant ($\equiv$ torque/angular displacement), assuming that a current of 1.0 mA produces an angular deflection of 15°.

27 Faraday's law of magnetic induction

What led me ... to the special theory of relativity was the conviction that the electromotive force acting on a body in motion in a magnetic field was nothing else but an electric field.

ALBERT EINSTEIN (1952)

Some day you will tax it.

MICHAEL FARADAY (On being asked by P. M. Gladstone what was the use of electricity)

27-1 Introduction

If a particle of charge q is at rest relative to an observer, then at a displacement $\mathbf{r}$ from the particle he measures the electric field

$$\mathbf{E} = \frac{q}{4\pi\epsilon_0 r^3}\mathbf{r} \tag{27-1}$$

Consider now this same particle, but this time as seen by an observer relative to whom it travels at the velocity $\mathbf{v}$. In addition to the electric field $\mathbf{E}$ this observer will also detect a $\mathbf{B}$-field given by (see Problem 9, Chapter 25)

$$\mathbf{B} = \frac{\mu_0 q}{4\pi r^3}\mathbf{v} \times \mathbf{r}$$

and which, by use of (27-1), may be cast into the equivalent form

$$\mathbf{B} = \mu_0\epsilon_0\mathbf{v} \times \mathbf{E} \qquad (27\text{-}2)$$

Bearing in mind the superposition principle and the fact that $\mathbf{v}$ is the velocity of the particle, we see that associated with an electric field source in motion there is, in general, both a magnetic as well as an electric field.

In 1839 Michael Faraday (1791–1867) first reported on his observations of a related effect, namely, that an *electric field is also, in general, associated with a magnetic source in motion*. That is, even though no electric charge may be present, an electric field comes into existence whenever there is relative motion between a magnetic source and the observer. Moreover, since a $\mathbf{B}$-field, which varies in time because the current which produces it does, is physically equivalent to the $\mathbf{B}$-field produced by a time-independent magnetic source in motion it follows that an electric field is also associated, in general, with time-dependent magnetic sources. The precise relation between this induced electric field and the properties of the causal, time-varying magnetic source is described quantitatively by a relation known as *Faraday's law*. The purpose of this chapter is to study this important physical law.

27-2   Experimental verification

In this section we describe a number of simple experiments which confirm the fact that an electric field is generally associated with any time-varying $\mathbf{B}$-field.

Consider first a loop of wire containing a galvanometer and suppose, as in Figure 27-1a, that the north pole of a permanent magnet approaches the loop. Under this circumstance we find that a current, which is directed as shown, flows in the loop. Similarly, if the north pole recedes from the loop, then as in Figure 27-1b the current flows in the opposite sense. Figures 27-1c and 27-1d show the analogous results obtained by use of the south pole of a

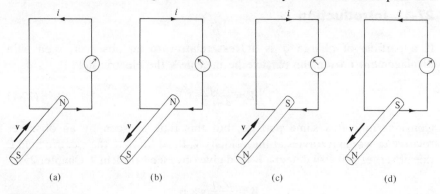

(a)   (b)   (c)   (d)

Figure 27-1

permanent magnet. Note that in each case the very existence of current in the loop implies that at least inside the wire there must exist an electric field. Further, experiment shows that if the relative motion between the magnet and the loop ceases, then so does the current. This implies that an electric field is associated only with the *motion* of a magnetic source and not with the source itself.

In connection with experiments such as those in Figure 27-1 there are three features worthy of special emphasis. First, the fact that the magnet is in motion while the loop is stationary in each case is not relevant. Were we to view the situation, say, in Figure 27-1a, from the point of view of an observer moving toward the loop at the velocity **v** (according to whom the magnet is stationary while the loop moves toward the magnet at the velocity $-\mathbf{v}$), we would find the current flow in the loop to be unaltered. A second feature is that the use of a permanent magnet is in no way restrictive. If in each of the cases in Figure 27-1 the magnet were replaced by the appropriate end of a coil of wire carrying a current, then precisely the same results would be obtained. In other words, when in motion, a wire carrying a current also produces an electric field. The third feature is that a force must be exerted on the magnets in Figure 27-1 in order to sustain the motion. Reference to the figure shows that the current induced in the loops produces, in each case, a **B**-field, which exerts a force that tends to oppose the motion of the magnet. Thus to sustain the given motion an external agent must exert a force and carry out positive work on the moving magnet.

Having established that an electric field is associated with the motion of magnetic sources, we now describe briefly how to verify the fact that an electric field is also produced by a stationary but *time-dependent* magnetic source. To this end consider, in Figure 27-2, two coplanar circular loops A and B. Suppose that B contains a galvanometer and that a current may be generated in A by closing a switch S. Experiment shows that immediately after S is closed there is a momentary flow of current around B and that this flow ceases shortly afterward. By contrast, the current around loop A becomes steady at the value $\mathscr{E}/R$. Similarly, when the switch S is subsequently opened, so that the current in A drops to zero, there is again a momentary flicker of current around B. This time it flows in the opposite

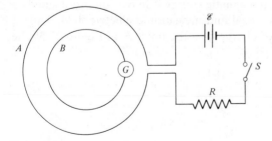

Figure 27-2

direction. It follows that since there is a change of current in *A* whenever *S* is opened or closed, an electric field must be associated with a *change* in current. But according to the law of Biot-Savart, a time-varying current will generate a time-varying **B**-field. Hence we conclude that an electric field is associated not only with electric charge and with the motion of magnetic sources, but also with a stationary but time-varying magnetic source.

27-3   The electric field produced by moving magnetic sources

Before proceeding with a study of Faraday's law in its most general form, let us consider first the special case of the electric field associated with moving magnetic sources. It will be assumed throughout this discussion that all velocities are small compared to the speed of light *c* ($\cong 3 \times 10^8$ m/s). This means that we shall concern ourselves only with those cases in which the relative velocity **v** between the observer and the magnetic source is small compared to *c*. For situations involving speeds $v \sim c$, the theory of relativity is required.

Consider a stationary magnetic source—the north pole of a magnetized rod in Figure 27-3a, for example—and let **B** represent the magnetic induction at some fixed point *P*. Figure 27-3b shows this same system, but this time as seen by an observer with respect to whom the magnetized rod has the velocity v_s perpendicular to the rod. If this observer carries out a sequence of experiments involving the measurements of forces on charged particles, he would find at point *P* the same magnetic induction **B** as above. In addition, however, he would also find that at point *P* there exists an electric field **E**, given by

$$\mathbf{E} = -\mathbf{v}_s \times \mathbf{B} \qquad (27\text{-}3)$$

Very generally, then: *If the source of a magnetic induction field **B** is moving at a velocity $\mathbf{v}_s$ relative to an observer, then this observer sees the original **B**-field as well as the electric field* $(-\mathbf{v}_s \times \mathbf{B})$. Note that two observers in relative motion will, in general, observe *different* electric fields.

It is worth emphasizing that this electric field **E** in (27-3) as seen by an observer with respect to whom a magnetic source is in motion is not simply a mathematical abstraction; it is a real and measurable electric field. Thus if the observer places a particle of charge *q* at the point *P* in Figure 27-3b, he

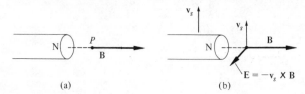

(a)   (b)

Figure 27-3

measures a force $\mathbf{F} = q\mathbf{E}$ on the particle. This situation is analogous to that in Newtonian mechanics, where in making use of accelerated coordinate systems it is necessary to introduce "fictitious" forces. As we saw there, these forces are very real and directly observable. However, electric forces arise even for magnetic sources in unaccelerated, *uniform* motion. Because of the fact that Newton's laws are invariant under Galilean transformations, no fictitious forces arise for such motions. By contrast, the electric field $\mathbf{E}$ in (27-3) comes into existence with all types of motion of a magnetic source, including uniform motion.

Now even though the validity of (27-3) rests mainly on experimental grounds, the existence of this electric field can also be justified on the basis of logical consistency. To see this, consider in Figure 27-4a a particle of charge q at rest at a point where there is a magnetic induction $\mathbf{B}$ produced, say, by a loop of wire carrying a current i. Since the particle is at rest and $\mathbf{E} = 0$ there is no force on the particle. Hence its acceleration vanishes. Let us now view this same physical situation, but this time as shown in Figure 27-4b as seen by an observer with respect to whom the loop and the particle have an upward velocity $\mathbf{v}_s$. According to this observer, the particle has a velocity $\mathbf{v}_s$ and is in a magnetic induction $\mathbf{B}$. Hence it experiences a magnetic force $\mathbf{F}_m$, given by

$$\mathbf{F}_m = q\,\mathbf{v}_s \times \mathbf{B}$$

and if there were no other force acting on it there would be a contradiction, since according to Newton's law the particle would accelerate. This seeming paradox is resolved by noting that the particle is also subject to the electric field $\mathbf{E}$ in (27-3). Thus it experiences also an electric force $\mathbf{F}_e$

$$\mathbf{F}_e = q\mathbf{E} = -q\,\mathbf{v}_s \times \mathbf{B}$$

and it is the sum $(\mathbf{F}_e + \mathbf{F}_m)$ that acts on the particle. On adding these two formulas together, we find that because of (27-3) the total force acting $(\mathbf{F}_m + \mathbf{F}_e)$ is indeed zero and, consistent with our physical expectations, the particle does not accelerate. Hence we see that if a magnetic source is in

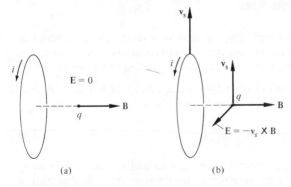

(a)   (b)

Figure 27-4

motion relative to an observer, an electric field of precisely the form in (27-3) must come into existence in order that inconsistencies such as this do not arise.

Example 27-1 Consider the uniform magnetic induction $\mathbf{B}_0$ of strength 0.02 tesla in a very long ideal solenoid. What is the electric field seen by an observer who travels at a velocity of 10 m/s:
(a) Along the axis of the solenoid?
(b) At right angles to the axis of the solenoid?

Solution
(a) Since the magnetic induction in an ideal solenoid is directed along its axis and since the cross product of two parallel vectors vanishes, it follows from (27-3) that, for this case,

$$\mathbf{E} = -\mathbf{v}_s \times \mathbf{B} = 0$$

Thus, as shown in Figure 27-5a, there are no electric field lines according to this observer.

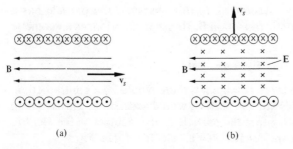

(a)   (b)

Figure 27-5

(b) Assuming that the observed velocity $\mathbf{v}_s$ of the solenoid is upward (see Figure 27-5b), it follows from (27-3) that the electric field $\mathbf{E}$ is perpendicular to and directed down into the plane of the diagram. The strength of this electric field is

$$E = |\mathbf{E}| = |-\mathbf{v}_s \times \mathbf{B}| = (10 \text{ m/s}) \times (0.02 \text{ T})$$
$$= 0.2 \text{ V/m}$$

Example 27-2 Consider an infinitely long, straight wire that carries a current i as seen by an observer, with respect to whom the wire travels at a velocity $\mathbf{v}_s$ along the direction of the current. Calculate the electric field according to this observer.

Solution The situation is shown in Figure 27-6. According to (25-6), the strength of the magnetic induction at a distance r from the wire is

$$B = \frac{\mu_0 i}{2\pi r} \tag{27-4}$$

and the **B**-field lines are circles, concentric with the wire and with a sense given by the right-hand rule. According to the definition of the cross product, the direction of the vector $(-\mathbf{v}_s \times \mathbf{B})$ is everywhere radial. Hence, by use of (27-4) and (27-3), we find

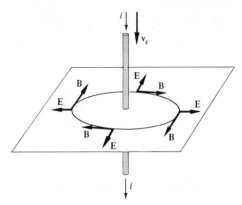

Figure 27-6

that

$$\mathbf{E} = -\mathbf{v}_s \times \mathbf{B} = \mathbf{e}\,\frac{\mu_0 v_s i}{2\pi r} \tag{27-5}$$

with **e** a unit vector in the radial direction.

According to this observer, in addition to the **B**-field there is also the radial electric field given in (27-5). Since the electric field associated with an infinite line charge, of charge per unit length λ, is $\mathbf{e}\lambda/2\pi\epsilon_0 r$ (see (20-6)), it follows on comparison with (27-5) that, in addition to the current i, this observer also sees on the wire a charge per unit length

$$\lambda = \mu_0\epsilon_0 v_s i \tag{27-6}$$

Note that this charge density λ and its associated electric field are very real and directly observable. Assuming the values $v_s = 10^3$ m/s and $i = 10$ amperes, we find by use of (27-6) that

$$\lambda = \mu_0\epsilon_0 v_s i$$
$$= (4\pi \times 10^{-7}\,\text{T-m/A}) \times (8.85 \times 10^{-12}\,\text{F/m}) \times (10^3\,\text{m/s}) \times (10\,\text{A})$$
$$= 1.1 \times 10^{-13}\,\text{C/m}$$

27-4 The motion of a conductor in a magnetic field

One of the very important applications of the electric field formula in (27-3) is to the problem of a conductor in motion relative to a magnetic source. The purpose of this section is to describe certain physical effects associated with this motion.

First, however, let us consider the case of a thin, conducting rod of length l traveling in a uniform **B**-field at a velocity **v** perpendicular to its axis. See Figure 27-7a. From the viewpoint of an observer at rest relative to the rod, the magnetic sources are traveling at the velocity $\mathbf{v}_s = -\mathbf{v}$. Hence, inside the

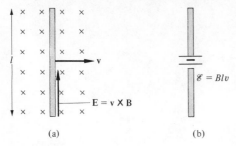

Figure 27-7

conductor there is an electric field, given in accordance with (27-3) by

$$\mathbf{E} = \mathbf{v} \times \mathbf{B} \tag{27-7}$$

(Note the sign!) As shown in Figure 27-7a, for the given directions of **B** and **v** this electric field is directed vertically upward along the rod and has a strength vB. In effect, then, there exists a uniform electric field of strength Bv inside the wire. But this electric field cannot be sustained! For the electrons inside the rod will respond to the field and rearrange themselves in a way so as to nullify its effects. Thus positive and negative charge will accumulate at the upper and lower end of the rod, respectively, until eventually the total electric field inside the rod vanishes. In effect then, as shown in Figure 27-7b, the ends of the rod become the two terminals of a seat of electromotive force with emf $\mathscr{E}$ given by

$$\mathscr{E} = Blv \tag{27-8}$$

and with the upper end of the rod the positive terminal. Note that since the rod is not connected to a closed conducting path, no current flows.

Alternatively, we may think of this redistribution of charge inside the rod in the following way. The free electrons in its interior share in the rod's motion and thus travel at the velocity **v** directed to the right in Figure 27-7a. Because of their negative charge $(-q)$, the force on these electrons $(-q\mathbf{v} \times \mathbf{B})$ is directed downward along the rod. Hence, just as above, negative charge accumulates at the lower end of the rod and a compensating positive charge appears at the other end.

Let us now reexamine this situation, but this time suppose as shown in Figure 27-8 that the moving conductor slides on two metallic rails and is thereby part of a closed conducting path. For simplicity assume that the moving conductor has negligible resistance while the rest of the current path has a constant resistance R. Since only the rod is in motion relative to the magnetic sources, it alone experiences the electric field in (27-7) and the emf in (27-8) is developed only across it. In effect, then, we have the situation shown in Figure 27-8b, of a battery of emf $\mathscr{E} = Blv$ connected across a resistor R. According to Ohm's law, the current i is

$$i = \frac{\mathscr{E}}{R} = \frac{Blv}{R} \tag{27-9}$$

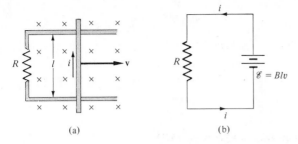

(a)   (b)

Figure 27-8

and the direction of this current is upward along the moving conductor in Figure 27-8a. According to (26-15), the external **B**-field exerts on the current in the moving rod a force **F**, which is directed to the left and has the magnitude

$$|\mathbf{F}| = ilB = \frac{l^2 B^2 v}{R} \tag{27-10}$$

where the second equality follows by use of (27-9). To keep the rod moving to the right at the given velocity **v** it is necessary to exert an external force of strength given in (27-10) but directed to the right.

Figure 27-9 illustrates the fact that current need not flow in a conductor moving through a **B**-field under all circumstances. If a rectangular conducting loop travels at some velocity **v** through a uniform **B**-field, directed as shown, then no current flows. This time an emf of strength Blv directed upward is developed in each vertical segment. Hence this situation is physically the same as that shown in Figure 27-9b, in which two identical batteries are connected in a way so that no current flows. By contrast, if the loop were to travel through an inhomogeneous **B**-field, say that in Figure 27-10a, with only one of the vertical wire segments in the **B**-field, then current would again flow. Figure 27-10b shows the equivalent circuit diagram for this case.

Figure 27-11 shows the general case of a thin loop of wire traveling at a velocity **v** through a **B**-field, with **v** and **B** at an arbitrary angle. If **B** is the magnetic induction at the position of an element dl of the loop, then according to (27-7) the electric field **E** in this element is $\mathbf{v} \times \mathbf{B}$. Hence the emf $d\mathscr{E}$

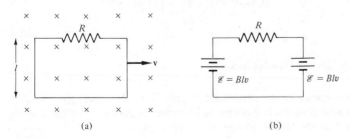

(a)   (b)

Figure 27-9

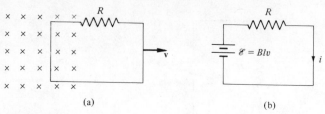

Figure 27-10

developed between its endpoints is (Section 23-2)

$$d\mathcal{E} = \mathbf{E} \cdot d\mathbf{l} = (\mathbf{v} \times \mathbf{B}) \cdot d\mathbf{l} \qquad (27\text{-}11)$$

and the emf $\mathcal{E}$ developed around the entire loop is obtained by adding together the contributions from all elements of the loop. The result is

$$\mathcal{E} = \oint (\mathbf{v} \times \mathbf{B}) \cdot d\mathbf{l} \qquad (27\text{-}12)$$

where the line integral is to be carried out over the entire loop. In general, different parts of the loop may travel at different speeds, and the magnetic induction may not be uniform around the loop. All such variations must be allowed for in evaluating the line integral in (27-12).

Example 27-3 If, in Figure 27-8, $B = 0.1$ tesla, $l = 10$ cm, $v = 15$ m/s, and $R = 10\,\Omega$, calculate:

(a) The emf developed between the ends of the rod.
(b) The current.
(c) The force required to keep the rod moving.

Solution
(a) Substitution into (27-8) yields

$$\mathcal{E} = Blv = (0.1 \text{ T}) \times (0.1 \text{ m}) \times (15 \text{ m/s}) = 0.15 \text{ V}$$

(b) The current is given by Ohm's law:

$$i = \frac{\mathcal{E}}{R} = \frac{0.15 \text{ V}}{10\,\Omega} = 1.5 \times 10^{-2} \text{ A}$$

(c) The force required to keep the rod moving is, according to (27-10),

$$F = \frac{l^2 B^2 v}{R} = \frac{(0.1 \text{ m})^2 \times (0.1 \text{ T})^2 \times (15 \text{ m/s})}{10\,\Omega}$$
$$= 1.5 \times 10^{-4} \text{ N}$$

The direction of this force is parallel to $\mathbf{v}$.

Example 27-4 Consider again the situation in Figure 27-8. Show that the rate at which the external agent must expend energy to keep the conductor traveling at the velocity $\mathbf{v}$ is the same as the power dissipated in the resistor.

Solution   The rate dW/dt at which the external agent carries out work is

$$\frac{dW}{dt} = \mathbf{F} \cdot \mathbf{v}$$

where $\mathbf{F}$ is the force exerted. Substituting the value for $|\mathbf{F}|$ in (27-10) and noting that $\mathbf{F}$ and $\mathbf{v}$ are parallel, we find that

$$\frac{dW}{dt} = \frac{l^2 B^2 v^2}{R} \qquad (27\text{-}13)$$

On the other hand, according to (27-9), the current i in the resistor is

$$i = \frac{Blv}{R}$$

and substitution into (27-13) yields

$$\frac{dW}{dt} = Ri^2$$

Finally, since Ri^2 is the power dissipated in the resistor, the desired result follows.

Example 27-5   A wire of length l is forced to travel at a velocity $\mathbf{v}$ along the direction of the current in a long, straight wire (see Figure 27-12). Calculate the emf developed between the ends of the moving wire.

Solution   Let $d\mathscr{E}$ represent the emf developed across an element of length dx of the moving wire at a distance x from the current. Making use of (27-11) and the fact that the magnetic induction $\mathbf{B}$ at this element is directed vertically down into the plane of the figure and has the strength $B = \mu_0 i / 2\pi x$, we find that

$$d\mathscr{E} = -v \frac{\mu_0 i}{2\pi x} \, dx \qquad (27\text{-}14)$$

The minus sign reflects the fact that the electric field in the moving conductor is directed toward the current. Integrating over all values from $x = a$ to $x = (a + l)$ we

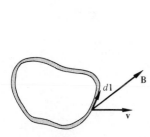

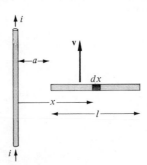

Figure 27-11   　　　　　　　　　　　　　　　　*Figure 27-12*

find for the emf across the wire

$$\mathscr{E} = \int d\mathscr{E} = \int_a^{a+l} \left(-\frac{v\mu_0 i \, dx}{2\pi x} \right)$$

$$= -\frac{v\mu_0 i}{2\pi} \int_a^{a+l} \frac{dx}{x} = -\frac{v\mu_0 i}{2\pi} \ln x \Big|_a^{a+l}$$

$$= -\frac{v\mu_0 i}{2\pi} \ln \left(1 + \frac{l}{a} \right)$$

Again, the minus sign reflects the fact that the end of the moving rod closest to the current is at the higher potential.

Example 27-6   A rectangular loop of wire of sides a and l is situated near a long, straight wire in which there flows a current i. Suppose, as shown in Figure 27-13, that the loop travels to the right with a velocity $\mathbf{v}$. Calculate the emf developed around the loop.

Solution   Since the magnetic induction $\mathbf{B}$ on the left segment of the loop has the magnitude $\mu_0 i / 2\pi x$ and is directed down into the plane of the figure, it follows (see (27-8)) that the emf $\mathscr{E}_L$ on this segment is directed upward and has the magnitude

$$\mathscr{E}_L = \frac{v l \mu_0 i}{2\pi x}$$

Similarly, the emf $\mathscr{E}_R$ on the right-hand segment is also directed upward and has the value

$$\mathscr{E}_R = \frac{v l \mu_0 i}{2\pi (x + a)}$$

Since the induced electric field $\mathbf{v} \times \mathbf{B}$ has no components along the upper and the lower segments of the loop, no emf is developed along these. Hence the total emf $\mathscr{E}$ around the loop is clockwise and has the value

$$\mathscr{E} = \mathscr{E}_L - \mathscr{E}_R = \frac{\mu_0 i l v}{2\pi} \left(\frac{1}{x} - \frac{1}{x + a} \right)$$

Thus we see that if a closed loop travels through an *inhomogeneous* $\mathbf{B}$-field an emf is developed around it. This is to be contrasted with the fact, established in the problems and exemplified in Figure 27-9, that no emf is developed around a rigid loop moving through a *uniform* $\mathbf{B}$-field.

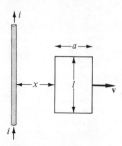

Figure 27-13

27-5 Lenz's law for moving conductors

In Example 27-4 it was established that the current through the moving conductor in Figure 27-8 is directed so that the external force that causes the conductor to move carries out positive work. In other words, the current is directed so that it tends to oppose the relative motion of the conductor and the magnetic sources. In 1834, Heinrich Lenz (1804–1865) showed that this property is very generally true. In his own words:

> If a constant current flows in the primary circuit A and if by the motion of A, or of the secondary circuit B, a current is induced in B, the direction of this induced current will be such that by its electromagnetic action on A it tends to oppose the relative motion of the two circuits.

The purpose of this section is to derive this law.

Suppose that the rigid loop of wire in Figure 27-11 has a resistance R and is forced to travel at a velocity $\mathbf{v}$ through a magnetic induction $\mathbf{B}$ under the action of an external force $\mathbf{F}$. In order to establish the validity of Lenz's law, it is necessary to prove that this force carries out positive work in this process. We shall now show that the rate $\mathbf{F} \cdot \mathbf{v}$ at which the external force carries out work is invariably positive. Lenz's law then follows as a direct consequence.

The emf $\mathcal{E}$ developed around the moving loop in Figure 27-11 is given by (27-12). Making use of the vector identity

$$(\mathbf{a} \times \mathbf{b}) \cdot \mathbf{c} = \mathbf{a} \cdot (\mathbf{b} \times \mathbf{c})$$

which is valid for any three vectors $\mathbf{a}$, $\mathbf{b}$, and $\mathbf{c}$, this formula for $\mathcal{E}$ may be expressed equivalently as

$$\mathcal{E} = \oint \mathbf{v} \cdot (\mathbf{B} \times d\mathbf{l}) = \mathbf{v} \cdot \oint \mathbf{B} \times d\mathbf{l} \qquad (27\text{-}15)$$

where the second equality follows since the loop is presumed to be rigid and hence each of its elements $d\mathbf{l}$ has the same velocity.

Now the force $\mathbf{F}_B$ *on* the loop due to the external $\mathbf{B}$-field is

$$\mathbf{F}_B = \oint i\, d\mathbf{l} \times \mathbf{B} = -i \oint \mathbf{B} \times d\mathbf{l} \qquad (27\text{-}16)$$

where the second equality follows from the vector identity in (10-3). On the other hand, the force $\mathbf{F}$ that the external agent must exert to keep the loop moving at the velocity $\mathbf{v}$ is equal and opposite to $\mathbf{F}_B$. Hence the rate $\mathbf{v} \cdot \mathbf{F}$ at which the external agent carries out work is

$$\mathbf{v} \cdot \mathbf{F} = -\mathbf{v} \cdot \mathbf{F}_B = -\mathbf{v} \cdot \left[-i \oint \mathbf{B} \times d\mathbf{l} \right]$$

$$= i\mathbf{v} \cdot \oint \mathbf{B} \times d\mathbf{l} = i\mathcal{E} \qquad (27\text{-}17)$$

$$= Ri^2 > 0$$

where the second equality follows from (27-16), the fourth from (27-15), and the fifth by use of Ohm's law. Since the quantity Ri^2 is inherently positive, it follows that so is the rate $\mathbf{v} \cdot \mathbf{F}$ at which work is carried out by the external agent. The validity of Lenz's law is thereby established.

It is interesting to note that, according to (27-17), the rate at which work is carried out by the external agent is the same as that at which heat is dissipated in the resistor. Thus, consistent with the ideas of energy conservation, the work carried out by the external agent is converted entirely to another form of energy.

Example 27-7   A small loop of wire has a resistance of $10\,\Omega$ and a force **F** of 0.1 newton is required to pull it at a velocity of 2.0 m/s through an inhomogeneous magnetic induction. Assuming that the direction of **F** is parallel to that of **v**:
  (a) What is the rate at which the external agent carries out work?
  (b) What current flows around the loop?
  (c) What emf is developed around the loop?

Solution
  (a) Using the given values for **F** and **v** we find that

$$\mathbf{v} \cdot \mathbf{F} = (2.0\text{ m/s}) \times (0.1\text{ N}) = 0.2\text{ W}$$

  (b) On equating this power to the ohmic heating losses in accordance with (27-17) we obtain

$$0.2\text{ W} = Ri^2 = (10\,\Omega)i^2$$

and this yields

$$i = 0.14\text{ A}$$

  (c) Making use of these values for R and i, we find by use of Ohm's law that

$$\mathscr{E} = Ri = (10\,\Omega) \times (0.14\text{ A}) = 1.4\text{ V}$$

27-6 Faraday's law

So far in this chapter we have considered only the electric field—and the associated emf in a conductor—due to the relative motion between an observer and magnetic sources. As noted previously, a **B**-field that varies in time due to such a motion is physically indistinguishable from one that varies in time for other reasons, for example, because the currents in the sources themselves vary in time. We might expect, therefore, that an electric field is very generally associated with all time-varying magnetic fields. This is indeed the case. The precise form of this relationship between a **B**-field that varies in time for any reason and its associated electric field is given by *Faraday's law*. The purpose of this section is to describe this all-important law.

Consider, in Figure 27-14, a region of space in which there is a time-varying **B**-field and let S be an arbitrary open surface that bounds a given closed curve l. As in the discussion of Ampère's law (Chapter 25), let us

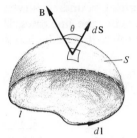

Figure 27-14

assign a sense of direction along the curve and determine a positive sense to the normal to S in accordance with the right-hand rule. (If the curve is grasped in the right hand with the fingers pointing along the sense of l, then the outstretched thumb will point along the positive sense of the normal to S.) The magnetic flux Φ_m through the surface S is defined by

$$\Phi_m = \int_S B \cos \theta \, dS = \int_S \mathbf{B} \cdot d\mathbf{S} \tag{27-18}$$

with $d\mathbf{S}$ a vectorial area element along the normal to S and with θ the angle between $\mathbf{B}$ and $d\mathbf{S}$. Now according to Gauss' law for magnetism, (25-11), the total magnetic flux out of every closed surface vanishes. Hence the magnetic flux Φ_m in (27-18) is the same for *all* open surfaces S bounded by the given curve l. In other words, the magnetic flux Φ_m is determined exclusively by the bounding curve itself and not by the particular surface S chosen.

Consider now a region of space in which there exist an electric field $\mathbf{E}$ and a magnetic induction field $\mathbf{B}$. Faraday's law states that for *all* closed curves l, $\mathbf{E}$ and $\mathbf{B}$ are related by

$$\oint_l \mathbf{E} \cdot d\mathbf{l} = -\frac{d}{dt} \int_S \mathbf{B} \cdot d\mathbf{S} \tag{27-19}$$

where the integral on the left is to be carried out around the closed curve l, and on the right the surface integral is over any open surface bounded by l. It is assumed that the positive sense of l and S are related as in Figure 27-14 by the right-hand rule. Implicit in (27-19) is the fact that this relation is valid for *all* closed curves l.

An alternate way of expressing Faraday's law is in terms of magnetic flux. Substituting (27-18) on the right-hand side of (27-19) we obtain

$$\oint_l \mathbf{E} \cdot d\mathbf{l} = -\frac{d}{dt} \Phi_m \tag{27-20}$$

which states in words:

The line integral of the electric field about any closed path l is equal to the negative of the time rate of change of magnetic flux through any surface bounded by l.

For the special, but frequently occurring, case where the bounding curve l is selected to coincide with a loop of a conducting wire, the line integral in (27-20) can be identified with the emf $\mathscr{E}$ developed around it. Faraday's law in this case assumes the form

$$\mathscr{E} = -\frac{d}{dt}\Phi_m \tag{27-21}$$

or, in words,

The emf developed around a closed conducting path is equal to the negative of the rate of change of magnetic flux through any surface bounded by that path.

In particular, the current around a closed loop in a time-varying **B**-field is the same as if a battery of emf $(-d\Phi_m/dt)$ were inserted in the loop.

It should be noted that just as Ampère's law in (25-12) does not by itself determine a unique **B**-field for a given current distribution, by the same token Faraday's law in (27-19) does not associate a unique electric field with a given time-dependent **B**-field. Nevertheless, Faraday's law is very useful in a practical sense and is one of the very important characterizations of the electric field. In particular, as noted above, if the closed loop l in (27-19) is selected to coincide with a closed conducting path, then the line integral in that relation represents the emf developed around this conductor. In many cases of practical importance, only this emf is of direct physical interest. And for these cases, according to (27-21), the significant quantity is the rate of change of magnetic flux through the conducting loop; the details of the electric field itself are at best only of peripheral interest.

As an illustration, consider again the loop of wire traveling through a uniform **B**-field in Figure 27-9. Because **B** is uniform, the magnetic flux through the loop does not vary in time. Hence, consistent with our previous analysis the emf around the loop vanishes. By the same token, in Figure 27-10, the loop travels through an inhomogeneous field and thus the magnetic flux through it varies in time. Hence current flows around the loop in this case.

Example 27-8 A circular conducting loop of radius a lies in a plane at right angles to a time-dependent but uniform magnetic induction, whose strength is

$$B(t) = \alpha + \beta t$$

where α and β are positive constants (see Figure 27-15). Calculate the emf developed around the loop.

Solution Assuming the sense for the bounding curve shown in the figure, the flux Φ_m through the planar surface bounded by the loop is

$$\Phi_m = \int_S \mathbf{B} \cdot d\mathbf{S} = B\pi a^2 = (\alpha + \beta t)\pi a^2$$

Substituting this result into (27-21) we find that

$$\mathscr{E} = -\frac{d\Phi_m}{dt} = -\frac{d}{dt}[(\alpha + \beta t)\pi a^2] = -\beta\pi a^2$$

The fact that this is negative (since $\beta > 0$) means that the emf developed in the loop is opposite to the sense assumed in Figure 27-15. This implies, in turn, that the resultant current in the loop is directed counterclockwise. If the flux had been decreasing, that is, if $\beta < 0$, then the current would have been directed clockwise.

Example 27-9 Consider, in Figure 27-16, the uniform but time-dependent magnetic induction $B(t)$ between the pole faces of an electromagnet. Calculate the strength of the electric field $E(t)$ in this region. Assume that the electric field lines are circles concentric with the axis of the electromagnet and that the magnetic induction is uniform.

Solution Let us apply Faraday's law to a circular loop of radius r with its center on the axis of symmetry of the electromagnet. Assuming that the sense of the bounding curve is as shown in the figure, we find

$$\oint \mathbf{E} \cdot d\mathbf{l} = 2\pi r E \qquad (27\text{-}22)$$

with E the strength of the electric field. Since the sense of the bounding curve is such that the normal to the surface bounded by it is downward, the magnetic flux through it is positive and given by

$$\Phi_m = \int \mathbf{B} \cdot d\mathbf{S} = B(t) \int dS = \pi r^2 B(t) \qquad (27\text{-}23)$$

where the second equality follows since the magnetic induction is uniform.
    Substituting (27-22) and (27-23) into Faraday's law, we find that

$$2\pi r E(t) = -\frac{d}{dt}[\pi r^2 B(t)] = -\pi r^2 \frac{dB}{dt}$$

or, in other words,

$$E = -\frac{r}{2}\frac{dB}{dt} \qquad (27\text{-}24)$$

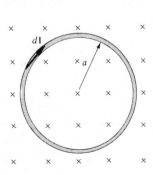

Figure 27-15

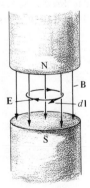

Figure 27-16

Thus, if the field decreases, so that $dB/dt < 0$, then the sense of the electric field lines is as shown in Figure 27-16. However, if the field increases, so that $dB/dt > 0$, then the electric field lines are oriented in the opposite sense.

The physical principle illustrated here is used in a practical way to produce very energetic electrons in a device called a *betatron*. The betatron was invented by Donald W. Kerst at the University of Illinois in 1941 and has been used to produce electron beams with energies as high as 300 MeV. If an electron is introduced into the region between the pole pieces in Figure 27-16, then because of the **B**-field it will orbit (approximately) in a circle, and because of the associated electric field in (27-24) it will also undergo a tangential acceleration. In this way its energy steadily rises to the desired value.

27-7 Applications of Faraday's law

In this section the general utility of Faraday's law in the form in (27-21) will be illustrated by applying it to a number of idealized physical situations.

Consider, in Figure 27-17a, a loop of wire in the presence of a time-dependent external magnetic induction $\mathbf{B}_0(t)$. If the vector $d\mathbf{l}$ represents the sense of the curve, it follows that the magnetic flux through the loop and associated with $\mathbf{B}_0$ is positive. Hence if $\mathbf{B}_0(t)$ is increasing, then $d\Phi_m/dt$ will be positive. The emf in the loop will then be negative, according to (27-21), and the current thus flows in a direction opposite to the sense of $d\mathbf{l}$. By contrast, if as shown in Figure 27-17b the external field $\mathbf{B}_0(t)$ is decreasing, then $d\Phi_m/dt$ will be negative. This time, then, the induced emf is positive, and the current will flow along $d\mathbf{l}$. Note that if $\mathbf{B}_i$ is the **B**-field due to the induced current i in the loop, then for $\mathbf{B}_0$ increasing, $\mathbf{B}_i$ is antiparallel to $\mathbf{B}_0$, and for $\mathbf{B}_0$ decreasing it is parallel to $\mathbf{B}_0$. In either case, the direction of the induced current is such as to produce a **B**-field that tends to *oppose* the *change* in magnetic flux through the loop. This leads to the alternate form of Lenz's law:

---

*The direction of the induced current in a loop of wire in the presence of a time-dependent **B**-field is such as to oppose changes in magnetic flux through it.*

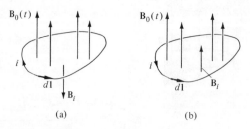

(a)　　　　　　　　(b)

Figure 27-17

Brief reflection shows that this form for Lenz's law is more general than that for moving conductors as derived in Section 27-5.

In working problems involving the application of (27-21) to particular physical situations it is often simplest to disregard the sign in that relation and to use it only to obtain the magnitude of the induced emf. The direction of the associated current flow can then be obtained by use of the above form of Lenz's law.

Example 27-10 Suppose the external field in Figure 27-15 varies in time as

$$B_0 = \alpha + \beta t$$

Determine the direction of the induced current in terms of the parameter β, assuming that $\alpha > 0$.

Solution For the assumed direction of $d\mathbf{l}$ in the figure, the flux through the loop will be positive (at $t = 0$) and an increasing function of time for $\beta > 0$. For this case, then, the induced field must be (perpendicular to and) out of the plane of the figure in accordance with Lenz's law. By the right-hand rule, then the current will be flowing counterclockwise, that is, in a direction opposite to $d\mathbf{l}$.

For the case $\beta < 0$, the flux will be decreasing. Here the induced current must flow clockwise, so the induced field will be parallel to the external field.

Example 27-11 Consider again a conducting rod sliding across two rails in a time-independent and uniform magnetic induction. See Figure 27-18. This time calculate the induced emf by use of (27-21).

Solution First, with regard to the direction of the induced emf and thus that of the current, we may reason as follows. Assuming a clockwise sense for the conducting path, we find that as the conducting rod moves to the right, the magnetic flux through the loop increases, since the area of the loop does. According to Lenz's law, then, the current in the loop will flow counterclockwise so that its associated magnetic induction will be directed opposite to that of the original magnetic induction $\mathbf{B}_0$.

With regard to the magnitude of the induced emf, we proceed as follows. The magnetic flux through the planar area bounded by the loop is $B_0 l x$, since the length and width of the loop are l and x, respectively. Substitution into (27-21) leads to

$$\mathscr{E} = -\frac{d\Phi_m}{dt} = -\frac{d}{dt}(B_0 l x)$$

$$= -B_0 l \frac{dx}{dt} = -B_0 l v$$

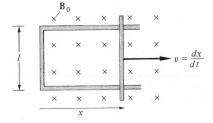

Figure 27-18

where the third equality follows since B_0 and l are constants, and the last since $v = dx/dt$. Note that except for the sign, which is due to our choice for the sense of the bounding curve, this result is identical to that in (27-8).

As this example illustrates, Faraday's law in the form in (27-21) can be utilized not only for the solution of problems involving explicitly time-dependent magnetic sources, but also for those involving magnetic sources in motion. On the other hand, the electric field formula in (27-3) is *not* generally applicable for stationary, but time-dependent, sources. It can be used only for the cases involving *static* magnetic sources in motion.

Example 27-12   A time-dependent current $i = i_0 \sin \omega t$ (i_0 and ω are constants) flows in a long, straight wire which lies in the plane of a nearby rectangular loop (see Figure 27-19). Calculate the emf induced in the loop.

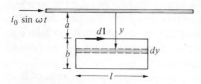

Figure 27-19

Solution   As shown, let us assign a clockwise sense to the loop. Then at an instant when the current flows to the right, the magnetic flux through the loop will be positive. It follows from Lenz's law that when the current in the long wire is increasing, the induced current will flow counterclockwise and conversely.

To calculate the emf, let $d\Phi_m$ be the flux through the infinitesimal area element of length l and width dy, and at a distance y below the long wire. Making use of (25-6), we find that

$$d\Phi_m = \mathbf{B} \cdot d\mathbf{S} = Bl\, dy$$

$$= \frac{\mu_0 l}{2\pi y} (i_0 \sin \omega t)\, dy$$

which, when integrated over all values of y from a to $(a+b)$, becomes

$$\Phi_m = \int d\Phi_m = \frac{\mu_0 l i_0 \sin \omega t}{2\pi} \int_a^{a+b} \frac{dy}{y}$$

$$= \frac{\mu_0 l i_0 \sin \omega t}{2\pi} \ln y \Big|_a^{a+b}$$

$$= \frac{\mu_0 l i_0 \sin \omega t}{2\pi} \ln \left(1 + \frac{b}{a}\right)$$

Finally, substitution into (27-21) yields the emf

$$\mathcal{E} = -\frac{d\Phi_m}{dt} = -\frac{\mu_0 l i_0}{2\pi} \ln \left(1 + \frac{b}{a}\right) \frac{d}{dt} \sin \omega t$$

$$= -\frac{\mu_0 l i_0 \omega}{2\pi} \ln \left(1 + \frac{b}{a}\right) \cos \omega t$$

The minus sign tells us that for times t for which $\cos \omega t > 0$ the direction of the current in the loop is counterclockwise; that is, opposite to the original positive sense assigned to the loop. For times t for which $\cos \omega t < 0$ the current flows in the opposite direction.

27-8 Alternating current generators

An important problem of considerable practical interest deals with the direct conversion of mechanical to electrical energy. The purpose of this section is to discuss the principles underlying the operation of a device known as an *alternating current generator* which can be used for this purpose.

Consider, in Figure 27-20, a planar coil of wire of area A, which is being rotated about an axis in its plane while in a uniform magnetic induction $\mathbf{B}$. Let us set up a stationary coordinate system with the axis of rotation along the z-axis and with $\mathbf{B}$ directed along the y-axis. If θ is the angle between the normal $\mathbf{n}$ to the plane of the loop and the direction of $\mathbf{B}$, then as the loop rotates at some angular velocity ω, the angle θ increases linearly in time as $\theta = \omega t$. As a result of the rotation, the magnetic flux through the loop varies in time, and therefore, according to Faraday's law, an emf is developed around it. In terms of Figure 27-20, this means that a potential difference will be developed between the two terminals a and b and thus if an electric network is connected across these terminals a current will flow.

To calculate the magnitude of the emf, we proceed as follows. Since $\mathbf{B}$ is constant and spatially uniform, the magnetic flux Φ_m through the loop at the instant shown in the figure is

$$\Phi_m = \int \mathbf{B} \cdot d\mathbf{S} = \int B \cos \theta \, dS$$
$$= BA \cos \omega t$$

where the second equality follows by use of the definition of the dot product,

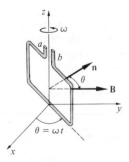

Figure 27-20

and the third since $\theta = \omega t$. Substitution into (27-21) leads to

$$\mathscr{E} = -\frac{d\Phi_m}{dt} = -\frac{d}{dt}(BA \cos \omega t) = -BA \frac{d}{dt} \cos \omega t = \omega BA \sin \omega t$$

$$\text{(27-25)}$$

As long as the loop continues to rotate, this emf will appear across the terminals a and b.

If there are N turns of wire in the loop, then, following the same steps as above, we find the more general formula

$$\mathscr{E} = N\omega BA \sin \omega t \qquad \text{(27-26)}$$

As a practical matter, the emf developed in a rotating loop can be utilized in an external circuit by use of two *slip rings*, which rotate with the loop (see Figure 27-21). As the slip rings rotate, electrical contact is made with the external circuit by means of two stationary brushes that are permanently in contact with the slip rings.

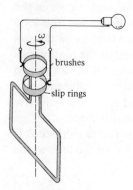

brushes

slip rings

Figure 27-21

When connected to a circuit so that a current flows, the external **B**-field exerts a torque on the rotating loop. According to Lenz's law the direction of this torque is such as to oppose the rotation. Hence an external agent must carry out work to keep the loop rotating.

Example 27-13 Suppose that a resistor of resistance R is connected across the terminals in Figure 27-20. If the coil has N turns each of area A, the magnetic induction is B, and ω is the angular velocity calculate the torque required to keep the loop rotating.

Solution Since the emf $\mathscr{E}$ across the resistor is given according to (27-26) by

$$\mathscr{E} = \omega NBA \sin \omega t$$

it follows from Ohm's law that the current i through the resistor is

$$i = \frac{\mathscr{E}}{R} = \frac{\omega NBA}{R} \sin \omega t$$

Making use of (26-22) which gives the torque on a loop carrying a current i, we find that

$$\tau = iNAB \sin \theta$$

$$= \frac{\omega (NAB)^2}{R} \sin^2 \omega t$$

where the first equality is obtained by making the substitution $\alpha \rightarrow \theta$. Note that, in accordance with Lenz's law, this torque is always nonnegative, thus reflecting the fact that the external agent who causes the loop to rotate must carry out positive work during the entire cycle. Note also that the rate at which the external agent carries out work, $\tau\omega$, is precisely the same as the rate Ri^2 at which heat is dissipated in the resistor. Hence all of the mechanical work is converted directly to heat in this case.

27-9 Summary of important formulas

Suppose that $\mathbf{B}$ is the magnetic induction associated with certain magnetic sources. Then from the viewpoint of an observer with respect to whom these magnetic sources have the velocity $\mathbf{v}_s$ there is, in addition to $\mathbf{B}$, an electric field $\mathbf{E}$, which at each point in space is

$$\mathbf{E} = -\mathbf{v}_s \times \mathbf{B} \qquad (27\text{-}3)$$

The emf $d\mathscr{E}$ developed between the ends of a thin, conducting element $d\mathbf{l}$ traveling at a velocity $\mathbf{v}$ is

$$d\mathscr{E} = \mathbf{E} \cdot d\mathbf{l} = (\mathbf{v} \times \mathbf{B}) \cdot d\mathbf{l} \qquad (27\text{-}11)$$

where $\mathbf{B}$ is the external (time-independent) $\mathbf{B}$-field.

Faraday's law states that for *all* closed curves l and regardless of what electric or magnetic sources may be present, $\mathbf{E}$ and $\mathbf{B}$ are related by

$$\oint_l \mathbf{E} \cdot d\mathbf{l} = -\frac{d}{dt} \int_s \mathbf{B} \cdot d\mathbf{S} \qquad (27\text{-}19)$$

where S is any open surface bounded by l. (See Figure 27-14 for the relation between the normal to S and sense of the curve l.) For the special case that l is selected to coincide with a loop of wire, (27-19) assumes the form

$$\mathscr{E} = -\frac{d\Phi_m}{dt} \qquad (27\text{-}21)$$

where $\mathscr{E}$ is the emf developed around the loop and Φ_m is the magnetic flux through it.

QUESTIONS

1. Consider the four situations in Figure 27-1. Show in each case by use of Lenz's law that the direction of the current is as shown.
2. Consider the situation in Figure 27-1a, but suppose this time that the magnet is stationary and the loop moves toward it. Find the direction of current by use of Lenz's law.
3. Consider a region of space where there is a uniform magnetic induction parallel to the z-axis of a certain coordinate system. Does an observer who is moving along the z-axis detect any electric field? If so, what is its direction?
4. Repeat Question 3, but this time suppose that the observer is moving along: (a) the positive x-axis; (b) the positive y-axis.
5. Consider the situation in Figure 27-2. What is the direction of the current in loop B at the instant that the switch S is closed? What is its direction when the switch is subsequently opened? Use Lenz's law.
6. A steady current flows in a circular loop. Find the direction of the associated electric field at a point on the axis of the loop as seen by an observer who is traveling at a velocity v:
   (a) Directed along the axis of the loop.
   (b) Directed at right angles to the axis of the loop.
   (c) Directed along a line which makes an angle α with the axis.
7. Repeat Question 6, but suppose that the observer has in addition a certain *acceleration* along v.
8. A circular loop has a resistance R and is pulled through a *uniform* magnetic induction, which is directed perpendicularly to the plane of the loop. Will there be any current

in the loop? Explain your answer in physical terms.
9. Is any force required to pull the loop of Question 8, assuming that it travels at a uniform velocity?
10. What is the direction of the induced emf in the lower segment of the rotating rectangular loop in Figure 27-20? What is it along the two vertical segments?
11. Consider, in Figure 27-22, a flat metal plate, which initially is at rest in an inhomogeneous magnetic induction B. Explain, in terms of induced currents, why it is necessary to exert a force in order to pull the plate to the right. (Such currents are known as *eddy currents*.)

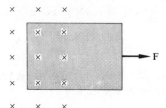

Figure 27-22

12. A small loop of wire is pulled at a velocity v between the pole pieces of a magnet as shown in Figure 27-23. What is the direction of the current in the loop just as it enters the region between the poles? What is the direction of this current on leaving this region?

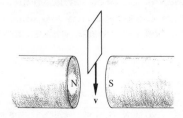

Figure 27-23

13. A particle of charge q (>0) is traveling at a velocity $\mathbf{v}$ in the plane of a small conducting loop. At an instant when it is located as shown in Figure 27-24, why is the magnetic flux through the loop increasing? What is the direction of the current in the loop?

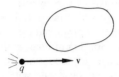

Figure 27-24

14. Consider the same situation as in Question 13, but this time as seen by an observer with respect to whom the particle is at rest so that the loop moves to the left at the velocity $\mathbf{v}$. Why must there be a current in the loop this time also?

15. Explain in physical terms why, in light of Question 14, an emf can be induced in a conducting loop which moves through an *electrostatic field* for which

$$\oint \mathbf{E} \cdot d\mathbf{l} = 0 \qquad \text{(electrostatic field)}$$

so that apparently no emf can be induced.

16. A segment of wire is pulled through a uniform magnetic induction at a velocity $\mathbf{v}$. Account for the emf developed between the ends of the wire in terms of the magnetic force on the charge carriers in the wire and their subsequent migrations.

17. Explain in physical terms the reason for the existence of an emf between the endpoints of a wire segment in a time-dependent magnetic induction $\mathbf{B}(t)$.

18. Suppose that the magnetic flux through a loop is increasing because a nonuniform $\mathbf{B}$-field perpendicular to its plane is rising. Explain physically why this is the same as if the magnetic induction were constant (in time) and the loop had a velocity in a certain direction.

19. Suppose that there is a loop of wire of radius r along the circular path in Figure 27-16. Would the electrons in the wire still be accelerated to high speeds? Explain.

20. Describe what physical properties an electromagnet—such as that in Figure 27-16—must have so that it can accelerate electrons, as in a betatron, in a circle of fixed radius r.

21. If a particle of charge q moves through a time-independent magnetic induction, its energy does not change. Explain why the same is generally *not* true for a time-dependent $\mathbf{B}$-field.

22. A particle of charge q is initially at rest in a time-dependent magnetic induction $\mathbf{B}(t)$. Explain why it will accelerate even though initially the magnetic force $q\mathbf{v} \times \mathbf{B}$ vanishes.

PROBLEMS

1. Consider a region of space in which there is a uniform $\mathbf{B}$-field of strength 0.2 tesla. Assuming that this field is oriented along the positive x-axis of a certain Cartesian coordinate system, find the electric field as seen by an observer who is traveling at:

   (a) A velocity of 10 m/s along the z-axis.
   (b) A velocity of 100 m/s along the positive x-axis.
   (c) A velocity of 5.0 m/s along the negative y-axis.

2. A long, tightly wound solenoid has 200 turns per meter and carries a

current of 5.0 amperes. Calculate the electric field as seen by an observer who is traveling at a velocity of 50 m/s: (a) along the axis of the solenoid; (b) at right angles to its axis.

3. Two very long, straight wires are parallel, carry equal and opposite currents of 10 amperes, and are separated by a distance of 50 cm.
   (a) What is the magnetic induction at a point midway between them?
   (b) What is the electric field at a point midway between the wires as seen by an observer who is traveling at a velocity of 30 m/s parallel to the wires?
   (c) Repeat (b) for an observer who is traveling at a velocity of 20 m/s in the plane of the wires and perpendicular to them.

4. Two parallel wires separated by a distance d carry equal and opposite currents i. Calculate the electric field $\mathbf{E}$ at the point P, which is at a distance x from one of the wires, as seen by an observer with respect to whom the wires are traveling at the velocity $\mathbf{v}$ directed as shown in Figure 27-25.

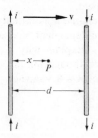

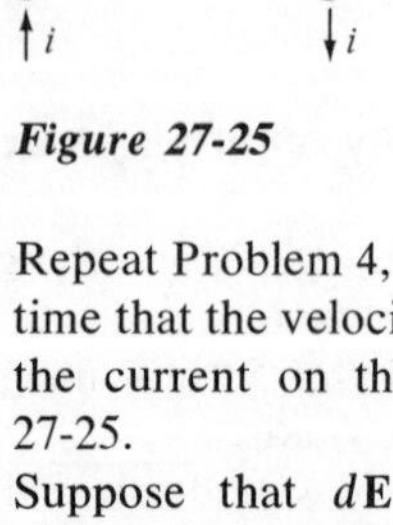

Figure 27-25

5. Repeat Problem 4, but suppose this time that the velocity $\mathbf{v}$ is parallel to the current on the left in Figure 27-25.

*6. Suppose that $d\mathbf{E}$ represents the electrostatic field produced by an infinitesimal element of charge dq.
   (a) Show that the magnetic induction $d\mathbf{B}$ as seen by an observer with respect to whom this charge element has a velocity $\mathbf{v}$ is

$$d\mathbf{B} = \mu_0 \epsilon_0 \mathbf{v} \times d\mathbf{E}$$

   (b) Making use of this result, show by integration that if $\mathbf{E}$ is the electric field produced by a certain distribution of charge, then the magnetic induction $\mathbf{B}$ as seen by an observer with respect to whom the charge has a velocity $\mathbf{v}$ is

$$\mathbf{B} = \epsilon_0 \mu_0 \mathbf{v} \times \mathbf{E}$$

   In Chapter 29, it will be seen that $\epsilon_0 \mu_0 = 1/c^2$, where c is the velocity of light. Hence $B \approx vE/c^2$ and is generally very small compared to E/c for speeds $v \ll c$.

7. Consider a sphere of radius a, which carries a uniform charge of density ρ_0. If the sphere is traveling at a velocity $\mathbf{v}$ find, by use of the results of Problem 6, the magnetic induction $\mathbf{B}$ inside and outside the sphere.

8. A circular coil of wire of radius a has N turns and carries a current i. Calculate the electric field $\mathbf{E}$ at a point P on the axis of the coil as seen by an observer with respect to whom the coil has a velocity $\mathbf{v}$ parallel to a diameter. Assume P is at a distance x from the plane of the loop on the side where $\mathbf{B}$ is directed toward the loop.

9. In Figure 27-7a, suppose that $B = 0.1$ tesla, $v = 10$ m/s, and $l = 20$ cm. (a) What is the electric field inside the rod? (b) What emf is developed between its ends? Which end acquires the positive charge?

10. In Figure 27-8a suppose that $v = 10$ m/s, $l = 10$ cm, $B = 0.5$ tesla, and

$R = 10\,\Omega$. Calculate (a) the emf in the circuit; (b) the current in the circuit; and (c) the power dissipated.

11. A runner whose height is 2.0 meters runs east at a speed of 20 km/hr perpendicular to the earth's **B**-field. Assuming that the horizontal component of **B** has a magnitude 5.0×10^{-5} tesla, what emf is developed between the top of his head and the soles of his feet?

12. Repeat Problem 11, but assume this time a "being" on a *neutron star* where the **B**-field has a strength of 10^9 tesla. Assume the same values for "its" height and speed as above.

13. If, for the physical situation in Figure 27-8a, $B = 0.5$ tesla, $l = 10$ cm, $v = 5$ m/s, and a current of 15 mA flows, calculate:
    (a) The resistance in the circuit.
    (b) The power dissipated.
    (c) The force required to keep the rod moving.
    (d) The power expended by the external agent and compare with your results to (b). Account for any differences.

14. If, for the physical situation in Figure 27-8a, $B = 0.1$ tesla, $l = 5$ cm, and $R = 2.0\,\Omega$, and a steady current of 2.0 mA flows, calculate (a) the velocity of the rod and (b) the rate at which the area of the loop increases.

15. A conducting rod of length l and resistance R slides down the conducting rails on the edges of an inclined plane of angle α as shown in Figure 27-26. Assume that a uniform magnetic induction **B** is directed vertically downward as shown in the figure, that the total electrical resistance is R, and that no frictional forces act. At an instant when the rod has the velocity **v**, as shown:
    (a) Show that the emf developed around the conducting path is $Blv \cos \alpha$.
    (b) Calculate the magnitude and direction of the current.

16. Consider again the situation in Figure 27-26 and suppose that the rod has a mass m.
    (a) Show that the magnetic force F on the rod is

    $$F = \frac{B^2 l^2 v}{R} \cos \alpha$$

    and find the direction of this force.
    (b) Show that the equation of motion for the rod is

    $$mg \sin \alpha - \frac{B^2 l^2 v \cos \alpha}{R} = m \frac{dv}{dt}$$

    (c) Find the terminal velocity.

17. A conducting rod of length l is pivoted at a point P and rotates with uniform angular velocity ω in a plane at right angles to a uniform induction **B** (see Figure 27-27).
    (a) Show that the emf $d\mathscr{E}$ developed across an infinitesimal

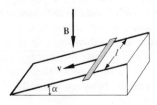

Figure 27-26

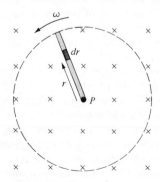

Figure 27-27

element of the rod of length dr at a distance r from P is

$$d\mathcal{E} = -\omega B r\, dr$$

Discuss the meaning of the minus sign in terms of the direction of the emf development between the ends of the rod.

(b) By integration, calculate the magnitude and the direction of the total emf developed between the ends of the rod.

18. A conducting rod of length l is pivoted at a point P and the other end slides along a circular conducting track with a uniform speed v. Assume that there is a uniform magnetic induction **B** directed as shown in Figure 27-28, that P and the track are connected by a resistor R, and that all other electrical resistance in the circuit is negligible.

Figure 27-28

(a) Calculate the magnitude and the direction of the emf developed in the circuit.

(b) What is the current through the resistor?

(c) How much power is dissipated in the resistor?

(d) What torque is required to keep the rod rotating?

19. A circular, conducting disk of radius a rotates with uniform angular velocity ω about its axis. Assuming the existence of a uniform magnetic induction **B** parallel to the

axis, show that the emf developed between the center of the disk and its rim is

$$\mathcal{E} = \frac{\omega a^2 B}{2}$$

(*Hint*: Think of the disk as a collection of rods, such as the one in Problem 17.) (*Note*: This device used for generating an emf was first studied by Faraday and is known as the *Faraday disk dynamo*.)

20. Consider again the physical situation described in Problem 19. Calculate the electric field at each point of the disk due to the charge separation that occurs. Assume the sense of rotation in Figure 27-28.

21. A conducting rod of length l is located near a long, straight wire, which carries a current i, as in Figure 27-29. If the rod is pulled at a velocity **v** along the direction of the current, calculate in terms of a, l, α, v, and i the emf developed between the ends of the rod.

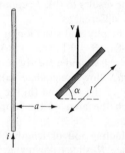

Figure 27-29

22. Consider again the situation in Figure 27-29, but suppose this time that the rod is pulled to the right with a velocity **v**. Assuming that at $t = 0$ the lower end of the rod is at the distance a from the current, calculate, as a function of time, the emf developed between the ends of the rod.

23. Suppose that the rotating current

loop in Figure 27-20 is a square of side l. At the instant depicted in the figure:

(a) What emf is developed along the bottom segment of the loop?

(b) What is the magnitude and direction of the emf developed in each of the vertical segments of the loop?

(c) Make use of (a) and (b) to calculate the total emf developed around the loop and compare your result with that in (27-25).

24. Consider, in Figure 27-30, a square loop of wire of length l and resistance R, which is forced to travel at the velocity $\mathbf{v}$ directed as shown. Assuming that at $t = 0$ it first enters the region of a uniform magnetic induction $\mathbf{B}$, and that $a > l$:

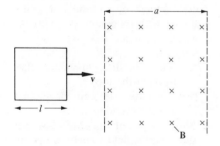

Figure 27-30

(a) Calculate the strength and the direction of the current in the loop as a function of time for the time intervals (1) $t \leqslant l/v$; (2) $l/v \leqslant t \leqslant a/v$; and (3) $t > a/v$.

(b) Calculate the force required to keep the loop moving at the constant speed v during each of the above time intervals.

25. Show by use of (27-12) that the emf developed around a rigid but closed conducting loop which moves at a velocity $\mathbf{v}$ in a uniform time-independent magnetic induction $\mathbf{B}$ vanishes. *Hint*: For this situation,

(27-12) may be written as

$$\mathcal{E} = (\mathbf{v} \times \mathbf{B}) \cdot \oint d\mathbf{l}$$

*26. Show explicitly that in the presence of a *time-dependent* $\mathbf{B}$-field, the work required to take a charged particle between any two points depends in general on the path connecting them. (*Hint*: Why is the work required to take a particle around a closed path not zero except in electrostatics?)

27. A conducting loop of radius 10 cm is in a uniform magnetic induction, which is directed as shown in Figure 27-31 and which increases at the constant rate of 0.02 T/s.

(a) What is the direction of the emf developed around the loop?

(b) What is the magnitude of the emf around the loop?

(c) If the resistance of the loop is 100 Ω, what current flows?

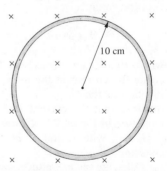

Figure 27-31

28. Consider again the situation in Figure 27-31, but suppose that this time the magnetic induction decreases at the rate of 0.2 T/s. Calculate the magnitude and the direction of the current in the loop.

29. A very long ideal solenoid of radius a has n turns per unit length and carries a current $i = i_0 + \alpha t$, where i_0 and α are positive constants.

(a) Calculate the magnetic flux

$\Phi_m(t)$ through a coaxial conducting loop of radius b ($>a$). See Figure 27-32.

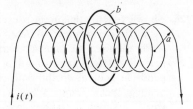

Figure 27-32

(b) Calculate the emf developed around the loop.

(c) If the loop has a resistance R, find the magnitude and the direction of the current in the loop.

30. A rectangular loop of wire of sides a and b has a resistance R and is subjected to a uniform magnetic induction perpendicular to its plane and varying in time in accordance with the formula

$$B = B_0 \cos \omega t$$

where B_0 and ω are constants. Calculate the current in the loop as a function of time.

31. Suppose that the current i in the long wire in Figure 27-19 varies as $i = i_0 - \alpha t$, with i_0 and α positive constants.

(a) Calculate the magnitude and direction of the emf developed in the rectangular loop.

(b) If the loop is cut at some point, what is the potential difference between the cut ends?

*32. A circular loop of wire of radius a is in a uniform magnetic induction $\mathbf{B}$ and is attached to a resistor R as shown in Figure 27-33.

(a) What is the magnetic flux through the circular loop?

(b) Suppose that the loop is quickly rotated in a short time interval Δt by 180° about the axis AC

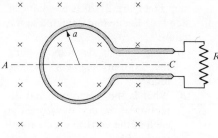

Figure 27-33

while the rest of the circuit maintains its original position. What is the direction of the current in the resistor R?

(c) Show that the total charge Q that flows through the resistor is

$$Q = \frac{2\pi a^2 B}{R}$$

*33. Suppose, in Figure 27-19, that the conducting loop is circular of radius a and with its center at a perpendicular distance b ($>a$) from the long wire.

(a) What is the magnetic flux through the loop?

(b) Calculate the emf developed around the loop.

*34. Show by use of Faraday's law that the electric field cannot drop abruptly to zero, in a direction perpendicular to itself. (*Hint:* Assume that, on the contrary, the electric field does drop abruptly to zero, evaluate Faraday's law for the rectangular path *abcd* in Figure 27-34, and consider the limit as the sides *ad* and *bc* become vanishingly small.)

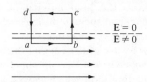

Figure 27-34

*35. Show by use of Faraday's law that even if magnetic monopoles existed, so that the magnetic flux out of a closed surface did not vanish, the relation

$$\frac{d}{dt}\oint_S \mathbf{B}\cdot d\mathbf{S} = 0$$

would still be valid.

*36. A circular loop of wire is in a time-dependent magnetic induction $\mathbf{B}(t)$ directed perpendicular to its plane. By starting with Faraday's law show that the direction of the emf developed in the coil is the same as that obtained by use of Lenz's law.

37. Consider, in Figure 27-35, a cylindrical region of radius a in which there is a uniform magnetic induction $\mathbf{B}(t)$, with $B(t) = B_0 - \alpha t$, with B_0 and α positive constants.

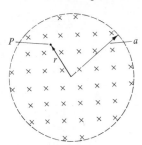

Figure 27-35

(a) What is the instantaneous force on a proton of charge q when it is at rest at the point P?

(b) What is the instantaneous acceleration of an electron if it is placed at rest at this point? Assume that $r = 50\,\text{cm}$ and $\alpha = 0.02\,\text{T/s}$.

38. Consider again the cylindrically symmetric magnetic induction $B(t) = B_0 - \alpha t$ in Figure 27-35. Suppose that a loop of wire of resistance R and radius $a/2$ is placed so

that it lies in the plane of the figure and is totally immersed in the uniform magnetic induction.

(a) Calculate the magnitude and the direction of the current in the loop.

(b) Suppose that the loop is replaced by one that is a square of side $a/2$. What changes, if any, are required for your answer to (a)?

*39. Show that the electric field $\mathbf{E}$ at any point inside the dotted circle in Figure 27-35 is

$$\mathbf{E} = \frac{1}{2}\mathbf{r} \times \frac{d\mathbf{B}}{dt}$$

where $\mathbf{B}$ is the spatially uniform magnetic induction and $\mathbf{r}$ is a vector from the center to the point under consideration. What is the equation of motion for a particle of charge q and mass m in this region?

*40. For the situation in Figure 27-18, suppose that the external field $\mathbf{B}(t)$ is not constant but varies in time and that the total resistance in the conducting path is R.

(a) Show that the emf developed around the closed conducting path is

$$\mathscr{E} = B(t)lv + xl\frac{dB}{dt}$$

(b) For the special case $B(t) = B_0 - \alpha t$, where B_0 and α are positive constants, calculate the force F required to keep the wire moving at the velocity v. Assume that at $t = 0$, $x = x_0$.

(c) Compare the power Fv expended by the external agent with that dissipated in the resistor and account for the difference in physical terms.

28 Inductance and magnetic materials

First we must inquire whether the elements are eternal or subject to generation and destruction; for when this question has been answered their number and character will be manifest.

ARISTOTLE

28-1 Introduction

In our study of dielectrics it was convenient to introduce first the notion of capacitance and then to use capacitors as a tool to measure the electric properties of insulators. Our study of magnetic materials will proceed in a similar way. This time we first introduce the concept of *inductance* and then show how the magnetic properties of matter can be measured by use of *inductors*.

28-2 Self-inductance

If a time-dependent current is generated in one of two neighboring conducting loops, then because of the changing magnetic flux, a current will, in general, also be induced in the other. Similarly, if a time-dependent current is generated in an isolated loop, the changing magnetic flux produces an additional emf around the loop itself. It is convenient to characterize this induced emf in a loop—whether it be due to its own changing current or to that in a neighboring one—in terms of certain quantities known as the *coefficients of inductance*. The purpose of this and Section 28-3 is to define these quantities.

Consider, in Figure 28-1, a loop of wire around which flows a time-dependent current $i \equiv i(t)$. Associated with this current, there will be a magnetic induction $\mathbf{B}(t)$, which in turn produces a certain magnetic flux Φ_m through the loop. According to (25-5), $\mathbf{B}(t)$ is directly proportional to the current. Hence we may write

$$\Phi_m = Li \qquad (28\text{-}1)$$

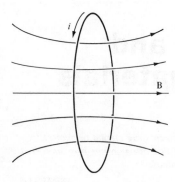

Figure 28-1

with L a certain proportionality constant, which is known as the *inductance* or the *self-inductance* of the loop. Note that L is independent of i; it depends only on the shape and size of the loop. Following convention we shall always take the positive sense of the normal to the loop to be related to the direction of the current by the right-hand rule. With this choice, the magnetic flux Φ_m in (28-1) will be positive, as will be the self-inductance L of the loop.

For the special case of a closely packed coil or an ideal solenoid for which the magnetic flux Φ_m through *each* of the N turns is the same, the total flux through the coil is $N\Phi_m$. For this case, (28-1) assumes the form

$$N\Phi_m = Li \qquad (28\text{-}2)$$

Note that in both (28-1) and (28-2) Φ_m is the flux through a *single* turn of the coil.

By use of the above definition of inductance, the emf across a coil of wire, or an *inductor*, as it is known, can be expressed directly in terms of L. Assuming that a current i flows around the loop, we find, by substituting (28-1) into Faraday's law, that

$$\mathscr{E} = -\frac{d\Phi_m}{dt} = -\frac{d}{dt}(Li)$$

and, since L is a time-independent constant, this may be expressed equivalently as

$$\mathscr{E} = -L\frac{di}{dt} \qquad (28\text{-}3)$$

In other words, the flow of a current $i(t)$ around a loop produces the same effect as if a "battery" of emf $(-L\,di/dt)$ had been introduced into the loop. Although (28-3) has been derived for a single loop, it is easy to confirm by use of (28-2) that it is equally valid for a closely packed coil or an ideal solenoid.

When used in electric circuits, the symbol

will always be used to represent an inductor. Circuits containing inductors will be discussed in Section 28-4.

According to the definition in (28-1), the unit of inductance is the $T\text{-}m^2/A = Wb/A$. It is convenient to define a unit of inductance called the *henry* (abbreviated H) by the relation

$$1\,H = 1\,Wb/A$$

The related units of the *millihenry* $mH = 10^{-3}\,H$ and the *microhenry* $\mu H = 10^{-6}\,H$ are also frequently used. This unit is named in honor of Joseph Henry (1797–1878), a contemporary of Faraday, who independently carried out studies of inductive effects associated with time-dependent currents.

Example 28-1   A very long, ideal solenoid has N turns and a length l. Calculate its inductance assuming that each turn is circular and has a radius a.

Solution   Assuming that a current i flows, the magnetic induction **B** is uniform and parallel to the axis of the coil. According to (25-9), its magnitude is

$$B = \frac{\mu_0 Ni}{l}$$

Hence, the magnetic flux Φ_m through a single turn of the coil is

$$\Phi_m = B\pi a^2 = \frac{\mu_0 iN\pi a^2}{l}$$

and substitution into (28-2) leads to

$$L = \frac{N\Phi_m}{i} = \frac{N\mu_0 iN\pi a^2}{li} = \frac{\mu_0 N^2 \pi a^2}{l} \qquad (28\text{-}4)$$

For example, if $a = 1\,cm$, $N = 10^3$, and $l = 20\,cm$, then

$$L = \frac{\mu_0 N^2 \pi a^2}{l} = \frac{(4\pi \times 10^{-7}\,\text{T-m/A}) \times (10^3)^2 \times \pi \times (10^{-2}\,m)^2}{0.2\,m} = 2.0 \times 10^{-3}\,H$$

Example 28-2 A toroid of inner radius a and outer radius b has N turns and a rectangular cross section of width c. Assuming that a current i flows, calculate:
  (a) The magnetic flux through a single turn of the toroid.
  (b) The self-inductance of the toroid.

Solution Figure 28-2a shows a top view of the toroid and Figure 28-2b a cross-sectional view without the wire.

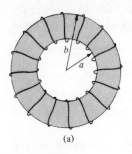
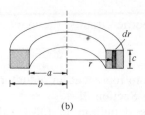

(a)   (b)

Figure 28-2

(a) According to the argument in Chapter 25, the lines of magnetic induction in a toroid are circles concentric with its axis. The magnitude of the **B**-field at a distance r ($a < r < b$) from its center is given by (25-15); that is,

$$B = \frac{\mu_0 N i}{2\pi r}$$

Hence the magnetic flux $d\Phi_m$ through a small rectangular area of width dr and height c, and at a distance r from the axis is

$$d\Phi_m = Bc\, dr = \frac{\mu_0 N i}{2\pi r} c\, dr$$

The flux Φ_m through a single turn is then obtained by integration

$$\Phi_m = \int d\Phi_m = \int_a^b \frac{\mu_0 N i}{2\pi r} c\, dr = \frac{\mu_0 N i c}{2\pi} \int_a^b \frac{dr}{r} = \frac{\mu_0 N i c}{2\pi} \ln r \Big|_a^b = \frac{\mu_0 N i c}{2\pi} \ln \frac{b}{a}$$

since $d[\ln r]/dr = 1/r$.
  (b) Substituting the above formula for Φ_m into (28-2) we obtain

$$L = \frac{N\Phi_m}{i} = \frac{\mu_0 N^2 c}{2\pi} \ln \frac{b}{a} \tag{28-5}$$

28-3 Mutual inductance

A time-dependent current in a coil will, in general, produce a changing magnetic flux not only through itself but also through any other nearby coil. To study this possibility consider, in Figure 28-3, two *closely packed* coils, which carry the currents i_1 and i_2 and consist of N_1 and N_2 turns of wire, respectively. Let Φ_{12} be the magnetic flux through a single turn of coil 1 due

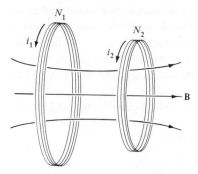

Figure 28-3

to the current i_2 in coil 2 and Φ_{21} the corresponding flux through a single turn of coil 2 due to i_1. It is important to note that Φ_{12}, for example, is *not* the total magnetic flux through a single turn of coil 1, since its own current i_1 produces a flux as well. The total magnetic flux through coil 1 due to i_2 is $N_1\Phi_{12}$, and, correspondingly, $N_2\Phi_{21}$ represents the total magnetic flux through coil 2 due to i_1.

Now as in the discussion of self-inductance, the magnetic fluxes $N_1\Phi_{12}$ and $N_2\Phi_{21}$ through the two coils will be proportional, respectively, to i_2 and i_1. Accordingly, we define the coefficients of mutual inductance M_{12} and M_{21} by

$$M_{21}i_1 = N_2\Phi_{21}$$
$$M_{12}i_2 = N_1\Phi_{12}$$

(28-6)

so that M_{12} and M_{21} are current-independent proportionality constants; they depend only on the geometric parameters characterizing the two loops and their relative separation and orientation. By contrast to the coefficient of self-inductance L, which characterizes the flux through a loop due to its own current, the mutual inductance M_{12} characterizes the flux through coil number 1 due to a current in a neighboring coil 2. Correspondingly, M_{21} represents the flux through coil 2 due to a unit current in coil 1.

We shall always assume in the following that each coil has been assigned a sense of direction such that a positive current in one produces a positive flux through the other. If, for example, the current i_1 in Figure 28-3 is positive, then if the sense of the other loop is along i_2, the flux Φ_{21} through the latter is positive. With this choice, the sign of Φ_{21} is always the same as i_1, and similarly for Φ_{12} and i_2. It follows then by reference to (28-6) that the coefficients of mutual inductance M_{12} and M_{21} are invariably positive.

Detailed calculations of the coefficients of mutual inductance for a variety of coils under a variety of relative positions show that the the two coefficients M_{12} and M_{21} are always the same. Hence we shall write

$$M = M_{12} = M_{21}$$

(28-7)

and the symbol M will be used to represent the mutual inductance of either coil. The relation in (28-7) is extremely useful in practice, and is a special case of a more general relation, which is known as a *reciprocity relation*. We emphasize that the validity of (28-7) is *not* dependent on the currents or the fluxes through the two coils. In general, the fluxes Φ_{12} and Φ_{21} will *not* be equal to each other despite (28-7).

The emf $\mathscr{E}_1$ developed in coil 1 due to the changing current in coil 2 may be calculated by use of Faraday's law. Since $N_1\Phi_{12}$ is the total flux through coil 1 due to the current i_2 in coil 2, it follows that

$$\mathscr{E}_1 = -\frac{d}{dt}(N_1\Phi_{12}) = -\frac{d}{dt}(M_{12}i_2)$$

$$= -M\frac{di_2}{dt} \tag{28-8}$$

where in the last equality we have used (28-7) and the fact that M is a time-independent constant. In a similar way, the emf $\mathscr{E}_2$ developed in coil 2 due to i_1 is

$$\mathscr{E}_2 = -M\frac{di_1}{dt} \tag{28-9}$$

Example 28-3 A large, closely packed coil of N_2 turns completely surrounds a very long, ideal solenoid of length l, radius a, and N_1 turns, as in Figure 28-4. Calculate the mutual inductance M between the coils.

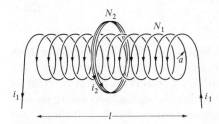

Figure 28-4

Solution Assuming a current i_1 in the solenoid, the magnetic induction **B** inside is uniform, parallel to its axis, and has the strength $B = \mu_0 N_1 i_1 / l$. Outside the solenoid the magnetic induction vanishes. The magnetic flux Φ_{21} through a single turn of the large coil is thus

$$\Phi_{21} = B\pi a^2 = \left(\frac{\mu_0 N_1 i_1}{l}\right)\pi a^2$$

and hence, by use of (28-6) and (28-7), the mutual inductance M is found to be

$$M = \frac{N_2\Phi_{21}}{i_1} = \frac{N_2}{i_1}\left(\frac{\mu_0 N_1 i_1}{l}\right)\pi a^2 = \frac{\mu_0 \pi a^2}{l}N_1 N_2$$

It is interesting to note that even though it is very difficult to calculate the flux Φ_{12} through the solenoid directly, by use of this formula for M and (28-7) the calculation is straightforward. For, according to (28-6),

$$\Phi_{12} = \frac{M_{12}i_2}{N_1} = \frac{Mi_2}{N_1} = \frac{\mu_0 \pi a^2 N_2}{l} i_2$$

where the final equality follows by use of the above formula for M. By contrast, the calculation of Φ_{12} by direct integration is extraordinarily complex!

Example 28-4 A rectangular loop of wire of dimensions l and $(d - 2a)$ lies in the plane of two very long, parallel wires, which comprise part of a circuit (see Figure 28-5). Calculate the mutual inductance between the loops.

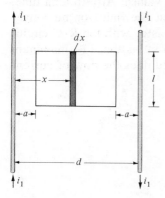

Figure 28-5

Solution Assuming that a current i_1 flows in the long wires, it follows that the flux $d\Phi_{21}$ through the small element of area of length l and thickness dx of the loop is

$$d\Phi_{21} = Bl \, dx$$

$$= \frac{\mu_0 i_1}{2\pi} \left(\frac{1}{x} + \frac{1}{d - x} \right) l \, dx$$

The total flux Φ_{21} through the small loop is found, by integration from $x = a$ to $x = d - a$, to be

$$\Phi_{21} = \int d\Phi_{21} = \frac{\mu_0 i_1 l}{2\pi} \int_a^{d-a} dx \left[\frac{1}{x} + \frac{1}{d-x} \right]$$

$$= \frac{\mu_0 i_1 l}{2\pi} [\ln x - \ln (d - x)] \Big|_a^{d-a}$$

$$= \frac{\mu_0 i_1 l}{\pi} \ln \left(\frac{d - a}{a} \right)$$

Finally, using the value $N_2 = 1$ and (28-6), we obtain

$$M = \frac{\Phi_{21}}{i_1} = \frac{\mu_0 l}{\pi} \ln \left(\frac{d - a}{a} \right)$$

28-4 The *R-L* circuit

Consider, in Figure 28-6, an inductor of inductance L in series with a resistor R, both connected across a battery of emf $\mathscr{E}$. Suppose that the circuit also contains a switch S, which is closed at the initial instant $t = 0$. According to Lenz's law, the inductor strives to oppose any change of magnetic flux through itself. Therefore, just after the switch is closed, an emf will be produced in the inductor and its sense will be such as to oppose, at least momentarily, the flow of electric current in the circuit. Initially, therefore, the current i through the battery will vanish. After a long time, however, the current becomes steady and then the inductor no longer influences the flow of current. This is to be contrasted with the R-C circuit analyzed in Section 23-6, where initially the current assumes its maximum value and only after the capacitor is fully charged does the flow of current cease.

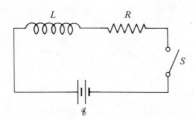

Figure 28-6

According to (28-3), an inductor through which flows a current i behaves as if it were a battery of emf $(-L\,di/dt)$. Hence the circuit equation for the R-L circuit is

$$Ri = \mathscr{E} - L\frac{di}{dt}$$

and this can be expressed equivalently as

$$L\frac{di}{dt} + Ri = \mathscr{E} \tag{28-10}$$

This is the basic equation for the R-L circuit. Its solution, subject to the initial condition $i = 0$ at $t = 0$, constitutes a complete description for the current at any time t.

To solve (28-10), let us first express it in the form

$$\frac{di}{i - \mathscr{E}/R} = -\frac{dt\,R}{L}$$

and then integrate:

$$\ln\left(\frac{\mathscr{E}}{R} - i\right) = -\frac{tR}{L} + \alpha$$

In order to satisfy the initial condition $i = 0$ at $t = 0$, the integration constant α must be $\ln(\mathscr{E}/R)$. Hence the final formula for the current in the *R-L* circuit is

$$i = \frac{\mathscr{E}}{R}(1 - e^{-Rt/L}) \tag{28-11}$$

Figure 28-7 shows a plot of i as a function of time. The current rises steadily from its initial value of zero and approaches its saturation value $\mathscr{E}/R$ asymptotically. In other words, after a long time, the current in the circuit assumes the value which it would have had if the inductance were *not* in the circuit. In this context by a "long time" is meant a period of time that is long compared to the *R-L* "time constant" τ_L, defined by

$$\tau_L = \frac{L}{R} \tag{28-12}$$

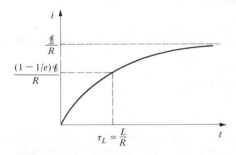

Figure 28-7

As shown in the figure τ_L represents the time interval from the closing of the switch to the instant at which the current has risen to $(1 - 1/e)$ of its final value, $\mathscr{E}/R$.

It is interesting to contrast the behavior of this *R-L* circuit with the corresponding behavior of the *R-C* circuit. In the latter the capacitor acts initially as a short, so that at first the current flows as if the capacitor were not present. After a long time, when the charge on the capacitor has built up, it acts as a large resistor and prevents the flow of current in the circuit. By contrast, in the *R-L* circuit, the inductor opposes any changes in magnetic flux through its coils so that initially it acts as a very large resistor, and stops any current flow. After a long time, however, the current achieves the steady value it would have if the inductance were not present, so that, effectively, it acts as a short.

Example 28-5 A 20-volt battery is suddenly connected across a 0.2-henry inductor that has a resistance of 100 Ω. Calculate:
(a) The *R-L* time constant.
(b) The potential drop across the inductor at $t = \tau_L$.
(c) The final current.

Solution

(a) Substituting the given values for R and L into (28-12) we obtain

$$\tau_L = \frac{L}{R} = \frac{0.2\text{ H}}{100\ \Omega} = 2.0 \times 10^{-3}\text{ s}$$

(b) The potential drop across the inductor is $(\mathcal{E} - Ri)$. Making use of (28-11) we obtain then, at $t = \tau_L$,

$$\mathcal{E} - Ri = \mathcal{E}e^{-t/\tau_L} = \frac{\mathcal{E}}{e} = 0.37 \times 20\text{ V} = 7.4\text{ V}$$

since $1/e \cong 0.37$ and the emf of the battery is 20 volts.

(c) After a long time, the current $i(\infty)$ in the circuit is the same as if the inductor were not present. Thus

$$i(\infty) = \frac{\mathcal{E}}{R} = \frac{20\text{ V}}{100\ \Omega} = 0.2\text{ A}$$

Example 28-6 Consider the two-loop network in Figure 28-8. Assuming that at $t = 0$ the switch S is closed, calculate:
(a) The initial current through the battery.
(b) The initial potential drop across the inductor.
(c) The final currents through R_1 and R_2.

Solution

(a) Just after the switch is closed, the inductor will not allow any current to pass and thus acts as an infinite resistor. Initially, then, the current will flow along the path $ABCDEFA$, and thus according to Ohm's law the current i_0 through the battery is

$$i_0 = \frac{\mathcal{E}}{R_1 + R_2}$$

(b) The potential drop V_L across the inductor must be the same as that across R_2. Making use of the result of (a), we find that

$$V_L(0) = R_2 i_0 = \frac{R_2 \mathcal{E}}{R_1 + R_2}$$

(c) After a long time, steady-state conditions prevail, and the potential drop across the inductor, $L\, di/dt$, will vanish. Thus, there will be no current through R_2 since the potential drop across R_2 must be the same as that across L. The final value for the current $i(\infty)$ through R_1 is therefore

$$i(\infty) = \frac{\mathcal{E}}{R_1}$$

Example 28-7 Consider the circuit in Figure 28-9 and suppose that at $t = 0$ the switch S is closed. Calculate:
(a) The initial current through the battery.
(b) The final current through the battery.
(c) The final charge Q_f on the capacitor.

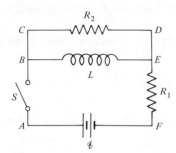

Figure 28-8

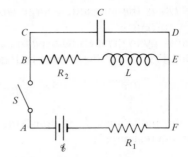

Figure 28-9

Solution

(a) Just as the switch is closed, the capacitor acts as a short, and the inductor behaves as if it were a resistor of infinite resistance. Hence, the initial current $i(0)$ travels along the path $ABCDEFA$, and since the total resistance in this path is R_1 we conclude that

$$i(0) = \frac{\mathscr{E}}{R_1}$$

(b) After a long time, when steady-state conditions prevail, the capacitor acts as an infinite resistor and the inductor acts as a short. Accordingly, the final current $i(\infty)$ follows the path $ABEFA$ and, since the total resistance along this path is $(R_1 + R_2)$, it has the value

$$i(\infty) = \frac{\mathscr{E}}{R_1 + R_2}$$

(c) Since in the final state the potential drop across the inductor vanishes, the potential across the capacitor must be the same as that across R_2. Hence

$$\frac{Q_f}{C} = R_2 i(\infty)$$

or, in other words, the final charge Q_f on the capacitor is

$$Q_f = CR_2 i(\infty) = \frac{CR_2\mathscr{E}}{R_1 + R_2}$$

where the final equality follows by use of the result of (b).

28-5 Magnetic energy

In Chapter 22 we found that associated with a charged capacitor there is the electrical energy $U_E = Q^2/2C$. The purpose of this section is to establish the analogous formula for inductors:

$$U_M = \frac{1}{2} Li^2 \tag{28-13}$$

where U_M is the magnetic energy stored in an inductor L through which flows a current i.

By analogy to the method used to derive (23-25), let us multiply the circuit equation (28-10) for the circuit in Figure 28-6 by the current i:

$$\mathscr{E}i = Ri^2 + iL\frac{di}{dt}$$

Since the inductance L is constant, and $d(i^2)/dt = 2i\,di/dt$, this may be written equivalently as

$$\mathscr{E}i = Ri^2 + \frac{d}{dt}\left(\frac{1}{2}Li^2\right)$$

Now the term $\mathscr{E}i$ represents the rate at which the battery carries out work on the other circuit elements. Furthermore, as we saw in the analogous derivation involving the R-C circuit, the term Ri^2 is the rate at which heat is dissipated in the resistor. It follows then from the ideas of energy conservation that the term $d(Li^2/2)/dt$ must represent the rate at which *energy is being stored in the magnetic field of the inductor*. The validity of (28-13) is thus established.

To illustrate, let us consider the special case of a long ideal solenoid of length l, radius a, and N turns. According to (28-4), the self-inductance L of the coil is $\mu_0 N^2 \pi a^2/l$. Hence, assuming a current i flows, we find that

$$U_M = \frac{1}{2}Li^2 = \frac{1}{2}\frac{\mu_0 N^2 \pi a^2}{l}i^2 \qquad (28\text{-}14)$$

Example 28-8   Consider the R-L circuit in Figure 28-6 for the special parameter values $\mathscr{E} = 100$ volts, $R = 50\ \Omega$, and $L = 0.1$ henry. Calculate the maximum value for the energy stored in the inductor.

Solution   Since the current in the circuit is a monotonically increasing function (see Figure 28-7), it follows by use of (28-13) that the maximum value for U_M occurs after a long time. Since the saturation value $i(\infty)$ for the current is $\mathscr{E}/R$, we obtain

$$U_M = \frac{1}{2}L[i(\infty)]^2 = \frac{1}{2}L\left(\frac{\mathscr{E}}{R}\right)^2 = \frac{1}{2}\times(0.1\ \text{H})\times\left(\frac{100\ \text{V}}{50\Omega}\right)^2 = 0.2\ \text{J}$$

Example 28-9   Confirm for the special case of an ideal solenoid, of N turns, length l, and radius a, that the *magnetic energy per unit volume* u_M may be expressed in the form

$$u_M = \frac{B^2}{2\mu_0} \qquad (28\text{-}15)$$

Solution   Since the B-field inside an ideal solenoid is $\mu_0 Ni/l$, (28-14) may be expressed as

$$U_M = \frac{1}{2\mu_0}\left(\frac{\mu_0 Ni}{l}\right)^2 l\pi a^2 = \frac{B^2 l\pi a^2}{2\mu_0}$$

and since the volume of the solenoid is $l\pi a^2$, (28-15) follows directly. Although here derived only for the ideal solenoid, (28-15) is found to be valid very generally.

28-6 Oscillations of the *L-C* circuit

An interesting illustration of the notion of electromagnetic energy deals with the discharge of a capacitor through an inductor. Consider, in Figure 28-10, a capacitor that has an initial charge Q_0 and suppose that at $t = 0$ it is connected across an inductor L. Although there is inevitably some resistance in any circuit, for the moment let us neglect it. If $i = i(t)$ represents the current in the circuit at any time t and $q(t)$ is the charge on the capacitor at this instant, then, the circuit equation is

$$\frac{1}{C}q = -L\frac{di}{dt} \qquad (28\text{-}16)$$

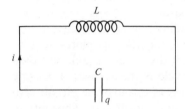

Figure 28-10

Furthermore, assuming that the right-hand plate of the capacitor has its charge q, the current is dq/dt, and therefore the circuit equation may be expressed equivalently as

$$\frac{d^2q}{dt^2} + \frac{1}{LC}q = 0 \qquad (28\text{-}17)$$

The solution of this relation, subject to the initial conditions $q(0) = Q_0$ and $i(0) = 0$, gives a formula for the charge on the capacitor at any time t.

To solve (28-17), let us recall the equation of motion for the simple harmonic oscillator in (6-20); that is,

$$\frac{d^2x}{dt^2} = -\omega^2 x \qquad (6\text{-}20)$$

and its solution in terms of the constants of integration A and α:

$$x(t) = A\cos(\omega t + \alpha) \qquad (6\text{-}21)$$

Comparison of (28-17) with (6-20) and (6-21) shows then that the solution of (28-17) is

$$q(t) = A\cos\left(\frac{t}{\sqrt{LC}} + \alpha\right)$$

where, just as in (6-21), A and α are integration constants. Imposing the initial conditions $q(0) = Q_0$ and $i(0) = 0$, we find that $A = Q_0$ and $\alpha = 0$. The solution of (28-17) is therefore

$$q(t) = Q_0\cos\omega t \qquad (28\text{-}18)$$

with ω, the *angular frequency of the oscillations*, defined by

$$\omega = \frac{1}{\sqrt{LC}} \qquad (28\text{-}19)$$

Differentiating (28-18), we obtain a formula for the current in the circuit:

$$i = \frac{dq}{dt} = -\omega Q_0 \sin \omega t \qquad (28\text{-}20)$$

The minus sign here is consistent with the fact that for $\omega t < \pi/2$, the capacitor discharges, and thus the direction of the current in the circuit is opposite to that assumed in the figure. (Recall that the right-hand plate of the capacitor was originally positive.)

The solutions in (28-18) through (28-20) show that the charge on the capacitor varies sinusoidally in time with period $P = 2\pi/\omega = 2\pi(LC)^{1/2}$. Immediately after being connected, the capacitor starts to discharge and continues to do so until, at time $\pi/2\omega$, $q = 0$. Thereafter its plates acquire charges of the opposite polarity, after which it discharges again, and then finally it regains its original charge Q_0. The cycle is then repeated. During this charging and discharging process, the current through the circuit also varies in time and increases, decreases, and reverses itself at the appropriate times. The current achieves its maximum value (ωQ_0) or minimum value $(-\omega Q_0)$ when the charge on the capacitor is zero, and vanishes when the charge on the capacitor is $\pm Q_0$.

It is of considerable interest to examine the results in (28-18) and (28-20) from the viewpoint of electric and magnetic energy. The electrical energy U_E stored in the capacitor at time t is $q^2/2C$, and thus

$$U_E = \frac{1}{2C} q^2 = \frac{Q_0^2}{2C} \cos^2 \omega t \qquad (28\text{-}21)$$

Similarly, the magnetic energy U_M stored in this stystem at time t is $\frac{1}{2}Li^2$, and hence

$$U_M = \frac{1}{2} Li^2 = \frac{1}{2} L \omega^2 Q_0^2 \sin^2 \omega t \qquad (28\text{-}22)$$

$$= \frac{Q_0^2}{2C} \sin^2 \omega t$$

where the final equality follows by use of the definition for ω in (28-19). Note that, in accordance with conservation of energy ideas, the total energy $U = U_E + U_M$ is a time-independent constant with the expected initial value $Q_0^2/2C$. (Recall that $\sin^2 \omega t + \cos^2 \omega t = 1$.) Figure 28-11 shows a plot of these formulas for U_E and U_M as functions of time. At $t = 0$, $U_M = 0$ and U_E has its maximum value $Q_0^2/2C$. As time goes on, U_E decreases as the capacitor discharges, while the magnetic energy U_M rises with the current. At the instant $t = \pi/2\omega$ the capacitor is completely discharged, and thus U_E vanishes, whereas the current achieves a maximum. The total energy U is

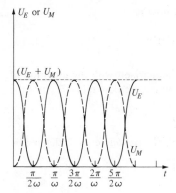

Figure 28-11

then all magnetic. After this the capacitor starts to be charged up with the opposite polarity until, at time $t = \pi/\omega$, the current falls to zero and the total energy is again stored entirely in the electric field of the capacitor. In this way, then, the total energy oscillates back and forth between the magnetic energy in the inductor and the electrical energy in the capacitor.

Example 28-10   A 0.1-μF capacitor has an initial charge of 2.0 μC and is connected across a 0.1-henry inductor. Calculate:
  (a) The total energy in the system.
  (b) The period of oscillation.
  (c) The maximum current flow in the circuit.

Solution
  (a) According to (28-21), the maximum energy stored in the capacitor U_E is

$$U_E = \frac{Q_0^2}{2C} = \frac{(2.0 \times 10^{-6}\,\text{C})^2}{2 \times (1.0 \times 10^{-7}\,\text{F})}$$

$$= 2.0 \times 10^{-5}\,\text{J}$$

  (b) Since the period P of the oscillation is $2\pi/\omega$, it follows, from (28-19), that

$$P = \frac{2\pi}{\omega} = 2\pi\sqrt{LC} = 2\pi[(0.1\,\text{H}) \times (1.0 \times 10^{-7}\,\text{F})]^{1/2}$$

$$= 6.3 \times 10^{-4}\,\text{s}$$

  (c) According to (28-20), the maximum current i_{max} is ωQ_0. Hence

$$i_{max} = \omega Q_0 = \frac{Q_0}{\sqrt{LC}} = \frac{2.0 \times 10^{-6}\,\text{C}}{[0.1\,\text{H} \times 1.0 \times 10^{-7}\,\text{F}]^{1/2}}$$

$$= 0.02\,\text{A}$$

Example 28-11   Suppose that a resistor R is connected in series with the other elements of an *L-C* circuit. Assuming again that the capacitor has an original charge Q_0, find the charge on the capacitor at time t.

Solution In place of (28-17), the circuit equation now is

$$L\frac{d^2q}{dt^2} + R\frac{dq}{dt} + \frac{1}{C}\,q = 0$$

since the potential drop across the resistor is Ri. Comparison with the equation of motion for a damped harmonic oscillator in (6-35) and its solution in (6-37) gives the solution

$$q(t) = Ae^{-Rt/2L}\cos{(\omega't + a)}$$

where, as in (6-37), A and α are integration constants. The parameter ω' is defined by

$$\omega' = \left[\frac{1}{LC} - \left(\frac{R}{2L}\right)^2\right]^{1/2}$$

and is real, if R is assumed to be small. It is left as an exercise to verify the validity of this solution and to calculate the values of A and α in terms of the initial conditions $q(0) = Q_0$ and $i(0) = 0$.

28-7 Diamagnetism, paramagnetism, and ferromagnetism

In studying dielectrics we found it convenient to make use of the uniform electric field produced in a parallel-plate capacitor. In a similar way, in order to study the magnetic properties of matter it is convenient to make use of the uniform magnetic induction **B** that exists in the interior of an ideal solenoid. In practice, measurements in the laboratory are usually carried out by use of a toroid. However, since the **B**-field in a toroid is not uniform it is much simpler conceptually to carry out the analysis by use of a solenoid.

Suppose that a very long ideal solenoid has a length l and a radius a, and consists of N turns. A measurement of the self-inductance L_0 of this coil leads, in accordance with (28-4), to the value

$$L_0 = \frac{\mu_0 N^2 \pi a^2}{l} \tag{28-23}$$

where the subscript on L_0 is to emphasize the fact that this formula is valid provided that no magnetizable matter is nearby.

Consider again this solenoid, but suppose now that it is wound on a long, homogeneous cylinder of matter of radius a and of length l. A measurement of the self-inductance of the coil under this circumstance yields a new value, L, which, in general, will differ from the value L_0. However, further experimentation shows that the dependency of L on the parameters N, l, and a is precisely that given in (28-23). In other words, L differs from L_0 only by an overall factor. By analogy to the definition of the dielectric constant κ in Section 22-8, we define the *relative permeability* κ_m of the sample by

$$L = \kappa_m L_0 \tag{28-24}$$

so that κ_m is the ratio of the inductances of the coil with and without matter being present. As will be seen in the following section, physically the parameter κ_m represents the ratio B/B_0, where, for a given current i in the coil, B is the **B**-field in the sample, and B_0 is the **B**-field in the solenoid after the sample is removed. A related parameter, χ_m, is also frequently used to characterize magnetic materials. It is known as the *magnetic susceptibility* and is defined by

$$\chi_m = \kappa_m - 1 \qquad (28\text{-}25)$$

Note that both κ_m and χ_m are dimensionless.

Before turning to a physical interpretation of magnetic susceptibility in the next section, let us summarize briefly the experimental situation. For the case of dielectrics, we found that the electric susceptibility χ_e is always positive. For magnetic materials the situation is not quite so simple. Experiments show that both positive and negative values for χ_m occur in nature.

It is convenient to classify magnetic substances into three main categories: *diamagnets*, *paramagnets*, and *ferromagnets*.[1]

A *diamagnetic material* is one whose magnetic susceptibility χ_m is very small and *negative*. A typical value is of the order of -10^{-5}. Table 28-1 lists (static) values of χ_m for a number of diamagnetic materials at room temperature. By contrast, a *paramagnetic material* is one for which the magnetic susceptibility, although still very small in magnitude, has a *positive* numerical value. In Table 28-1 values for χ_m for several paramagnetic substances are also listed.

Table 28-1 Magnetic susceptibilities at room temperature

Diamagnetic materials		Paramagnetic materials	
Substance	χ_m	Substance	χ_m
Bi	-1.7×10^{-5}	Al	2.5×10^{-5}
Cd	-2.3×10^{-6}	Ca	1.4×10^{-5}
C (diamond)	-2.2×10^{-5}	O_2 (gas)	1.8×10^{-6}
Cu	-1.1×10^{-6}	Pt	2.0×10^{-4}
Hg	-2.4×10^{-5}	Th	1.6×10^{-6}
Pb	-1.6×10^{-5}	W	3.5×10^{-6}

The third class of magnetic materials of particular interest consists of *ferromagnets*. Generally speaking, at room temperature most substances are either diamagnetic or paramagnetic. For these the magnetic susceptibility is generally independent of the strength of the current in the coils used to produce the magnetic behavior. By contrast, the elements Fe, Co, Ni, Gd, and Dy, and some of their alloys and oxides, are ferromagnetic and exhibit an altogether different behavior. On measuring the inductance of a coil

---

[1] A more complete listing would also include ferrimagnets and antiferromagnets.

wound around a ferromagnet, we find for χ_m large positive values, typically of the order of several hundred to several thousand. That is, the magnetic susceptibility of ferromagnets is 7 to 8 orders of magnitude larger than that for paramagnetic substances! Furthermore, not only do the values of χ_m, as so determined, vary with temperature, but they also depend on the strength of the current in the coils. In addition, as will be discussed in the next section, ferromagnets exhibit a phenomenon known as *hysteresis*, one of whose manifestations is a dependency of χ_m on the history of the specimen. The fact that χ_m for a ferromagnet varies with the current in the coils implies, for example, that the inductance L of such an inductor depends not only on the geometric parameters characterizing it but on the current through it as well.

To sum up, then, by measuring the self-inductance of a coil wrapped around a sample of material we may ascertain its magnetic susceptibility χ_m in accordance with (28-24) and (28-25). Materials for which χ_m has a small negative value (independent of the current) are known as diamagnets. Paramagnets are identical to diamagnets in many of their observable properties except for the fact that χ_m is small and positive. Finally, ferromagnets are materials for which χ_m is very large ($\sim 10^3$) and positive. For a ferromagnetic material χ_m varies, in general, with the strength of the current through the surrounding coils. Moreover, associated with each ferromagnetic material there is a critical temperature, known as the *Curie-point*, above which the material becomes paramagnetic. For example, the Curie point of iron is 1043 K. This means that if iron is heated to a temperature above 1043 K it loses all of its ferromagnetic properties and behaves as an ordinary paramagnetic substance. Similarly, the ferromagnetic element gadolinium (Gd) becomes paramagnetic at 289 K.

28-8 Magnetization currents

Consider a long, ideal solenoid of N turns and length l, and carrying a current i. According to (28-2), the magnetic flux (assuming no magnetizable bodies nearby) Φ_m° through a single turn of a coil is

$$\Phi_m^\circ = \frac{L_0 i}{N} \tag{28-26}$$

with L_0 the inductance of the coil.

Consider again this same coil, but suppose this time it is wrapped around a cylinder of matter of magnetic susceptibility χ_m (see Figure 28-12). The flux Φ_m through a single turn of the coil is now

$$\Phi_m = \frac{Li}{N} = \frac{(1 + \chi_m)L_0 i}{N}$$

$$= (1 + \chi_m)\Phi_m^\circ \tag{28-27}$$

where the second equality follows by use of (28-24) and (28-25), and the last from (28-26). Thus, for paramagnetic and ferromagnetic materials (for which $\chi_m > 0$), the magnetic flux Φ_m through the coil is increased whereas for diamagnetic materials (for which $\chi_m < 0$) it is decreased. It is convenient to think of this increase—or decrease, for the case of a diamagnetic substance— of magnetic flux through the sample in terms of a *magnetization current* i_m flowing on its surface. Figure 28-12 illustrates the case for $\chi_m > 0$, for which the direction of flow of i_m is along that of the current i through the coils. For diamagnetic materials for which $\Phi_m < \Phi_m^\circ$, the direction of this magnetization current is in the opposite direction. Note that in both cases the total **B**-field inside the sample is the vector sum of the fields produced by each of the currents: i and i_m. This way of characterizing **B**-fields inside matter in terms of i_m is particularly useful for ferromagnetic substances which may exhibit magnetic behavior even when the external current i vanishes.

To discuss the physical mechanism that underlies these magnetization currents, let us consider first a homogeneous paramagnetic substance. Studies in quantum mechanics have shown that for these, the constituent molecules may be thought of as small magnetic dipoles. That is, the molecules of a paramagnetic substance behave as if they consisted of small current loops, which under normal circumstances are oriented randomly, so there is no net magnetic effect due to them. Consider now, in Figure 28-13, a cross section of such a paramagnetic material inside a long solenoid around which flows a certain current i. The microscopic current loops in the sample find themselves in a certain magnetic induction field $\mathbf{B}_0$ and, in accordance with (26-24), they experience a torque, which tends to line them up with their normals parallel to $\mathbf{B}_0$. Reference to the figure shows that when they are lined up, these currents due to neighboring loops cancel in the interior of the sample. Hence there is no magnetization current in this region. However, this cancellation cannot take place at the surface. Effectively, therefore, a certain magnetization current i_m flows on the surface along the direction of the causal current i. The total **B**-field inside the sample is due to this total current $(i + i_m)$ on its surface.

With regard to ferromagnetic materials, the situation is very similar, although more complex in detail. If the sample is initially unmagnetized, then its microscopic current loops are oriented at random as for the case of a

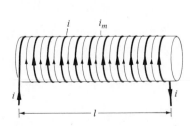

Figure 28-12

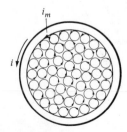

Figure 28-13

paramagnet. If the sample is then placed into an external **B**-field, these magnetic dipoles, or small current loops, again tend to line up with the external **B**-field, just as for the case of paramagnetic substance, but this time in a much more organized way. Indeed, for ferromagnetic materials, for which $\chi_m \gg 1$, the surface current i_m is very much greater than that in the surrounding coils, and thus the torque experienced by a given dipole is due mainly to the **B**-field produced by its neighbors. Moreover, the collective effect of these torques between the dipoles is so strong that for temperatures below the Curie point, they will remain lined up even after the external current is removed. From a macroscopic point of view, they thereby constitute a permanent magnet. This behavior is to be contrasted with that in a paramagnetic material, whose magnetization current vanishes as the causal current does.

To see the kinds of effects observed in laboratory experiments with ferromagnets consider, in Figure 28-14, a plot of the **B**-field inside a ferromagnetic sample as a function of the current i in the surrounding coils. Suppose that originally the sample is unmagnetized and is thus at the point A in the figure. As the current is turned on and gradually increased, the **B**-field inside the sample is found to increase steadily and to saturate at a maximum value at C. As the current is subsequently decreased, instead of retracing the original curve AC backward the **B**-field is found to follow along a different path CD, associated with consistently higher values for $|\mathbf{B}|$. In particular, even when the current through the coils vanishes, the **B**-field in the sample at point D does not! In other words, the sample has now become a *permanent magnet*. As noted above, once a ferromagnetic specimen has been magnetized, an external **B**-field is not required to keep the dipoles aligned; the fields produced by the individual dipoles are sufficient for this purpose. As the sense of the current around the sample is reversed, the **B**-field variations now follow curve DEF, and a second reversal of the current then leads to an increasing **B**-field along path FGH, and finally to saturation at C. Again, at point G there is a residual field, but now with a sense opposite to that at D. It

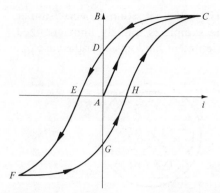

Figure 28-14

is because of these residual fields at D and G that one often says that the magnetic susceptibility of ferromagnets depend on their past history. The curve AC in the figure is usually referred to as the magnetization curve, and the closed loop $CDEFGHC$ is called a *hysteresis loop*.

Finally, let us consider diamagnetic substances, for which $\chi_m < 0$ and for which the magnetization current i_m flows in a direction opposite to the current in the coils. Consider, in Figure 28-15a, an electron in a circular orbit perpendicular to a uniform magnetic induction $\mathbf{B}_0$. Because of this $\mathbf{B}$-field, the electron experiences the force $q\mathbf{v} \times \mathbf{B}$, and therefore, because of its negative charge it will orbit about a field line in the sense shown. However, because of its negative charge, the electron produces a circular current i with sense opposite to that of its velocity. Hence the magnetic induction $\mathbf{B}_e$ due to the current associated with this orbiting electron will be directed opposite to the causal field $\mathbf{B}_0$.

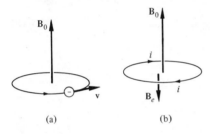

(a)   (b)

Figure 28-15

Consider now a sample of a diamagnetic material, such as bismuth, in a solenoid that carries a current i. The free electrons in their motions through the lattice will behave as small current loops just as for the case of paramagnets. But this time the magnetization current on the surface of the sample will be *opposite* to that of the external current i. Hence, the effective magnet flux through the sample is decreased and in accordance with (28-27) the magnetic susceptibility for diamagnetic materials is negative.

To summarize, then, in a paramagnetic material the constituent molecules can be thought of as small current loops, which in response to the torque produced by an external $\mathbf{B}$-field tend to line up along that field. A ferromagnet is one for which this lining up is very much more organized than for a paramagnet. Finally, in a diamagnet the electrons can be thought of as orbiting in circles about the lines of magnetic induction, thereby producing a $\mathbf{B}$-field directed opposite to the external one. It should be noted that more detailed studies show that most substances have diamagnetic as well as paramagnetic properties and that the observed magnetic behavior of any substance depends on which one of these dominates.

†28-9 The magnetization vector

In discussing dielectrics we found it convenient to define a vector **P** to represent the number of electric dipoles per unit volume in the sample. In a similar way, for magnetic materials it is convenient to define a *magnetization vector* **M** as the number of magnetic dipoles per unit volume in the material.

Consider, in Figure 28-16, a region of space in which there are magnetized bodies. Let l be an arbitrary, directed curve in this region, and let i_m represent the net magnetization current, which flows along the positive sense (defined by the right-hand rule) through an open surface bounded by l. The curve l itself may be entirely inside the material or partially inside and partially outside. In terms of i_m, define the *magnetization vector* **M** associated with this material by the requirement

$$\oint_l \mathbf{M} \cdot d\mathbf{l} = i_m \qquad (28\text{-}28)$$

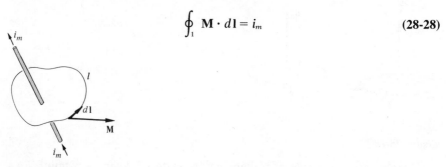

Figure 28-16

where i_m is the magnetization current flowing through any open surface bounded by the curve l, and the line integral is around l. It is implicit here that this relation is valid for *all* closed curves l. In the space outside of material bodies $i_m = 0$, and, consistent with (28-28), we shall assume that $\mathbf{M} \equiv 0$ in such regions. For the same reason that Ampère's law, (25-12), does not, in general, associate a unique **B**-field with a given current distribution, (28-28) does not determine a unique **M** for a given i_m either. Nevertheless, this definition suffices for our needs.

To determine the relation between **M** and the magnetic susceptibility χ_m of a material consider again, as in Figure 28-12, an ideal solenoid wrapped around a magnetic substance. If a current i is sent through the coil, a certain magnetization current i_m will flow on the surface of the sample. Let us define the *magnetic field* **H** inside the sample so that it satisfies the relation

$$\oint_l \mathbf{H} \cdot d\mathbf{l} = i \qquad (28\text{-}29)$$

where, as in Ampère's law, the line integral is around a closed curve and i is the net current flowing through an open surface S bounded by that curve, with the positive sense given by the right-hand rule. To understand the significance of **H**, let us compare (28-29) with Ampère's law,

$$\oint_l \mathbf{B} \cdot d\mathbf{l} = \mu_0(i + i_m) \qquad \text{(28-30)}$$

where the form of the right-hand side follows since the *total* current flowing through a surface bounded by l is $(i + i_m)$. Comparing this with (28-29), we see that the magnetic field **H** for the situation in Figure 28-12 is the same as the product of μ_0^{-1} and the **B**-*field in the absence of magnetization current*. In other words, if no matter is present, so that $i_m = 0$, then up to a factor μ_0 the two fields **H** and **B** are identical.

There is a simple algebraic relation between **B**, **H**, and **M**, which may be obtained as follows. Substituting for the currents i and i_m in (28-30) by use of (28-28) and (28-29) and assuming the same curve l in all three integrals, we obtain

$$\oint \mathbf{B} \cdot d\mathbf{l} = \mu_0(i + i_m)$$

$$= \mu_0 \left\{ \oint \mathbf{H} \cdot d\mathbf{l} + \oint \mathbf{M} \cdot d\mathbf{l} \right\}$$

Hence, since the curve l is arbitrary, it follows that[2]

$$\mathbf{B} = \mu_0(\mathbf{H} + \mathbf{M}) \qquad \text{(28-31)}$$

Although derived only for the system in Figure 28-12, this relation is applicable very generally and is basic to all discussions of the magnetic behavior of matter.

We shall now derive the relation between **M** and χ_m. For the situation in Figure 28-12 the flux Φ_m through a single turn of the coil is

$$\Phi_m = \int_S \mathbf{B} \cdot d\mathbf{S}$$

where the integral goes over a surface bounded by a turn. Correspondingly, the magnetic flux Φ_m° through a single turn of the coil in the absence of the sample is

$$\Phi_m^\circ = \int_S \mu_0 \mathbf{H} \cdot d\mathbf{S}$$

since the magnetic induction **B** is the same as $\mu_0 \mathbf{H}$ in the absence of matter. Substituting these formulas for Φ_m and Φ_m° into (28-27), we find, assuming a "linear material" for which χ_m is independent of current and therefore of **H**, that

$$\int_S \mathbf{B} \cdot d\mathbf{S} = \mu_0(1 + \chi_m) \int_S \mathbf{H} \cdot d\mathbf{S}$$

---

[2]Strictly speaking, since $\oint d\mathbf{l} = 0$ for any closed curve l (Example 26-7), this argument only shows that the vector $[\mathbf{B} - \mu_0(\mathbf{H} + \mathbf{M})]$ is a constant. However, since $\mathbf{M} = 0$ outside of matter, and since for the ideal solenoid $\mathbf{H} = \mathbf{B} = 0$ outside, it follows that this constant is zero.

Since the surface S is arbitrary, this is equivalent to

$$\mathbf{B} = \mu_0(1 + \chi_m)\mathbf{H} \tag{28-32}$$

and combining this with (28-31), we obtain the sought-for formula

$$\mathbf{M} = \chi_m \mathbf{H} \tag{28-33}$$

for the magnetization $\mathbf{M}$ inside the sample.

The value for χ_m, as we saw above, can be determined by a measurement of inductance. Hence, since $\mathbf{H}$ is known independently it follows that $\mathbf{M}$ itself can be measured. The definition in (28-28) then enables us to calculate the magnetization currents i_m associated with the given sample.

Example 28-12 Suppose that a current of 2.0 amperes flows in the coils in Figure 28-12 and that the sample is iron, for which $\chi_m = 200$. Assuming that there are 10^4 turns per meter and that (28-32) and (28-33) are applicable, calculate:
(a) The magnetic field strength $\mathbf{H}$ in the sample.
(b) The magnetization vector $\mathbf{M}$.
(c) The magnetic induction $\mathbf{B}$.

Solution
(a) In the absence of the sample $\mathbf{B} = \mu_0\mathbf{H}$. Since $\mathbf{B}$ is uniform and parallel to the axis of the coil and has a strength $\mu_0 ni$, it follows that the magnitude of the magnetic field is

$$H = ni = (10^4/\text{m}) \times 2.0\ \text{A} = 2.0 \times 10^4\ \text{A/m}$$

(b) Using the given value $\chi_m = 200$ and the above value for H, we find by use of (28-33) that

$$M = \chi_m H = (200) \times (2.0 \times 10^4\ \text{A/m})$$
$$= 4.0 \times 10^6\ \text{A/m}$$

(c) The substitution of these values into (28-32) leads to

$$B = \mu_0(1 + \chi_m)H = (4\pi \times 10^{-7}\ \text{T-m/A}) \times (201) \times (2.0 \times 10^4\ \text{A/m}) = 5.0\ \text{T}$$

The directions of $\mathbf{B}$, $\mathbf{H}$, and $\mathbf{M}$ are all parallel to the axis of the coil, and with their sense given by the right-hand rule.

It is interesting to note that the effect of the ferromagnetic core here is to increase the $\mathbf{B}$-field inside the solenoid by the factor $(1 + \chi_m) \approx 200$. The iron core thus effectively decreases the current required to produce a given $\mathbf{B}$-field by a factor $(1 + \chi_m)$. Equivalently, the magnetic induction $\mu_0\mathbf{H}$ produced by the external current i is negligible compared to the $\mathbf{B}$-field inside the sample. Thus, we can think of a material characterized by a very large χ_m as one which tends to pull $\mathbf{B}$-lines into its interior. This feature plays an important role in variety of applications; for example, in the design of galvanometers, as we saw in Section 26-7.

28-10 Summary of important formulas

The self-inductance L of a loop is defined by

$$\Phi_m = Li \tag{28-1}$$

where Φ_m is the total flux through the loop when it carries a current i. The magnetic energy U_M stored in the loop when it carries a current i is

$$U_M = \frac{Li^2}{2} \tag{28-13}$$

If L_0 is the self-inductance of a coil, then the inductance L of the same coil when wound around a homogeneous sample is given by

$$L = \kappa_m L_0 \tag{28-24}$$

where κ_m is the relative permeability of the sample.

QUESTIONS

1. Define or describe briefly what is meant by the following terms: (a) mutual inductance; (b) relative permeability; (c) magnetic susceptibility; and (d) magnetization vector.

2. What important feature (or features) distinguishes diamagnetic, paramagnetic, and ferromagnetic materials from each other?

3. Making use of (28-3), explain why the rules for combining inductors in series and parallel are the same as those for combining resistors.

4. Two identical loops of wire, each of one turn, are placed near each other so that they almost coincide. If L is the self-inductance of either loop, why must the mutual inductance of the pair also have the value L?

5. Associated with any real circuit there is, in general, some resistance. Explain in physical terms why it is that associated with any circuit there is also inevitably some inductance. For what type of circuits would this effect be a serious problem?

6. Is it possible for the self-inductance of a coil to have a negative value? What about the mutual inductance of two coils? Explain.

7. Explain in physical terms why and under what circumstances an inductor in a circuit behaves as an infinite resistor. Does the magnetic energy $Li^2/2$ stored in the inductor have anything to do with this feature?

8. Explain in terms of magnetic energy why and under what circumstances an inductor in a circuit behaves as a short; that is, as a resistor with zero resistance.

9. In more advanced studies, it is shown that a formula for the mutual inductance M between two coils is

$$M = \mu_0 \oint_{l_1} \oint_{l_2} \frac{d\mathbf{l}_1 \cdot d\mathbf{l}_2}{r}$$

where the symbols are as defined in Figure 28-17. Show by use of this formula that the reciprocity law in (28-7) is valid.

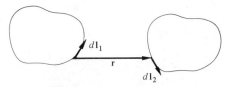

Figure 28-17

10. Consider the formula in (28-4) for the self-inductance of an ideal solenoid. Consider the limiting case $N \to \infty$, $l \to \infty$, with the ratio $N/l = n$ fixed. Give a physical interpretation for the fact that L diverges in this limit, and explain why this is not a practical problem.

11. Explain in terms of magnetic energy, why it is not possible to alter suddenly the current in a circuit that contains an inductance. Can you

think of a practical use for this feature?

12. Suppose that an ammeter were inserted in the R-L circuit in Figure 28-6. Describe the motion of its pointer as a function of time if (a) $\tau_L = 1$ ms; (b) $\tau_L = 5$ seconds. (Assume that a full-scale reading on the ammeter corresponds to the maximum current in the circuit.)

13. If a voltmeter is connected across the inductor in Figure 28-6, describe the motion of the pointer if (a) $\tau_L = 1$ ms; (b) $\tau_L = 5$ seconds. (Assume that $\mathscr{E}$ is the full-scale deflection on the voltmeter.)

14. An unmagnetized iron nail is brought near a permanent magnet. Why will the nail be attracted to the magnet? If a second nail is brought near the first, what happens?

15. Explain in microscopic terms the distinction between paramagnetism and diamagnetism. What would happen to a diamagnetic needle suspended in a magnetic field? Contrast this behavior with that of a paramagnetic needle.

16. Consider a magnetized iron rod. Explain in microscopic terms why, if

this rod is cut in half, a new north and south pole appear at the cut surfaces. Could an isolated magnetic pole be obtained in this way? Explain.

17. Because isolated magnetic poles do not exist, the flux of **B** out of any closed surface vanishes. Why is this not true, in general, for the flux of **H** out of a closed surface? Under what circumstances is the relation

$$\oint_S \mathbf{H} \cdot d\mathbf{S} = 0$$

true?

18. What happens to a ferromagnet when its temperature is raised above the Curie point?

19. Suppose a long, thin rod is inserted into an inductor in an R-L circuit. If the time constant of the circuit is found to increase slightly, is the rod diamagnetic, ferromagnetic, or paramagnetic? Explain.

20. A coil of wire carrying a current i is wrapped around a long, iron rod. Explain why most of the lines of magnetic induction will be in the interior of the rod even if the coil is of relatively short length.

PROBLEMS

1. A long, ideal solenoid is 2 cm in diameter and has 40 turns/cm.
   (a) What is the self-inductance per meter of this coil?
   (b) If a current of 1.5 amperes flows through the coil, what is the magnetic flux in a single turn?
   (c) What is the total flux through the turns in 0.5 meter of the coil?

2. A long, ideal solenoid has 2×10^3 turns and an inductance of 2.0 henry.
   (a) What is the total magnetic flux

through the solenoid if a current of 0.3 ampere flows?
   (b) If the solenoid has a cross sectional area of 2 cm², what is the strength of the magnetic induction inside the coil when this current flows?

3. A toroid has a square cross section, an inner radius of 10 cm, and an outer radius of 12 cm. If it has 300 turns:
   (a) Calculate the self-inductance.
   (b) Calculate the magnetic flux through a single turn if a steady current of 3.0 amperes flows.

*4. A toroid of inner radius a and outer radius b has a circular cross section of radius $(b - a)/2$ and N turns. Calculate the self-inductance L of this coil. (*Hint*: Use the method in Example 28-2.)

5. Consider the formula for the self-inductance of a toroid in (28-5). Show that in the limit as the radius of the toroid becomes very large (while its cross-sectional area is fixed) the formula for its inductance per unit length reduces to that for a very long, straight solenoid.

6. A coil of 10 turns is wrapped around a very long, ideal solenoid of self-inductance 5.0 mH and of 10^3 turns.

(a) If a current i flows in the solenoid, what is the magnetic flux through a single turn of the *coil*?

(b) What is the mutual inductance between the coil and the solenoid?

7. A coil of 100 turns is wrapped around a long, ideal solenoid of self-inductance 2.0 henry and having 20,000 turns.

(a) What is the mutual inductance between the coil and the solenoid?

(b) If a current of 10 amperes flows in the coil, what is the magnetic flux through a single turn of the solenoid? What is the total flux through the entire solenoid?

8. Suppose the toroid in Figure 28-18 has 100 turns, an inner radius of 20 cm, an outer radius of 25 cm, and a square cross section.

(a) What is the self-inductance of the toroid?

(b) What is the mutual inductance between the toroid and the loop?

(c) If a current of 2.0×10^{-2} ampere flows in the toroid what is the magnetic flux through the loop?

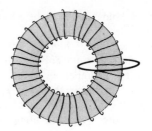

Figure 28-18

*9. Consider the situation in Figure 28-5. If a current i flows in the rectangular loop, what is the total magnetic flux through the area bounded by the two parallel wires? Assume that there is no current flowing in the long wires. (*Hint*: Use the formula for M in Example 28-4.)

10. A 10-volt battery is suddenly connected across a 10-mH inductor and a 100-Ω resistor in series.

(a) What is the initial current in the circuit?

(b) What is the initial potential across the inductor?

(c) What is the initial value for the time rate of change of the current?

11. Consider the same circuit as in Problem 10.

(a) What is the time constant associated with this circuit?

(b) What current flows in the circuit at the instant $t = \tau_L$?

(c) What is the potential drop across the resistor at $t = 10\tau_L$?

12. What value for the inductance is required in an R-L circuit so that for a resistance of 75 Ω the time constant will be 2.0 ms? If the emf of the battery is 100 volts, how much energy is stored in the inductor under steady-state conditions?

13. A certain R-L circuit has a time constant τ_L. Show that the time $t_{1/2}$ that must elapse for the current to achieve half of its final steady-state value is $t_{1/2} = \tau_L \ln 2$.

14. A man holds in his right hand a long, ideal solenoid of self-inductance 0.2 henry, around which flows the current $i = \alpha + \beta t$, with $\alpha = 10$ amperes and $\beta = -10$ A/s. Suppose that for $t \leqslant 1$ second, the direction of the current through the coils is along his fingers.
    (a) What is the magnitude of the emf induced in the solenoid at $t = 0.5$ second?
    (b) Does the man's thumb point in the direction of increasing or decreasing potential?

15. Two inductors L_1 and L_2 are very far away from each other.
    (a) Show that the equivalent inductance L if they are connected in series is $L = L_1 + L_2$.
    (b) Show that the equivalent inductance if they are in parallel is
    $$L = \frac{L_1 L_2}{L_1 + L_2}$$

16. Consider the circuits in Figure 28-19 and suppose that M is the mutual inductance between the inductors. Show that the circuit equations are:
    $$R_1 i_1 + L_1 \frac{di_1}{dt} + M \frac{di_2}{dt} = \mathscr{E}$$
    $$R_2 i_2 + L_2 \frac{di_2}{dt} + M \frac{di_1}{dt} = 0$$
    What are the values for i_1 and i_2 after a long time when steady-state conditions have been reached?

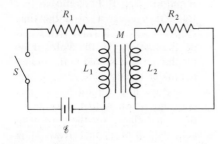

R_1 R_2 M L_1 L_2 S $\mathscr{E}$

Figure 28-19

*17. Using the method of Section 28-5 and the results of Problem 16, verify that the magnetic energy U_M stored in the circuits in Figure 28-19 is
    $$U_M = \frac{1}{2} L_1 i_1^2 + \frac{1}{2} L_2 i_2^2 + M i_1 i_2$$

18. Consider two coaxial ideal solenoids of radii a and b ($> a$), respectively. Suppose that the outer one has N_2 turns and the inner one N_1 turns, and that they have the same length l.
    (a) Show that the B-field at a distance r from the axis has the magnitude
    $$B = \frac{\mu_0}{l} \begin{cases} (N_1 i_1 + N_2 i_2) & r \leqslant a \\ N_2 i_2 & a \leqslant r \leqslant b \\ 0 & r \geqslant b \end{cases}$$
    where i_1 and i_2 are the currents in the inner and outer coils, respectively.
    (b) Calculate the magnetic energy U_M by use of (28-15).
    (c) Using your result to (b) and Problem 17, show that the mutual inductance M of these solenoids agrees with that in Example 28-3.

19. If a voltmeter were connected across the inductor in Figure 28-6, show that at time t after the switch is closed it would read $\mathscr{E} e^{-t/\tau_L}$.

*20. Two inductors L_1 and L_2 are connected in series in a way so that their mutual inductance is M. Show that the inductance L of the equivalent inductor is
    $$L = L_1 + L_2 + 2M$$

*21. If the two inductors of Problem 20 are connected in parallel, show that the equivalent inductance is
    $$L = \frac{L_1 L_2 - M^2}{L_1 + L_2 - 2M}$$

22. Two inductors, each having a self-inductance of 0.4 henry, are connected in parallel across a 20-volt battery.
    (a) If the final steady current through the battery is 2.0 amperes, what is the resistance in the circuit?
    (b) If the R-L time constant is found to be 0.05 second, what is the equivalent inductance?
    (c) Making use of the result of Problem 21, calculate the mutual inductance of the two coils.

23. Consider the circuit in Figure 28-20 and suppose that the switch S is closed at $t = 0$.

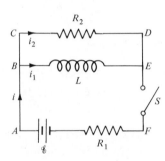

Figure 28-20

    (a) Describe the path followed by the current immediately after the switch is closed.
    (b) What is the path followed by the current after a long time?

24. Assuming in Figure 28-20 the parameter values $\mathscr{E} = 20$ volts, $R_1 = 10\,\Omega$, $R_2 = 50\,\Omega$, and $L = 50\,mH$, calculate:
    (a) The initial values for i, i_1, and i_2.
    (b) The values for i, i_1, and i_2 after a long time.
    (c) The maximum value for the magnetic energy stored in the inductor.

25. Consider the circuit in Figure 28-21 and suppose that the parameters have the values $\mathscr{E} = 50$ volts, $R_1 =$

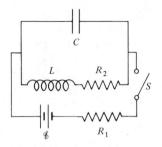

Figure 28-21

$50\,\Omega$, $R_2 = 10\,\Omega$, $L = 2.0$ henry, and $C = 5\,\mu F$. If the switch is closed at $t = 0$:
    (a) What is the initial current through the battery?
    (b) What is the current through the battery after a long time?
    (c) What is the final charge on the capacitor?

26. For the circuit in Figure 28-21 calculate:
    (a) The final value for the magnetic energy stored in the inductor.
    (b) The final electric energy stored in the capacitor.
    (c) The rate at which energy is dissipated in R_1 under steady-state conditions.
    Use the parameter values in Problem 25.

27. For the circuit in Figure 28-22 assume the parameter values $\mathscr{E} = 40$ volts, $R_1 = 10\,\Omega$, $R_2 = 60\,\Omega$, $R_3 = 30\,\Omega$, $R_4 = 20\,\Omega$, $C = 2\mu F$, $L_1 = 1.0$

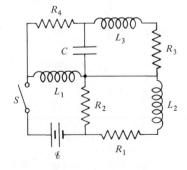

Figure 28-22

henry, $L_2 = 2.0$ henry, $L_3 = 3.0$ henry, and calculate:

(a) The initial value of the current through the battery.

(b) The final current through R_2.

(c) The final current through R_1.

(d) The final current through the battery.

28. For the physical situation described in Problem 27 calculate the final energy stored in each inductor and in the capacitor.

29. A capacitor of capacitance $2.0~\mu F$ has an initial charge of $5.0~\mu C$ and is connected across a 2.0-mH inductor.

(a) What is the angular frequency of the oscillations of this circuit?

(b) What is the period?

(c) What is the maximum energy stored in the inductor?

*30. A 2.0-μF capacitor with an original charge of $2.0~\mu C$ is connected across a 20-Ω resistor and a 2.0-mH inductance, as in Figure 28-23. By reference to the results of Example 28-11, calculate:

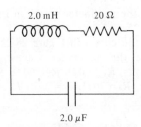

2.0 mH   20 Ω

2.0 μF

Figure 28-23

(a) The angular frequency of oscillations if the resistor were not present.

(b) The actual angular frequency ω'.

(c) The total energy dissipated in the resistor.

31. A time-dependent emf $\mathscr{E}\cos\omega_0 t$ ($\mathscr{E}, \omega_0$ positive constants) is con-

nected across an L-C combination, as in Figure 28-24.

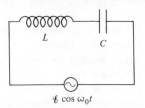

L C

$\mathscr{E}\cos\omega_0 t$

Figure 28-24

(a) Show that if $q(t)$ is the charge on the capacitor at time t, then the circuit equation may be written

$$L\frac{d^2q}{dt^2} + \frac{q}{C} = \mathscr{E}\cos\omega_0 t$$

(b) Verify that

$$q(t) = A\cos(\omega t + \phi)$$
$$+ \frac{(\cos\omega_0 t)\mathscr{E}/L}{(\omega^2 - \omega_0^2)}$$

where $\omega = 1/(LC)^{1/2}$ and A and ϕ are integration constants, is a solution of (a).

(c) Explain in physical terms what happens as the external frequency ω_0 becomes equal to the *resonant* frequency $(LC)^{-1/2}$ of this L-C combination.

*32. Consider again the circuit in Figure 28-24, but suppose that this time a resistor R is in series with the inductor.

(a) Write down the circuit equation in terms of the charge q on the capacitor and its time derivatives.

(b) Verify that a solution of this equation is

$$q(t) = \frac{\mathscr{E}\cos(\omega_0 t + \alpha)}{\left[\left(\omega_0^2 L - \frac{1}{C}\right)^2 + R^2\omega_0^2\right]^{1/2}}$$

where α satisfies

$$\sin \alpha = \frac{-R\omega_0}{\left[\left(\omega_0^2 L - \frac{1}{C}\right)^2 + R^2\omega_0^2\right]^{1/2}}$$

33. Making use of the results of Problem 32 for the particular choice of $L = 1.0$ mH and $C = 1.0$ μF, make a plot of the maximum potential drop V_{max} across the resistor as a function of ω_0 for the following choices: **(a)** $R = 1\ \Omega$; **(b)** $R = 100\ \Omega$; and **(c)** $R = 10^3\ \Omega$. Explain by reference to your graphs why this circuit can be used as a frequency selector, for example, in a radio.

†34. Show by use of (28-28) that if **M** is the magnetization vector associated with a given magnetization current i_m, then $(\mathbf{M} + \mathbf{M}_0)$, where $\mathbf{M}_0$ is an *arbitrary* but constant vector, also satisfies this relation. How do we resolve this ambiguity in practice?

†35. Show by use of (28-32) that for a sample for which $\chi_m = 500$, the ratio of the strength of the magnetic induction inside the sample to that outside is approximately 500. It is for this reason that we often say that if a ferromagnetic material is introduced into a region containing a magnetic induction it tends to "suck" the field lines into its interior.

†36. A steel rod having a length of 0.5 meter has a relative permeability of 1000 and is initially unmagnetized. If we uniformly wrap around it 100 turns of wire, what current is required so that the magnetic induction inside the iron is 1.5 tesla? Calculate also the strength of **M** and **H** inside the iron.

†37. Figure 28-25 shows the basic elements of a *transformer*. N_1 turns of one coil, called the *primary* coil, are wound around a ring of high permeability such as iron. A coil, called

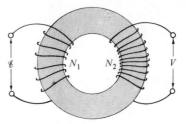

Figure 28-25

the *secondary*, having N_2 turns is also wound about the ring.

(a) If a current i flows in the primary, why is it that the magnetic flux through a single turn of either coil is the same to a high degree of approximation.

(b) Show that if a time-dependent emf $\mathcal{E}$ is applied to the primary, then the voltage V across the secondary is

$$V = \frac{\mathcal{E}N_2}{N_1}$$

(*Note*: If $N_1 < N_2$, then $V > \mathcal{E}$, and the transformer is said to be a *step-up* transformer. Conversely, if $N_1 > N_2$, then it is a *step-down* transformer.)

†38. (a) By constructing an appropriate Gaussian surface, part of which lies in the interior of magnetizable matter, show by use of Gauss' law for magnetism that the normal components of **B** are continuous across the interface.

(b) Show by use of Ampère's law that because of magnetization currents, the tangential components of **B** are *not* in general continuous across the surface of a magnetizable material.

†39. By arguments similar to those in Problem 38 show that the tangential components of **H** are continuous, but that the normal components of **H** may be discontinuous, across the interface between two magnetic materials.

29 Maxwell's equations and electromagnetic waves

*One cannot escape the feeling that these mathematical formulas
[Maxwell's equations] have an independent existence and an intelligence
of their own, that they are wiser than we are, wiser even than their
discoverers, that we get more out of them than we put into them.*

HEINRICH R. HERTZ (1857–1894)

Who will then explain the explanation!

BYRON

29-1   Introduction

For the past ten chapters we have been studying the electric and magnetic
fields produced by charges at rest and in motion. The important results of
this study are the two laws of Gauss (one for electricity and one for
magnetism) and the laws of Ampère and of Faraday. The laws governing
specialized phenomena, such as those associated with electrostatics and
magnetostatics, are all derivable from these four fundamental laws. Except
for the addition to Ampère's law of a term known as the *displacement
current* which we shall discuss in Section 29-3, these four relations comprise
Maxwell's equations. When supplemented by certain empirical relations

such as Ohm's law, Maxwell's equations constitute a complete and quantitative description of *all* electromagnetic phenomena.

29-2   Recapitulation

As a preliminary to a discussion of Maxwell's equations and electromagnetic waves, in this section we review briefly the four basic relations of electromagnetism.

First, there is *Gauss' law* for electricity. This states that the total electric flux out of every closed surface S is the same as the product of $1/\epsilon_0$ and the *total* charge $q(S)$ *inside* that surface. In mathematical terms, then,

$$\oint_S \mathbf{E} \cdot d\mathbf{S} = \frac{1}{\epsilon_0} q(S) \qquad \text{(Gauss)} \qquad (29\text{-}1)$$

where $\mathbf{E}$ is the electric field and, as in Figure 29-1a, the surface S is closed.

Because of the fact that magnetic monopoles—that is, isolated north or south magnetic poles—do not exist, the corresponding *Gauss' law for magnetism* is simpler and states that the total magnetic flux out of every closed surface vanishes. In mathematical terms, we may express this as

$$\oint_S \mathbf{B} \cdot d\mathbf{S} = 0 \qquad \text{(Gauss)} \qquad (29\text{-}2)$$

where $\mathbf{B}$ is the magnetic induction and where again, as in Figure 29-1a, the surface S is presumed to be closed. Note that (29-2) is valid regardless of what currents or permanent magnets may be present inside or outside of the closed surface S.

A third important property of the electromagnetic field is given by *Faraday's law*. This states that the electromotive force developed about any closed path l is the same as the negative of the rate of change of magnetic

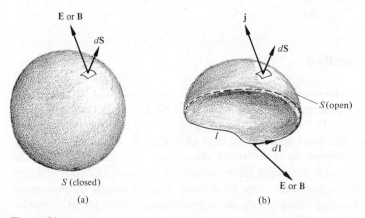

E or B

dS

S (closed)

(a)

j

dS

S(open)

l

dl

E or B

(b)

Figure 29-1

flux through every open surface S bounded by l. In the notation defined in Figure 29-1b, Faraday's law is

$$\oint_l \mathbf{E} \cdot d\mathbf{l} = -\frac{d}{dt} \int_S \mathbf{B} \cdot d\mathbf{S} \qquad \text{(Faraday)} \qquad (29\text{-}3)$$

where the sense of the bounding curve and that of the outward normal to the surface S are related by the right-hand rule (Section 27-6).

Finally, there is *Ampère's law*. This may be expressed by the formula

$$\oint_l \mathbf{B} \cdot d\mathbf{l} = \mu_0 \int_S \mathbf{j} \cdot d\mathbf{S} \qquad \text{(Ampère)} \qquad (29\text{-}4)$$

where, as in Figure 29-1b, l is a directed, closed curve that bounds an arbitrary, but open, surface S. The current density $\mathbf{j}$ in this formula represents the vector sum of *all* currents, whether they be associated with the actual motion of charge or with magnetization currents inside or on the surface of matter. As for Faraday's law, the outward normal to S is assumed to be related to the sense of l by the right-hand rule.

It is worth emphasizing that all other relations, such as the equations of electrostatics or of magnetostatics or the law of Biot-Savart, are direct consequences of (29-1) through (29-4). For the phenomena of electrostatics, for example, $\mathbf{B}$ and $\mathbf{j}$ both vanish. Hence, (29-3) reduces for this case to

$$\oint_l \mathbf{E} \cdot d\mathbf{l} = 0$$

and the combination of this and Gauss' law, (29-1), constitute a complete specification of the electrostatic field. Similarly, the $\mathbf{B}$-field associated with a static current distribution may be calculated directly by use of (29-2) and (29-4). The law of Biot-Savart follows from these as a special case for currents in thin wires.

To summarize, then, the scope of the basic relations in (29-1) through (29-4) encompasses all phenomena of electromagnetism that we have studied up to now. However, these relations are *not* Maxwell's equations. Maxwell's equations result from (29-1) through (29-4) only after a certain inconsistency, associated mainly with Ampère's law, is removed.

29-3 Maxwell's equations

Maxwell was the first to point out that there is a difficulty with Ampère's law in (29-4) if it is assumed to be valid for *all* closed paths l. The details of his argument are somewhat complex and will be briefly touched upon in Section 29-10. In outline, Maxwell showed that Ampère's law in (29-4) leads to certain unphysical predictions if it is assumed to be valid for all closed curves. Specifically, he established that Ampère's law predicts that the net current flowing out of *every* closed surface must vanish. But it is easy to see that this prediction is false. For if, as shown in Figure 29-2, one plate of a capacitor is surrounded by a closed surface S (the dashed surface in the figure), then during the time that the capacitor is being charged up, there is a definite flow of current into S. That is, since charge is accumulating on the plate of the capacitor, there must be a net current flow onto the plate and thus into S. Therefore, to be consistent with experiment, Maxwell reasoned, that Ampère's law in (29-4) cannot be correct as it stands.

Bearing this difficulty in mind, and making use of arguments of the type discussed in Section 29-10, Maxwell proposed adding the term

$$\epsilon_0\mu_0 \frac{d}{dt}\int_S \mathbf{E}\cdot d\mathbf{S}$$

to the right-hand side of Ampère's law. With this modification, Ampère's law becomes

$$\oint_l \mathbf{B}\cdot d\mathbf{l} = \mu_0 \int_S \mathbf{j}\cdot d\mathbf{S} + \mu_0\epsilon_0 \frac{d}{dt}\int_S \mathbf{E}\cdot d\mathbf{S} \qquad (29\text{-}5)$$

where, as before, the sense of the surface S and its bounding curve l are related by the right-hand rule (Figure 29-1b). And it is this relation in (29-5) and the two laws of Gauss and that of Faraday in (29-1) through (29-3) that collectively are known as Maxwell's equations.

The above added term, which makes (29-5) differ from Ampère's law, is known as the *displacement current*, and it fell to Heinrich R. Hertz (1857–1894) to establish its undisputable physical reality by means of experiments involving electromagnetic waves. However, nearly two decades were to elapse from the time of Maxwell's original proposal until Hertz succeeded in demonstrating that—consistent with (29-5), Faraday's law, and the two laws of Gauss—electromagnetic waves do indeed exist. It is interesting to note in this connection that (29-1) through (29-4)—that is, Maxwell's equations without the displacement current—do *not* admit solutions corresponding to electromagnetic waves.

As noted earlier, our understanding of all electromagnetic phenomena

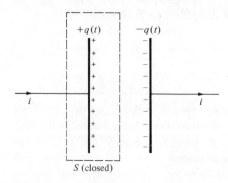

Figure 29-2

including those described in the preceding ten chapters is based on Maxwell's equations. Moreover, as will be shown in the following sections of this chapter, the existence and the properties of electromagnetic waves also follow from these relations. Maxwell's equations, in other words, describe not only the full range of ordinary electromagnetic phenomena, but also those of visible and ultraviolet light, radio waves, infrared radiation, X rays, and so forth. Hertz's observation that they "are wiser even than their discoverers" can hardly be classed as a metaphor.

Example 29-1 In the presence of material bodies characterized by a dipole moment per unit volume **P** and a magnetization **M**, Maxwell's equations consist of the two relations

$$\oint_S \mathbf{D} \cdot d\mathbf{S} = q_t(S)$$

$$\oint_l \mathbf{H} \cdot d\mathbf{l} = \int_S \mathbf{j}_t \cdot d\mathbf{S} + \frac{d}{dt} \int_S \mathbf{D} \cdot d\mathbf{S} \qquad (29\text{-}6)$$

plus Gauss' law for magnetism, (29-2), and Faraday's law, (29-3). Here **D** is the displacement vector as defined in (22-26), **H** is the magnetic field as defined in (28-29) and q_t and $\mathbf{j}_t$ are all charges and currents except those induced in dielectric and magnetic materials. Express (29-6) directly in terms of **E** and **B**, assuming that S and l lie in the interior of a homogeneous, isotropic, and linear material.

Solution According to (22-31), for a linear material the displacement vector **D** is related to **E** via

$$\mathbf{D} = \kappa \epsilon_0 \mathbf{E}$$

with κ the dielectric constant. Similarly, according to (28-25) and (28-32), **B** and **H** are related by

$$\mathbf{B} = \mu_0 \kappa_m \mathbf{H}$$

with κ_m the relative permeability. Substitution into (29-6) leads to

$$\oint_S \mathbf{E} \cdot d\mathbf{S} = \frac{1}{\kappa \epsilon_0} q_t(S)$$

$$\oint_l \mathbf{B} \cdot d\mathbf{l} = \mu_0 \kappa_m \int_S \mathbf{j}_t \cdot d\mathbf{S} + \epsilon_0 \mu_0 \kappa \kappa_m \frac{d}{dt} \int_S \mathbf{E} \cdot d\mathbf{S} \qquad (29\text{-}7)$$

where κ and κ_m have been taken out from under the integral signs since they are constant. Note that these relations reduce to (29-1) and (29-5), respectively, if the parameters $\kappa \epsilon_0$ and $\kappa_m \mu_0$ are replaced by ϵ_0 and μ_0, respectively.

29-4 The solutions of Maxwell's equations in free space

Let us consider Maxwell's equations—(29-1), (29-2), (29-3), and (29-5)—in a region of space very far away from any charges and currents. For simplicity, assume for the moment that no dielectrics or magnetizable bodies are nearby. Setting $q(S) = 0$, $\mathbf{j} = 0$, Maxwell's equations assume the form

$$\oint_S \mathbf{E} \cdot d\mathbf{S} = 0$$

$$\oint_S \mathbf{B} \cdot d\mathbf{S} = 0$$

(29-8)

$$\oint_l \mathbf{E} \cdot d\mathbf{l} = -\frac{d}{dt} \int_S \mathbf{B} \cdot d\mathbf{S}$$

$$\oint_l \mathbf{B} \cdot d\mathbf{l} = \epsilon_0 \mu_0 \frac{d}{dt} \int_S \mathbf{E} \cdot d\mathbf{S}$$

Note the symmetrical way that $\mathbf{E}$ and $\mathbf{B}$ appear in these relations; this feature is a direct consequence of the introduction of the displacement current. Note also that the solution $\mathbf{E} = \mathbf{B} = 0$ satisfies these equations. However, there are nonzero solutions as well. These correspond to electromagnetic waves and are the the subject matter of the remainder of this chapter.

Let us describe some general properties of these equations. First, they are *homogeneous* in $\mathbf{E}$ and $\mathbf{B}$, in that if $\mathbf{E}$ and $\mathbf{B}$ are two fields that satisfy (29-8) then so do the fields $\alpha \mathbf{E}$ and $\alpha \mathbf{B}$, for α an arbitrary constant. For the special choice $\alpha = 0$, we see that consistent with our previous studies, in the absence of charges and currents the electric and magnetic fields may vanish. Second, they are *linear*; that is, if $(\mathbf{E}_1, \mathbf{B}_1)$ and $(\mathbf{E}_2, \mathbf{B}_2)$ separately satisfy (29-8), then so do $(\alpha \mathbf{E}_1 + \beta \mathbf{E}_2)$, $(\alpha \mathbf{B}_1 + \beta \mathbf{B}_2)$ for arbitrary constants α and β. This linearity property is particularly important and it is related directly to the *superposition principle* for the electric field and the magnetic field. Note that according to this superposition principle *any* linear combination of solutions of (29-8) is also a solution.

To display a particular nonzero solution, let $\mathbf{E}_0$ be an arbitrary, but constant, vector with dimensions of electric field. In terms of $\mathbf{E}_0$ define a Cartesian coordinate system with $\mathbf{E}_0$ along the x-axis, and let $\mathbf{i}$, $\mathbf{j}$, and $\mathbf{k}$ form a triad of orthogonal unit vectors along its axes (Figure 29-3a). If h is an arbitrary function of a *single variable*, then:

The electromagnetic fields defined by

$$\mathbf{E} = \mathbf{i} E_0 h(z - ct)$$

(29-9)

$$\mathbf{B} = \mathbf{j} \frac{E_0}{c} h(z - ct)$$

satisfy the charge-current-free Maxwellian equations in (29-8), *provided that the parameter c is defined by*

$$c = (\mu_0 \epsilon_0)^{-1/2}$$

(29-10)

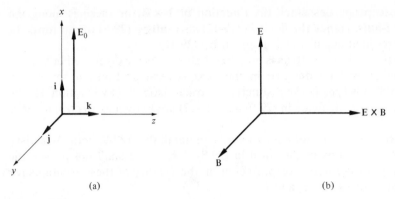

Figure 29-3

In other words, the electromagnetic fields in (29-9) satisfy (29-8) for *any* constant E_0 and *any* function h, provided that the parameter c is defined by (29-10). Inserting the known values $\epsilon_0 = (4\pi \times 9.0 \times 10^9 \text{ N-m}^2/\text{C}^2)^{-1}$ and $\mu_0 = 4\pi \times 10^{-7}$ T-m/A, we find that

$$c = 3.0 \times 10^8 \text{ m/s}$$

which is precisely the speed of light in a vacuum! More generally, as we shall see in Section 29-5, the parameter c is the speed not only of visible light but of *all* electromagnetic waves in a vacuum.

Although the solution in (29-9) has been defined in terms of the coordinate system in Figure 29-3a, the significant features of this solution are actually independent of the system chosen. The following properties deserve particular emphasis:

1. **E** and **B** are mutually perpendicular.
2. **E** and **B** vary spatially only along the direction of **E** × **B**. If for example, in Figure 29-3b **E** is along the *x*-axis and **B** is along the *y*-axis, then **E** and **B** are functions only of *z* and *not* of *x* and *y*.
3. The space-time variation of both **E** and **B** is described in its entirety by the single variable $(z - ct)$, where *z* is the coordinate along the direction of **E** × **B**.
4. The ratio of the magnitudes of **E** and **B** is

$$\frac{|\mathbf{E}|}{|\mathbf{B}|} = c = (\mu_0 \epsilon_0)^{-1/2} \tag{29-11}$$

and is thus a *universal constant* whose value can be determined in terms of purely electric and magnetic quantities.

For example, the electromagnetic fields **E** and **B** defined by

$$\mathbf{E} = \mathbf{i} E_0 h(z + ct)$$

$$\mathbf{B} = -\mathbf{j} \frac{E_0}{c} h(z + ct) \tag{29-12}$$

have these properties since the direction of $\mathbf{E} \times \mathbf{B}$ for these is along the *negative* z-axis. Hence the fields in (29-12) also satisfy (29-8) for arbitrary E_0 and h, provided again that c is given by (29-10).

Of particular interest to us is property 3 above. As we discussed in Section 18-4 and, as will be detailed in the next several sections, a space-time variation of this type is *the* distinctive characteristic of wave motion. Hence the electromagnetic fields in (29-9) and (29-12) are known as *electromagnetic waves*.

Unfortunately, it is not easy to verify in detail that (29-9) actually satisfy Maxwell's equations in the form in (29-8). Hence we shall not pursue the matter any further here except to confirm the validity of these solutions for particular choices for S and l.

Example 29-2   Verify that (29-9) satisfy the first two relations in (29-8) for the special case where the closed surface is a cube of side L, as in Figure 29-4.

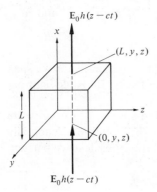

Figure 29-4

Solution   Since the direction of $\mathbf{E}$ is along the x-axis, it follows that there is no contribution to the flux integral $\oint_S \mathbf{E} \cdot d\mathbf{S}$ from the front or back face or from either of the side faces. (For each element of area $d\mathbf{S}$ of each of these faces, $\mathbf{E} \cdot d\mathbf{S} = 0$.) The flux out of an element of area $dy\,dz$ of the top face with coordinates (L, y, z) is $E_0 h(z - ct)dy\,dz$. Correspondingly, the *outward* flux through the element of the bottom face with coordinates $(0, y, z)$ and of area $dy\,dz$ is $-E_0 h(z - ct)dy\,dz$, since $\mathbf{E}$ and $d\mathbf{S}$ are antiparallel on this face. Hence

$$\oint_S \mathbf{E} \cdot d\mathbf{S} = \int_0^L dy \int_0^L dz\, E_0 h(z - ct) - \int_0^L dy \int_0^L dz\, E_0 h(z - ct) = 0$$

thus confirming the validity of the first of (29-9) for this special case.

The argument that the magnetic flux out of the cube in Figure 29-4 vanishes follows along the same lines. This time we find an exact cancellation between the flux out of the front and back faces of the cube. The flux out of each of the remaining faces vanishes identically since for each of these $\mathbf{B} \cdot d\mathbf{S} = 0$.

Example 29-3   Show that (29-9) satisfies the fourth relation of (29-8) for the special choice where S is a square of side L in the y-z plane (see Figure 29-5).

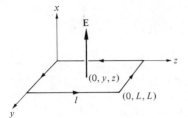

Figure 29-5

Solution   Assuming that the positive sense of $d\mathbf{S}$ is along the positive x-axis, the electric flux through the surface is:

$$\int \mathbf{E} \cdot d\mathbf{S} = \int_0^L \int_0^L (\mathbf{i} \, dy \, dz) \cdot (\mathbf{i} E_0 h(z - ct)) = \int_0^L dy \int_0^L dz \, E_0 h(z - ct)$$

$$= E_0 L \int_0^L dz \, h(z - ct)$$

since $d\mathbf{S} = \mathbf{i} \, dy \, dz$ in this case, and $\mathbf{i} \cdot \mathbf{i} = 1$. Similarly, since $\mathbf{B}$ has a component only along the y-axis, we obtain for the line integral of $\mathbf{B}$ around the perimeter of the square

$$\oint \mathbf{B} \cdot d\mathbf{l} = \int_0^L (\mathbf{j} \, dy) \cdot \left[\mathbf{j} \frac{E_0}{c} h(-ct) \right] + \int_L^0 (\mathbf{j} \, dy) \cdot \left[\mathbf{j} \frac{E_0}{c} h(L - ct) \right]$$

$$= \frac{L E_0}{c} [h(-ct) - h(L - ct)]$$

since $\mathbf{j} \cdot \mathbf{j} = 1$ and since along the left side of the square $z = 0$ while on the right $z = L$. If we now substitute these formulas into the fourth relation in (29-8) there results

$$\frac{L E_0}{c} [h(-ct) - h(L - ct)] = \epsilon_0 \mu_0 \frac{d}{dt} E_0 L \int_0^L dz \, h(z - ct)$$

$$= \epsilon_0 \mu_0 E_0 L \int_0^L dz \frac{d}{dt} h(z - ct)$$

$$= \epsilon_0 \mu_0 E_0 L \int_0^L dz \left[-c \frac{d}{dz} h(z - ct) \right]$$

$$= -\epsilon_0 \mu_0 E_0 L c \, h(z - ct) \Big|_0^L$$

$$= \epsilon_0 \mu_0 E_0 L c [h(-ct) - h(L - ct)]$$

where the third equality follows from the identity

$$\frac{d}{dx} f(x + y) = \frac{d}{dy} f(x + y)$$

which is valid for any function f. (See Example 18-5 in this connection.) Finally, then, because of (29-10) this is an identity for all values for the parameters E_0 and L and for any choice of the function h. The validity of (29-9) for the particular surface in Figure 29-5 is thus confirmed.

29-5 The propagation of electromagnetic waves

In order to understand better the physical significance of the solutions in (29-9) and (29-12) let us review some of the important ideas developed in Chapter 18 for wave motion on a string. Consider a string stretched along the x-axis of a certain coordinate system and under a tension T_0 (Figure 18-7). If a disturbance is generated somewhere along the string and propagates, say, along the positive sense of the x-axis, then the displacement $y = y(x, t)$ at time t of an element of the string at the point x may be expressed as

$$y = h(x - ut) \tag{18-5}$$

Correspondingly, for a wave traveling along the negative sense of the x-axis, the displacement is of the form $y = h(x + ut)$. In these formulas, h is an arbitrary function of a single variable and the parameter u characterizes the speed of the wave along the string. (Its value is given by $u = (T_0/\mu)^{1/2}$, where μ is the mass per unit length of the string.) As shown in Figure 18-3, the actual displacements of the constituent particles of the string are perpendicular to the direction of propagation of the wave and thus the wave is *transverse*. By contrast, for a *longitudinal* wave, such as a sound wave in air, the motions of the constituent particles are along the direction of propagation of the wave. Finally, since the motions of the constituent particles in Figure 18-6a are along the y-axis, the wave is also said to be *plane polarized* or *linearly polarized* along this axis. Thus, the wave shown in Figure 18-6b is linearly polarized along the z-axis. As exemplified in Figure 18-6c, not all waves on a string need be plane polarized.

Let us now return to the solution in (29-9) of the charge-current-free Maxwell equations. In light of the preceding discussion, we see that this solution represents a wave of some type traveling at the speed c along the positive sense of the z-axis. We shall refer to it as an *electromagnetic wave*. The "displacement" of the constituents of the wave are the fields **E** and **B** in (29-9) themselves, and since these fields are perpendicular to the direction of propagation of the wave it follows that electromagnetic waves are *transverse*. It is customary to define the direction of polarization of the wave by the direction of the electric vector. Therefore, the particular electromagnetic wave defined by (29-9) is linearly polarized along the x-axis. Similarly, the fields in (29-12) correspond to a transverse wave polarized along the x-axis and traveling along the negative sense of the z-axis.

Very generally, then, the charge-current-free Maxwellian equations in (29-8) admit wave solutions corresponding to transverse electromagnetic waves traveling at the speed c. The electric and magnetic fields **E** and **B** of the wave are perpendicular to each other and to the direction of propagation of the wave, $\mathbf{E} \times \mathbf{B}$. The ratio $|\mathbf{E}|/|\mathbf{B}|$ for all electromagnetic waves traveling through empty space is the constant c, defined by (29-10).

Although the two classes of solutions in (29-9) and (29-12) correspond to linearly polarized electromagnetic waves, unpolarized solutions of (29-8) also exist. This follows most directly from the fact that linear combinations of solutions of (29-8) are also solutions. For example, the electromagnetic fields given by

$$\mathbf{E} = E_0[\mathbf{i}h_1(z - ct) + \mathbf{j}h_2(z + ct)]$$

$$\mathbf{B} = \frac{E_0}{c}[\mathbf{j}h_1(z - ct) + \mathbf{i}h_2(z + ct)]$$

for h_1 and h_2 arbitrary functions satisfy (29-8) and correspond to an unpolarized electromagnetic wave.

Figure 29-6 shows at $t = 0$ an instantaneous picture of a polarized electromagnetic wave pulse associated with a particular function $h(z)$. The electric field vector of the wave lies in the x-z plane, and the corresponding magnetic field is in the y-z plane. Except for a scale factor, the envelopes of the vectors for these two fields are the same. In the course of time, as the wave travels along the positive sense of the z-axis, the equation for the envelope of the electric vector evolves as $x = h(z - ct)$ and similarly for $\mathbf{B}$. Consistent with our discussion of traveling waves in Chapter 18, we can visualize the propagation of the pulse in Figure 29-6 by imagining the entire pattern to travel along the z-axis at the speed c.

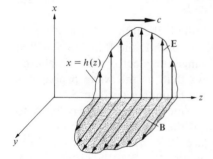

Figure 29-6

Example 29-4 The *index of refraction n* of a substance is defined as the ratio of the speed c of light in a vacuum to its speed v when traveling through that substance:

$$n = \frac{c}{v} \tag{29-13}$$

Calculate the index of refraction of a material characterized by a dielectric constant κ and a relative permeability κ_m.

Solution According to (29-7), in the absence of "true" charges and currents ($q_t = 0$; $j_t = 0$), the first, second, and third relations of (29-8) still apply, whereas the fourth must be replaced by

$$\oint_l \mathbf{B} \cdot d\mathbf{l} = \epsilon_0\mu_0\kappa\kappa_m \frac{d}{dt} \int_S \mathbf{E} \cdot d\mathbf{S}$$

Hence, comparing this relation with (29-8) through (29-10) we see that the speed v of waves in this medium is given by

$$v = [\epsilon_0 \mu_0 \kappa \kappa_m]^{-1/2} = \frac{c}{(\kappa \kappa_m)^{1/2}}$$

since $c = (\epsilon_0 \mu_0)^{-1/2}$. Finally, comparison with (29-13) leads to

$$n = (\kappa \kappa_m)^{1/2} \tag{29-14}$$

29-6 The Poynting vector

As for other types of wave motion, a flow of energy is generally associated with the propagation of an electromagnetic wave. The purpose of this section is to define a quantity known as the *Poynting vector*, in terms of which the flow of electromagnetic energy is conveniently described.

Consider, in Figure 29-7, a region of space through which an electromagnetic wave is propagating, and let S be a closed surface in this region. If $\mathbf{E}$ and $\mathbf{B}$ are the fields associated with the wave, then according to (28-15) and the result of Problem 26, Chapter 22, the total energy U inside S is

$$U = \frac{\epsilon_0}{2} \int_V \mathbf{E}^2 \, dV + \frac{1}{2\mu_0} \int_V \mathbf{B}^2 \, dV \tag{29-15}$$

where in both integrals the region of integration is the total volume V inside S. As the wave propagates along, the values of both $\mathbf{E}$ and $\mathbf{B}$ inside S will in general change, and associated with this change the electromagnetic energy U inside S will also be modified. It follows that, as the wave propagates along, energy will in general flow across the bounding surface S.

To characterize this energy flow, let us define the *Poynting vector* $\mathbf{N}$ to represent the magnitude and the direction of the energy flow per unit area per unit time. This means that if, in Figure 29-7, $d\mathbf{S}$ is a vectorial area element of the closed surface S, then the quantity $\mathbf{N} \cdot d\mathbf{S}$ is the outward flow of electromagnetic energy per unit time through $d\mathbf{S}$. In particular, the total flow of energy per unit time out of S is the surface integral

$$\oint_S \mathbf{N} \cdot d\mathbf{S}$$

with the integration extending over the entire closed surface S.

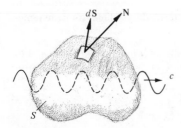

Figure 29-7

Now according to the ideas of energy conservation, if there is an outward flow of energy from S the total electromagnetic energy inside must change. Hence follows the basic relation

$$\frac{d}{dt} U + \oint_S \mathbf{N} \cdot d\mathbf{S} = 0 \qquad (29\text{-}16)$$

with U defined in (29-15). In physical terms, this states that as an electromagnetic wave propagates along, at each instant the rate at which the energy inside a given surface increases plus the outward flow of energy per unit time through the surface must vanish. Or equivalently, the rate at which the energy increases inside a given surface is numerically equal to the flow of energy per unit time *into* that surface.

By use of the formula for U in (29-15) and the fact that (29-16) is applicable for all closed surfaces S, an explicit formula for the Poynting vector in terms of $\mathbf{E}$ and $\mathbf{B}$ can be derived. The result is

$$\mathbf{N} = \frac{1}{\mu_0} \mathbf{E} \times \mathbf{B} \qquad (29\text{-}17)$$

and below we shall confirm the validity of this formula for a special case. Note that the direction of the Poynting vector is along the propagation direction $\mathbf{E} \times \mathbf{B}$ of the electromagnetic wave. Thus the wave defined by (29-9) propagates along the positive z-axis, which is along the direction of $\mathbf{N}$. Similarly, for the wave in (29-12), $\mathbf{E} \times \mathbf{B}$ is along the negative z-axis, consistent with the fact that the associated wave travels along this direction also.

It is interesting to note that momentum is also associated with the flow of electromagnetic energy in a wave. Detailed studies of Maxwell's equations show that if $\mathbf{E}$ and $\mathbf{B}$ are the fields at a given point in space, then the momentum density, or the momentum per unit volume $\mathbf{P}_v$ in that region, is

$$\mathbf{P}_v = \frac{1}{c^2} \mathbf{N} \qquad (29\text{-}18)$$

with $\mathbf{N}$ given in (29-17). A number of applications of this formula are given in the problems. One important physical prediction of (29-18) is that if an electromagnetic wave falls on matter it exerts mechanical pressure on it. This is known as *radiation pressure* and is the basis for a number of suggested proposals for space-propulsion systems that utilize the radiant energy output of the sun.

Example 29-5 Confirm the validity of (29-17) for the solution in (29-9) by using for S the cubical surface in Figure 29-4.

Solution Substituting the given forms for $\mathbf{E}$ and $\mathbf{B}$ into (29-17), we find that

$$\mathbf{N} = \frac{1}{\mu_0} \mathbf{E} \times \mathbf{B} = \frac{E_0^2}{\mu_0 c} h^2 (z - ct)\mathbf{k}$$

with $\mathbf{k} = \mathbf{i} \times \mathbf{j}$ a unit vector along the z-axis. Accordingly, the flow of energy per unit time $\mathbf{N} \cdot d\mathbf{S}$ out of the top, the bottom, and the front and back faces of the cube vanishes. Hence

$$\oint \mathbf{N} \cdot d\mathbf{S} = \frac{E_0^2}{\mu_0 c} \left[\int_0^L \int_0^L dx\, dy\, h^2(L - ct) - \int_0^L \int_0^L dx\, dy\, h^2(-ct) \right]$$

$$= \frac{L^2 E_0^2}{\mu_0 c} [h^2(L - ct) - h^2(-ct)] \tag{29-19}$$

since along the right face $z = L$ and on the left $z = 0$. Similarly, the substitution of the given forms for $\mathbf{E}$ and $\mathbf{B}$ into (29-15) leads to

$$U = \int_0^L \int_0^L \int_0^L dx\, dy\, dz \left[\frac{\epsilon_0}{2} E_0^2 h^2(z - ct) + \frac{1}{2\mu_0} \frac{E_0^2}{c^2} h^2(z - ct) \right]$$

$$= \epsilon_0 E_0^2 L^2 \int_0^L dz\, h^2(z - ct)$$

where the second equality follows from (29-10) and the fact that the integrand is independent of x and y. Making use of the same method as in Example 29-3 we find that

$$\frac{dU}{dt} = \frac{d}{dt} \epsilon_0 E_0^2 L^2 \int_0^L dz\, h^2(z - ct) = \epsilon_0 E_0^2 L^2 \int_0^L dz\, \frac{d}{dt} h^2(z - ct)$$

$$= -\epsilon_0 E_0^2 L^2 \int_0^L dz\, c \frac{d}{dz} h^2(z - ct) = -\epsilon_0 c E_0^2 L^2 h^2(z - ct) \Big|_0^L$$

$$= -\epsilon_0 c E_0^2 L^2 [h^2(L - ct) - h^2(-ct)]$$

The validity of (29-17) for this special case then follows by comparison with (29-19) and the relation $\epsilon_0 c = 1/\mu_0 c$.

29-7   Monochromatic waves

A *monochromatic* electromagnetic wave is analogous to a sinusoidal wave as defined in Chapter 18, and is one for which the function h in (29-9) or (29-12) is either a sine function or a cosine function. The term "monochromatic" refers to the fact that for visible light a sinusoidal wave is one composed of a single color, or frequency.

To be specific, let us make use of a sine function. Comparison with (18-8) shows then that for a monochromatic wave, (29-9) may be expressed as

$$\mathbf{E} = \mathbf{i} E_0 \sin(kz - \omega t)$$

$$\mathbf{B} = \mathbf{j} \frac{E_0}{c} \sin(kz - \omega t) \tag{29-20}$$

where k and ω are two parameters, known respectively as the *wave number* and the *angular frequency* of the wave. Since the space-time variation of an electromagnetic wave can only be in terms of the variable $(z - ct)$, or

$(z + ct)$, it follows that ω and k must be related by (compare (18-9))

$$\omega = kc \tag{29-21}$$

Hence only one of ω or k is required to specify the wave. Figure 29-8 shows an instantaneous picture of the linearly polarized electromagnetic wave defined by the fields in (29-20). The temporal evolution of the wave may be visualized by imagining the given pattern to be moving along the positive sense of the z-axis at the speed c.

The *wavelength* λ of a monochromatic wave is defined as the distance between two neighboring maxima, or minima, of the wave at a fixed time t; see Figure 29-8. As in the derivation of (18-10), it follows that λ and k are related by

$$\lambda = \frac{2\pi}{k} \tag{29-22}$$

and thus a knowledge of either one is tantamount to a knowledge of the other.

A fourth parameter of interest in connection with a monochromatic wave is its frequency ν. This is defined so that $1/\nu$ is the period of the wave, and thus $1/\nu$ represents the minimum positive time interval after which the wave repeats its form at each point of space. Substitution of this definition into (29-20) leads to the identification $\nu = \omega/2\pi$, and combining this with (29-22) enables us to express (29-21) equivalently as

$$\lambda\nu = c \tag{29-23}$$

To summarize, then, a monochromatic electromagnetic wave, such as the one described by (29-20) is specified by two independent parameters. One of these is its *amplitude E_0*, which, as we shall see in the next section, determines the amount of energy transported by the wave. The second is usually selected to be its wavelength λ or its frequency ν. These are related by (29-23).

Figure 29-9 shows diagrammatically the wavelength intervals and the associated frequency ranges for all known types of electromagnetic radiation. The totality of these radiations constitute the *electromagnetic spec-*

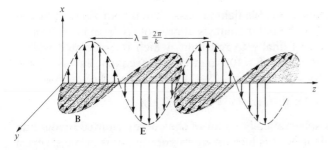

Figure 29-8

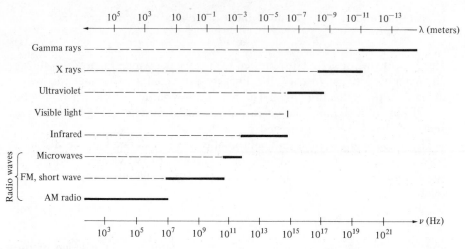

Figure 29-9

trum. At the long-wavelength end of this spectrum are *radio waves*, which are characterized by wavelengths typically of the order of 10 meters and longer, while at the other extreme we find *gamma rays*, with wavelengths of the order of 10^{-11} meter and shorter. Between these two extremes the spectrum is composed of *X rays, ultraviolet light, visible light*, and *infrared radiation*. Note that in the figure λ increases towards the left, while ν increases towards the right. The unit of frequency is the hertz (Hz), which has been defined in Chapter 18 as 1 cycle per second. Thus since $c = 3.0 \times 10^8$ m/s it follows from (29-23) that the frequency associated with a radio wave of wavelength $\lambda = 20$ meters is

$$\nu = \frac{c}{\lambda} = \frac{3.0 \times 10^8 \, \text{m/s}}{20 \, \text{m}} = 1.5 \times 10^7 \, \text{Hz}$$

Of particular interest to us is the portion of the electromagnetic spectrum that consists of visible light. The wavelengths of electromagnetic waves to which the human eye is sensitive lie along the very restricted interval

$$4.5 \times 10^{-7} \, \text{m} \leqslant \lambda \leqslant 6.8 \times 10^{-7} \, \text{m}$$

The shortest wavelengths of visible light are associated with violet light and for increasing values of λ, the color changes continuously from blue through green, yellow, orange, and finally to red light, which is associated with a wavelength of approximately 6.8×10^{-7} meter. The frequencies associated with visible light are found by use of (29-23) to lie along the interval

$$6.7 \times 10^{14} \, \text{Hz} \geqslant \nu \geqslant 4.4 \times 10^{14} \, \text{Hz}$$

Figure 29-10 shows schematically a plot of the energy emitted by our sun as a function of wavelength. It is interesting to note that this curve shows a peak at wavelengths that correspond to visible light. That is, the spectrum of

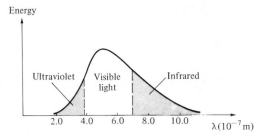

Figure 29-10

the radiation emitted by our sun peaks at precisely those wavelengths where the eye is most sensitive. By a happy coincidence (!) the earth's atmosphere also happens to be transparent for these same wavelengths. Almost all of the ultraviolet light emitted by the sun and much of that radiated in the infrared is absorbed by the earth's atmosphere. Several decades ago it was found that the atmosphere is also transparent to radiation of wavelengths in the range from 10^{-2} to 10^2 meters. This very important discovery ushered in the era of *radio astronomy*. Fortunately, the atmosphere absorbs radiation of all wavelengths except for radio waves and visible light. Life, as we know it on earth, would not be possible if the atmosphere were transparent to ultraviolet light or X rays.

Example 29-6 Consider yellow light of wavelength 5.7×10^{-7} meter. Calculate:
  (a) Its frequency.
  (b) Its period.
  (c) The wave number associated with it.

Solution
  (a) Solving (29-23) for ν and substituting the given values for c and λ, we find that

$$\nu = \frac{c}{\lambda} = \frac{3.0 \times 10^8 \text{ m/s}}{5.7 \times 10^{-7} \text{ m}} = 5.3 \times 10^{14} \text{ Hz}$$

  (b) The period T is the reciprocal of ν; hence

$$T = \frac{1}{\nu} = \frac{1}{5.3 \times 10^{14} \text{ Hz}} = 1.9 \times 10^{-15} \text{ s}$$

  (c) Substituting the given value for λ into (29-22), we obtain

$$k = \frac{2\pi}{\lambda} = \frac{2\pi}{5.7 \times 10^{-7} \text{ m}} = 1.1 \times 10^7 /\text{m}$$

Example 29-7 An electromagnetic wave of wavelength λ_0 and of frequency $\nu_0 = c/\lambda_0$ originally traveling in free space enters a block of matter characterized by a dielectric constant κ. Assuming that $\kappa_m \approx 1$:
  (a) What is the speed of the wave in the material?

(b) Explain physically why the frequency of the wave must be the same inside and outside of the material.

(c) Find the wavelength of the wave inside the material.

Solution

(a) According to (29-14), the index of refraction n of the material is $\sqrt{\kappa}$ since we have assumed that $\kappa_m \approx 1$. Hence substitution into (29-13) yields for the velocity v of the wave in the medium

$$v = \frac{c}{\sqrt{\kappa}}$$

(b) Since the incident electromagnetic fields oscillate at the frequency ν_0, the induced dipole moment inside the material will also oscillate at this same frequency. It follows that the total electromagnetic field inside the material—which is the sum of the external fields plus those produced by the dipoles—will also oscillate with the same frequency ν_0. Hence the frequency of the wave inside and outside of this material is the same.

(c) The relation, $\nu\lambda = c$ in (29-23) is very generally valid for any wave. Thus, since the velocity of the wave in the medium is $c/\sqrt{\kappa}$, while its frequency according to (b) is ν_0, it follows that the wavelength λ_m inside the medium is

$$\lambda_m = \frac{v}{\nu_0} = \frac{c}{\nu_0\sqrt{\kappa}} = \frac{\lambda_0}{\sqrt{\kappa}}$$

where in the final equality we have used the fact that the free-space wavelength $\lambda_0 = c/\nu_0$. Generally $\kappa > 1$, so the wavelength λ_m inside the medium is less than the wavelength λ_0 in free space. Equivalently, this result may be expressed in terms of the index of refraction

$$\lambda_m = \frac{\lambda_0}{n} \qquad (29\text{-}24)$$

since $n = \sqrt{\kappa}$ in this case.

For example, if light of wavelength 5.5×10^{-7} meter in free space enters a medium of refractive index $n = 1.5$, its wavelength in the medium, according to (29-24), is

$$\lambda_m = \frac{\lambda_0}{n} = \frac{5.5 \times 10^{-7}\,\text{m}}{1.5} = 3.7 \times 10^{-7}\,\text{m}$$

and this no longer corresponds to visible light. The velocity of light v in this medium is

$$v = \frac{c}{n} = \frac{3.0 \times 10^8\,\text{m/s}}{1.5} = 2.0 \times 10^8\,\text{m/s}$$

29-8   Energy propagation in a monochromatic wave

According to Figure 29-9, the frequency of visible light is of the order of 10^{15} Hz. This means that both the electric and the magnetic vectors of the wave reverse themselves about 10^{15} times each second. Since this is very large compared to the frequencies associated with the motion of macroscopic matter, including notably those associated with our sensory organs, it follows that when light enters our eye what we observe is radiation averaged over many cycles. Hence for purposes of discussing the energy flow associated with visible light, as well as certain other parts of the electromagnetic spectrum, only the time average of the Poynting vector is of physical interest. The purpose of this section is to show that this time average $\bar{N}$ of the magnitude of the Poynting vector for a monochromatic wave is

$$\bar{N} = \frac{1}{2} \sqrt{\frac{\epsilon_0}{\mu_0}} E_0^2 \qquad (29\text{-}25)$$

where E_0 is the amplitude of the wave.

To this end consider the monochromatic wave characterized by the amplitude E_0 and wave number k in (29-20). According to (29-21) through (29-23), this may be expressed equivalently as

$$\mathbf{E} = \mathbf{i} E_0 \sin \left[\frac{2\pi}{\lambda} (z - ct) \right]$$

$$\mathbf{B} = \mathbf{j} \frac{E_0}{c} \sin \left[\frac{2\pi}{\lambda} (z - ct) \right]$$

Substitution into (29-17) shows that the direction of energy flow is along the z-axis and that the associated magnitude N of the Poynting vector is

$$N = \left| \frac{1}{\mu_0} \mathbf{E} \times \mathbf{B} \right| = \frac{E_0^2}{\mu_0 c} \sin^2 \left[\frac{2\pi}{\lambda} (z - ct) \right]$$

Note that, in general, N varies with both position and time.

To calculate the time average $\bar{N}$ of the Poynting vector $\mathbf{N}$, let us make use of the fact that the average value $\bar{f}$ of a function $f(t)$ over a fixed interval $a < t < b$ is defined by

$$\bar{f} = \frac{1}{(b-a)} \int_a^b f(t) \, dt$$

Applying this definition to the Poynting vector, we find that the average value of N over a period $T = 1/\nu$ is

$$\bar{N} = \nu \int_0^{1/\nu} N \, dt = \frac{E_0^2}{\mu_0 c} \nu \int_0^{1/\nu} \sin^2 \left[\frac{2\pi}{\lambda} (z - ct) \right] dt \qquad (29\text{-}26)$$

In the problems it will be established that the value of this integral is $1/2\nu$ and is thus independent of spatial variable z. Hence, since $\epsilon_0 \mu_0 c^2 = 1$, the relation in (29-25) follows.

It is also shown in the problems that for a wave composed of more than one frequency, the right-hand side of (29-25) must be replaced by a sum of terms, one for each frequency. In this connection it should be noted that (29-25) is independent of λ and is thus the same for all frequencies. Hence, the ratio of the energies associated with two waves of different frequencies is independent of the frequencies involved; it depends only on the ratio of the squares of the amplitudes of the two waves.

It should be emphasized that (29-25) applies only to physical situations for which the observer is so far from the source that (29-20) is applicable "everywhere." More realistically, as will be shown in the following section, $\bar{N}$ varies inversely with the square of the distance from the source. For example, since the planet Jupiter is about 5.2 times as far away from the sun as is the earth, it follows that the solar energy incident per unit time on a unit area of Jupiter is $(1/5.2)^2 = 0.037$ of that falling on the earth. In applying (29-25) to practical situations this feature must be kept in mind.

Example 29-8 A 1-mW helium-neon laser sends out a continuous beam of red light. Assuming that the beam has a 0.2-cm^2 cross section and does not appreciably diverge, calculate:
(a) The average energy flux $\bar{N}$.
(b) The amplitude E_0 of the associated wave.
(c) The strength of the magnetic induction B_0 of the wave.

Solution
(a) Since the output of the laser is 1 mW and since the area of the beam is $0.2 \text{ cm}^2 = 2.0 \times 10^{-5} \text{ m}^2$, the flux or the energy per unit area per unit time, $\bar{N}$, is

$$\bar{N} = \frac{1.0 \times 10^{-3} \text{ W}}{2.0 \times 10^{-5} \text{ m}^2} = 50 \text{ W/m}^2$$

(b) Solving (29-25) for E_0, and substituting the above value for $\bar{N}$, we obtain

$$E_0 = \left[2\bar{N} \sqrt{\frac{\mu_0}{\epsilon_0}} \right]^{1/2} = \left[2 \times 50 \text{ W/m}^2 \times \left(\frac{4\pi \times 10^{-7} \text{ T-m/A}}{8.9 \times 10^{-12} \text{ C}^2/\text{N-m}^2} \right)^{1/2} \right]^{1/2}$$
$$= 1.9 \times 10^2 \text{ V/m}$$

(c) Since for an electromagnetic wave the amplitude B_0 of the **B**-field is E_0/c, it follows that

$$B_0 = \frac{E_0}{c} = \frac{1.9 \times 10^2 \text{ V/m}}{3 \times 10^8 \text{ m/s}} = 6.3 \times 10^{-7} \text{ T}$$

†29-9 The sources of electromagnetic waves

Having demonstrated that the charge-current-free Maxwell equations admit solutions corresponding to electromagnetic waves, in this section we briefly consider the problem of how such waves are generated.

In our previous studies of Maxwell's equations with the displacement current term omitted, we have seen that the electromagnetic fields associated with various charge-current distributions go to zero very rapidly far away from these sources. Specifically, if r is the separation distance between the observation point and the source, then both **E** and **B** fall off as

$1/r^2$ as r becomes very large. In turn, this means that the Poynting vector $\mathbf{N}$ in (29-17) falls off as $1/r^4$ and, as will be seen below, no electromagnetic waves are associated with such fields. However, if the displacement current term is included in Maxwell's equations, then under certain circumstances there are solutions of these equations that at large separation distances from the source fall off only as $1/r$. For these solutions—which are known as the *radiation fields*—the Poynting vector varies as $1/r^2$, and associated with these fields there is an unambiguous flow of energy.

Consider, in Figure 29-11, a charge-current source located at the origin of a certain coordinate system. Assuming that for very large distances r from this source both $\mathbf{E}$ and $\mathbf{B}$ vary as $1/r$, it follows from (29-17) that

$$\mathbf{N} \cong \mathbf{N}_0 \frac{1}{r^2} \qquad (r \to \infty) \qquad \text{(29-27)}$$

with $\mathbf{N}_0$ a vector independent of r. In general, however, $\mathbf{N}_0$ will vary with the angles (θ, ϕ) as defined in the figure. Now in terms of these angles, an area element $d\mathbf{S}$ of the sphere is

$$d\mathbf{S} = \mathbf{e} r^2 \sin\theta \, d\theta \, d\phi$$

with $\mathbf{e}$ a unit vector along the radial direction. Hence the electromagnetic energy per unit time that passes through $d\mathbf{S}$ is (for $r \to \infty$)

$$\mathbf{N} \cdot d\mathbf{S} = \left(\mathbf{N}_0 \frac{1}{r^2} \right) \cdot (\mathbf{e} r^2 \sin\theta \, d\theta \, d\phi)$$

$$= (\mathbf{N}_0 \cdot \mathbf{e}) \sin\theta \, d\theta \, d\phi \qquad \text{(29-28)}$$

and this is independent of r. Thus we have the result:

Regardless of the separation distance between the observer and the source, provided $\mathbf{E}$ and $\mathbf{B}$ both vary as $1/r$ for large r, electromagnetic energy will be radiated by the source unless $|\mathbf{N}_0| \equiv 0$.

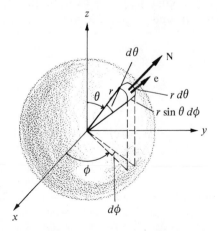

Figure 29-11

The *total* rate dW/dt at which energy is radiated by the source may be obtained by integrating (29-28) over the entire sphere. The result is

$$\frac{dW}{dt} = \int_0^{2\pi} d\phi \int_0^{\pi} \sin\theta\, d\theta\, \mathbf{N}_0 \cdot \mathbf{e} \qquad (29\text{-}29)$$

To illustrate the types of charge-current distributions that radiate electromagnetic energy, consider in Figure 29-12 a particle of charge q which has an acceleration $\mathbf{a}$. For distances r very far from the particle, the electric and magnetic fields are found to be

$$\mathbf{E} = \frac{q}{4\pi\epsilon_0 c^2 r^3}\, \mathbf{r} \times (\mathbf{r} \times \mathbf{a})$$

$$\mathbf{B} = \frac{q}{4\pi\epsilon_0 c^3 r^2}\, \mathbf{a} \times \mathbf{r} \qquad (29\text{-}30)$$

provided that the speed of the particle is small compared to c. Substituting these formulas into (29-27), we find that

$$\mathbf{N}_0 = \mathbf{e}\,\frac{q^2}{16\pi^2\epsilon_0}\frac{a^2}{c^3}\sin^2\theta$$

with θ as defined in the figure. The energy radiated by the particle vanishes along the direction of the acceleration, and is a maximum in a direction perpendicular to $\mathbf{a}$. The rate at which the particle radiates energy in all directions is found by substitution into (29-29) to be

$$\frac{dW}{dt} = \frac{q^2 a^2}{6\pi\epsilon_0 c^3} \qquad (29\text{-}31)$$

Note the important feature that if $a = 0$, then no energy is radiated. In other words, in order for a charged particle to radiate energy it is *necessary that it accelerate.*

Correspondingly, in order for a macroscopic current distribution to radiate electromagnetic energy, it is necessary that the current vary in time. For example, if a current $i = i_0 \cos\omega t$ flows around a current loop of area A,

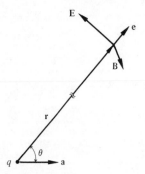

Figure 29-12

then the rate at which energy is radiated is found to be

$$\frac{dW}{dt} = \frac{1}{3} \sqrt{\frac{\mu_0}{\epsilon_0}} (i_0 A)^2 \left(\frac{\omega}{c}\right)^4$$

Example 29-9 At what rate is energy being radiated by a proton in a cyclotron at an instant when it is undergoing a centripetal acceleration of 2.0×10^{15} m/s²?

Solution Substituting the given data into (29-31), we obtain

$$\frac{dW}{dt} = \frac{2}{3} \frac{q^2}{4\pi\epsilon_0} \frac{a^2}{c^3}$$

$$= \frac{(0.67) \times (1.6 \times 10^{-19}\ \text{C})^2 \times (9.0 \times 10^9\ \text{N-m}^2/\text{C}^2) \times (2.0 \times 10^{15}\ \text{m/s}^2)^2}{(3.0 \times 10^8\ \text{m/s})^3}$$

$$= 2.3 \times 10^{-23}\ \text{W}$$

†29-10 The displacement current

The purpose of this section is to consider in a quantitative way the difficulty associated with Ampère's law and to show how this is resolved by the introduction of the displacement current.

Imagine, in Figure 29-13, applying Ampère's law to a sequence of closed curves $l_1, l_2, \ldots$, whose lengths, as in the figure, approach zero. In this limit, suppose that the sequence of surfaces $S_1, S_2, \ldots$ bounded by these curves approaches a *finite* closed surface S. Since the magnetic induction **B** is generally finite, it follows that the sequence of numbers

$$\oint_{l_1} \mathbf{B} \cdot d\mathbf{l}_1, \quad \oint_{l_2} \mathbf{B} \cdot d\mathbf{l}_2, \quad \ldots$$

tends to zero because the sequence of path lengths does. Accordingly, in this limit the left-hand side of (29-4) vanishes, and since the limiting surface S is closed, we find

$$\oint_S \mathbf{j} \cdot d\mathbf{S} = 0 \qquad (29\text{-}32)$$

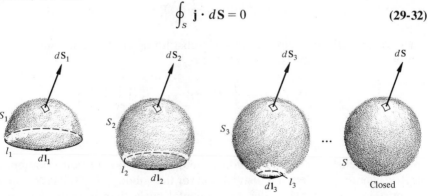

Figure 29-13

But the integral in this relation represents the net flow of current outward through the *closed* surface S. Hence, we conclude that Ampère's law implies of necessity that *the net current flow out of every closed surface must vanish.*

But this cannot always be true! Consider again, in Figure 29-2, a capacitor that is being charged up by a current i, and let S be a surface that encloses the positively charged plate in its entirety. The rate at which the charge on the plate increases is dq/dt, so the current flowing *out* of the closed surface is $-dq/dt$. Hence, for this case,

$$\oint_S \mathbf{j} \cdot d\mathbf{S} = -\frac{dq}{dt}$$

and this is in direct contradiction with the implication of Ampère's law in (29-32).

Based on arguments of this type, Maxwell reasoned that Ampère's law in the form in (29-4) could not be correct. According to the ideas of charge conservation, the outward flow of current from a closed surface is zero only if the rate of change of charge inside that surface vanishes. More generally, the correct form of (29-32) must be

$$\oint_S \mathbf{j} \cdot d\mathbf{S} = -\frac{d}{dt} q(S) \qquad (29\text{-}33)$$

with $q(S)$ the total charge inside S. Hence, to be consistent with experiment it is necessary to modify Ampère's law so that its limiting form for a closed surface does *not* lead to the generally incorrect (29-32), but rather to the correct version in (29-33).

Let us now confirm that the modified form of Ampère's law in (29-5) does not suffer from this difficulty. To this end, let us apply (29-5) to the sequence of surfaces in Figure 29-13. As in the above derivation of (29-32), the left-hand side of (29-5) vanishes in this limit so that this time there results

$$\oint_S \mathbf{j} \cdot d\mathbf{S} = -\epsilon_0 \frac{d}{dt} \oint_S \mathbf{E} \cdot d\mathbf{S}$$

where S is an arbitrary closed surface. Substituting on the right-hand side by use of Gauss' law, (29-1), we find that

$$\oint_S \mathbf{j} \cdot d\mathbf{S} = -\epsilon_0 \frac{d}{dt} \oint_S \mathbf{E} \cdot d\mathbf{S} = -\epsilon_0 \frac{d}{dt} \left[\frac{1}{\epsilon_0} q(S) \right]$$

$$= -\frac{d}{dt} q(S)$$

and this, at least, is consistent with (29-33) and the law of conservation of electric charge. The detailed justification for the validity of (29-5) is founded, of course, on a much broader experimental basis than is implied by this calculation.

29-11   Summary of important formulas

Maxwell's equations are

$$\oint_S \mathbf{E} \cdot d\mathbf{S} = \frac{1}{\epsilon_0} q(S) \qquad (29\text{-}1)$$

$$\oint_S \mathbf{B} \cdot d\mathbf{S} = 0 \qquad (29\text{-}2)$$

$$\oint_l \mathbf{E} \cdot d\mathbf{l} = -\frac{d}{dt} \int_S \mathbf{B} \cdot d\mathbf{S} \qquad (29\text{-}3)$$

$$\oint_l \mathbf{B} \cdot d\mathbf{l} = \mu_0 \int_S \mathbf{j} \cdot d\mathbf{S} + \epsilon_0 \mu_0 \frac{d}{dt} \int_S \mathbf{E} \cdot d\mathbf{S} \qquad (29\text{-}5)$$

where the symbols S and l are as defined in Figure 29-1. The quantity $q(S)$ represents the *total* charge inside the closed surface S and $\mathbf{j}$ is the current density.

In regions of space far from charges and currents, Maxwell's equations admit wave solutions of the form

$$\mathbf{E} = \mathbf{i}E_0 h(z - ct)$$
$$\mathbf{B} = \mathbf{j}\frac{E_0}{c} h(z - ct) \qquad (29\text{-}9)$$

with h an arbitrary function and c defined by

$$c = (\epsilon_0 \mu_0)^{-1/2} \qquad (29\text{-}10)$$

Associated with these waves there is an energy flow given by the *Poynting vector*

$$\mathbf{N} = \frac{1}{\mu_0} \mathbf{E} \times \mathbf{B} \qquad (29\text{-}17)$$

which represents the rate of energy flow per unit area along the direction of propagation of the wave. For a monochromatic wave, the time average $\bar{N}$ of the magnitude of the Poynting vector is independent of wavelength and is

$$\bar{N} = \frac{1}{2} \sqrt{\frac{\epsilon_0}{\mu_0}} E_0^2 \qquad (29\text{-}25)$$

QUESTIONS

1. Define or describe briefly what is meant by the following terms: (a) displacement current; (b) Maxwell's equations; (c) index of refraction; and (d) electromagnetic spectrum.

2. By starting with Maxwell's equations, state what restrictions you must impose on the various physical quantities $\mathbf{E}$, $\mathbf{B}$, $\mathbf{j}$, and ρ so that the equations of electrostatics result.

3. Repeat Question 2, but this time for the equations governing the relation between time-independent currents and their associated magnetic fields.

4. Explain in physical terms why the displacement current term in (29-5) does not appreciably alter the **B**-field associated with a slowly varying current distribution.

5. Under what circumstances is Kirchhoff's first rule—which states that the algebraic sum of the currents approaching a junction must vanish—consistent with Maxwell's equations?

6. Does the argument in Section 29-10 yield a unique value for the displacement current? Give an example of a different term that could be added to Ampère's law so that (29-33) is still satisfied.

7. In Chapter 18, we defined wave motion as "the propagation of motion or of energy through a medium without an associated propagation of matter." Describe in what respects this definition applies to an electromagnetic wave and in what respects it does not.

8. In our study of dielectrics it was noted that the dielectric constant for ordinary water is 80. According to (29-14), it follows that the velocity v of visible light in water would be

$$v = \frac{c}{\sqrt{\kappa}} \cong \frac{c}{9}$$

Explain in physical terms why it is that the measured velocity is actually about $(3/4)c$. (*Hint:* Think of the physical meaning of dielectric constant for a time-dependent electric field.)

9. State a typical value for the wavelength of each of the following types of radiation: (a) visible light; (b) X rays; (c) ultraviolet light; and (d) infrared radiation.

10. State a typical value for the frequency associated with each of the following: (a) visible light; (b) gamma rays; (c) radio waves; and (d) ultraviolet light.

11. What is the frequency of an AM radio station in your community? What is the wavelength of the radiation associated with this frequency?

12. Consider an electromagnetic wave characterized by the electric field

$$\mathbf{E} = \mathbf{j} E_0 h (x - ct)$$

What is the direction of propagation of the wave? Along what direction does the associated **B**-field vector point?

13. What is the direction of the electric field associated with an electromagnetic wave if the **B**-field of the wave has the form

$$\mathbf{B} = -\mathbf{k} \frac{E_0}{c} h (y + ct)$$

14. Consider an electromagnetic wave characterized by the electric field

$$\mathbf{E} = \mathbf{i} E_0 \cos (kz - \omega t + \alpha)$$

where E_0, k, ω, and α are fixed constants.

(a) What is the amplitude of this wave?

(b) Why is the speed of propagation ω / k?

(c) Need the velocity ω / k be equal to c? Explain.

15. Does an electromagnetic wave always propagate along a straight line? Explain.

16. According to classical ideas, a hydrogen atom consists of a proton and an electron orbiting about it. Why does the electron have to accelerate in this orbit? Show that, in light of (29-31), this atom is unstable; that is, show that the electron would have to radiate energy and thus gradually spiral into the proton. (*Note:* This was one of several incorrect deduc-

tions of the classical laws, which argued against the validity of classical physics, and which led eventually to the development of quantum mechanics.)

17. An electromagnetic wave propagates at the speed c in a given coordinate system. A second observer who is traveling at a speed $u = c/2$ along the propagation direction also observes that the velocity of this wave is c. Can you account for this "paradox" in terms of classical ideas? (*Note*: It was this type of failure of the ideas of the Galilean transformation that helped to bring forth the theory of relativity.)

18. An electron is traveling at a uniform speed $u \sim c$ through a medium of index of refraction n. If u is greater than the speed of light c/n in this medium it is observed that the electron emits electromagnetic radiation known as *Cerenkov radiation*. Give a physical explanation for the existence of this radiation.

19. An electron traveling at a speed $u \sim c$ traverses the interior of the eyeball of a human. In light of Question 18, explain why he will observe a light flash, even though his eyes are closed.

20. Two electromagnetic waves are traveling in mutually perpendicular directions in free speace. If the waves overlap in some region, do you expect them to modify each other in any way? Explain.

21. An electromagnetic wave enters a block of matter of dielectric constant κ and of index of refraction $n = \sqrt{\kappa}$. Explain in microscopic terms why the wave is modified inside the block. Can you account for the fact that the speed of the wave is less than that in free space?

22. Describe the circumstances under which a charged particle will radiate electromagnetic energy. What must happen to the mechanical energy of the particle? To its velocity? To its momentum?

23. In light of Question 22, discuss why do you think that momentum must be associated with an electromagnetic wave.

24. What is the direction of the momentum in an electromagnetic wave traveling radially outward from the sun? Will you experience a force when sunlight shines on you? Explain.

PROBLEMS

1. The sensitivity of a healthy human eye is greatest for green-yellow light of wavelength 5.5×10^{-7} meter. What is the frequency of this light? What is the associated wave number?

2. What is the wavelength of a beam of X rays that are characterized by a frequency of 2.5×10^{19} Hz? What is the angular frequency of these X rays?

3. A monochromatic electromagnetic wave is characterized by a wavelength of 6.0×10^{-7} meter in free space.
   (a) What is the frequency of this wave?
   (b) What is the angular frequency of the wave?
   (c) What is the wave number?

4. Suppose that the wave in Problem 3 enters a region of space containing a nonmagnetic material of dielectric constant $\kappa = 4$.
   (a) What is the index of refraction of this medium?

(b) What is the velocity of the wave in this material?

(c) What is the wavelength inside this matter?

5. A monochromatic wave has associated with it an average energy flux of 12 W/m^2.

(a) What is the strength of the electric field of the wave?

(b) What is the strength of the magnetic induction?

(c) Can the frequency of this wave be determined from the given data? Explain.

6. Calculate the strength of the electric and magnetic fields in the radiation reaching us from the sun, assuming for the "solar constant" (the average amount of energy per unit time incident normally on unit area just outside the earth's atmosphere) a value of $1.37 \times 10^3 \text{ W/m}^2$ and assuming the sun to be a monochromatic source. Neglect atmospheric absorption.

7. Using the data in Problem 6, calculate the energy flux incident on 1 square meter of Saturn and of Pluto, assuming that these planets are, respectively, 9.2 and 40 times as far away from the sun as is the earth.

8. A monochromatic point source radiates energy in all directions at the average rate of 250 watts.

(a) What is the average flux of energy at a point 2 meters from the source?

(b) What is the amplitude of the electric field at this point?

(c) What is the magnitude of the magnetic induction at a point 5 meters from the source?

9. A monochromatic source emits visible light with a wavelength (in air) of 5.5×10^{-7} meter. What is the wavelength of this light in water? (Assume an index refraction for water of 4/3.) What wavelength

would an underwater observer see?

10. Show that for an electromagnetic wave, the energy density $\epsilon_0 E^2/2$ associated with the electric field is the same as the magnetic energy density $B^2/2\mu_0$.

11. Suppose the parallel-plate capacitor in an R-C circuit has plates of area A and separation distance d. If $\mathcal{E}$ is the emf of the battery and current first starts to flow at $t = 0$:

(a) Show that the electric field E between the plates is

$$E = \frac{\mathcal{E}}{d}[1 - e^{-t/RC}]$$

with $C = \epsilon_0 A/d$ the capacitance.

(b) Calculate the value of the quantity $\epsilon_0 dE/dt$ associated with this field.

*12. Making use of the results of Problem 11 show explicitly that the quantity

$$\oint_S \left[\mathbf{j} + \epsilon_0 \frac{d\mathbf{E}}{dt} \right] \cdot d\mathbf{S}$$

vanishes for the closed surface S in Figure 29-2.

13. Show that the electric field $\mathbf{E}$ of an electromagnetic wave propagating along the z-axis satisfies the wave equation

$$\frac{\partial^2}{\partial z^2} \mathbf{E} = \frac{1}{c^2} \frac{\partial^2}{\partial t^2} \mathbf{E}$$

Will the $\mathbf{B}$-field of an electromagnetic field satisfy this relation? Prove your answer.

14. Consider, in Figure 29-14, a segment

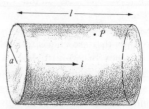

Figure 29-14

of length l of a very long wire of resistance R and radius a, through which there flows a uniform current i.

(a) What is the magnetic induction **B** at a point P on the surface of the wire?

(b) Show that the magnitude of the electric field at point P is Ri/l.

(c) What is the magnitude and the direction of the Poynting vector at this point?

(d) Show thus that the rate at which energy flows out of this surface is $-Ri^2$ and give a physical interpretation of this result.

15. Establish the relation

$$\int_0^{1/\nu} \sin^2\left[\frac{2\pi}{\lambda}(z - ct)\right] dt = \frac{1}{2\nu}$$

and thus show the equivalence of (29-25) and (29-26).

16. Consider two monochromatic waves of frequencies ν_1 and ν_2 and of amplitudes E_{01} and E_{02} traveling along the same direction. Show that the time average of the Poynting vector $\bar{N}$ is

$$\bar{N} = \frac{1}{2}\sqrt{\frac{\epsilon_0}{\mu_0}}(E_{01}^2 + E_{02}^2)$$

17. Starting with (29-18), show that the time average of the magnitude of the momentum density $\mathbf{P}_v$ of a monochromatic wave is

$$\bar{P}_v = \frac{\epsilon_0 E_0^2}{2c}$$

where E_0 is the amplitude of the wave.

18. Consider, in Figure 29-15, a monochromatic electromagnetic wave of amplitude E_0 traveling perpendicular to a material surface of area A.

(a) Using the result of Problem 17, calculate the time average of the momentum inside the volume $Ac\,\Delta t$ in the figure.

(b) Assuming that all of this radia-

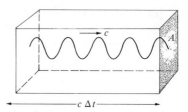

Figure 29-15

tion is absorbed at the surface during the time interval Δt, show that the "radiation pressure" P_R on the surface is

$$P_R = \frac{1}{2}\epsilon_0 E_0^2 = \frac{1}{c}\bar{N}$$

(*Hint:* What is the change in momentum per unit time of the radiation?)

(c) If the radiation is reflected by the surface, show that the radiation pressure is

$$P_R = \frac{2}{c}\bar{N}$$

19. How large must the "reflecting sail" on a spaceship of total mass 5.0×10^3 kg be so that it will accelerate radially outward from the sun at 1.0 m/s². Make use of the result of Problem 18 and assume an incident solar flux of 1.0 kW/m². Neglect the gravitational attraction of the sun.

20. Using the value for the solar constant in Problem 6 and the result of Problem 18, calculate the force on the earth due to radiation pressure from the sun. Assume the earth to be a perfectly absorbing flat disk of radius 6.4×10^3 km with its plane perpendicular to the sun's rays. Compare this with the gravitational attraction between the earth and the sun.

21. If a charged particle travels through a medium of refractive index n at a speed v greater than the speed of light in the medium it emits elec-

tromagnetic waves (Cerenkov radiation) on a conical surface of half angle θ determined by

$$\cos \theta = \frac{c/n}{v}$$

See Figure 29-16 and compare with Figures 18-18 and 18-19. For an electron traveling through a medium of refractive index $n = 1.5$, calculate the emission angle θ for $v = 2.0 \times 10^8$ m/s and for $v = 2.9 \times 10^8$ m/s.

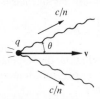

Figure 29-16

*22. Consider, in Figure 29-17, a monochromatic electromagnetic wave characterized by the angular frequency ω and wave number k $(= \omega/c)$ traveling along the z-axis of a certain coordinate system. Let ω' and k' be the corresponding parameters for this same wave as seen by a primed observer who travels along the z-axis at the velocity u. Assuming the validity of the Lorentz transformation in (3-33) and that the phases of the wave $(kz - \omega t)$ for one observer and

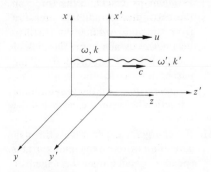

Figure 29-17

$(k'z' - \omega' t')$ for the other are the same, derive the Doppler shift formula

$$\nu' = \nu \left[\frac{1 - u/c}{1 + u/c} \right]^{1/2}$$

where $\nu' = \omega'/2\pi$ is the observed frequency for the primed observer. (*Hint*: Why must the relation $\omega' = k'c$ be valid for all choices for u?)

23. Using the results of Problem 22, show that if a monochromatic source emits radiation of wavelength λ, then to an observer traveling away from the source at a velocity u the radiation has the wavelength λ', given by

$$\lambda' = \lambda \left[\frac{1 + u/c}{1 - u/c} \right]^{1/2}$$

24. Light having a wavelength of 5.5×10^{-7} meter leaves a distant galaxy which is receding from us at a velocity of $c/2$. Making use of the result of Problem 23, calculate:
   (a) The original frequency of this light.
   (b) The wavelength of this light when it reaches earth.
   (c) The observed wavelength if the galaxy were traveling toward us at a velocity of $c/2$.

25. A certain line in the spectrum of hydrogen has a wavelength, as seen on earth, of 4.35×10^{-7} meter. When this line is observed in the spectrum from a distant galaxy, it is found to have the value of 6.60×10^{-7} meter. What is the velocity of recession of this galaxy from us? What would the observed wavelength of this line be if the galaxy were approaching us at this velocity? (*Hint*: Use the Doppler shift formula in Problem 23.)

26. For a galaxy that recedes from us at a velocity of $0.99c$, what would be the observed wavelength of the 21-cm line in the spectrum of hydrogen emitted by this galaxy?

27. A spaceship that is approaching earth at a speed of $0.2c$ sends out a flash of light with a wavelength of 4.5×10^{-7} meter. What is the wavelength of this light as seen on earth?

28. Consider the solution in (29-9) for the charge-current-free Maxwellian equations.
    (a) Show that if the first relation in (29-8) is satisfied by these solutions for all closed surfaces, then so is the second.
    (b) Show that if the fourth relation in (29-8) is satisfied by these solutions for all closed curves l, then so is the third.

29. Verify that the flux out of a closed sphere of radius r and centered at the origin due to the electric field

$$\mathbf{E} = \mathbf{i}E_0 h(z - ct)$$

vanishes. (*Hint*: In terms of the angles θ and ϕ defined in Figure 29-11, the area element $d\mathbf{S}$ is given by

$$d\mathbf{S} = \mathbf{e}r^2 \sin\theta \, d\theta \, d\phi$$

where $\mathbf{e}$ is a unit vector in the radial direction.)

30. For the rectangular path in Figure 29-18 verify that the fields in (29-9)

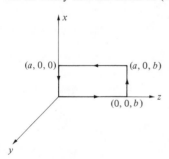

Figure 29-18

satisfy the third relation of (29-8) for all choices of the function h.

†31. A particle of charge q is moving along the trajectory

$$x = a_0 t^2 + bt$$

where a_0 and b are constants. Calculate, by use of (29-31), the rate at which it radiates energy.

†32. A particle of charge q oscillates with simple harmonic motion of angular frequency ω and amplitude A. Calculate the total energy it radiates during one complete oscillation.

†33. A particle of charge q and mass m orbits at right angles to a uniform B-field at a velocity v ($\ll c$).
    (a) What is its acceleration?
    (b) Show that the rate of change dW/dt of its energy is

$$\frac{dW}{dt} = -\frac{q^4 B^2}{3\pi\epsilon_0 m^3 c^3} W$$

34. Making use of the result of Problem 33, show that the fractional rate of energy loss dW/W during one period is

$$\frac{2}{3}\frac{q^3 B}{\epsilon_0 c^3 m^2}$$

and make a numerical estimate of this loss for a proton moving in B-field of strength 1.0 tesla.

35. For an electron traveling at a velocity of $c/137$ about a proton (a hydrogen atom) in a circular orbit of radius $a = 5.3 \times 10^{-11}$ meter calculate the total energy radiated during one complete orbit. Assume that the velocity and the orbital radius are not appreciably modified by the radiation.

30 The properties of light

We all know what light is; but it is not easy to tell what it is.

SAMUEL JOHNSON (According to
Boswell)

30-1 Introduction

Visible light is the radiation associated with that part of the electromagnetic spectrum to which the human eye is receptive. Since visual perceptions play such an important role in man's daily activities, it is not surprising that the study of light, or *optics*, as it is known, began early in his history. For the earliest recorded observations of optical phenomena we are indebted to the ancient Greeks. However, it was not until well into the seventeenth century that definitive theories of the nature of light were developed.

One of the earliest quantitative attempts to describe light was made by Isaac Newton, who developed a *corpuscular theory*. Basic to this theory is the idea that the light emitted by a source consists of streams of corpuscles, which stream out in rectilinear paths in all directions from the source. One of the important implications of this theory is that the velocity of light in a dense medium, such as water, is larger than is its velocity in a less dense one such as air. As we shall see, this is *not* consistent with experiment. Mainly because of this and related incorrect predictions, the corpuscular theory of Newton was eventually abandoned. But not until the nineteenth century!

In 1670 Christian Huygens (1629–1695), who was a contemporary of Newton, proposed a competing *wave theory* of light. Although this theory was very appealing in a variety of respects, it did not receive wide acceptance for well over a century. One of the main arguments against it was that if light were indeed a wave motion, then it should exhibit wavelike phenomena, such as diffraction and interference. Since at that time diffraction effects, such as the bending of light rays around corners, had not been observed, the conclusion that light could not be a wave motion was inescapable. Today we know that the wavelenght of light is of the order of 5.0×10^{-7} meter and is thus small compared with the dimensions of macroscopic bodies. In order to observe diffraction phenomena, it is necessary to use an apparatus, such as a slit system, whose dimensions are comparable to the wavelength of visible light. Hence it was not until well over one hundred years after Huygens that the wave nature of light was unambiguously established.

Thomas Young (1773–1829) finally vindicated Huygens. He showed that when two coherent beams of light fall onto the same area of a viewing screen, there appear alternating dark and bright bands characteristic of wavelike interference phenomena. For the first time, then, man had clear evidence for the wave nature of light. Subsequently, Jean B. Foucault (1819–1868) succeeded in measuring the velocity of light in water. Contrary to the predictions of Newton, he found that the measured speed was less in water than in air. The ideas of Newton seemed to be dead. Moreover, with the wave nature of light thus established, Augustin J. Fresnel (1788–1827), assuming that light was a wave motion in the "ether," derived formulas for the intensity ratios for the reflected and the transmitted light resulting from a light beam incident on the interface between two media. These relations are known as *Fresnel's equations*. Although derived on the basis of the incorrect ether hypothesis, they have been found to be consistent with experiment nevertheless.

As we saw in Chapter 29, our views of the nature of light were modified again in the nineteenth century, when James Clerk Maxwell enunciated his equations and demonstrated that they admitted wavelike solutions corresponding to waves that travel in free space at the velocity of light c. His prediction for this speed $c = (\epsilon_0 \mu_0)^{-1/2} \cong 3.0 \times 10^8$ m/s is in excellent agreement with other measurements for the speed of light. In one bold stroke, then, Maxwell showed that *all* quantitative predictions of the entire field of optics could be deduced directly from electromagnetic theory. Further, he showed that these results for visible light were applicable to a much wider range of phenomena—that is, to the entire electromagnetic spectrum, of which visible light is but a very small part.

Finally, it is interesting to note that even though the Newton–Huygens controversy regarding the nature of light seemed to have been firmly settled by Maxwell in the nineteenth century, the matter was again reopened during the early part of our own century. In an effort to account for certain experimental results associated with *blackbody radiation* and the *photo-electric effect*, Max Planck (1858–1947) and Albert Einstein (1879–1955) hypothesized that under suitable circumstances electromagnetic waves take on the attributes of streams of corpuscles. These corpuscles are today called *photons*. Planck and Einstein suggested that a monochromatic electromagnetic wave of frequency ν and thus of wavelength $\lambda = c/\nu$ must consist of a stream of photons, each of energy E and of momentum p, with

$$E = h\nu \qquad p = \frac{h}{\lambda} \qquad\qquad (30\text{-}1)$$

The constant h is known as *Planck's constant*; it has the measured value

$$h \cong 6.63 \times 10^{-34} \, \text{J-s} \qquad\qquad (30\text{-}2)$$

Today we know that light and electromagnetic waves in general are capable of exhibiting both particlelike and wavelike characteristics; and that the particular aspect exhibited in a given case depends on the details of the prevailing physical situation. Certain experiments will bring out the particulate aspects of the wave, and others its wavelike features. In the following we shall be concerned only with those phenomena which exhibit the wave aspects of light. Some experiments that bring out the corpuscular nature of light—as well as the wavelike nature of elementary particles, such as electrons or protons—will be briefly discussed in Chapter 34.

We begin our study of optics in this chapter by considering first the phenomena and laws of *geometrical optics*. This study involves the reflection and refraction of light from surfaces whose dimensions are large compared to a wavelength. Under these circumstances, the fact that light is a wave motion plays only a peripheral role. Then, we shall examine certain of the phenomena of diffraction that arise when light interacts with objects of dimensions comparable to the wavelength of light. This is the field of *physical optics*. The fact that light is a wave motion plays a crucial role in the understanding of the phenomena of this field.

30-2   Measurements of the velocity of light

Prior to the time of Maxwell it was not known that light was an electromagnetic wave whose velocity c could be ascertained indirectly in terms of the electromagnetic parameters ϵ_0 and μ_0. Consequently, during this earlier period much effort was expended on direct measurements of this fundamental constant. The purpose of this section is to examine briefly some of these earlier attempts.

The first recorded effort to obtain a value for the speed of light was carried out by Galileo during the seventeenth century. Galileo and an assistant, each in possession of a lantern with a removable dark slide, stationed themselves on two nearby hills about a kilometer apart. The experiment was then initiated by Galileo's removing the dark slide from his lantern so that a beam of light went out towards his assistant. On seeing this light, the assistant immediately removed the slide from his own lantern, thereby sending a beam of light back toward Galileo. The time interval measured from the instant that Galileo first removed his slide to that when he saw the light from his assistant's lantern is thus a measure of the velocity of light. Unfortunately, this rather direct experiment could not succeed. Assuming that the distance d between the two lanterns is 1 km, and using the fact that the speed of light is 3.0×10^8 m/s, we find that the time interval τ for the light to go successively between the two lanterns is

$$\tau = \frac{2d}{c} = \frac{2 \times 1.0 \, \text{km}}{3.0 \times 10^8 \, \text{m/s}} = 6.7 \times 10^{-6} \, \text{s}$$

Since this is imperceptibly small compared to a man's reaction time, a sensible value for the speed of light could not be found this way.

A second attempt to obtain a value for the speed of light achieved more success and was made in 1675 by the Danish mathematician and astronomer, Ole Roemer (1644–1710). The planet Jupiter is about five times farther away from the sun than is the earth, and orbits about the sun with a period of 11.86 years. In turn, twelve moons are in orbit about Jupiter. The innermost of these is called Io and has a period $P \cong 1.77$ days = 42.5 hours. While making measurements of this period, Roemer observed a systematic variation in its value throughout the course of the year. During that part of the year when the earth, due to its own orbital motion, recedes from Jupiter, he found the observed periods to be, on the average, larger than the values measured during the remainder of the year, when the earth approaches Jupiter. To understand how a value for c can be deduced from these data, consider this astronomical situation in Figure 30-1 (not to scale). Suppose that, at $t = 0$, Io enters Jupiter's shadow at point A. Because of the fact that the velocity of light is finite, news of this eclipse will reach earth, located at this instant at E_1, a certain time t_1 later. About 42.5 hours later, when the earth is at E_2, Io

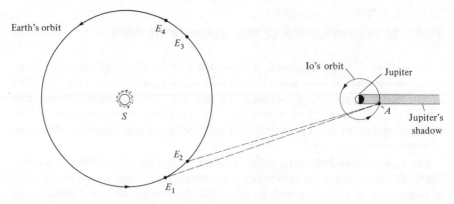

Figure 30-1

will again be eclipsed by Jupiter at A, and this event will be seen on earth at a later time, t_2. The *measured* value for the period of Io is thus $(t_2 - t_1)$. However, because of the fact that the earth travels in its orbit at a velocity of 30 km/s $\cong 10^{-4}\, c$, the time it takes for the eclipse to be seen on earth the second time, when it is at E_2, is less than when it was at E_1. In other words, because of the finiteness of the speed of light, and the fact that in 42.5 hours the earth moves closer to Jupiter by the approximate distance

$$(30\ \text{km/s}) \times (42.5 \times 3600\ \text{s}) \cong 4.6 \times 10^6\ \text{km}$$

the observed period $(t_2 - t_1)$ is less than the true value by about

$$\frac{4.6 \times 10^6\ \text{km}}{3.0 \times 10^8\ \text{m/s}} \cong 15\ \text{s}$$

Similarly, when the earth travels from E_3 to E_4 on the other side of its orbit, the measured period will exceed the actual one by about 15 seconds.

Analyzing these data in more detail, Roemer concluded that the time required for light to travel a distance equal to the diameter of the earth's orbit is about 22 min. Since the radius of this orbit is 1.5×10^8 km, this yields for the speed of light the approximate value

$$\frac{2 \times 1.5 \times 10^8\ \text{km}}{22 \times 60\ \text{s}} = 2.3 \times 10^8\ \text{m/s}$$

This is very close to the presently accepted value of 3.0×10^8 m/s.

A second astronomical determination of c was subsequently carried out by James Bradley (1692–1762). Bradley observed that during the course of a year a star located near the axis of the earth's orbit will appear to traverse a small circular path with a sense opposite to that of the earth's orbital motion (Figure 30-2). This phenomenon is called *aberration*.

Figure 30-3 illustrates schematically the principle underlying Bradley's measurement for the case of a star vertically above the earth's orbit. If the earth were not moving in orbit about the sun, then to a terrestrial observer the star would always be in a fixed position vertically above the orbital plane. However, the earth does travel at a certain velocity **v** about the sun. Hence, when it is at E_1, to an observer on earth the velocity of the light from

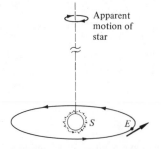

Figure 30-2

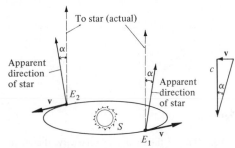

Figure 30-3

the star has a component v directed opposite to the earth's velocity. As shown in the figure, the apparent direction of the star thus makes an angle α with its actual direction, where[1]

$$\tan \alpha = \frac{v}{c}$$

Correspondingly, when the earth is on the opposite side of its orbit at E_2 the star will appear to be displaced by the angle α in the opposite direction. Hence during the course of a year the star will appear to orbit in a circle, or more accurately in an ellipse, of angular diameter 2α. Making use of the known value $v \cong 3.0 \times 10^4$ m/s and the observed value for α ($\cong 20.6''$), he obtained a value very close to the one accepted today.

An important and successful terrestrial experiment to measure c was pioneered in 1849 by Armand H. L. Fizeau (1819–1896). Figure 30-4 is a schematic diagram of the apparatus he employed. The key to this measurement is a toothed wheel, which can be set into rapid rotation about its axis. The actual measurement is initiated when the light emitted by a source S goes through a lens L_1 and is reflected toward the left by a half-silvered mirror M_1, which partially reflects and partially transmits the incident light. Assuming that at this instant the position of the wheel is such that a gap is present, the incident light will travel successively through the lenses L_2 and L_3 and reach the mirror M_2. From there it will be reflected backward and retrace its path toward the wheel. Now if the angular velocity of the wheel is such that by the time the light has traveled this distance of length $2l$ its further passage is blocked by a tooth in the wheel, then the observer O will not see any light. On the other hand, if the wheel rotates sufficiently

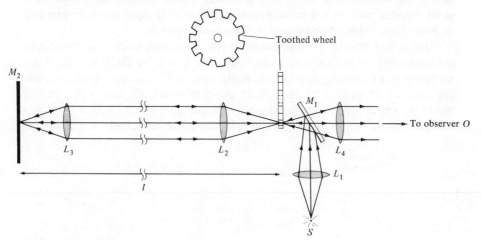

Figure 30-4

[1]According to the theory of relativity, the correct formula is: $\tan \alpha = v/(c^2 - v^2)^{1/2}$ and this reduces to Bradley's formula for $v \ll c$.

slowly, then the original light which went through a gap in the wheel will be reflected back from M_2, go through this same gap, be partially transmitted by M_1, and finally be observed at O. Thus, knowing the distance l and the angular velocity of the wheel at which no light is seen by the observer, Fizeau could measure the speed of light. Making use of a wheel with 720 teeth, and a distance $l \cong 8$ km, he found that when the wheel rotated at about 600 revolutions per minute the observer at O saw no light. He arrived at the value $c \cong 3.1 \times 10^8$ m/s.

This important experimental determination of c by Fizeau, has been followed by a number of other measurements. As a result of an ever-improving sequence of these experiments the accepted value for the velocity of light today[2] is

$$c = 2.997925 \times 10^8 \text{ m/s } (\pm 300 \text{ m/s}) \tag{30-3}$$

For most purposes the approximation $c \cong 3.0 \times 10^8$ m/s is sufficiently accurate.

30-3 Index of refraction

The value for c quoted in (30-3) refers to the speed of light only in a vacuum. The speed of light in a transparent medium, such as air, water, or glass, is, generally speaking, less than c. If v is the measured speed of light in a material medium, the *refractive index* or the *index of refraction*, n, of that medium is defined by

$$n = \frac{c}{v} \tag{30-4}$$

In the following we shall be concerned only with *nondispersive* materials, for which $n > 1$. The speed of light in such a material is always smaller than is its speed in a vacuum.

Table 30-1 lists values for the refractive index for typical substances. For

Table 30-1   Index of refraction at 293 K for $\lambda = 5.893 \times 10^{-7}$ m

Substance	n
Air (1 atm)	1.0003
Benzene	1.50
Crown glass	1.52
Diamond	2.42
Fused quartz	1.46
Water	1.33

[2] For a more recent and accurate value see K. M. Evenson et al., *Phys. Rev. Lett.* **29**, 1346 (1972).

most practical purposes the refractive index for air is unity, and for the other materials listed n ranges from a value of 1.33 for water to 2.42 for diamond. As we shall see below, the index of refraction varies slightly with the frequency of the light ν or equivalently with its wavelength λ given by (29-23):

$$\lambda = \frac{c}{\nu} \tag{30-5}$$

The values for n in the table correspond to the light associated with an intense line in the spectrum of sodium, which has the wavelength $\lambda = 5.893 \times 10^{-7}$ meter.

In discussions of optics one often makes use of the units of length of the micron (abbreviated μ) and the angstrom (abbreviated Å). These are defined by

$$1 \text{ micron} = 1 \ \mu = 10^{-6} \text{ m}$$
$$1 \text{ angstrom} = 1 \text{ Å} = 10^{-10} \text{ m} \tag{30-6}$$

Thus the wavelength of yellow light, $\lambda = 5893$ Å, as used in the table corresponds to a wavelength of 5.893×10^{-7} meter $= 0.5893 \ \mu$. The micrometer (which is equivalent to the micron) and its abbreviation μm (see Table 1-1) are rarely used in optics.

In order to discuss the physical meaning of refractive index let us recall the relation (29-14)

$$n = (\kappa \kappa_m)^{1/2}$$

for the index of refraction n of a material characterized by a dielectric constant κ and relative permeability κ_m. As it stands, this formula is applicable throughout the electromagnetic spectrum and is not restricted to visible light. However, experiment shows that for optical frequencies ($\nu \sim 6 \times 10^{14}$ Hz), the relative permeability κ_m of most materials is, to a high degree of approximation, unity. Hence, for visible light,

$$n = \sqrt{\kappa} \tag{30-7}$$

and this suggests the following physical picture for the index of refraction. If a monochromatic electromagnetic wave of an optical frequency ν enters a dielectric medium, the electric field of the wave causes the constituent dipoles of the medium to oscillate at the same frequency ν. The resultant motion of charge produces, in turn, a secondary electromagnetic wave, also at the same frequency ν. According to the superposition property of the electromagnetic fields, the total wave traveling through the medium is the sum of the original wave and that produced by the oscillations of the dipoles. And it is the resultant *total* wave that travels through the medium at the velocity $v = c/n$, with n given in (30-7).

Based on this physical picture, the reason why the index of refraction might vary with frequency is readily apparent. For the efficiency with which the dipoles of a given medium oscillate generally varies with the frequency of the incident wave. It follows from (30-7) that the index of refraction will also vary with the wavelength or the frequency of the incident wave. Figure 30-5 shows a plot of the observed values of the refractive index of crown glass as a function of wavelength at a fixed temperature. Note that this variation is *not* very large. Indeed, experiment shows that for many substances this variation can be safely neglected and we shall assume this to be so in much of the following.

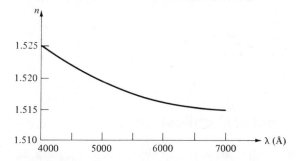

Figure 30-5

In a similar way, the above physical picture for refractive index also suggests and makes plausible the fact that n might vary with temperature. By way of example, in Table 30-2 we list the values of n for water at a wavelength of 0.5893 μ at several temperatures. Again, experiment shows that for many substances the temperature variation of n can (within reasonable limits) be safely neglected and we shall assume this to be so in the following.

Table 30-2   Index of refraction of water for $\lambda = 5893$ Å

T (°C)	15	25	35	45	55
n	1.3338	1.3329	1.3316	1.3301	1.3285

Example 30-1   A monochromatic wave of yellow light of wavelength 5893 Å is traveling in free space.
(a) What is the frequency of this wave?
(b) What is its velocity in a medium for which $n = 1.52$?
(c) What is its wavelength in the medium?

Solution
(a) Substituting the given value for λ into (30-5) we find that

$$\nu = \frac{c}{\lambda} = \frac{3.0 \times 10^8 \text{ m/s}}{5893 \text{ Å}} = \frac{3.0 \times 10^8 \text{ m/s}}{5.893 \times 10^{-7} \text{ m}}$$

$$= 5.1 \times 10^{14} \text{ Hz}$$

where in the last equality only two significant figures have been kept.

(b) Making use of the definition of n in (30-4) we find, for the wave speed v in the medium,

$$v = \frac{c}{n} = \frac{3.0 \times 10^8 \text{ m/s}}{1.52} = 2.0 \times 10^8 \text{ m/s}$$

(c) Since the frequency of the wave is the same in free space as in the medium, it follows that the wavelength λ_m in the medium is less than its value λ_0 in free space by the factor n^{-1}. See (29-24). Hence

$$\lambda_m = \frac{\lambda_0}{n} = \frac{5893 \text{ Å}}{1.52} = 3900 \text{ Å}$$

to two significant figures.

30-4   The laws of geometrical optics

If a beam of visible light falls on the surface of an object such as a piece of wood, the reflected light goes out in all directions, and the object can be seen from several directions. This type of reflection is called *diffuse reflection* and is exemplified in Figure 30-6a. Diffuse reflection arises if the normal irregularities associated with the surfaces of most objects are very large compared to the wavelength of the light being used. If a beam of light falls on a very smooth or highly polished surface—that is, one for which the ir-regularities are small compared to the wavelength of light—then the reflected light goes out as a beam in a single direction. This type of reflection is called *specular* and is shown in Figure 30-6b. For example, the reflection of light from the wall or floor of a room is generally diffuse, whereas that reflected from a mirror is specular. In the remainder of this chapter we shall be concerned only with specular reflection. Hence it will be assumed that the irregularities of all surfaces of interest are small compared to the wavelength of visible light, and thus in the following the word "reflection" shall always mean "specular reflection."

Consider, in Figure 30-7, a beam of monochromatic light, which ap-proaches, at an angle θ_1 with respect to the normal, the planar interface

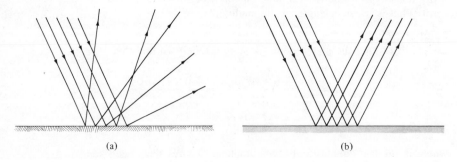

(a)   (b)

Figure 30-6

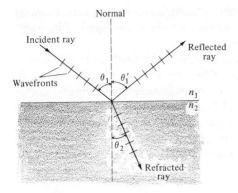

Figure 30-7

between two media of respective indices of refraction n_1 and n_2. The incoming beam is called the *incident ray*, and planes at right angles to the direction of propagation of this ray are called the *wavefronts* associated with the wave. Experiment shows that, as a consequence of the incident ray's striking the interface, there arise two additional waves. One of these propagates backward into the same medium as the incident wave and is called the *reflected wave*. It propagates along the direction of the *reflected ray* at some angle θ_1' with respect to the normal. The second wave arising at the interface propagates into the lower medium and it is known as the *refracted wave*. It propagates along the direction of the *refracted ray*, which makes a certain angle θ_2 with respect to the normal. Associated with the reflected and the refracted waves there are also wavefronts, and these are represented in the figure by short lines perpendicular to the respective rays. Since the electric and magnetic vectors of an electromagnetic wave are perpendicular to the direction of propagation of the wave, it follows that the wavefronts of the various waves are the planes containing the electric and the magnetic vectors.

Experiment as well as detailed studies of Maxwell's equations show that the intensities and the directions of propagation of the incident, the reflected, and the refracted rays in Figure 30-7 are related to each other in a certain way. The laws of *geometrical optics* relate the directions of propagation of these three rays and are:

1. The incident, the reflected, and the refracted rays all lie in a single plane. This plane is called the *plane of incidence* and can be thought of as being determined by the incident ray and the normal to the interface.

2. The *angle of reflection* θ_1' is equal to the *angle of incidence* θ_1:

$$\theta_1 = \theta_1' \tag{30-8}$$

3. The angle of incidence θ_1 and the *angle of refraction* θ_2 are related by *Snell's law*:

$$n_1 \sin \theta_1 = n_2 \sin \theta_2 \tag{30-9}$$

Although we shall be concerned with these three laws of geometrical optics only for visible light, it should be noted that they are applicable very generally for all electromagnetic waves.

It is interesting that although these laws can be derived from Maxwell's theory, they were known much earlier. Thus the law of reflection in (30-8) was known to Euclid and the law of refraction was determined empirically in the seventeenth century by Willebroad Snell (1591–1626). Snell's law was also derived from Newton's corpuscular theory by René Descartes (1596–1650). However, the fact that these laws are applicable to the entire electromagnetic spectrum was not known until late in the nineteenth century.

30-5   Applications

In this section we apply the laws of geometrical optics to a number of particular physical situations.

Example 30-2   A directed beam of light from a source S strikes a mirror (that is, an interface for which most of the incident light is reflected) at a point A and at an angle of 60° with respect to the normal N_1; see Figure 30-8. Assuming that there is a second mirror inclined at an angle of 130° with respect to the first, find the three angles θ_1, θ_2, and θ_3 as defined in the figure. Assume that N_2 is the normal to the second mirror.

Solution   According to the law of reflection, the angle of incidence is equal to the angle of reflection. Hence

$$\theta_1 = 60°$$

To obtain θ_2 and θ_3, note that in triangle ABC the measure of angle B is 130° while that for angle CAB is $90° - \theta_1 = 30°$. Hence, since the sum of the angles of a triangle is 180°, it follows that

$$\theta_2 = 180° - 130° - 30° = 20°$$

Finally, since N_2 is the normal to the second mirror, a second application of the law of reflection yields

$$\theta_3 = 70°$$

Example 30-3   A ray of light traveling in air strikes, at an angle of incidence of 45°, the planar surface of a piece of glass of index $n = 1.5$. Calculate the angle of refraction θ (Figure 30-9).

Solution   Since the refractive index of air is unity, Snell's law in (30-9) takes the form

$$\sin 45° = 1.5 \sin \theta$$

where we have made the substitutions $n_1 = 1$, $n_2 = 1.5$, $\theta_1 = 45°$, and $\theta_2 = \theta$. Hence

$$\sin \theta = \frac{\sin 45°}{1.5} = 0.47$$

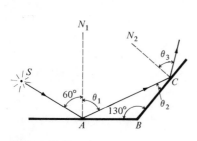

Figure 30-8

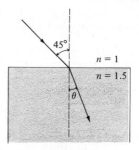

Figure 30-9

which, by reference to a table of trigonometric functions, is equivalent to

$$\theta \cong 28°$$

Example 30-4 A beam of light is produced by a source S, which is under water and at a vertical distance d below the surface, as in Figure 30-10 (not to scale). Assuming almost normal incidence so that the angles of incidence and refraction θ_1 and θ_2, respectively, are sufficiently small so that $\sin \theta_1 \cong \tan \theta_1 \cong \theta_1$, and similarly for θ_2, find the depth y below the surface where the source appears to be to the observer O. Assume that $n = 4/3$.

Solution Since the incident ray is under water and the refracted ray is in air, according to Snell's law the angles θ_1 and θ_2 are related by

$$\frac{4}{3} \sin \theta_1 = \sin \theta_2 \qquad \qquad \textbf{(30-10)}$$

By hypothesis, the angles θ_1 and θ_2 are presumed small, and thus

$$\sin \theta_1 \cong \tan \theta_1 = \frac{a}{d}$$

$$\sin \theta_2 \cong \tan \theta_2 = \frac{a}{y}$$

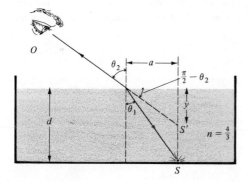

Figure 30-10

Substitution into (30-10) then yields

$$\frac{4a}{3d} = \frac{a}{y}$$

or, equivalently,

$$y = \frac{3}{4}d$$

In other words, to the observer O, the depth of the water appears to be only three fourths of its actual depth.

More generally, if a flat pan or a lake or a stream is filled to a depth d with a transparent substance of refractive index n, then its apparent depth to an observer looking perpendicularly down on the surface is d/n.

Example 30-5   Consider, in Figure 30-11, a ray incident, at an angle θ_1, on a face of a prism of vertex angle A and of index of refraction n. Find the angle of deviation δ of this ray as it traverses the prism in terms of the parameters n, θ_1, and A.

Solution   From the fact that the exterior angle of a triangle is equal to the sum of the remote interior angles it follows that

$$\delta = \theta_1 - \theta_1' + \theta_2 - \theta_2'$$

Further, since the sum of the angles θ_1' and θ_2' is supplementary to angle B and the latter is the supplement of angle A we have the additional relation

$$A = \theta_1' + \theta_2'$$

Hence applying Snell's law at the two refractive surfaces

$$\sin \theta_1 = n \sin \theta_1'$$
$$\sin \theta_2 = n \sin \theta_2'$$

and eliminating the angles θ_1', θ_2, and θ_2' from among the above four relations we obtain finally

$$\delta = \theta_1 - A + \sin^{-1}[(n^2 - \sin^2 \theta_1)^{1/2} \sin A - \cos A \sin \theta_1]$$

An analysis of this formula shows that for any given prism there is a value for the angle of incidence θ_1 for which this deviation angle δ is a minimum. It is left as an

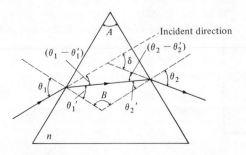

Figure 30-11

exercise to show that this minimum value for δ, call it δ_m, satisfies

$$n \sin \frac{A}{2} = \sin \frac{A + \delta_m}{2} \tag{30-11}$$

and that the value of the angle of incidence, call it $\overline{\theta}_1$, at which this occurs is

$$\overline{\theta}_1 = \sin^{-1}\left[n \sin \frac{A}{2} \right]$$

This formula for the angle of minimum deviation δ_m in (30-11) is often used when it is desired to measure the refractive index of transparent solids. The minimum angle of deviation δ_m for any given prism can be measured by varying the angle of incidence θ_1 and observing the varying direction of the transmitted ray. Combining the value of δ_m so obtained with a knowledge of the angle A we can obtain by use of (30-11), a measured value for n. By varying the color or the wavelength of the incident light, a dispersion curve such as that in Figure 30-5 can be constructed.

30-6   Total internal reflection

Suppose, in Figure 30-7, that the incident ray is in air, and therefore $n_1 = 1$. Then, the angle θ_2 of the refracted ray will be less than the angle of incidence θ_1, or, in other words, the refracted ray is bent toward the normal. For according to Snell's law

$$\sin \theta_1 = n_2 \sin \theta_2 \tag{30-12}$$

and, since $\sin \theta$ is an increasing function of θ for $0 \leqslant \theta \leqslant 90°$ we find, assuming $n_2 > 1$ that, except at normal incidence,

$$\theta_2 < \theta_1$$

Imagine now an experiment in which the angle of incidence θ_1 is increased to its maximum value of $90°$. The angle of refraction will also increase and will approach a certain maximum value, call it θ_m. Substituting the value $\theta_1 = 90°$ into (30-12) we find that the maximum angle of refraction θ_m satisfies

$$\sin \theta_m = \frac{1}{n_2} \tag{30-13}$$

In other words, regardless of the angle of incidence, all refracted light rays coming from above the surface in Figure 30-7 are confined to the interior of a cone of half-angle θ_m as defined in (30-13). For example, since the index of refraction of water is 4/3, it follows that if you are under water and look upward, all of the light coming to you from above is confined to the interior of a cone of half-angle $\theta_m = \sin^{-1} 0.75 = 49°$.

Consider now the situation depicted in Figure 30-12 in which a ray of light originally travels in a medium of refractive index $n_1 > 1$ and then enters a second medium, such as air, which has an index of unity. If θ_1 is, as before, the angle of incidence and θ_2 the corresponding angle of refraction, then

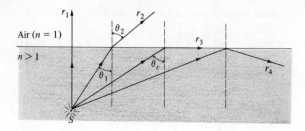

Figure 30-12

according to Snell's law

$$n_1 \sin \theta_1 = \sin \theta_2 \tag{30-14}$$

This time, as exemplified by the ray r_2 in Figure 30-12, the angles θ_1 and θ_2 are related by

$$\theta_2 > \theta_1$$

In other words, for this case, the angle of refraction (except for the normal ray r_1) is invariably larger than is the angle of incidence and thus the refracted rays are all bent away from the normal.

Imagine now an experiment involving light going from such a refractive medium into air. As the angle of incidence θ_1 is increased, it will eventually achieve a certain value, θ_c, known as the *critical angle* for which the angle of refraction $\theta_2 = 90°$. This is represented by the ray r_3 in the figure. Substituting this value $\theta_2 = 90°$ into (30-14) we find for θ_c

$$\sin \theta_c = \frac{1}{n_1} \tag{30-15}$$

The physical significance of θ_c is that for all angles of incidence θ_1 in the range $\theta_c < \theta_1 < 90°$ there exists no real angle of refraction θ_2 that satisfies Snell's law. Hence there can be no refracted ray for such angles. According to the discussion in Chapter 29, however, there is a flow of energy associated with the incident, the reflected and the refracted ray. To be consistent with the law of energy conservation, then, it follows that for $\theta_1 > \theta_c$, for which there is no refracted ray, *all* of the energy incident on the interface must be reflected back into the medium. This phenomenon is therefore called *total internal reflection*. Although there is generally some reflection at all angles, only for $\theta_1 > \theta_c$ is the reflection total.

It should be emphasized that total internal reflection is associated only with a beam of light going from a medium of higher to one of lower index of refraction. Generally speaking, a refracted ray is always associated with a light ray that is incident from the lower index side onto the interface between two media.

Example 30-6 Calculate the critical angle θ_c associated with an interface between air and:
 (a) Water, of refractive index 1.33.
 (b) Glass, of refractive index 1.5.
 (c) Diamond, of refractive index 2.5.

Solution In each case, making use of (30-15), we find that
 (a) For water,

$$\theta_c = \sin^{-1} 0.75 = 49°$$

 (b) For glass,

$$\theta_c = \sin^{-1} 0.67 = 42°$$

 (c) For diamond,

$$\theta_c = \sin^{-1} 0.4 = 24°$$

Example 30-7 A beam of light is incident from the glass side of the interface between water ($n = 4/3$) and glass ($n = 3/2$), as in Figure 30-13. For what angles of incidence will this beam experience total internal reflection?

Solution If θ_1 is the angle of incidence and θ_2 the angle of refraction, then, according to Snell's law,

$$\frac{3}{2} \sin \theta_1 = \frac{4}{3} \sin \theta_2$$

The critical angle θ_c is defined, as above, to be the value of θ_1 for which $\theta_2 = 90°$. Hence

$$\frac{3}{2} \sin \theta_c = \frac{4}{3}$$

or, equivalently,

$$\theta_c = \sin^{-1} \frac{8}{9} = 63°$$

Thus, total internal reflection takes place for all incident angles θ in the range

$$63° \le \theta \le 90°$$

Example 30-8 A beam of light is incident on a 45°-45°-90° prism, of index 1.5; see Figure 30-14. Describe what happens to the beam if it is incident normally on:
 (a) A leg of the prism as in Figure 30-14a.
 (b) The hypotenuse of the prism as in Figure 30-14b.

Solution Since the refractive index is 1.5 it follows that with respect to air the critical angle θ_c is

$$\theta_c = \sin^{-1} \frac{1}{n} = \sin^{-1} 0.67 = 42°$$

(a) If the beam is incident normally on a leg of the prism, then the angle of incidence at point A is 45°. Since this exceeds the critical angle, it follows that the incident beam will be totally internally reflected. Accordingly, the ray follows the

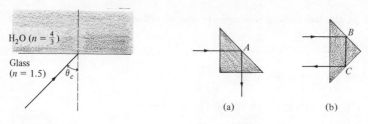

Figure 30-13   **Figure 30-14**

path indicated in Figure 30-14a. In effect, the direction of the beam is turned through 90°.

(b) In a similar way, if the incident beam strikes the hypotenuse of the prism as in Figure 30-14b, then at both B and C its angle of incidence is 45°. This exceeds the critical angle $\theta_c = 42°$, so the beam undergoes total internal reflection at both of these points. In effect, the direction of the beam is reversed.

30-7   Fresnel's equations

Consider again Figure 30-7, in which a light beam strikes the interface between the media of refractive indices n_1 and n_2. Maxwell's equations enable us to predict not only the directions of the reflected and the refracted beams, but also their intensities relative to that of the incident beam. The amplitudes of these relative intensities satisfy certain relations known as *Fresnel's equations*.

As we saw in Chapter 29, a monochromatic electromagnetic wave traveling in free space, is characterized uniquely by its amplitude E_0 and wavelength λ. (See (29-20) through (29-22).) The average flow of energy $\bar{N}$ per unit area per unit time associated with such a wave when traveling in free space is given by

$$\bar{N} = \frac{1}{2} \sqrt{\frac{\epsilon_0}{\mu_0}} E_0{}^2 \qquad (29\text{-}25)$$

and thus depends exclusively on the amplitude E_0. More generally, if an electromagnetic wave travels in a (nonmagnetic) medium of index of refraction n, then the argument used in Example 29-4 shows that the relation (29-25) is still applicable, provided that the factor ϵ_0 is replaced by $\epsilon_0 \kappa$. Hence, introducing the symbol I to represent the energy flow per unit area per unit time associated with a monochromatic wave traveling through a medium of refractive index n, we find by use of (29-14) that

$$I = \frac{1}{2} \sqrt{\frac{\epsilon_0}{\mu_0}} n E_0{}^2 \qquad (30\text{-}16)$$

with E_0 the amplitude of the electric field vector of the wave. In the following, we shall use the term *intensity* for the quantity I given by (30-16).

Consider, in Figure 30-15, an electromagnetic wave of amplitude E_0

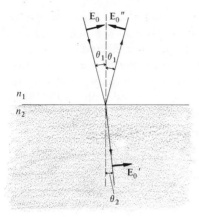

Figure 30-15

traveling in a medium of refractive index n_1 and striking at an angle θ_1 the interface with a second medium of refractive index n_2. Let E_0'' be the corresponding amplitude of the reflected wave and E_0' that of the refracted wave. The directions of the electric vectors for these three waves must be perpendicular to the respective propagation directions, and for the directions of the electric fields shown in the figure the associated **B** vectors are perpendicular to and out of the plane of the diagram. According to (30-16), the energy flow associated with the incident, the reflected, and the refracted rays are completely specified in terms of the respective amplitudes E_0, E_0', and E_0''.

The relations between these three amplitudes E_0, E_0', and E_0'' are in general fairly complex and thus we shall consider in detail only the special case of near-normal incidence. Here the angle of incidence satisfies the relation $\theta_1 \ll 1$ and thus Snell's law has the form

$$n_2\theta_2 = n_1\theta_1$$

In the problems it will be verified that for these angles, the ratios E_0''/E_0 of the reflected to the incident amplitude and E_0'/E_0 of the refracted to the incident amplitudes are given by

$$\frac{E_0''}{E_0} = \frac{n_2 - n_1}{n_1 + n_2} \qquad \frac{E_0'}{E_0} = \frac{2n_1}{n_1 + n_2} \qquad \text{(30-17)}$$

These are *Fresnel's equations* for this special case of normal incidence. Substitution into (30-16) yields, for the reflected intensity I'' and the refracted intensity I',

$$I'' = I_0\left(\frac{n_2 - n_1}{n_2 + n_1}\right)^2$$

$$I' = I_0\frac{4n_1n_2}{(n_1 + n_2)^2} \qquad \text{(30-18)}$$

where I_0 is the intensity of the incident light and where, for example, the second equality follows since, according to (30-16), $I'/I_0 = n_2 E_0'^2/n_1 E_0^2$.

For the special case in which $n_2 = n_1$, we expect no reflected light since, in effect, there is no interface. Substitution of the value $n_2 = n_1$ shows that, consistent with this expectation, the relations in (30-18) imply that all of the incident light is transmitted: ($I'' = 0$; $I' = I_0$). If the two relations in (30-18) are added together, we find after some algebra that

$$I'' + I' = I_0$$

so that regardless of the values for n_2 and n_1 the sum of the reflected and refracted intensities is equal to the incident intensity. In this sense the intensity ratios in (30-18) are consistent with the ideas of energy conservation, as they must be.

More generally, if the angle of incidence θ_1 is not small, then if the light is polarized with the electric vectors in the plane of incidence, as in Figure 30-15 (p-polarization), then (30-17) must be replaced by

$$\frac{E_0''}{E_0} = \frac{\tan(\theta_1 - \theta_2)}{\tan(\theta_1 + \theta_2)} \qquad \frac{E_0'}{E_0} = \frac{2\cos\theta_1 \sin\theta_2}{\sin(\theta_1 + \theta_2)\cos(\theta_1 - \theta_2)} \qquad \textbf{(30-19)}$$

with θ_2 given by Snell's law. Correspondingly, if the light is polarized with the electric vectors perpendicular to the plane of incidence (s-polarization), then the ratios of these amplitudes are

$$\frac{E_0''}{E_0} = \frac{\sin(\theta_1 - \theta_2)}{\sin(\theta_1 + \theta_2)} \qquad \frac{E_0'}{E_0} = \frac{2\cos\theta_1 \sin\theta_2}{\sin(\theta_1 + \theta_2)} \qquad \textbf{(30-20)}$$

These relations in (30-19) and (30-20) are known as Fresnel's equations.

It is established in the problems that, for normal incidence ($\theta_1 \approx 0$ and $\theta_2 = n_1\theta_1/n_2$), both (30-19) and (30-20) reduce to (30-17). This must follow since the two cases of the incident ray being polarized in, or perpendicular to, the plane of incidence are physically indistinguishable in this case.

Example 30-9 A beam of light traveling in air is incident on a glass plate of refractive index 1.5. If *all* of the light is transmitted, and the electric vector lies in the plane of incidence, as in Figure 30-15, what is the angle of incidence θ_1?

Solution Reference to (30-19) shows that since $\tan 90° = \infty$ the reflected amplitude E_0'' will vanish if $\theta_1 + \theta_2 = 90°$. Substituting this datum and the values $n_1 = 1$ and $n_2 = 1.5$ into Snell's law, we find that

$$\sin\theta_1 = n_2 \sin\theta_2 = 1.5 \sin(90° - \theta_1) = 1.5\cos\theta_1$$

Hence, $\tan\theta_1 = 1.5$, and this yields

$$\theta_1 = 56°$$

As will be discussed in Chapter 33, this angle of incidence at which no reflection takes place is known as *Brewster's angle*.

30-8 Summary of important formulas

If a light beam traveling in a medium of refractive index n_1 is incident at an angle θ_1 on the interface between it and a second medium of refractive index n_2, then some of the light will be reflected at the same angle θ_1 and the rest will be refracted into the second medium at an angle θ_2. The angle of refraction θ_2 is given by Snell's law:

$$n_1 \sin \theta_1 = n_2 \sin \theta_2 \qquad\qquad (30\text{-}9)$$

Near normal incidence ($\theta_1 \cong \theta_2 n_2/n_1$) the intensities associated with the incident beam, I_0, the transmitted beam I', and the reflected beam I'' are related by

$$I'' = I_0 \left(\frac{n_2 - n_1}{n_2 + n_1}\right)^2$$

$$(30\text{-}18)$$

$$I' = I_0 \frac{4n_1 n_2}{(n_2 + n_1)^2}$$

QUESTIONS

1. Define or describe briefly what is meant by the following terms: (a) aberration of starlight; (b) index of refraction; (c) critical angle; (d) Fresnel's equations; and (e) specular reflection.
2. State an experimental fact that is in contradiction with Newton's corpuscular theory. Are the laws of reflection and refraction consistent with this theory?
3. In what way or ways are the photons of Einstein and Planck different from Newton's corpuscles? In what way or ways are they similar?
4. Define what is meant by the "wavelength" of light and make use of your definition to distinguish between the phenomena of geometrical and of physical optics.
5. What is the physical basis for the difference between diffuse and specular reflection?
6. Light from a flashlight is reflected from a mirror in a dark room. Exp-

lain why, even though the reflection is specular, the image of the light can be seen from various parts of the room.
7. How far apart would Galileo and his assistant have had to be in order that the time for the light to travel back and forth between them be 1 second? How does this compare to the distance between the earth and the moon?
8. Figure 30-16 shows a curved, transparent rod made of a material of high

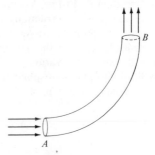

Figure 30-16

refractive index. By use of the notion of total internal reflection and Fresnel's equations, explain why it is that if light is admitted at one end, *A*, of the rod, an appreciable fraction of this light might emerge at the other end, *B*. (*Note*: When used in this way, the rod is often referred to as a *light pipe*, and the field of study concerned with the optical properties of a collection of flexible light pipes is known as *fiber optics*.)

9. Is the concept of refractive index restricted to visible light or is it more generally applicable? Explain.

10. Can Snell's law be applied to X rays? To ultraviolet light? Explain.

11. The wavelength of green light is 5600 Å in air and its wavelength in water is 4200 Å. If you view this light while your head is under water, would it appear to be green or violet? Explain.

12. On what basic physical laws are the laws of reflection and refraction based? How were they originally discovered?

13. By reference to the graph in Figure 30-5, determine whether or not the speed of light of wavelength 6000 Å in crown glass is greater or less than the speed of light of wavelength 4000 Å in the same medium.

14. Suppose that white light—that is, light containing a mixture of all visible wavelengths—is incident on the prism in Figure 30-11. Assuming that the index of a refraction is a decreasing function of wavelength, the various colors in the incident beam will come out at various angles θ_2. Will violet light correspond to a smaller or larger angle of deviation δ?

15. You are given a prism of a certain index *n* and of vertex angle 30°, as in Figure 30-11. Describe how you would measure the variation of *n* with wavelength by its use.

16. Light is incident from air onto the face of a medium of refractive index 2. What is the maximum possible angle of refraction?

17. What is the critical angle associated with the system in Question 16? Is there a relation between the critical angle and the angle computed in Question 16? Explain.

18. Describe what you would see if you were under water and looked vertically upward. Assume that $n = 4/3$ and that the surface of the water is smooth.

19. Explain why a swimming pool that is actually 2.5 meters deep appears to be only about 2.0 meters deep to an observer looking perpendicularly down on the surface.

20. In terms of the properties of the microscopic constituents of a dielectric, what is the physical mechanism that underlies the phenomena of reflection and refraction? Would reflection and refraction take place for a light beam incident on a transparent magnetic material?

21. Consider a monochromatic light beam incident normally on the interface between two media. Assuming that $n_1 \neq n_2$, is it possible for all of the incident light to be transmitted? For it all to be reflected?

22. A light beam consisting of more than one color is incident from air onto a plate of crown glass, whose refractive index is plotted in Figure 30-5. Explain why there is a separation of the colors in the refracted beam but not in the reflected beam.

PROBLEMS

1. **(a)** Calculate the time it takes for light to reach us from the sun. Assume the earth's orbital radius to be 1.5×10^8 km.
   (b) What are the maximum and minimum times it takes light to reach us from the planet Jupiter? Assume that Jupiter's orbit has a radius of 7.8×10^8 km.

2. Consider a beam of orange light of wavelength 6500 Å.
   (a) What is its frequency?
   (b) What is its wavelength when traveling in crown glass of index 1.52?
   (c) What is its velocity in this medium?

3. If the speed of light in a certain medium is 2.0×10^8 m/s what is the refractive index of the medium? What is the wavelength of this light in the medium if its free-space wavelength is 4500 Å?

4. Light of wavelength 5000 Å enters a medium where its wavelength is found to be 3500 Å.
   (a) What is the refractive index of the medium?
   (b) What is the frequency of the light?
   (c) What is the dielectric constant of the medium at this frequency?

5. A beam of light is incident at an angle θ onto the surface of a plane mirror. If the mirror is rotated by an angle α about an axis in its plane and perpendicular to the plane of incidence, by what angle is the reflected beam rotated?

6. By use of the graph in Figure 30-5 calculate the ratio of the speed of light in crown glass for light of wavelength 7.0×10^{-7} meter to light of wavelength 4.0×10^{-7} meter.

7. In Figure 30-17 suppose that light

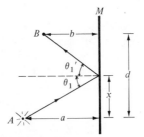

Figure 30-17

goes from point A to point B after reflection by mirror M. Show that if the path of the ray is such that the angle of incidence is equal to the angle of reflection, then the time required for the light to travel from A to B is a minimum. (*Hint*: Show that the time $t(x)$ to go via the indicated path is

$$t(x) = \frac{1}{c}\{[a^2 + x^2]^{1/2} + [(d-x)^2 + b^2]^{1/2}\}$$

and find the value of x for which this is minimum.) (*Note*: This and the result of Problem 8 represent a derivation of the laws of geometrical optics by use of *Fermat's principle*, which states that the actual path— out of the totality of possible paths—followed by a light beam is that one which takes the *least time*.)

8. Consider, in Figure 30-18, a beam of light that goes from point A to point B inside a medium of refractive index n. If the parameters a, b, and d are as defined in the figure:
   (a) Show that the time t required to go from A to B is

$$t = \frac{1}{c}\{[a^2 + x^2]^{1/2} + n[b^2 + (d-x)^2]^{1/2}\}$$

   (b) Find that value of x for which the time t is a minimum and

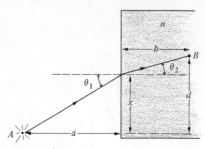

Figure 30-18

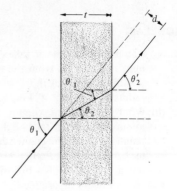

Figure 30-19

show that at this condition the angles θ_1 and θ_2 are related by Snell's law.

9. A beam of light originates at a point below the surface of a tank of water. Assuming that $n = 1.33$, calculate the angle the emerging beam makes with respect to the vertical if the angle of incidence is: (a) 30°; (b) 45°; (c) 75°.

10. For the physical situation described in Problem 9 calculate the angle of incidence for which the emerging beam will come out horizontally. What happens if the angle of incidence is increased beyond this value?

11. A plate of glass (index 1.5) is under water (index 1.33).
    (a) If light is originally traveling in the water and is incident at an angle of 30°, what is the angle of refraction?
    (b) If a light beam is originally traveling in the glass and is incident at an angle of 60°, what is the angle of refraction as the beam enters the water?

12. A beam of light is incident at an angle θ_1 onto a flat glass slab of thickness t and refractive index n; see Figure 30-19.
    (a) What is the value of θ_2 in terms of n and θ_1?
    (b) Why must $\theta_1' = \theta_2$ and $\theta_2' = \theta_1$?
    (c) Show that the lateral deviation d of the beam after the second

refraction is

$$d = t \sin \theta_1 \left[1 - \frac{\cos \theta_1}{(n^2 - \sin^2 \theta_1)^{1/2}} \right]$$

and evaluate this for small θ_1.

13. A ray of light is incident at an angle of 30° on a glass plate 1 cm thick. If the lateral displacement between the incident and transmitted beams is 0.2 cm, calculate the index of refraction of the plate by use of the results of Problem 12.

14. Show that if a beam of light is incident from air on a surface of refractive index n at an angle θ_0, so that the reflected and refracted beams are perpendicular to each other, then

$$\tan \theta_0 = n$$

15. Two parallel beams of light go through the prism in Figure 30-11 of vertex angle $A = 30°$. If the refractive index of the prism for one of the beams is 1.50 and for the other it is 1.60, and if the beams enter at an angle of incidence corresponding to the minimum deviation for the former, calculate:
    (a) The deviation δ of each beam.
    (b) The angle between the two beams when they emerge from the prism.

16. Show that the angle of minimum

deviation δ_m for a prism of vertex angle A is given by (30-11).

17. Consider again, in Figure 30-14a, a beam of light incident normally onto one of the shorter sides of a 45°-45°-90° prism. What is the minimum value for the refractive index so that the beam undergoes total internal reflection? Assume the prism to be in air.

18. Repeat Problem 17, but assume this time that on the surface of the prism at A there is a thin layer of water of refractive index 1.33.

19. Light is incident at an angle θ_1 on the face of a cube of glass of index 1.5; see Figure 30-20. Find the maximum value for the angle θ_1 so that the light is totally internally reflected at point A on an adjacent face. Assume that the cube is in water of refractive index 1.33.

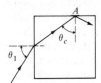

Figure 30-20

20. At what angle, with respect to the vertical, must a fish look in order to see a fisherman sitting on a distant shore? Assume that $n = 1.33$.

21. Generalize (30-15) to the case that the second medium is not air but has an index of refraction n_2 and show explicitly that here

$$\sin \theta_c = \frac{n_2}{n_1}$$

22. Two parallel light beams are incident from air onto the large face of a 45°-45°-90° prism of refractive index 1.30; see Figure 30-21. Calculate the angle between the refracted light beams when they emerge from the prism.

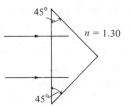

Figure 30-21

23. A light beam is incident normally from air onto a glass plate of refractive index 1.5. What fraction of the incident energy (expressed as a percent) enters the plate? What happens to the remainder?

24. Show explicitly, by use of (30-18), that the sum of the reflected and refracted intensities is numerically equal to the incident intensity.

25. Suppose that a light beam is incident from air perpendicularly onto the face of a certain transparent material. Calculate its refractive index if 50 percent of the light is reflected.

26. Show explicitly that for near-normal incidence ($\theta_1 \cong n_2\theta_2/n_1 \cong 0$), both (30-19) and (30-20) reduce to (30-17).

27. (a) Assuming near-normal incidence in Figure 30-15, show that the requirement that the tangential components of the electric field be continuous implies that

$$E_0 - E_0'' = E_0'$$

(b) Similarly, assuming that the tangential components of $\mathbf{H}$ ($\equiv \mathbf{B}/\mu_0$ since we assume that the media are not magnetic) are continuous, show that at normal incidence

$$n_1 E_0 + n_1 E_0'' = n_2 E_0'$$

(c) By combining the results of (a) and (b) confirm the validity of (30-17).

*28. Using the method of Problem 27, establish the validity of (30-19). Make use of the vectors in Figure 30-15 but do not assume the angles θ_1 and θ_2 to be small.

29. Show that the amplitudes in both (30-19) and (30-20) satisfy the relation

$$n_1 \cos \theta_1 E_0{}^2 = n_1 \cos \theta_1 E_0{}''^2$$
$$+ n_2 \cos \theta_2 E_0{}'^2$$

What is the physical significance of this relation?

*30. Consider, in Figure 30-22, a very narrow, rectangular path ABCDA, which straddles the interface be-

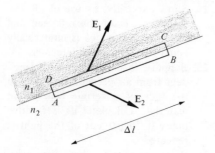

Figure 30-22

tween two media of refractive indices n_1 and n_2. Assume in the following that the length Δl is small but finite, and that the width of the rectangle tends to zero.

(a) Show that the tangential components of the electric field on the two sides of the interface must be the same by applying Faraday's law to the closed path ABCDA. (See Problem 34, Chapter 27.)

(b) Similarly, by applying another of Maxwell's equations to the path ABCDA show that the tangential components of H ($\equiv B/\mu_0$ in the nonmagnetic case) are continuous. (See Problem 39, Chapter 28.)

*31. Using methods analogous to those in Problem 30, show by use of Gauss' law that the normal components of D ($\equiv \kappa \epsilon_0 E$) are continuous across the interface between two media. (*Hint*: Evaluate Gauss' law for a very short pillbox, which straddles the interface. See also Problem 38, Chapter 28.)

31 Mirrors and lenses

The beholding of the light is itself a more excellent and fairer thing than all the uses of it.

FRANCIS BACON

31-1 Introduction

In Chapter 30 we studied the laws which govern the reflection and the refraction of a collimated beam of light rays at the interface between two media. We shall now extend these studies to the more usual and complex situations involving the simultaneous reflections and refractions of large collections of such light beams.

Except under controlled laboratory conditions, the light waves emitted by a source normally consist of a superposition of many light beams, which go radially out in all directions from the source. It is for this reason, for example, that when an object—such as this page—is illuminated by a light source, it can be seen from many vantage points. The incident light is diffusely scattered so that each point of the object becomes, in effect, a point source from which light rays go out in all directions. Our perception of the object is determined by the particular subset of these rays which enters the eyes. Figure 31-1a, for example, shows some of the light rays that emanate from a point source S. Note that the light rays which enter the eyes of the observers O_1 and O_2 are distinct, so that each observer perceives S from his

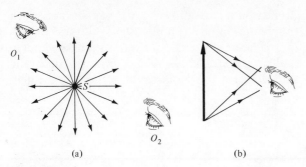

Figure 31-1

own point of view. Similarly, Figure 31-1b shows some of the light rays seen by an observer viewing an illuminated object. Although only the rays that go to the observer's eyes from the top and bottom of the arrow are shown, it is understood in such diagrams that light rays from intermediate points of the arrow enter his eyes as well. In other words, the observer perceives all points of the object, but for the sake of diagrammatic simplicity not all of the rays involved have been drawn.

Consider, in Figure 31-2, a point source S in air at a certain distance from the flat surface of a refracting medium. Each ray from S will, upon striking the interface, give rise to both a reflected and a refracted ray whose directions may be calculated from the laws of geometrical optics. An important question that arises in this connection is whether in addition to this predictable behavior of the individual light beams there are any significant regularities displayed by the *totality* of these reflected and refracted rays. The main purpose of this chapter is to establish an affirmative answer to this question by describing some of the quantitative laws that govern the behavior of such *collections* of light rays.

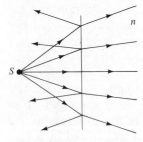

Figure 31-2

31-2 Plane mirrors

A *mirror* is a highly polished surface having the property that most of the light incident on it is reflected, and very little is therefore transmitted through it. Ordinary household or laboratory mirrors can be made by evaporating a thin, metallic film on a smooth glass or Pyrex surface. Such mirrors are nearly 100 percent reflecting. Although we shall confine ourselves here to an analysis of reflections by such a mirror, it should be kept in mind that the results obtained are also applicable to ordinary dielectric surfaces, with the only exception that for the latter *not* all of the incident light is reflected.

Consider, in Figure 31-3, a point object O at a perpendicular distance s from a plane mirror MM'. As shown, some of the light rays that go radially outward from O will strike the mirror and be reflected backward in accordance with the law of reflection. An observer, such as O_1, looking into the mirror will see the object by virtue of some of these reflected rays which enter his eyes. To him, the object will *appear* to be at an *image point I*, which is at a certain distance s' *behind* the mirror. Most interesting is the fact that to *all* observers looking into the mirror the object appears to be at the same point I, which is thus known as the *image* of the object. This point, which lies along the perpendicular from O to the mirror, is precisely as far behind the mirror as the object is in front. Hence the object distance s and the image distance s' in the figure are equal.

An image of a point object of this type whose rays do not actually intersect in a point, but only appear to do so, is said to be a *virtual image.* It follows by reference to Figure 31-3 that all images formed in plane mirrors must be virtual. By contrast, for a concave spherical mirror, as will be seen in Section 31-3, it is very much possible for the light rays from a point object to intersect in a real point after reflection in the mirror. This type of an image is known as a *real image.* Note that for a real image there is an actual flow of energy through the image point, whereas no such flow is associated with a virtual image.

To establish the equality of s and s' consider, in Figure 31-4, an arbitrary ray OB from the object, which makes an angle θ with the ray OA perpendicular to the mirror. On striking the mirror, the ray OA is reflected back on itself, whereas the ray OB is reflected along the direction BC, which, according to the laws of geometrical optics, also makes an angle θ

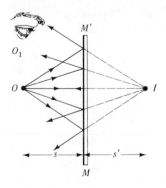

Figure 31-3

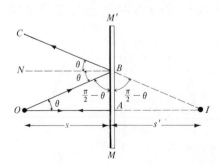

Figure 31-4

with the normal NB to the mirror. Let us extend the reflected rays AO and BC backward through the mirror and let them intersect at a point I at a certain distance s' behind the mirror. The two triangles OAB and ABI have the common side AB and by construction $\angle OAB = \angle IAB = 90°$ and $\angle OBA = \angle IBA = 90° - \theta$. Hence these two triangles are congruent, and therefore

$$s = s'$$

That is, regardless of the value for θ, the backward extension of the reflected ray BC intersects the corresponding extension of the perpendicular ray AO at the point I, which is precisely as far behind the mirror as the object is in front. Thus, the totality of all reflected rays appears to come from this image point I, and to *any* observer who sees only the reflected light the object appears to be at the image point I.

Figure 31-5 illustrates image formation of an object of finite dimensions in a plane mirror. For each point o of the object which is at a perpendicular distance s from the mirror there is a corresponding image point i the same distance behind it. Hence as shown in the figure by the dotted arrow, the image is precisely of the same size as the object but tilted by the angle α in the opposite sense.

Figure 31-6 shows the reflection of a complete three-dimensional object in a plane mirror. Again for each object point o in front of the mirror there is an image point i an equal distance behind it, with the line through o and i perpendicular to the mirror. Note that the sense of the z-axis of the image is opposite to that of the object, but that the senses of the x- and y-axes are unaltered. We describe this by saying that the *sense of front and back* for the image is opposite to that for the object. If you look into a mirror and move your right hand, then, consistent with the above arguments, you will find that the *left hand* of the image is the one that moves. Accordingly, we also say that the sense of left and right for the image in a plane mirror is opposite to that of the object.

Figure 31-7 illustrates the problem of image formation for an object O, which lies between two mirrors at right angles to each other. This time, as illustrated by the reflected rays, there will be three images: I_1, I_2, and I_3. Note that for I_2 the ray from O undergoes *two* reflections before entering

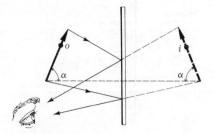

Figure 31-5

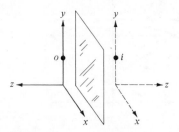

Figure 31-6

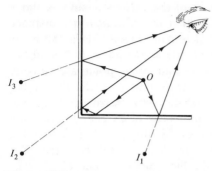

Figure 31-7

the observer's eye, so the sense of front and back for I_2 is the same as that for the object. The rays for the images I_1 and I_3 involve only one reflection, and thus for these the sense of front and back is opposite to that for the object. These features can be readily confirmed by looking into two mirrors at right angles. If you raise your right hand, you will find that each of the images on the left and the right will raise the left hand, while for the middle image the right hand will be raised.

31-3 Spherical mirrors

Besides the plane mirror, there are a number of other mirror geometries which are of physical and practical interest. One of these is the spherical mirror, and is the subject of this section and the next. We shall find that under certain conditions image formation also can take place in spherical mirrors.

Consider in Figure 31-8 an object O at a distance s from a *concave* spherical mirror of radius R. Let the point C represent the *center of curvature*—that is, the center of the spherical surface of which the mirror is a part—and define the line through O and C as the *axis* of the mirror. The intersection of this axis with the mirror defines its vertex V and, for reasons to be given below, we define the *focal point* F of the mirror as the point midway between V and C. Thus the focal point F is at a distance $R/2$ from both the vertex and the center of curvature.

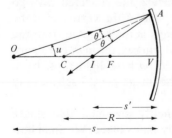

Figure 31-8

Figure 31-9 illustrates how the definitions of the various quantities above may be extended to situations other than that of an object at a distance greater than R from a concave spherical mirror. Figure 31-9a shows an object at a distance $s < R$ from a concave mirror of radius R. The above definition for the mirror axis, the vertex, and focal point are directly applicable to this case. The main difference is that now the object lies between the vertex and the center of curvature of the mirror. Correspondingly, Figure 31-9b shows how these definitions can be extended to the case of a *convex* mirror. The important difference for this case is that both the center of curvature C and the focal point F are here *behind* the mirror. The axis of the mirror is still defined to be the line joining O and C.

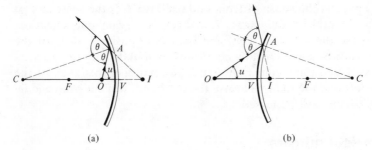

(a)   (b)

Figure 31-9

Let us now consider any one of the situations in Figures 31-8 and 31-9, and follow a typical ray OA from O, which makes an angle u with the axis and strikes the mirror at a point A. Since the radius vector CA is perpendicular to the mirror, it follows from the laws of geometrical optics that CA will bisect the angle between the incident and the reflected rays. The ray OV is itself perpendicular to the mirror and is thus reflected back on itself. It follows that if an image is formed in the spherical mirror, it must be at the point of intersection of the mirror axis OV and the reflected ray AI or its backward extension. For the situations in Figure 31-9 this image point I is analogous to that for a plane mirror and is *behind* the mirror; that is, the light rays are reflected by the mirror and *appear* to the observer to originate at the point I *behind* the mirror. These images are thus *virtual*. By contrast, for the situation in Figure 31-8 the image is *real*. Here the reflected rays go through the image point I, which is on the same side of the vertex as is the object. Hence for this case the observed light rays actually come from the image point I, and there is, in general, a flow of energy through the image point. By contrast, although energy is associated with all light rays, for a virtual image no energy actually goes through the image point itself; it only appears to do so.

So far in this discussion it has been implicitly assumed that image formation—whether it be real or virtual—actually takes place. We shall now establish that for *paraxial rays*, that is, rays for which the angle u in Figures

31-8 and 31-9 is sufficiently small that $u \cong \tan u \cong \sin u$, an image is actually formed in each case. Moreover, it will be shown that for a concave mirror the object distance s and the image distance s' are related by

$$\frac{1}{s} + \frac{1}{s'} = \frac{2}{R} \qquad \text{(concave)} \qquad (31\text{-}1)$$

while for a convex mirror the corresponding formula is

$$\frac{1}{s} + \frac{1}{s'} = -\frac{2}{R} \qquad \text{(convex)} \qquad (31\text{-}2)$$

where a negative value for s' means that the image is virtual and thus located a distance $|s'|$ behind the mirror.

Alternatively, we may express (31-1) and (31-2) in the form

$$\frac{1}{s} + \frac{1}{s'} = \frac{1}{f} \qquad (31\text{-}3)$$

where f is the *focal length* of the mirror and has the value $+R/2$ for a concave mirror and $-R/2$ for a convex one. Physically, f is the image distance associated with an object at infinity. As illustrated in Figure 31-10a, for the case of a concave mirror the incoming parallel rays from a very distant object are brought together at the focal point F of the mirror. Correspondingly, as shown in Figure 31-10b, the reflected rays from a point object at the focal point of a concave mirror are parallel to the mirror axis. The analogous result for a convex mirror is shown in Figure 31-13 and will be considered in Section 31-4.

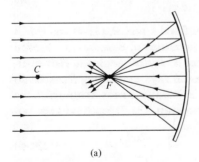
(a)

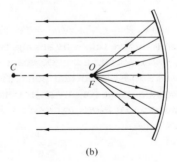
(b)

Figure 31-10

To establish (31-1) suppose that $s > R/2$, and consider in Figure 31-11 an arbitrary ray OA from the object, which makes an angle u with the axis. Define the angles α, β, and θ as in the figure, and let h be the length of the perpendicular from A to the axis and ϵ the distance from its foot to the vertex V. Since the exterior angle of a triangle is the sum of the remote interior angles it follows that

$$\beta = \alpha + \theta \qquad \alpha = u + \theta$$

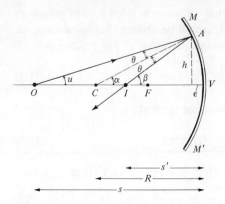

Figure 31-11

Eliminating the angle θ, we obtain

$$u + \beta = 2\alpha \tag{31-4}$$

Moreover, reference to the figure shows that

$$\tan u = \frac{h}{s - \epsilon}$$

$$\tan \beta = \frac{h}{s' - \epsilon} \tag{31-5}$$

$$\tan \alpha = \frac{h}{R - \epsilon}$$

$$\epsilon = R(1 - \cos \alpha)$$

since $s = \overline{OV}$, $R = \overline{CV}$, and $s' = \overline{IV}$. Now for fixed s and R, (31-4) and (31-5) represent only five relations among the six quantities u, β, α, h, ϵ, s'. Hence they do *not* associate a unique image distance s' for *all* angles u. However, and this is a most important point, the paraxial rays emanating from O do form a unique image for any object distance s. For these rays, with which we shall concern ourselves exclusively from now on, (31-5) may be approximated by

$$u = \frac{h}{s} \qquad \beta = \frac{h}{s'} \qquad \alpha = \frac{h}{R} \qquad \epsilon = \frac{R\alpha^2}{2} \cong 0 \tag{31-6}$$

since if u is small, so that $u \approx \tan u \approx \sin u$, then so are each of the remaining angles α, β, and θ. Substituting the first three relations of (31-6) into (31-4), we find, after canceling the common factor h, the desired relation in (31-1).

The validity of (31-1) for $s < R/2$ as well as (31-2) can be established by analogous arguments. Details will be found in the problems.

Figure 31-12a shows a plot of the image distance s' as a function of s for a concave mirror, as predicted by (31-1). Note that if $s > R/2$, then $s' > 0$, and the image is real, whereas if $s < R/2$, then $s' < 0$ and, as confirmed in Figure

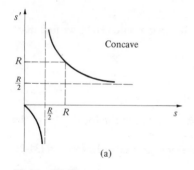

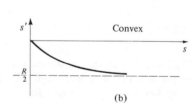

(a)

(b)

Figure 31-12

31-9a, the image is virtual. If the object is at the focal point F, so that $s = R/2$, then according to the graph $s' = \infty$. Hence, as shown in Figure 31-10b, the reflected rays are parallel to the mirror axis in this case. Figure 31-12b shows the corresponding plot of s' against s for a convex mirror. As can be confirmed by reference to Figure 31-9b, the image is virtual for any object distance in this case. Hence, as shown in the graph, only negative values for s' occur.

31-4 Applications

In this section we apply (31-1) and (31-2) to several concrete physical situations.

Example 31-1 A point object is at a distance of 20 cm from a spherical mirror of radius 10 cm. What is the nature and location of the image if the mirror is (a) Concave? (b) Convex?

Solution
(a) Here we are given the values $s = 20$ cm and $R = 10$ cm. Substitution into (31-1) yields

$$s' = \frac{20}{3}\,\text{cm} = 6.7\,\text{cm}$$

Since $s' > 0$, the image is real and on the same side of the mirror as the object.
(b) For the convex mirror, we must use (31-2). Substituting the value $s = 20$ cm, we find that

$$s' = -4.0\,\text{cm}$$

so the image is virtual this time.

Example 31-2 Parallel rays are incident on a spherical mirror having a radius of 0.5 meter. What is the nature and location of the image formed by these rays if the mirror is: (a) Concave? (b) Convex?

Solution

(a) Substituting the given values, $s = \infty$ and $R = 0.5$ meter, into (31-1), we find that

$$\frac{1}{\infty} + \frac{1}{s'} = \frac{2}{0.5 \text{ m}}$$

that is,

$$s' = 0.25 \text{ m}$$

The image is real and, as shown in Figure 31-10a, it is located at the focal point of the mirror at a distance $R/2$ from the vertex.

(b) Substituting $s = \infty$ and $R = 0.5$ meter into the convex mirror formula in (31-2), we find in a similar way that

$$s' = -0.25 \text{ m}$$

and this is shown in Figure 31-13. Note that the image is virtual and located a focal length $(R/2)$ *behind* the mirror.

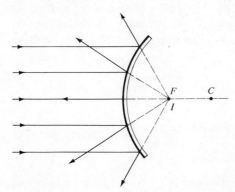

Figure 31-13

Example 31-3   A point object O is at a distance $(3/4)R$ from a concave mirror of radius R; see Figure 31-14.

(a) What is the nature and location of the image?

(b) How can an observer tell which of O and I is the object and which is the image?

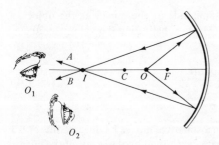

Figure 31-14

Solution

(a) The substitution of the given values for s and R into (31-1) yields

$$\frac{1}{(3/4)R} + \frac{1}{s'} = \frac{2}{R}$$

and thus

$$s' = +\frac{3}{2}R$$

so the image is real.

(b) A person viewing the object O directly will see it regardless of which observation point he happens to choose. This follows since a point object is one that radiates light in all directions. On the other hand, because of the finite extent of the mirror, the reflected rays are confined—as shown in Figure 31-14—to the interior of the cone AIB. In other words, the observer O_1 will see the image, since some of the reflected rays will enter his eyes. On the other hand, an observer O_2 looking at this same point I in space but from outside of the cone AIB will see nothing, since the reflected rays that would enter his eyes if the mirror were large enough simply are not reflected. Thus by changing points of view we can tell the object and the image apart. That is, there are always positions of the observer for which the image is invisible!

31-5   Finite objects

Consider, in Figure 31-15, an object of finite size—say an arrow of length y—at a distance s from and perpendicular to the axis of a concave spherical mirror of radius R. The image of the point A at the base of the arrow will occur at the point A', which is on the axis and at a distance s' from the vertex as determined by (31-1). The object distance of the point B, which represents the head of the arrow, is slightly greater than is the object distance s for the point A, but for simplicity let us neglect this difference by confining ourselves to small objects. In general, if the object is not small, the image will undergo a distortion known as *curvature of field*. As is implied in the figure, we shall ignore these and other *spherical aberrations*, as they are

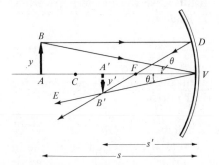

Figure 31-15

referred to, so that the image, $A'B'$, will always be assumed to be parallel to the object, AB.

The position of the image B' of the point B may be found by drawing any two of the following four rays. The two of these that have been drawn in the figure are (1) the ray BD parallel to the mirror axis, which after reflection goes through the focal point F; and (2) the ray BV to the vertex of the mirror, whose reflection VE makes the same angle θ with the axis that the incident ray does. The remaining two rays, which have not been included in the figure, are (3) the ray BC through the center of curvature, which will be reflected back on itself; and (4) the ray BF through the focal point, which after reflection will be parallel to the axis. (See Figure 31-16.) The intersection of any two of these reflected rays will determine a unique point B' at a certain distance y' below the axis. The perpendicular from B' to the axis will intersect the axis at the point A', which is the image of the point A. All other rays drawn from points on the arrow between A and B will be imaged between points A' and B', and thus there is no need to draw any additional rays.

Consider the two right triangles ABV and $VA'B'$. They are similar to each other since their angles are equal in pairs. Hence we obtain

$$\frac{y}{y'} = \frac{s}{s'} \tag{31-7}$$

By use of this formula and (31-1) the size of the image y' for any given object can be readily determined.

In connection with a discussion of the reflection of objects of finite size in a spherical mirror it is convenient to define the *lateral magnification, m,* of the object to be the ratio of the size of the image to that of the object. The sign of m is by definition negative if the image is inverted relative to the object, and positive otherwise. For the situation in Figure 31-15 with a concave mirror, this means that

$$m = -\frac{y'}{y} \tag{31-8}$$

where the minus sign is required since in this case the image is inverted. Substitution for the ratio y'/y from (31-7) leads to

$$m = -\frac{s'}{s} \tag{31-9}$$

which together with (31-1) and (31-2) represent the basic tools necessary for the analysis of image formation in spherical mirrors.

Although (31-9) has been derived only for the special situation shown in Figure 31–15—that is, where the object is at a distance $s > R$ from a concave spherical mirror and for which the image is real and inverted—it is true in general. That is, regardless of whether the mirror is concave or convex or whether the image is real or virtual or whether it is erect or inverted, (31-9) is invariably correct. The validity of this feature is illustrated in Figure 31-16 for various special cases. Figure 31-16a shows an object AB at a point between the focal point F and the center of curvature C. Its image is found by drawing the ray BD parallel to the axis, which after reflection goes through the focal point, and the ray BF through the focal point F, which after reflection is parallel to the axis. The intersection of these two rays as well as the ray CB through the center of curvature determines the image point B'. Note that consistent with (31-9) the image is inverted and larger than the object. Similarly, in Figure 31-16b, where the object distance is less than the focal distance, we see that the image is virtual and erect. This means that the image distance s' is negative and thus, consistent with (31-9), the magnification is positive. Finally, Figure 31-16c shows the case of an object in front of a convex mirror. The image is virtual, thus $s' < 0$; and therefore, consistent with (31-9), the lateral magnification m is positive. Hence the image is erect, as shown.

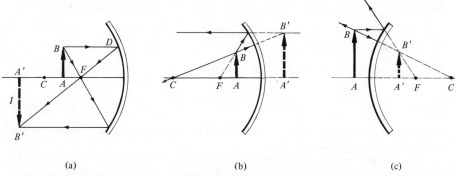

(a)   (b)   (c)

Figure 31-16

Example 31-4 An object of height 2.0 cm is perpendicular to the axis and at a distance of 5.0 cm from a concave spherical mirror of radius 15 cm.
  (a) Locate and describe the nature of the image.
  (b) Calculate the lateral magnification.
  (c) What is the size of the image?

Solution
  (a) Substituting the given data, $R = 15$ cm and $s = 5.0$ cm, into (31-1) yields

$$\frac{1}{5\,\text{cm}} + \frac{1}{s'} = \frac{2}{15\,\text{cm}}$$

so that $s' = -15$ cm. Hence the image is virtual and at a distance of 15 cm behind the mirror.
  (b) Substituting the above values for s and s' into (31-9) yields for m

$$m = -\frac{s'}{s} = -\frac{-15\,\text{cm}}{5\,\text{cm}}$$

$$= +3$$

Since this is positive, it follows that the image is erect.

(c) Since $m = +3$, it follows by use of (31-8) that the image is three times as large as the object. Since the object has a height of 2 cm, the height of the image is 6 cm.

Example 31-5 An object is at a distance s from a spherical mirror of radius R. If the object is displaced by a small amount ds, show that the corresponding displacement ds' of the image is

$$\frac{ds'}{ds} = -m^2$$

where m is the lateral magnification, and describe this result qualitatively.

Solution Let us start with the basic relation

$$\frac{1}{s} + \frac{1}{s'} = \pm \frac{2}{R}$$

where the upper sign is for a concave mirror and the lower for a convex one. Taking differentials on both sides of this relation we find, since R is constant, that

$$-\frac{ds}{s^2} - \frac{ds'}{s'^2} = 0$$

Solving for the ratio ds'/ds, we thus obtain

$$\frac{ds'}{ds} = -\left(\frac{s'}{s}\right)^2 = -m^2$$

where the last equality follows by (31-9). The quantity ds'/ds is known as the *longitudinal magnification*.

Even though the lateral magnification m can be either positive or negative, the ratio ds'/ds is negative in all cases. This means that if an object is brought closer to a mirror—that is, if $ds < 0$—then $ds' > 0$, so that if the image is real, then it recedes from the mirror while if it is virtual it approaches the mirror. Contrariwise, if $ds > 0$, so that the object recedes from the mirror, then $ds' < 0$, so that the image either approaches or recedes from the mirror depending on whether the image is real or virtual.

31-6 Image formation by spherical refracting surfaces

We now turn from the problem of image formation in mirrors, and for the remainder of this chapter consider the related problem of the formation of images by light rays refracted at the spherical interface between two media.

Figure 31-17 shows a point object O at a distance s_1 from the spherical interface of radius R between two media of refractive indices n_1 and n_2, respectively. The refracting surface is assumed to be convex in this case. The perpendicular ray OC from the object to the center of curvature C cuts the surface at the vertex V and defines the *axis* of the system. A second ray OA, which makes an arbitrary angle u with respect to the axis, has also been

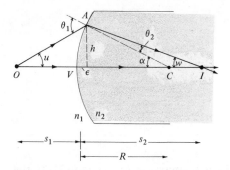

Figure 31-17

drawn in the figure. Assuming that $n_2 > n_1$, this ray will be refracted toward the normal into the second medium and, as shown, it will intersect the axis at a point I at a certain distance s_2 from the vertex. The angle of incidence θ_1 and the angle of refraction θ_2 in the figure are defined with respect to the normal to the surface at A, which is the radius CA from the center of curvature C.

Just as for spherical mirrors, it will be established below that for *paraxial rays* image formation also takes place for spherical refracting surfaces. For the specific case shown in the figure, with the object in the n_1-medium on the convex side and at a distance s_1 from the vertex, an image will be formed at a distance s_2 from the vertex given by

$$\frac{n_1}{s_1} + \frac{n_2}{s_2} = \frac{n_2 - n_1}{R} \quad \text{(convex)} \quad \text{(31-10)}$$

If s_2, is positive then the image is real and is formed in the n_2-medium to the right of the interface, as shown in Figure 31-17. On the other hand, if $s_2 < 0$, then, as shown in Figure 31-18, a virtual image is formed in the same medium as is the object. Correspondingly, for the case of the concave interface, as in Figure 31-19, the relation between the object and the image distance is

$$\frac{n_1}{s_1} + \frac{n_2}{s_2} = -\frac{n_2 - n_1}{R} \quad \text{(concave)} \quad \text{(31-11)}$$

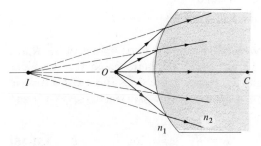

Figure 31-18

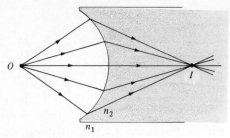

Figure 31-19

Note that, just as for the spherical mirror formulas in (31-1) and (31-2), the relations (31-10) and (31-11) may be obtained from each other by the substitution $R \rightleftarrows -R$. The use of these formulas makes possible the calculation of the image distance s_2 for any given object distance s_1. Positive values for s_2 imply a real image, as in Figures 31-17 and 31-19, and for these the image is on the opposite side of the interface from the object. Figure 31-18 shows the case of a virtual image, for which s_2 is negative, and here the image is in the same medium as is the object.

To derive (31-10), let us define the various angles u, θ_1, θ_2, α, and w, and the distances s_1, s_2, h, and ϵ as in Figure 31-17. Since the exterior angle of a triangle is the sum of the remote interior angles, it follows that

$$\alpha = w + \theta_2 \qquad \theta_1 = \alpha + u \qquad \text{(31-12)}$$

while according to Snell's law,

$$n_1 \sin \theta_1 = n_2 \sin \theta_2 \qquad \text{(31-13)}$$

Reference to the figure makes it possible to establish the additional relations

$$\tan u = \frac{h}{s_1 + \epsilon}$$

$$\tan w = \frac{h}{s_2 - \epsilon} \qquad \text{(31-14)}$$

$$h = R \sin \alpha$$

$$\epsilon = R(1 - \cos \alpha)$$

Again, as for the case of the spherical mirror, for fixed values of s_1, R, n_1, and n_2, these seven relations do not yield a unique value for the eight quantities α, θ_1, θ_2, u, w, h, ϵ, and s_2. Hence there does *not* exist a unique image distance s_2 for all values of u. However, for paraxial rays, (31-13) and (31-14) become

$$n_1 \theta_1 = n_2 \theta_2 \qquad u = \frac{h}{s_1} \qquad w = \frac{h}{s_2} \qquad h = R\alpha \qquad \epsilon \simeq 0 \qquad \text{(31-15)}$$

and, substituting for θ_1 and θ_2 into the first of these by use of (31-12), we obtain

$$n_1(\alpha + u) = n_2(\alpha - w)$$

or

$$n_1 u + n_2 w = \alpha(n_2 - n_1)$$

Finally, eliminating the angles u, w, and α by use of (31-15), and canceling the common factor h, the sought-for relation in (31-10) results. The validity of (31-10) for a virtual image, as in Figure 31-18, can be established in the same way.

To derive (31-11) we may now proceed as follows. Reference to Figures 31-17 and 31-19 shows that if, for a given value for s_1, the object is placed at the image point (of a real image), then the resultant image occurs at the old object point. That is, if an object is placed at the point I in either of these figures, its image occurs at the point O. For the effect of exchanging the image and the object is to reverse the sense of direction of all light rays, and the laws of optics work just as well for the reversed rays as they do for the original ones. For the case of a mirror, this reversal means that $s' \rightleftarrows s$ and (31-1) and (31-2) are invariant under this exchange, as they must be. In the present case the interchange of the object and image means that $s_1 \rightleftarrows s_2$ and $n_1 \rightleftarrows n_2$, and making these substitutions in (31-10) we are led directly to (31-11). In other words, both (31-10) and (31-11) are invariant under the simultaneous replacements $s_1 \rightleftarrows s_2$ and $n_1 \rightleftarrows n_2$ and $R \rightarrow -R$. Therefore provided negative values for R are used for concave surfaces, (31-10) is very generally applicable to both concave and convex surfaces.

Graphs of (31-10) for the special case $n_2 > n_1$ are shown in Figure 31-20a and for $n_2 < n_1$ in Figure 31-20b. Note the similarity between these graphs and the corresponding ones for a mirror in Figure 31-12. For the case $n_1 < n_2$, if the object is placed a distance $Rn_1/(n_2 - n_1)$ from the vertex, the refracted rays are parallel to the axis and thus are imaged at infinity.

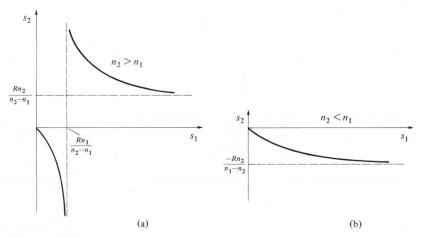

(a)   (b)

Figure 31-20

Example 31-6 A bundle of parallel rays are traveling in air and are incident from the left onto a spherical surface of glass ($n = 4/3$) of radius 10 cm. Describe the nature and location of the image if the surface is:
(a) Convex, as in Figure 31-17.
(b) Concave, as in Figure 31-19.

Solution
(a) The substitution of the given values $s_1 = \infty$, $R = +10$ cm, $n_1 = 1$, $n_2 = 4/3$ into (31-10) yields

$$\frac{1}{\infty} + \frac{4/3}{s_2} = \frac{(4/3) - 1}{10 \text{ cm}}$$

that is,

$$s_2 = +40 \text{ cm}$$

Hence the image is real and on the right-hand side of the interface.

(b) For this case, the parameter values are $s_1 = \infty$, $R = -10$ cm, $n_1 = 1$, and $n_2 = 4/3$. Substitution into (31-10) leads to

$$\frac{1}{\infty} + \frac{4/3}{s_2} = -\frac{(4/3) - 1}{10 \text{ cm}}$$

and this yields

$$s_2 = -40 \text{ cm}$$

Thus, after being refracted, the rays diverge and appear to come from the virtual image at a point 40 cm to the left of the interface.

Example 31-7 A point source of light is on the bottom of a tank of water ($n = 1.33$) of depth h. What is the apparent depth h' as seen by an observer who is in air and views the light at normal incidence?

Solution Since the observer views the source at normal incidence, only paraxial rays are involved and we may use (31-10). The surface of the water has zero curvature, so the parameter values are

$$R = \infty \qquad n_1 = \frac{4}{3} \qquad n_2 = 1 \qquad s_1 = h$$

Substitution into (31-10) yields

$$\frac{4/3}{h} + \frac{1}{h'} = -\frac{(4/3) - 1}{\infty}$$

and, consistent with the result in Example 30-4, this leads to

$$h' = -\frac{3}{4} h$$

31-7 The lensmaker's equation

Consider, in Figure 31-21, an object O on the *axis* of and at a distance s from a *thin lens* of refractive index n. Let R_1 and R_2 be the radii of curvature of its left and right faces, respectively, and, to be specific, suppose that their respective centers of curvature C_1 and C_2 are located as shown. The line joining C_1 and C_2 is called the *axis* of the lens. And the lens is "thin" provided that its maximum thickness is small compared to its two radii of curvature. In the following we shall be concerned only with thin lenses.

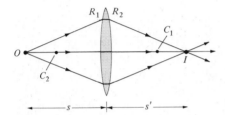

Figure 31-21

The main purpose of this section is to establish that the distance s' of the image I formed by a thin lens in air is related to the object distance s by

$$\frac{1}{s} + \frac{1}{s'} = \frac{1}{f}$$ (31-16)

where f is the "focal length" of the lens and is given by the lensmaker's equation

$$\frac{1}{f} = (n - 1)\left[\frac{1}{R_1} + \frac{1}{R_2}\right]$$ (31-17)

In utilizing this formula for the focal length f, the parameter R_1 is to be taken to be positive if, as in Figure 31-21, C_1 is to the right of the first refracting surface, and negative otherwise. Similarly, R_2 is positive if, as in the same figure, C_2 is to the *left* of the lens, and negative otherwise. Hence the focal length f can be positive or negative depending on the positions of the centers of curvature. The term *positive* or *converging lens* is used to characterize a lens with a positive focal length, and similarly a *negative* or a *diverging* lens is one whose focal length is negative. The particular lens shown in Figure 31-21 is known as a *double convex* lens and it has positive focal length. By contrast, the lens in Figure 31-22b is a *double concave* lens and since for it both R_1 and R_2 in (31-17) are negative, so is its focal length. Similarly the *plano-convex* and the *plano-concave* lenses in Figure 31-23 have positive and negative focal lengths, respectively.

The significance of the term "focal length" for the quantity represented by the symbol f in (31-16) and (31-17) can be seen by reference to Figure 31-22. Part (a) shows a beam of light rays incident on and parallel to the axis of a

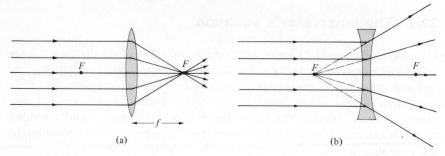

Figure 31-22

double convex lens, with its positive focal length. Since the object for these rays is at infinity, we find on setting $s = \infty$ in (31-16) that the image distance s' is equal to the focal length f. In other words, the focal length f is the distance from the positive lens of an image formed by an object at infinity. Equivalently, since an interchange of the object and its image can be achieved by reversing the direction of the light rays, the focal length is the distance from a positive lens at which a point source must be placed so that the emerging rays on the other side of the lens are parallel to the axis. Because of the inherent symmetry under the exchange $R_1 \rightleftarrows R_2$ in the lensmaker's equation in (31-17), the focal length f is the same on the two sides of the lens. Hence, as shown in the figure, the two focal points F of the lens are at the same distance f on opposite sides of the lens. Figure 31-22b shows a parallel beam incident on a double concave lens with its negative focal length. This time, setting $s = \infty$ in (31-16), we find that $s' = f$, but now f is negative! Therefore the image distance is negative and, as shown, the image is virtual and on the same side of the lens as is the object. For this case, the rays emerging from the lens diverge and thus appear to come from the side of the lens on which the rays are incident.

Comparison of (31-16) with the corresponding formulas for the spherical mirror in (31-1) and (31-2) shows that the graph in Figure 31-12 is also a graph of (31-16), provided that $R/2$ is interpreted as the focal length f of the lens. The important distinction is that for the mirror a positive image distance s' corresponds to a real image on the *same* side of the mirror as is the object.

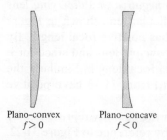

Plano–convex
$f > 0$

Plano–concave
$f < 0$

Figure 31-23

For a thin lens, on the other hand, a positive image distance means that the image is on the *opposite* side of the lens from the object.

To establish the validity of (31-16) and (31-17), let us consider the situation in Figure 31-21 and apply (31-10) and (31-11) at the two surfaces. For the refraction of the light rays from O at the first surface, the appropriate parameter values to substitute into (31-10) are $n_1 = 1$, $n_2 = n$, $s_1 = s$, $s_2 = s_2$, and $R = R_1$. Hence

$$\frac{1}{s} + \frac{n}{s_2} = \frac{n-1}{R_1} \tag{31-18}$$

Since the lens is presumed thin, it follows that the object distance for the refraction of these rays at the second surface is $-s_2$. Hence the appropriate parameter values to be used in (31-11) for the refraction of the light rays at the second surface are $s_1 = -s_2$, $s_2 = s'$, $n_1 = n$, $n_2 = 1$, and $R = R_2$. Hence

$$\frac{n}{-s_2} + \frac{1}{s'} = -\frac{(1-n)}{R_2} \tag{31-19}$$

Finally, eliminating s_2 between (31-18) and (31-19) and noting the formula for f in (31-17), we obtain the basic relation in (31-16).

Example 31-8 Assuming that the curved surface of a plano-convex lens (Figure 31-23) has a radius of curvature of 10 cm and that the refractive index of glass is 1.5, calculate the focal length of the lens.

Solution Assuming to be specific that the object is on the flat side of the lens, the radii of the lens are $R_1 = \infty$ and $R_2 = +10$ cm. Substituting these values into the lensmaker's equation, we obtain

$$\frac{1}{f} = (n-1)\left(\frac{1}{R_1} + \frac{1}{R_2}\right)$$

$$= (1.5-1)\left(\frac{1}{\infty} + \frac{1}{10\ \text{cm}}\right)$$

$$= 0.05/\text{cm}$$

whence

$$f = +20\ \text{cm}$$

The positive sign means that the lens is converging, so that, for example, a beam of parallel rays will be focused to a real image on the opposite side of the lens and at a distance 20 cm from it.

Example 31-9 Suppose that for the double concave lens in Figure 31-22b the two radii of curvature are -10 cm and -20 cm and that the refractive index of glass is 1.5. Calculate the focal length of this lens.

Solution This time the values for both R_1 and R_2 are negative. Substitution into the lensmaker's equation leads to

$$\frac{1}{f} = (n-1)\left(\frac{1}{R_1} + \frac{1}{R_2}\right)$$

$$= (1.5-1)\left(-\frac{1}{10\text{ cm}} - \frac{1}{20\text{ cm}}\right)$$

$$= -\frac{3}{40\text{ cm}}$$

Hence

$$f = -13.3\text{ cm}$$

Example 31-10 A parallel beam of light rays is incident on a glass sphere of radius 6 cm and of refractive index 1.5. Calculate the value of y where the rays come to a focus; see Figure 31-24.

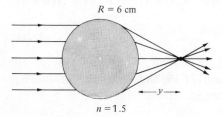

$R = 6$ cm

$n = 1.5$

Figure 31-24

Solution Since the lens is *not thin*, (31-16) and (31-17) are *not* applicable. Let us therefore proceed as in the derivation of these formulas themselves.

At the first surface, the appropriate parameter values are $s_1 = \infty$, $n_1 = 1$, $n_2 = 1.5$, and $R = 6$ cm. Substituting these into (31-10), we find that

$$0 + \frac{1.5}{s_2} = \frac{1.5-1}{6\text{ cm}}$$

so that $s_2 = 18$ cm. At the second surface, the corresponding parameter values are $s_1 = 12\text{ cm} - 18\text{ cm} = -6\text{ cm}$, $s_2 = y$, $n_1 = 1.5$, $n_2 = 1$, and $R = -6$ cm since the distance between the curved surfaces is 12 cm. A second application of (31-10) thus yields

$$-\frac{1.5}{6\text{ cm}} + \frac{1}{y} = \frac{(1-1.5)}{-6\text{ cm}}$$

Solving for y we obtain

$$y = 3\text{ cm}$$

31-8 Images formed by thin lenses

Consider, in Figure 31-25, an object AB, of height y, perpendicular to the axis of and at a distance s from a thin, converging lens of focal length f. Assuming y to be sufficiently small so that any optical distortions can be

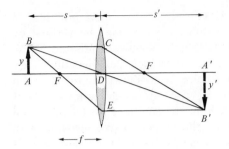

Figure 31-25

neglected, the image $A'B'$ of the object will be formed at a certain distance s' from the lens in accordance with (31-16). The lateral magnification m of the image is defined by

$$m = -\frac{y'}{y} \tag{31-20}$$

where y' is the size of the image. The purpose of the minus sign in this relation is so that m will be negative if, as in Figure 31-25, the image is inverted, and positive if the image is erect. This definition is the same as that for a spherical mirror in (31-8).

To determine the height y' of the image, it is necessary to find the image B' of the object point B. (The location of the image A' of the base of the object can be found directly from (31-16).) Figure 31-25 shows three rays, BCB', BDB', and BEB', the intersection of any two of which determine the image point B'. The ray BC is drawn parallel to the axis of the lens and thus, after being refracted in the lens, it will go through the focal point F as shown. Correspondingly, if the ray BE is drawn through the focal point F on the same side of the lens as the object, then after refraction it will be parallel to the lens axis. Finally, the ray BD through the center of the lens will emerge on the other side without being deviated. This follows from the result of Problem 12 in Chapter 30 and its associated Figure 30-19. The fact that all three of these rays (as well as all others that originate at B) must come together at the single point B' may be used as a consistency check when making use of this graphical method—illustrated in the figure—to determine the image formed by a thin lens.

Proceeding with the calculation of the height y' of the image, we note that in Figure 31-25 the triangles ABD and $A'B'D$ are similar. For they are both right triangles with the two angles $\angle BDA$ and $\angle B'DA'$ vertical angles and thus equal to each other. Hence

$$\frac{s}{s'} = \frac{y}{y'}$$

and substituting this into (31-20) we find that the lateral magnification m becomes

$$m = -\frac{s'}{s} \tag{31-21}$$

Even though this relation has been derived only for the converging lens in Figure 31-25, its validity extends to all thin lenses, as may be verified by applying the above argument to the situations in Figures 31-26 and 31-27.

By contrast to the situation in Figure 31-25, for which the object distance s exceeds the focal length f of the lens, Figure 31-26 shows the ray diagram for the two other possibilities: $s < f$ and $s = f$. As shown in Figure 31-26a, if the object is within a focal length of the lens, the emerging rays diverge and thus the image is virtual and larger than the object. Also consistent with (31-21), since $s' < 0$ for a virtual image, m is positive and the image is erect as shown. When the object is at the focal point itself, according to (31-16) $s' = \infty$. Substitution into (31-21) thus leads to $m = -\infty$, and hence, as implied by the parallel rays on the right side of the lens in Figure 31-26b, the image is inverted and "infinitely" large in this case.

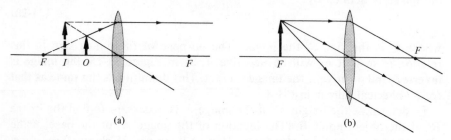

(a)   (b)

Figure 31-26

Figure 31-27 shows the analogous situation for the images formed by a thin, divergent lens with its negative focal length. Since only virtual images are possible for such a lens, the image distance s' is always negative. Hence, according to (31-21), m must be positive. As shown, therefore, for all three cases— $s > |f|$, $s < |f|$, and $s = |f|$—the image is virtual and erect.

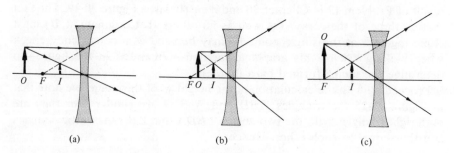

(a)   (b)   (c)

Figure 31-27

Example 31-11 Two thin lenses of focal lengths f_1 and f_2 are separated by a distance t; see Figure 31-28. Calculate the equivalent "focal length" of the combination.

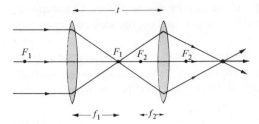

Figure 31-28

Solution   To solve this problem, let us assume that parallel rays are incident on the lens of focal length f_1 and that the lenses are a distance t apart. By definition of focal length, these incident rays will form an image at a distance f_1 from the f_1-lens or, equivalently, at a distance $(t - f_1)$ from the second lens. On substituting this value $(t - f_1)$ for the object distance into (31-16) we obtain

$$\frac{1}{t - f_1} + \frac{1}{s'} = \frac{1}{f_2}$$

where s' is the image distance from the second lens. Calling this distance s' from the second lens, where the incident parallel rays come to a focus, the focal length f of the combination, we find that

$$\frac{1}{f} = \frac{1}{f_2} + \frac{1}{f_1 - t} \tag{31-22}$$

Thus by simply adjusting the distance t between the two lenses, it is possible to vary the focal length of this lens combination. The special case $t = 0$ corresponds to a "thin lens" of focal length f given by

$$\frac{1}{f} = \frac{1}{f_1} + \frac{1}{f_2}$$

Note that if both lenses are convergent, then the focal length of the combination f is less than the focal lengths f_1 and f_2 of the constituent lenses.

31-9   Summary of important formulas

An object at a distance s from a concave spherical mirror of radius R will be imaged at a distance s' given by

$$\frac{1}{s} + \frac{1}{s'} = \frac{2}{R} \tag{31-1}$$

and the lateral magnification m of the object will be

$$m = -\frac{s'}{s} \tag{31-9}$$

For a convex mirror the radius R in (31-1) must be replaced by its negative, $-R$. In these formulas, a positive value for s' means that the image is *real* and *inverted*, and on the same side of the mirror as the object. Correspondingly, for $s' < 0$ the image is *virtual* and *erect*, and on the side of the mirror opposite to that of the object.

If an object is in a medium of refractive index n_1 and at a distance s_1 from a convex spherical interface with a medium of refractive index n_2, then an image will be formed at a distance s_2 from the interface, where

$$\frac{n_1}{s_1} + \frac{n_2}{s_2} = \frac{n_2 - n_1}{R} \qquad (31\text{-}10)$$

For a concave surface, the factor R in this formula must be replaced by $-R$. A positive value for s_2 means that a real image is formed in the n_2-medium, whereas a negative value for s_2 is associated with a virtual image in the n_1-medium.

If s is the distance of an object from a thin lens, then an image will be formed at a distance s', where

$$\frac{1}{s} + \frac{1}{s'} = \frac{1}{f} \qquad (31\text{-}16)$$

In this formula f is the focal length of the lens and is given by the lensmaker's equation

$$\frac{1}{f} = (n - 1)\left[\frac{1}{R_1} + \frac{1}{R_2}\right] \qquad (31\text{-}17)$$

where n is the refractive index of the lens, and R_1 and R_2 are positive provided that the centers of curvature of the two surfaces are oriented as in Figure 31-21. If either center of curvature is different than in this figure, the corresponding factor R_1 or R_2, or both, must be replaced by its negative. Positive values of s' correspond to a real image on the side of the lens opposite to the object, and negative values for s' imply that the image is virtual and on the same side of the lens as the object. The lateral magnification of a thin lens is given by the same formula as that for a spherical mirror in (31-9).

QUESTIONS

1. Define or describe briefly what is meant by the following terms: (a) virtual image; (b) lateral magnification; (c) paraxial rays; (d) thin lens; (e) lensmaker's equation; and (f) converging lens.

2. A small object is pushed toward a plane mirror. Describe the motion of the image.

3. A point object lies somewhere between two plane mirrors which are at an angle of 60° to each other. Explain why you will see five images.

4. Consider two plane mirrors hinged along an edge and making an angle of 45°. Locate all seven images formed by a small object placed at the angle bisector of the mirrors. Characterize each image in terms of the number of reflections it undergoes.

5. What is the distinction between a real and a virtual image? Can a vir-

tual image be photographed? Can it be projected onto a screen?

6. A small object in front of a concave spherical mirror produces a real image. In view of the symmetry under exchange of s and s' in (31-1), how can you tell the difference between the object and the image only by viewing the light rays?

7. What are the approximations made in deriving the formula in (31-1) for a spherical mirror? Under what circumstances would this formula be *invalid*?

8. A point object is initially very far away from a concave spherical mirror of radius R. Describe the motion of the image as the object is brought along the axis to a final resting position at an object distance of $R/4$.

9. Repeat Question 8, but suppose this time that the mirror is convex.

10. Suppose that the object in Question 8 were a small pencil with its point directed perpendicular to the axis. Describe the variation in the size and orientation of the pencil image as the pencil is brought to within a distance of $R/4$ from the mirror.

11. Repeat Question 10, but this time assume that the mirror is convex.

12. Which would be a better choice for the side-mounted rear view mirror of an automobile: a concave mirror or a convex one? Explain your answer.

13. Imagine looking into a mirror which is very slightly curved and seeing a virtual, erect, and enlarged image of your face. Can you conclude from this observation that the mirror is concave? Explain.

14. In the problems it is shown that, for a concave *parabolic* mirror, if light rays, which are parallel to the axis of the mirror, are incident on it, then they come to a focus at a certain point on the axis. Explain in what way this differs from the corresponding case of parallel light rays

incident on a concave *spherical* mirror.

15. What is the distinction between a *converging* and a *diverging* thin lens? Which one always produces a virtual image? Does one of these always yield a real image?

16. Where in front of a thin lens must an object be placed so that the image has a lateral magnification of -1? Must the lens be *converging*? Explain.

17. Why does the focal length of a thin lens increase if the lens is under water? Is this an important consideration for swimmers who wear contact lenses?

18. A thin, converging glass lens is placed into a transparent liquid whose refractive index exceeds that of glass. Explain why the lens becomes diverging under these circumstances! What would happen to the lens if it were double-concave?

19. Is it possible for the focal length of a thin lens to vary with the color of the light? Explain.

20. Explain in what sense the object and the image of a thin, converging lens are interchangeable. How could one tell the object and the image apart by viewing only?

21. Suppose you are suspended at a distance of 20 meters from the center of a large spherical room defined by a large spherical mirror having a radius of 50 meters. What would you see?

22. Light from the sun is focused by a thin, converging lens onto a piece of paper. Explain why the paper catches fire. Would this work with a thin, diverging lens? Explain.

23. A "zoom lens" is one whose focal length can be varied by means of a mechanical or electronic adjustment. Show how such a lens can be constructed by use of two or more thin lenses.

PROBLEMS

1. Consider an object 20 cm from a plane mirror. If you view the image from a distance of 50 cm directly behind the object (that is, 70 cm from the mirror), for what distance must you focus your eyes?
2. A small object is midway between two parallel, plane mirrors separated by a distance D.
   (a) Show that there are images in both mirrors at separation distances nD ($n = 1, 2, ...$) from the object.
   (b) Draw a ray diagram to show how the image in one of the mirrors at a distance of $2D$ from the object is formed.
3. A concave spherical mirror used for shaving has a radius of 30 cm.
   (a) Where is the image, if the shaver's face is 10 cm from the mirror?
   (b) What is the magnification?
   (c) Is the image real or virtual? Erect or inverted?
4. A man looks into a convex spherical mirror of radius 30 cm. If his face is 10 cm from the vertex of the mirror:
   (a) Where does he see his image?
   (b) What is the magnification?
5. A reflecting telescope has a concave spherical mirror with a radius of 1.5 meters.
   (a) Where is the image of a rocket that is at a distance of 100 km from the telescope?
   (b) What is the magnification?
   (c) Assuming that the rocket is a sphere with a radius of 20 meters, what is the radius of its image?
6. A small object is 10 cm in length and is oriented perpendicular to the axis of a concave spherical mirror of radius 80 cm. If the object is

20 cm from the mirror:
   (a) Where is the image? Is it real or virtual?
   (b) What is the magnification?
   (c) What is the size of the image? Is it erect or inverted?
7. Repeat Problem 6, but this time suppose that the mirror is convex.
8. An object is at a distance s from the vertex of a concave spherical mirror of radius R.
   (a) If D represents the *sum* of the object and image distance (that is, $D = s + s'$), show that

$$D = \frac{s^2}{s - (R/2)}$$

   (b) For what value of s is D minimum? Consider only object distances greater than $R/2$.
9. Repeat both parts of Problem 8, but this time for a convex mirror. (*Note*: The formula for D in (a) must be modified for this case.)
10. Show that the lateral magnification m of an object that is at a distance s from a concave spherical mirror of radius R is given by

$$m = -\frac{R}{2s - R}$$

What is the analogous formula for a convex mirror? What is the largest magnification achievable for this latter case?
11. An object of length l lies along the axis of a concave spherical mirror of radius R. If the nearer end of the object is at a distance s from the vertex of the mirror and if its length is very small compared to R, show that the length l' of the image is

$$l' = l\left(\frac{R}{2s - R}\right)$$

Assume that $s > R/2$.

12. Consider in Figure 31-29 an object O at a distance s from a *convex* spherical mirror of radius R. Follow the procedure used to derive (31-1) for a concave mirror and thus establish the validity of (31-2).

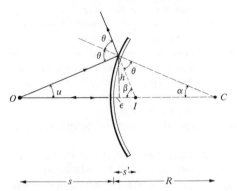

Figure 31-29

*13. Figure 31-30 shows a parabolic mirror whose equation in the given coordinate system is

$$y = \frac{a}{2} x^2$$

Consider an incoming ray traveling parallel to the y-axis along the line $x = x_0$ so that it strikes the mirror at the point $(x_0, ax_0^2/2)$.

   (a) Show that the angle of incidence θ satisfies

$$\tan \theta = ax_0$$

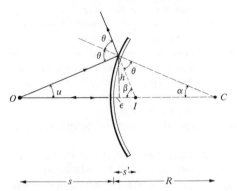

Figure 31-30

   (b) Show that the equation for the reflected ray is

$$y = -(x - x_0) \cot 2\theta + \frac{a}{2} x_0^2$$

   (c) Show that this reflected ray intersects the y-axis at a distance $1/2a$ above the origin. Explain why this result implies that the point with coordinates $(0, 1/2a)$ is the focal point of this parabolic mirror.

14. (a) An object of height 10 cm is perpendicular to the axis of a concave spherical mirror of radius 50 cm and is at a distance 60 cm from its vertex. Make a drawing of the situation to some scale, and determine the location and size of the image graphically. (*Hint:* Any ray through the center of curvature of the mirror is reflected back on itself, whereas any ray that goes through the focal point is reflected parallel to the axis.)

   (b) Repeat (a), but this time assume that the object is 40 cm from the mirror.

15. Repeat (a) and (b) of Problem 14, but suppose this time that the mirror is convex.

16. It is desired to project the image of an object 3 cm high onto a screen 40 meters from a spherical concave mirror so that the size of the image is 80 cm.

   (a) What lateral magnification is required?

   (b) What must be the object distance s?

   (c) What must be the radius of curvature of the mirror?

17. Show that for a plane mirror the *longitudinal magnification* (ds'/ds) is the same for all object distances, and determine its numerical value.

18. An object 15 cm high is in air and at a distance of 30 cm from a convex spherical glass surface ($n = 1.5$) of radius 60 cm. Find the position of the image and state whether it is real or virtual and whether it is erect or inverted.

19. A goldfish is 5 cm long and is at the center of a spherical bowl of radius 40 cm, which is full of water ($n = 1.33$). Neglect refraction due to the bowl itself.
    (a) Where does the fish appear to be to an external observer?
    (b) Does the fish appear to be erect or inverted? (Assume that the fish is perpendicular to the line from the observer's eye to the center.)

20. A point source O is at the bottom of a tank of water of depth h and refractive index 4/3. Suspended at a distance d above the tank is a plane mirror parallel to the surface of the water; see Figure 31-31. Where would an observer who looks into the mirror see the image of the object?

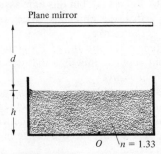

Figure 31-31

*21. Repeat Problem 20, but suppose that this time the plane mirror is replaced by a concave spherical one of radius R and that d represents the distance from the vertex of the mirror to the surface of the water. Assume, $d + h = R/2$.

22. A bundle of parallel light rays traveling in glass ($n = 1.5$) are incident at its spherical interface with water ($n = 1.33$). Describe the nature and location of the image if:
    (a) The surface is concave of radius R.
    (b) The surface is convex of radius R.

23. Repeat both parts of Problem 22, but suppose that this time the original light rays are traveling in water and are refracted by the spherical glass surface.

24. Suppose that in Figure 31-17 the object O has a height y_1 and is perpendicular to the axis. If y_2 is the corresponding size of the image, show that the lateral magnification defined as $m = -y_2/y_1$ is

$$m = -\frac{n_1 s_2}{n_2 s_1}$$

25. A glass rod of refractive index 1.5 has a length of 60 cm and has at its ends hemispherical surfaces, each of radius 10 cm. An object of height 2 cm is placed at a distance of 30 cm from one end, as shown in Figure 31-32.

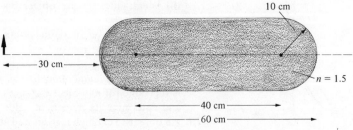

Figure 31-32

(a) Where is the image I_1 located after refraction by the left surface?

(b) What is the nature and size of this image I_1? (*Hint*: Use the result of Problem 24.)

(c) Find the location of the final image I_2, which is formed by the refraction of the light from I_1, considered as an object for the second surface.

(d) What is the nature and magnification of I_2 relative to the original object?

26. Consider the rod in Figure 31-32, but suppose that this time light rays parallel to its axis are incident on its left surface.

(a) Where does the light come to a focus after the first refraction?

(b) What is the location of the final image after two refractions?

*27. Parallel beams of light are incident on a glass sphere of radius R and index 1.5. If, as shown in Figure 31-33, a plane mirror is at a distance $(3/2)R$ from the sphere, calculate the location of the final image. (*Hint*: See Example 31-10.)

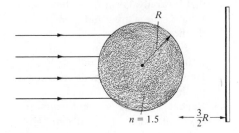

Figure 31-33

*28. Consider again the situation in Figure 31-33, but suppose that this time an object is located at a distance $3R$ from the left surface of the sphere.

(a) Find the location of the image after the first refraction through the surface of the sphere.

(b) Find the final position of the image.

29. A plano-convex thin lens is made of glass ($n = 1.5$) and the curved surface has a radius of 10 cm.

(a) What is the focal length of the lens?

(b) Repeat (a), but suppose that this time the lens is plano-concave with the same radius of curvature.

(c) Is the lens in (a) diverging or converging?

30. A double convex thin lens made of glass ($n = 1.5$) has a focal length of 30 cm. If the radius of one of the surfaces is 30 cm, what is the radius of the other surface?

*31. Show that if f_0 is the focal length of a thin lens in air and f' is the corresponding focal length when the lens is in a medium of refractive index n', then

$$f' = f_0 \frac{n'(n-1)}{(n-n')}$$

where n is the refractive index of the material of the lens. (*Hint*: Modify the arguments used to derive the lensmaker's equation.)

32. A thin, converging lens has a focal length of 20 cm. Find the location and magnification associated with an image whose object distance is: (a) 30 cm; (b) 20 cm; (c) 10 cm. In which of these cases is the image real?

33. An object is at a distance of 12 cm from a thin, converging lens of focal length 10 cm. (a) How far from the lens must a screen be placed so that the image is focused on it? (b) How large is the image if the object is perpendicular to the axis and has a size of 4 cm?

34. An object is at a distance d from a screen. Show that if a thin converging lens of focal length $f (< d/4)$ is placed at either of the distances

$$\frac{d}{2} \pm \frac{1}{2}[d(d-4f)]^{1/2}$$

from the object, then the image will come to a focus on the screen. What is the ratio of the lateral magnification in the two cases?

35. Show that if an object is at a distance $(x+f)$ from a thin lens of focal length f, then its image is at a distance $(x'+f)$, where

$$xx' = f^2$$

(*Note*: This form of the thin lens equation is known as the *Newtonian form* and is to be contrasted with the *Gaussian form* in (31-16).)

36. An object is at a distance s from a thin lens of focal length f.
  (a) Show that the distance d between the object and the image is

  $$d = \frac{s^2}{s-f}$$

  (b) Show that for a converging lens the distance between the object and any real images formed is always greater than $4f$.
  (c) At what object distance is the minimum in (b) achieved?

37. When an object having a height of 3 cm is placed 8 cm in front of a thin, converging lens, an image is formed on a screen 12 cm behind the lens. Suppose that the lens is moved 2 cm closer to the object. (a) Which way must we move the screen so that the image is still in focus? (b) What is the final size of the image?

38. An object 2 cm high is at a distance of 10 cm from a converging lens of focal length 4 cm. Make a drawing to scale and, by drawing appropriate rays from the top of the object, determine the size and location of the image. Compare your graphical values with the calculated ones.

39. Repeat Problem 38 but suppose that this time the lens is diverging and has a focal length of −4 cm. Assume that the object is 4 cm from the lens.

40. Two converging lenses each of focal length 10 cm are separated by a distance of 50 cm. An object 5 cm in height is placed to the left of the lenses and 15 cm from one of the lenses and 65 cm from the other.
  (a) Where is the image I_1 due to the first lens?
  (b) Where is the final image I_2 located?
  (c) What are the sizes of I_1 and I_2? Is the final image erect or inverted?

41. Consider the lens system of Problem 40, but suppose that this time the more remote lens is moved so that it is only 35 cm from the other. What is the nature and position of the final image?

42. In Figure 31-28 suppose that $f_1 = +10$ cm, $f_2 = +10$ cm, and $t = 50$ cm. How far from the second lens does the light come to a focus?

43. Repeat Problem 42, but suppose that this time $f_1 = -10$ cm, $f_2 = +10$ cm, and $t = 50$ cm.

44. Figure 31-34 shows the basic ele-

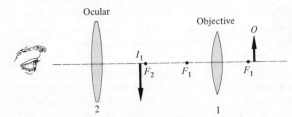

Figure 31-34

ments of the compound microscope. The object O is placed just outside the focal point of the objective lens and the associated image falls just inside the focal length of the ocular lens. Explain why the observer will see an enlarged image of the object. Will the observed image be real or virtual? Erect or inverted?

45. The basic elements of a telescope are two converging lenses: an *objective* lens (1) of focal length f_1 and an *ocular* lens (2) of focal length f_2. Unlike the microscope in Figure 31-34, for a telescope the lenses are separated by the distance $(f_1 + f_2)$. Explain, by constructing a ray diagram for an object at infinity, how the telescope works. Show also that the magnification M of a telescope is

$$M = -\frac{f_1}{f_2}$$

and explain the significance of the minus sign.

46. An object is at a distance $(a + b)$ from a plane mirror. Suppose that a thin, converging lens of focal length f $(f < a; f < b)$ is placed between the object and the mirror and at a distance a from the former. Find the nature and the location of the final image assuming that $b > af/(a - f)$. Justify your answer by constructing an appropriate ray diagram.

*47. A point object O is in front of a concave spherical mirror of radius R, and forms a real image I. Prove

that the time it takes for any two paraxial rays to go from O to I is the same. What is the relation between this result and Fermat's principle as described in Problems 7 and 8 of Chapter 30? (*Hint*: Make use of the geometry of Figure 31-11 and show that for paraxial rays the time it takes the ray OA to travel the path OAI is independent of the angle u.)

*48. Suppose, in Figure 31-17, that an object O is near an interface between two media of refractive indices n_1 and n_2, and a *real* image I is formed. Prove that *all* paraxial rays take the same time to go from O to I regardless of the values of R, n_1, and n_2. How is this result related to Fermat's principle?

49. By making use of the results of Problem 48 or of Fermat's principle, show that if a real image I is formed of an object O by a thin, positive lens, then the time taken in going from O to I is the same for *all* paraxial rays.

50. Suppose, in Figure 31-35, that parallel rays are incident at an angle u with respect to the axis of a thin, positive lens of focal length f. Show that the rays will converge to a point image I at a distance f behind the lens. The locus of the images formed in this way by various incident angles u defines the *focal plane* of the lens. (*Hint*: Consider the incident rays that go through the focal point and through the center of the lens.)

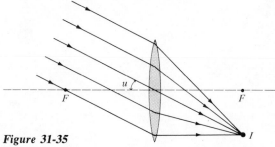

Figure 31-35

32 Interference and diffraction

Where there is a great deal of light, the shadows are deeper.

GOETHE

32-1   Introduction

The fact that light is a wave motion did not play an important role in our formulation of the laws of *geometrical optics* in Chapter 30. The reason for this is that these laws are applicable only to physical situations involving the interaction of light with objects whose dimensions are large compared to a wavelength. Under these conditions the fact that light is a wave motion plays only a very minor role.

By contrast, in this chapter we shall consider the entirely new set of phenomena that arises when light interacts with objects whose linear dimensions are comparable to a wavelength. This is the field of *physical optics*. The new physical effects here are: (1) the *diffraction* of light as manifested by the bending of light around the edges of sharp obstructions; and (2) the *interference* of two light beams for which the observed intensity is *not* the sum of the intensities of the separate beams. As we shall see, diffraction and interference are two very closely related phenomena. Although Christian Huygens suggested the possibility of the diffraction and the interference of visible light as early as 1690, it was not until 1803 that

Thomas Young (1773–1829) succeeded in establishing the validity of this hypothesis in the laboratory. His key experiment was simplicity itself. In his own words:

> I made a small hole in a window shutter and covered it with a piece of thick paper, which I perforated with a fine needle.... I placed a small looking-glass without the window shutter in such a position as to reflect the sun's light, in a direction nearly horizontal, upon the opposite wall.... I brought into the sunbeam a slip of card about one thirtieth of an inch in breadth and observed its shadow, either on the wall or on other cards held at different distances. Besides the fringes of colors on each side of the shadow, the shadow itself was divided by similar parallel fringes of smaller dimensions, differing in number according to the distance at which the shadow was observed but leaving the middle of the shadow always white. Now these fringes were the joint effects of the portions of light passing on each side of the slip of card, and inflected, or rather diffracted into the shadow....

32-2 Qualitative aspects of diffraction

The term diffraction refers to the experimental fact that light rays when passing around the sharp edge of an object will spread out, to some extent, into the region of its shadow. Because the wavelength of visible light is of the order of 5×10^{-5} cm and is thus small compared to the dimensions of ordinary bodies this effect is difficult to observe except under controlled laboratory conditions.

Consider, in Figure 32-1, a parallel beam of light rays incident at right angles onto a barrier AB, out of which has been cut an aperture H. Suppose first that the linear dimensions, a, of the aperture are of the order of a centimeter and thus very large compared to a wavelength. According to the ideas of geometrical optics, the incident light rays that strike the barrier are either absorbed or scattered by it, whereas those that are incident at the opening will go through undeviated. Thus we might expect that a bright spot will appear at EF, and that the remainder of the screen—that is, CE and DF—will remain dark. Indeed, provided that $a \gg \lambda$, these expectations are substantially in accord with observations. However, even for this case in which the linear dimensions of the aperture are very large compared to a wavelength, a careful examination of the edges of the bright spot at E and F shows that the line of demarcation between the bright spot and the shadow on the screen is not very sharp. But rather, very close scrutiny shows that near E and F there are some dark bands inside the bright region and some bright bands in the region of the shadow.

Imagine now repeating the above experiment, but with successively smaller apertures. The dividing line between the bright and the dark region on the screen will become less and less sharp until a situation such as that in Figure 32-2 is reached. Here the incoming light is incident on a tiny opening

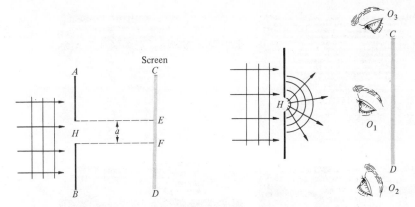

Figure 32-1   **Figure 32-2**

whose linear dimensions are small compared to the wavelength of visible light. The pattern of illumination on the screen under this circumstance is completely at variance with the ideas of geometrical optics. As shown in the figure, the observed pattern is best described as one that results if spherical waves—that is, waves whose surfaces of constant phase, or *wavefronts*, are spherical surfaces—emanate radially outward from the aperture. If geometrical optics were applicable to this situation, then only the observer O_1 who is lined up with the aperature H would see any light. In fact, we find that the light emanating from H is also seen by various other observers, such as O_2 and O_3, thus confirming that the light from H spreads out after going through the opening in the barrier. This phenomenon of the spreading out of a light beam after passing through an aperture of dimension of the order of 10^{-6} meter and smaller is an illustration of the *diffraction* of light; it is a direct and unambiguous manifestation of the fact that light is a wave motion.

Figure 32-3 illustrates the kinds of patterns observed for the analogous case of water waves in a ripple tank passing through an opening. Note that if the size of the opening is very large compared to a wavelength (the distance between neighboring crests or troughs), there is relatively little diffraction. On the other hand, for a narrow opening, the emerging rays are the radii of circles centered at the opening.

A qualitative way of thinking of the phenomenon of diffraction can be obtained by use of a rule known in honor of its discoverer as the *Huygens construction*. According to this rule, the wavefront of a wave at any instant $(t + \Delta t)$ may be obtained from that at a previous instant t by thinking of each point of the earlier wavefront as being a point source of outgoing spherical waves. The envelope of these spherical waves is then the wavefront at the later time $(t + \Delta t)$. By way of example, Figure 32-4a shows a portion AB of an "infinite" wavefront at some instant t. The associated wavefront $A'B'$ at the later instant $(t + \Delta t)$ is the envelope, or the common tangent plane in this case,

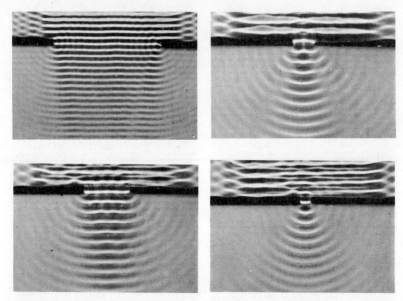

Figure 32-3 *Water waves being diffracted through slits of varying sizes in a ripple tank. Note how the diffraction pattern widens as the size of the slit is decreased down to a wavelength. (Courtesy of Project Physics, Holt, Rinehart and Winston, New York.)*

of the spherical wavefronts, each of radius $c\,\Delta t$ and emanating from the various points of the original wavefront AB. It is evident that, consistent with our expectations, the new wavefront $A'B'$, as given in this way by the Huygens construction, is parallel to AB and at a distance $c\,\Delta t$ from it. Figure 32-4b illustrates the corresponding situation of a wave of infinite extent approaching an aperture H. Applying the Huygens construction to the finite wavefront which is at the aperture at time t, we may construct the new wavefront at the later instant $(t + \Delta t)$. Note that only near the center of the aperture is the transmitted wavefront parallel to the original one. At the edges the transmitted light has been diffracted.

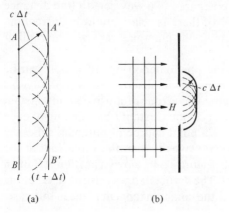

(a)   (b)

Figure 32-4

Despite its historical interest and simplicity, the Huygens construction will not be used in the following quantitative description of diffraction. Instead, we shall take the point of view that light is an electromagnetic wave and then establish that diffraction and interference are properties of all such waves.

32-3 A point source

By way of introduction to a quantitative study of interference, in this section we review briefly some of the important parameters associated with electromagnetic waves emitted by a single source.

Consider, in Figure 32-5, a monochromatic point source S, which radiates electromagnetic waves of wavelength λ. As shown, very close to the source the wavefronts of the outgoing radiation are spherical, but to an observer O who is very far from S these wavefronts will appear to be plane waves. If E_0 is the amplitude of the observed waves, then, according to (29-20) through (29-23), the electric field E in the radiation seen by the observer O is

$$E = E_0 \sin \left[\frac{2\pi}{\lambda} (z - ct) - \alpha \right] \tag{32-1}$$

where $c = 3.0 \times 10^8$ m/s is the speed of light and α is a constant phase factor. In writing down this formula, we have assumed that the source is at the origin of the coordinate system and that the observer is at a remote distance z along the z-axis. Since the $\mathbf{E}$ and $\mathbf{B}$ vectors in an electromagnetic wave are always perpendicular to each other and to the propagation direction of the wave, a knowledge of only the magnitude of the electric field will be needed in the following. The magnetic induction $\mathbf{B}$ associated with the wave in (32-1) is given by (29-20), but with the argument of the sine function the same as that in (32-1).

The argument of the sine function in (32-1), that is, the factor $[2\pi(z - ct)/\lambda - \alpha]$, is known as the *phase* of the wave. All points along a given wavefront have the same phase. In particular, for a plane wave all points of the wave which lie in a fixed plane at right angles to the propagation direction of the wave are characterized by the same phase.

The magnitude N of the Poynting vector—that is, the flow of energy per

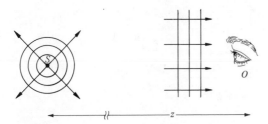

Figure 32-5

unit area per unit time received by the observer—is obtained by substituting
(32-1) and the associated formula for **B** into (29-17). The result is

$$N = \sqrt{\frac{\epsilon_0}{\mu_0}} E_0^2 \sin^2 \left[\frac{2\pi}{\lambda} (z - ct) - \alpha \right] \tag{32-2}$$

As in Chapter 29, only the average value I_0 of this quantity over a period λ/c
is of physical interest. Proceeding, therefore, as in the analogous derivation
of (29-25), we find for the average intensity

$$I_0 = \frac{1}{2} \sqrt{\frac{\epsilon_0}{\mu_0}} E_0^2 \tag{32-3}$$

Note that I_0 is independent of both the phase α and the wavelength λ of the
electromagnetic wave.

32-4   Interference

Consider, in Figure 32-6, two identical point sources S_1 and S_2, each emitting
electromagnetic waves of the same amplitude and of the wavelength λ. As
for the single-source case considered above, suppose the observer O is very
far away from the sources, and to simplify matters let us assume that S_1, S_2,
and O are along a straight line. We define the *path difference* between two
sources and the given observer to be the difference in the optical paths
between the sources and the observer. For the situation in Figure 32-6 this
means that since S_1, S_2, and O are collinear, the distance Δ between the
sources is itself the path difference. In other words, since the distance
between S_1 and O is z while that between S_2 and O is $(z - \Delta)$, the path
difference is $z - (z - \Delta) = \Delta$. In the following we shall be concerned mainly
with physical situations for which the path difference Δ is of the order of the
wavelength of light. By contrast, the distance z will always be assumed to be
so large that the wavefronts seen by the observer are planes, so that in general
$|z| \gg \Delta$.

Let us now calculate the average intensity at the position of the observer.
According to the superposition principle, the electric field E in the wave
seen by the observer is the sum of the fields produced by each of the sources
separately. Hence, taking note of (32-1) and neglecting effects associated

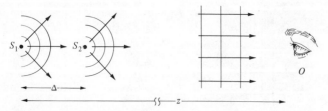

Figure 32-6

with the vector nature of the electric field, we find that

$$E = E_0 \sin\left[\frac{2\pi}{\lambda}(z - ct) - \alpha_1\right] + E_0 \sin\left[\frac{2\pi}{\lambda}(z - \Delta - ct) - \alpha_2\right] \quad \text{(32-4)}$$

with E_0 the observed electric field amplitudes, and α_1 and α_2 certain phase factors. Note that even though $\Delta \ll |z|$, the path difference Δ cannot be neglected compared to z in this formula.

To explore the possibility of interference, suppose first that the two sources are independent of each other. In this case there is no definite phase relation between them or, in other words, the phases α_1 and α_2 in (32-4) are *random*. We say in this case that the sources are *incoherent*. For incoherent sources the observed intensity I is the sum of the intensities due to each source. Hence, since S_1 and S_2 have the same amplitude E_0, it follows that if they are incoherent, then

$$I = 2I_0 \quad \text{(incoherent sources)} \quad \text{(32-5)}$$

with I_0 given by (32-3). Incoherent sources, such as the light coming from different points on this page, *do not interfere* with each other.

By contrast, if for some reason the phase difference $(\alpha_1 - \alpha_2)$ is constant over long periods of time, then we say that the two sources are *coherent*. Assuming for simplicity that $\alpha_1 = \alpha_2$, it will be established below that the average intensity I produced by two coherent sources is

$$I = 4I_0 \cos^2\frac{\delta}{2} \quad \text{(coherent sources)} \quad \text{(32-6)}$$

with I_0 defined by (32-3). The new quantity δ, which will play an important role in the following, is defined by

$$\delta = \frac{2\pi}{\lambda}\Delta \quad \text{(32-7)}$$

and is known as the *phase difference* between the coherent sources.

According to (32-6), the observed intensity I for coherent sources can vary from a maximum of $4I_0$ to a minimum of zero, depending on the path difference Δ between the sources. If Δ is an integral number of wavelengths, for example, λ or 2λ, then the phase difference δ is an integral multiple of 2π and, according to (32-6), I achieves its maximum value of $4I_0$. We say in this case that *constructive interference* has taken place. On the other hand, if Δ is an odd integral multiple of $\lambda/2$, such as $\lambda/2, 3\lambda/2, \ldots$, then δ is an odd multiple of π and I achieves its minimum value of zero. In this case we say that *destructive interference* has taken place.

Figure 32-7 shows a graph of (32-6) as a function of the phase difference δ. Note that, as shown above, at $\delta = \pi, 3\pi, 5\pi, \ldots$, the interference is destructive, whereas at $\delta = 0, 2\pi, 4\pi, \ldots$ it is constructive. The horizontal dotted line at $I = 2I_0$ represents the corresponding plot for the case of incoherent sources.

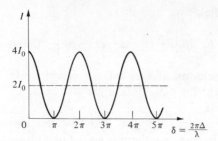

Figure 32-7

It is important and interesting to note that even for the case of complete destructive interference there is no contradiction with the principle of energy conservation. For even though the path difference Δ between the sources relative to the observer O in Figure 32-6 is an odd integral multiple of half of a wavelength, there are other observation points relative to which Δ is an integral multiple of a wavelength. For example, to an observer who views the sources S_1 and S_2 along the perpendicular bisector of the line joining them, the path difference of the sources relative to him vanishes. Hence, according to (32-6) and (32-7), he would see the maximum intensity $I = 4I_0$. As we might expect, the total intensity radiated by the two sources in *all* directions is precisely the same as that which would be radiated if they were incoherent.

Figure 32-8 shows the analogous interference pattern when water waves pass through neighboring holes in a ripple tank. Note that there are directions along which the amplitudes of the wave are a maximum and others along which they are minimum. The path difference associated with the former is invariably an integral multiple of λ and with the latter an odd multiple of $\lambda/2$.

To derive (32-6) let us make use of the trigonometric identity

$$\sin A + \sin B = 2 \cos \left(\frac{A - B}{2}\right) \sin \left(\frac{A + B}{2}\right)$$

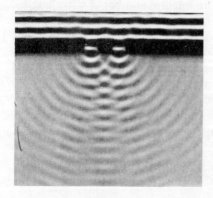

Figure 32-8 An interference pattern produced in a ripple tank by the diffraction of water waves from two nearby slits. (Courtesy Project Physics, Holt, Rinehart and Winston, New York.)

to reexpress the formula for E in (32-4) in the form

$$E = 2E_0 \cos \frac{\delta}{2} \sin \left[\frac{2\pi}{\lambda}(z - ct) - \frac{\delta}{2} - \left(\frac{\alpha_1 + \alpha_2}{2} \right) \right]$$

Here δ is the phase difference defined in (32-7), and use has been made of the fact that the sources are coherent and that $\alpha_1 = \alpha_2$. Comparing this form for E with that in (32-1), we see that it corresponds to the radiation emitted by a single source with amplitude $2E_0 \cos (\delta/2)$. Hence substitution into (32-3) yields the desired formula in (32-6).

The formal derivation of (32-5) for the radiation emitted by two incoherent sources proceeds in a similar way. It is left as an exercise to confirm that (32-5) results if an averaging process over the random phases α_1 and α_2 is carried out.

Example 32-1 Two point sources are a distance 1.0×10^{-6} meter apart and are emitting light of wavelength 5500 Å. Assuming that the observed intensity when either source is acting alone is 2.0 W/m², and that the observer is along the line joining the sources, calculate:

(a) The observed intensity if the sources are incoherent.

(b) The observed intensity if they are coherent.

(c) The maximum intensity possible for these coherent sources and the smallest additional distance they must be separated to achieve this maximum.

Solution

(a) The substitution of the given value $I_0 = 2.0$ W/m² into (32-5) yields for the incoherent case

$$I = 2I_0 = 4.0 \text{ W/m}^2$$

(b) For the coherent case, we must use (32-6). According to (32-7) the phase difference δ is

$$\delta = \frac{2\pi}{\lambda} \Delta = \frac{2\pi \times 1.0 \times 10^{-6} \text{ m}}{5.5 \times 10^{-7} \text{ m}} = 3.6\pi \text{ rad}$$

and hence

$$I = 4I_0 \cos^2 \frac{\delta}{2} = (4 \times 2.0 \text{ W/m}^2) \times \cos^2 1.8\pi = 5.2 \text{ W/m}^2$$

(c) The intensity will be maximum at the value $4I_0 = 8.0$ W/m² when the conditions for constructive interference are satisfied. This will take place when the path difference Δ is an integral number of wavelengths. Initially the distance between the sources is

$$1.0 \times 10^{-6} \text{ m} = 1.8 \times 5.5 \times 10^{-7} \text{ m} = 1.8\lambda$$

Hence they must be separated an additional distance of

$$(2 - 1.8)\lambda = 0.2\lambda = 0.2 \times 5.5 \times 10^{-7} \text{ m} = 1.1 \times 10^{-7} \text{ m}$$

to achieve this maximum.

Example 32-2 Suppose that the two sources in Figure 32-6 are emitting white light—that is, light of all wavelengths. If the separation distance between the sources is the minimum one for which blue light of wavelength 4500 Å suffers destructive interference, what wavelength will appear with maximum intensity?

Solution Reference to the graph in Figure 32-7 shows that for the blue light the phase difference δ must be π. Thus, according to (32-7) the value of Δ must be

$$\Delta = \frac{\lambda\delta}{2\pi} = \frac{4500\pi \text{ Å}}{2\pi} = 2250 \text{ Å}$$

And since a maximum in intensity I will occur when Δ is an integral number of wavelengths, it follows that the wavelength of the light associated with this maximum is

$$\lambda = 2250 \text{ Å}$$

In other words, radiation at this wavelength in the ultraviolet region will be at a maximum intensity for the given separation distance between the sources.

Example 32-3 Generalize (32-6) to the case of N identical coherent sources S_1, $S_2, \ldots, S_N$ located at the points with z-coordinates of $0, \Delta, 2\Delta, \ldots, (N-1)\Delta$, respectively. Assume that the observer is at a remote point along the z-axis and that $|z| \gg (N-1)\Delta$, so that he sees plane waves.

Solution Since the sources are coherent, the generalization of (32-4) for N sources is

$$E = E_0 \sum_{i=0}^{N-1} \sin \left[\frac{2\pi}{\lambda}(z - ct - i\Delta) + \alpha \right]$$

provided that the phases α are all the same. Making use of the following trigonometric identity (derived in Problem 13)

$$\sum_{i=0}^{N-1} \sin (x + i\theta) = \frac{\sin (N\theta/2)}{\sin (\theta/2)} \sin \left[x + \frac{(N-1)\theta}{2}\right] \tag{32-8}$$

this formula for E may be reexpressed in the form

$$E = E_0 \frac{\sin (N\pi\Delta/\lambda)}{\sin (\pi\Delta/\lambda)} \sin \left\{\frac{2\pi}{\lambda}\left[z - ct - \left(\frac{N-1}{2}\right)\Delta\right] + \alpha\right\}$$

Comparison with (32-1) and (32-3) thus yields for the observed intensity the result

$$I = I_0 \frac{\sin^2 [N\pi\Delta/\lambda]}{\sin^2 [\pi\Delta/\lambda]} \tag{32-9}$$

with I_0 the intensity which would be observed if only one source were radiating. It is left as an exercise to confirm that for $N = 2$ this reduces to (32-6), as it must.

32-5 Young's two-slit experiment

In order to verify experimentally the existence of interference phenomena as predicted by (32-6) and (32-7), it is necessary to have available coherent sources. In the next two sections we shall describe several methods that have been used in the past to produce such sources. These methods, although of considerable historical and conceptual interest, are no longer as widely used as they used to be. The discovery of the *laser*, which is a device capable of producing a very intense and highly coherent light beam, has made them obsolete.

The classic method used to demonstrate interference was devised by Thomas Young. Figure 32-9 shows the essential features of his experimental setup. Light from a line source S (perpendicular to the plane of the figure) is incident through a converging lens L onto a barrier AB out of which have been cut two slits S_1 and S_2. Assuming that S is in the focal plane of the lens, the cylindrical wavefronts emanating from S will approach the barrier as plane waves with wavefronts parallel to AB. (Equivalently, the source S and the lens L may be replaced by a laser beam.) Since, by definition, all points of a wavefront have the same phase, it follows that the waves that emerge from the slits will be coherent and, if allowed to fall on a screen, will produce an interference pattern. As is implied in the figure, the screen is assumed to be far enough away from the slits so that the waves incident on the screen are plane waves. For only then can the analysis of Section 32-4 be used. Alternatively, if the waves coming from S_1 and S_2 are first refracted through a thin, converging lens, then the screen can be moved up to coincide with the focal plane of this lens. (See Problem 50 in Chapter 31 and the corresponding analysis of the diffraction grating in Section 32-7.)

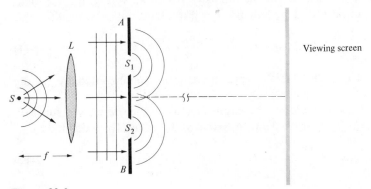

Figure 32-9

To analyze the interference pattern on the screen, let d represent the distance between the slits and D ($\gg d$) the distance from the slits to the screen. Draw a reference line perpendicular to and bisecting the line joining the slits and let it intersect the screen at P_0, as in Figure 32-10. Any point P on the screen can then be described in terms of its distance x from P_0 or, equivalently, in terms of the angle θ between the lines going from the midpoint of the slits to P_0 and to P.

Consider now the two rays S_1P and S_2P, which go from the slits to the point P. At S_1 and S_2 they are in phase, so their path difference, on arriving at

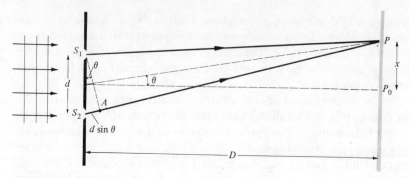

Figure 32-10

P, is the difference $(S_1P - S_2P)$. Since D is presumed to be very large and d very small, it follows that the rays S_1P and S_2P are essentially parallel. Hence the path difference Δ is effectively the distance S_2A, with A determined by drawing a perpendicular from S_1 to the ray S_2P. Reference to the figure shows that the angle between S_1S_2 and S_1A is also θ. Therefore the path difference between the two rays S_1P and S_2P is $d \sin \theta$ and substitution into (32-7) yields

$$\delta = \frac{2\pi d}{\lambda} \sin \theta \qquad (32\text{-}10)$$

The corresponding formula for the intensity $I(\theta)$ observed on the screen at the angle θ can be found by substitution into (32-6) with the result

$$I(\theta) = 4I_0 \cos^2 \left[\frac{\pi d \sin \theta}{\lambda} \right] \qquad (32\text{-}11)$$

where $4I_0$ is a constant which represents the maximum intensity on the screen.

Figure 32-11 shows a plot of the intensity $I(\theta)$ on the screen as a function of the quantity $\delta/2 = (\pi d \sin \theta)/\lambda$. Note that at the values for θ defined by

$$\sin \theta_m = \frac{m\lambda}{d} \qquad (m = 0, \pm 1, \pm 2, \dots) \qquad \text{(maxima)} \qquad (32\text{-}12)$$

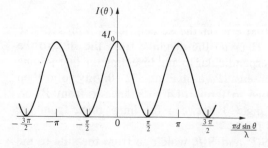

Figure 32-11

the intensity in the pattern is a maximum. The bright line corresponding to $m = 0$, which occurs at the center of the screen at P_0, is known as the *zeroth-order* fringe, and the neighboring lines at $\pm \theta_1$ corresponding to $m = \pm 1$ are called the first-order fringes, and so forth. The intensity maximum corresponding to θ_m appears at that position on the screen for which the path difference from S_1 and S_2 is $m\lambda$. Between these intensity maxima there are dark bands associated with positions on the screen for which the path difference is an odd integral multiple of $\lambda/2$. These minima occur for values of the angle θ given by

$$\sin \theta_m = \frac{(m + \frac{1}{2})\lambda}{d} \qquad (m = 0, \pm 1, \pm 2, \ldots) \qquad \text{(minima)} \qquad \textbf{(32-13)}$$

It is shown in the problems that for small values of θ, for which $\theta \cong \sin \theta$, the separation distance between the mth-order maximum at a distance x_m above P_0 on the screen in Figure 32-10 and its neighboring maximum at x_{m+1} is constant and is given by

$$x_{m+1} - x_m = \frac{\lambda D}{d} \qquad \textbf{(32-14)}$$

This means that if we carry out an experiment with fixed values for D and d, a measurement of the distance $(x_{m+1} - x_m)$ determines in effect the wavelength λ of the radiation emitted by the source. Thus by simply measuring the distance between neighboring maxima we can determine the wavelength of the light involved. This is the first practical method for measuring the wavelength of visible light that we have come across up to this point.

As will be seen in Section 32-8 and Problem 33, because of diffraction effects the number of maxima that will appear on the screen in Figure 32-10 is severely limited. In general, only a small number of bright maxima are visible, and these appear near the center of the screen at the point P_0. The intensities of the remaining ones drop off sharply with increasing distance from P_0.

Example 32-4 For the double-slit arrangement in Figure 32-10, suppose that the slits are a distance of 0.03 mm apart, that the screen is at a distance of 50 cm from the slits, and that the wavelength of the incident light is 4000 Å.
 (a) What value of θ corresponds to the first-order maximum?
 (b) What value of θ corresponds to the seventh minimum?
 (c) What is the spacing on the screen between successive maxima for the first few orders?

Solution
 (a) Setting $m = 1$ in (32-12), we find that

$$\sin \theta_1 = \frac{4.0 \times 10^{-7} \text{ m}}{3.0 \times 10^{-5} \text{ m}} = 1.3 \times 10^{-2}$$

and since this angle is small it follows that

$$\sin \theta_1 \cong \theta_1 = 0.013 \text{ rad}$$

(b) Setting $m = 6$ in (32-13), we find that

$$\sin \theta_6 = \left(6 + \frac{1}{2}\right)\frac{\lambda}{d} = 6.5 \times \frac{4.0 \times 10^{-7}\,\text{m}}{3.0 \times 10^{-5}\,\text{m}}$$

$$= 8.6 \times 10^{-2}$$

and hence

$$\sin \theta_6 \cong \theta_6 = 0.086\,\text{rad}$$

(c) It is apparent from the answers to (a) and (b) that for the first several orders the angles θ_1, θ_2, ... are small enough that (32-14) is applicable. Hence

$$x_{m+1} - x_m = \frac{\lambda D}{d} = \frac{(4.0 \times 10^{-7}\,\text{m}) \times (0.5\,\text{m})}{3.0 \times 10^{-5}\,\text{m}}$$

$$= 6.7\,\text{mm}$$

and this is large enough to be easily observable.

32-6 Fresnel's double mirror and biprism

Besides the apparatus used by Thomas Young, various other experimental setups have been suggested from time to time to produce interference effects. The purpose of this section is to describe two of these, which were originally proposed by Fresnel.

Figure 32-12a shows an apparatus known as *Fresnel's double mirror*. It consists of a small, nearly monochromatic, source S near two planar mirrors M_1 and M_2, which make a small angle α with each other. The virtual images S_1 and S_2 of the source S in the two mirrors act as two coherent sources separated by a certain distance d. Hence if any pairs of parallel rays, such as A and B, from these sources are brought together on a screen they will interfere. One important advantage of this apparatus over Young's setup in Figure 32-9 is that the distance d between S_1 and S_2 can be easily varied, within certain limits, simply by changing the angle α between the mirrors.

A second method for producing an interference pattern is by use of a

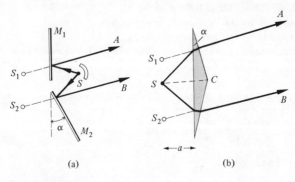

(a)   (b)

Figure 32-12

small angle prism known as a *Fresnel biprism* (see Figure 32-12b). If a point monochromatic source S is placed on the axis SC of, and at a distance a from, the prism, then when viewed from the other side, light rays appear to come from two virtual sources S_1 and S_2 located symmetrically with respect to S. In effect, then, as above, S_1 and S_2 constitute two coherent sources whose emitted radiation can be made to interfere. It is shown in the problems that for this case the distance d between S_1 and S_2 is

$$d = 2a\alpha(n - 1) \tag{32-15}$$

with n the refractive index of the prism and α the angle defined in the figure. Making use of this value for d and defining the angle θ as the angle made by the parallel rays S_1A and S_2B with the axis SC of the prism, we can obtain an explicit formula for the spatial variation of the interference pattern by substitution into (32-10) and (32-11).

32-7 The diffraction grating

Consider, in Figure 32-13, plane waves of wavelength λ normally incident on a flat barrier out of which have been cut a large number N of very thin parallel slits, each separated from its nearest neighbors by a distance d. If N is sufficiently large, say of the order of 5×10^3 per centimeter of barrier, then this system is known as a *grating* or a *diffraction grating*. The number of slits per unit length of the grating is known as the *grating constant*.

As shown in the figure, suppose that on the other side of the grating there are a thin, converging lens of focal length f and a screen in the focal plane of the lens. To determine the rays which are brought to a focus at a point P on the screen, draw a line from P through the center C of the lens and let θ be the angle this ray makes with the lens axis. It follows from the properties of thin lenses that all rays which emerge from the slits and are parallel to CP will come to a focus at P. The situation here is very analogous to that analyzed in Section 32-5, where it was assumed that the screen was very far

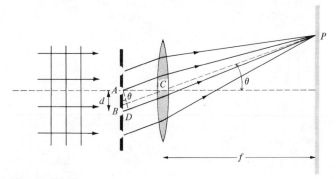

Figure 32-13

away from the slits. The function of the lens, as noted there, is simply to bring the parallel rays to a focus.

Now since the radiation incident on the grating in Figure 32-13 is a plane wave parallel to the grating, it follows that each slit may be considered to be a source of radiation that is coherent and therefore in phase with the radiation coming from the other slits. In particular, the rays that emanate from two neighboring slits A and B and are focused on the screen at P start out in phase at the points A and B, respectively. The common perpendicular AD to these two rays defines the wavefront of the wave which will be focused at P. Hence the time required for the light to travel from A through the lens to P is precisely the same as that required for the light to travel from D through the lens to P. This means that BD is the path difference Δ from the "sources" at A and B to the observer at P. Moreover, this argument can be repeated for all slits of the grating, and therefore this value for Δ is also the path difference relative to P of the radiation emanating from all neighboring slits. Reference to the figure shows that, just as for the two-slit case, this path difference Δ is

$$\Delta = d \sin \theta \tag{32-16}$$

The problem of obtaining the radiation pattern formed by a diffraction grating is thus that of calculating the radiation emitted by N coherent sources each of which has this same path difference Δ relative to its neighbor. But this has been solved in Example 32-3. Hence, substituting (32-16) into (32-9), we find that the intensity I at the point P on the screen in Figure 32-13 is

$$I = I_0 \frac{\sin^2 [(N\pi d \sin \theta)/\lambda]}{\sin^2 [(\pi d \sin \theta)/\lambda]} \tag{32-17}$$

The constant I_0 may be thought of as the intensity that would be observed if only one slit were open. It is left as an exercise to show that for the special case $N = 2$, this is precisely equivalent to (32-11), as it must be.

Figure 32-14 shows a plot of the formula for I in (32-17) as a function of the quantity $(\pi d \sin \theta)/\lambda$ for a value of $N \geqslant 10$. (For common gratings a more typical value is 5×10^3.) Note that there is a series of very sharp peaks,

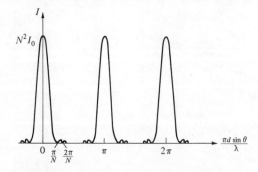

Figure 32-14

each of height $N^2 I_0$ and each with a width of approximately $2\pi/N$. The separation distance between peaks corresponds, according to (32-16), to the path differences Δ being an integral multiple of λ. On either side of these sharp peaks there are additional maxima and minima, but these are generally of the order I_0 and thus of no significance except for the lowest values of N.

In terms of the angle θ, which characterizes the point P on the screen in Figure 32-13, we may summarize the relation in (32-17) in the following way. For a given wavelength λ, the observed intensity I achieves its maximum of $N^2 I_0$ at those values θ_m given by

$$d \sin \theta_m = m\lambda \qquad (m = 0, 1, 2, \ldots) \tag{32-18}$$

For at these values the ratio on the right-hand side of (32-17) assumes the form

$$\frac{\sin^2 Nm\pi}{\sin^2 m\pi}$$

which is indeterminate since N and m are both integers, and thus both the numerator and the denominator vanish. However, this ratio may be evaluated by a limiting process and as shown in the problems the result may be expressed as

$$\lim_{x \to m} \frac{\sin^2 N\pi x}{\sin^2 \pi x} = N^2$$

for m an integer. The line corresponding to $m = 0$, is known as the *central maximum*, and is observed at $\theta = 0$ for *all wavelengths*. The line corresponding to $m = 1$ is called the *first-order* maximum, that at $m = 2$ is the *second-order* maximum, and so forth. The associated angles θ_1 and θ_2 may be obtained from (32-18) by the substitution of $m = 1$ and $m = 2$, respectively. Note that since $|\sin \theta| \le 1$, the maximum order observable is the greatest integer not exceeding the ratio d/λ.

Diffraction gratings play an extremely important role in optical spectroscopy. According to (32-18), if the spacing d for a grating is given, the measurement of the wavelength of a line is reduced to that of making a measurement of an appropriate angle. A variety of optical gratings is available commercially, including both reflection and transmission gratings.

Although it has been implicitly assumed in all of the above that the radiation of interest was visible light, it should be noted that interference effects can be produced for electromagnetic radiation of any wavelength. Indeed, in 1913 Max von Laue (1879–1960) suggested that a crystalline solid might act as a (reflection) grating for X rays. As we know today, the wavelengths of X rays are of the order of 10^{-7} cm to 10^{-8} cm, and this is comparable to the interatomic spacing expected in crystals. The first experiments along these lines—which are known today as *X-ray diffraction experiments*—were carried out shortly thereafter by Friedrich and Kipping. They confirmed not only the fact that X rays form a part of the electromagnetic spectrum, but also that the atoms in some solids are arranged in regular, crystal-like arrays. This phenomenon of X-ray diffraction is still used today as an important tool in studying the properties of both organic and inorganic crystals.

Example 32-5   Light of wavelength 6000 Å falls on a diffraction grating having 5000 lines/cm.

(a) Calculate the grating spacing.

(b) At what angle θ_1 does the first-order line occur?

(c) If a lens of focal length 12 cm is used to focus the light on a screen, how far is it from the central maximum to the first order?

Solution

(a) Since there are 5000 lines/cm, it follows that the distance d between the lines is

$$d = \frac{1}{5 \times 10^3/\text{cm}} = 2.0 \times 10^{-4} \text{ cm}$$

(b) Use of the value $m = 1$ in (32-18) yields for θ_1

$$\sin \theta_1 = \frac{\lambda}{d} = \frac{6.0 \times 10^{-7} \text{ m}}{2.0 \times 10^{-6} \text{ m}} = 0.30$$

whence

$$\theta_1 \cong 17.5°$$

(c) Reference to Figure 32-13 shows that

$$\tan \theta_1 = \frac{x}{f}$$

where x is the distance along the screen to the first-order maximum from the central maximum and f is the focal length of the lens. Using the known values for θ_1 and f we find that

$$x = f \tan \theta_1 = (12 \text{ cm}) \times \tan 17.5°$$
$$= (12 \text{ cm}) \times 0.31$$
$$= 3.7 \text{ cm}$$

Example 32-6   Light of wavelength 4500 Å is incident on a grating of 6000 lines/cm. What orders are observed?

Solution   Solving (32-18) for m we find that

$$m = (\sin \theta_m) \frac{d}{\lambda} = \sin \theta_m \frac{1.67 \times 10^{-4} \text{ cm}}{4.5 \times 10^{-5} \text{ cm}}$$
$$= 3.7 \sin \theta_m$$

Since $|\sin \theta| \leq 1$, it follows that only for the values $m = 0, 1, 2,$ and 3 does there exist an angle θ_m satisfying this relation. Hence, besides the central maximum, only the first-, second-, and third-order maxima will be observed.

32-8   Fraunhofer diffraction

On passing through an aperture in a barrier, a beam of light will spread out, to some extent, into the region where we would normally expect to see a shadow. As noted earlier, this is an example of the diffraction of light, or the property of light of bending around sharp edges. The distinction between diffraction and interference is not very precise and for our purposes a diffraction pattern may be thought of as the pattern formed by the interference of a very large number of coherent beams.

Historically, the two classes of diffraction phenomena that have been analyzed in depth are *Fresnel diffraction* and *Fraunhofer diffraction*. Figure 32-15 shows how a Fraunhofer diffraction pattern may be produced. Light from a line (or point) source S is rendered parallel by a positive lens L_1 and the resultant plane waves are incident on a barrier AB out of which has been cut a slit of width b. After the light emerges from the slit, the resultant rays are brought to a focus on a screen or a photographic plate CD, which is located in the focal plane of a second positive lens L_2. Note that waves which arrive at the barrier AB, as well as those that arrive at the screen CD, are plane waves. This feature is a main characteristic of Fraunhofer diffraction. If either of the lenses is removed so that waves that approach either the slit or the screen are spherical or cylindrical waves, then the observed pattern is said to be due to Fresnel diffraction. In Fraunhofer diffraction the distances from the source to the slit and from the slit to the screen are both effectively infinite. Fresnel diffraction takes place if either or both of these distances are finite. For reasons of simplicity, we shall consider here only Fraunhofer diffraction by a single slit.

Consider, in Figure 32-16, plane waves incident on a slit of width b and suppose that the resultant diffracted rays are brought to a focus on a screen by a lens L. In accordance with the Huygens principle, each point of the slit may be thought of as being a source of outgoing spherical waves. Moreover, as set up, these sources are coherent. To calculate the actual intensity $I(\theta)$ at

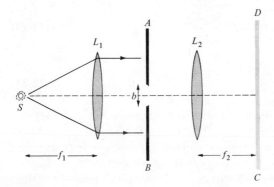

Figure 32-15

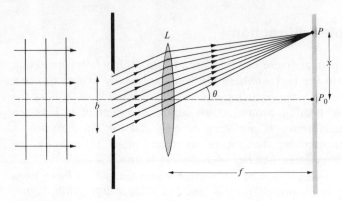

Figure 32-16

the point P on the screen, we may suppose, therefore, that in the region of the slit there are N equally spaced and coherent point (or line) sources and that we are interested in the resultant interference pattern for very large N. The distance d between these sources is given in terms of the slit width b by

$$d = \frac{b}{N} \tag{32-19}$$

and thus according to (32-17), the intensity $I_N(\theta)$ for N sources may be expressed by

$$I_N = \frac{I_m}{N^2} \frac{\sin^2 [(N\pi d \sin \theta)/\lambda]}{\sin^2 [(\pi d \sin \theta)/\lambda]}$$

where $I_m \equiv I_0 N^2$ represents the maximum intensity of the principle maximum (Figure 32-14). Applying this formula to the situation in Figure 32-16, we obtain by use of (32-19)

$$I_N = \frac{I_m}{N^2} \frac{\sin^2 [(\pi b \sin \theta)/\lambda]}{\sin^2 [(\pi b \sin \theta)/N\lambda]}$$

Passing to the limit $N \to \infty$, we find that the intensity $I(\theta)$ for the Fraunhofer diffraction pattern is given by

$$I(\theta) = I_m \frac{\sin^2 \beta}{\beta^2} \tag{32-20}$$

where $\beta \equiv \beta(\theta)$ is defined by

$$\beta = \frac{\pi b \sin \theta}{\lambda} \tag{32-21}$$

and we have used the fact that as N becomes very large, $N^2 \sin^2 (x/N) \to x^2$.

Figure 32-17 shows a plot of $I(\theta)$ as a function of $\beta = (\pi b \sin \theta)/\lambda$. Since the ratio $(\sin x)/x$ approaches unity as x tends to zero, it follows from (32-20) that at the value $\beta = 0$, $I(\theta)$ achieves its maximum value of I_m. The intensity function $I(\theta)$ vanishes whenever β assumes an integral multiple value of π. Hence, since $I(\theta)$ is nonnegative, it follows that there are

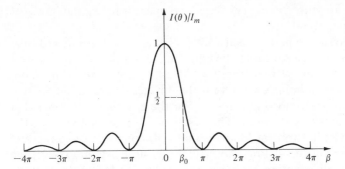

Figure 32-17

additional maxima between each pair of these zeros. To a rough approximation we may assume that these maxima lie midway between these zeros—
that is, at the values $\beta \cong 3\pi/2, 5\pi/2, 7\pi/2, \ldots$. Substituting these values for
β into (32-20), we find that at these secondary maxima $I(\theta)/I_m$ assumes the
values $(2/3\pi)^2, (2/5\pi)^2, (2/7\pi)^2, \ldots$; or, equivalently, the values 0.045, 0.016,
0.0083, $\ldots$. Thus the large central maximum, for which the ratio $I(\theta)/I_m$ is
unity, dominates the entire diffraction pattern! Nevertheless, the first
maximum at $\beta \cong 3/2\pi$ has an intensity of about 4.5 percent of that of the
central maximum and is thus frequently visible.

In order to describe the width of the central maximum in Figure 32-17, it is
convenient to define the parameter β_0 as the value of β for which the
intensity has dropped to half of its maximum value. Substituting the values
$\beta = \beta_0$ and $I = I_m/2$ into (32-20) yields

$$\sin^2 \beta_0 = \frac{\beta_0{}^2}{2} \qquad \text{(32-22)}$$

so that

$$\beta_0 \cong 1.4 \text{ rad} = 80°$$

The associated angle θ_0 may be found by substitution into (32-21).

A second angle of interest in this connection is the angle θ_1, defined as the
position on the screen where the intensity of the central maximum has
dropped to zero. Since the value for β associated with θ_1 is π, we find by
substitution into (32-21) that

$$\sin \theta_1 = \frac{\lambda}{b} \qquad \text{(32-23)}$$

Physically, we may understand the significance of θ_1 in the following way. If
light were not a wave, then the light coming through the aperture in Figure
32-16 would obey the laws of geometrical optics and come to a focus at the
point P_0 on the screen. However, light *is* a wave phenomenon. Hence this
point image is diffracted out until it has an angular diameter given approximately by twice the angle θ_1 in (32-23). In the limit of very large b, the width
of the central maximum shrinks to zero, corresponding to a point image.

That is, for an aperture size very large compared to λ, the laws of geometrical optics prevail.

At the other extreme, where the slit width b is very small compared to λ, the intensity on the screen appears to be uniform. This is the situation shown in Figure 32-2. In the analysis of the two-slit diffraction pattern in Figure 32-9 and in the corresponding analysis for the diffraction grating in Section 32-7, it was implicitly assumed that $b \ll \lambda$. For the case $b \sim \lambda$ a somewhat more complex analysis is required. The result of such an analysis for the two-slit case is given in Problem 33 and is that $I(\theta)$ consists of two factors: one of these is the single-slit Fraunhofer pattern $(\sin^2\beta)/\beta^2$ in (32-20); the other is the two-slit distribution $\cos^2[(\pi d \sin \theta)/\lambda]$ in (32-11). In the analysis carried out in Section 32-5, it was implicitly assumed that the slits were very narrow $(b \ll \lambda)$, so that $(\sin^2 \beta)/\beta^2 \cong 1$ and the first of these factors could be neglected. However, for large enough angles this is no longer true. It is for this reason that the two-slit pattern observed on the screen in Figure 32-10 extends out only to small angles, for which diffraction effects are negligible.

Example 32-7 A slit of width 0.02 mm is illuminated normally by plane monochromatic light waves of wavelength 6000 Å. A thin, positive lens of focal length 50 cm is used to project the diffraction pattern onto a screen (see Figure 32-16).

(a) What is the angular separation from the center of the principal maximum to the first minimum?

(b) At what value of θ has the intensity of the central maximum decreased to half its maximum value?

(c) How wide is the central maximum as observed on the screen?

Solution

(a) The required value of θ corresponds to $\beta = \pi$, and is given by the value θ_1 in (32-23). Substituting the known values for λ and b, we find that

$$\sin \theta_1 = \frac{\lambda}{b} = \frac{6.0 \times 10^{-5} \text{ cm}}{2.0 \times 10^{-3} \text{ cm}} = 0.03$$

Thus, we may replace $\sin \theta_1$ by θ_1 and conclude that

$$\theta_1 = 0.03 \text{ rad} = 1.7°$$

(b) This angle is the angle θ_0 in (32-22). Hence by use of (32-21) it follows that

$$\sin \theta_0 \cong \theta_0 = \frac{1.4\lambda}{\pi b} = \frac{1.4}{\pi} \times 0.03 = 0.013 \text{ rad}$$

(c) Reference to Figure 32-16 shows that the distance x_1 to the screen associated with the angle θ_1 is

$$x_1 = f \tan \theta_1 \cong f\theta_1 = (50 \text{ cm}) \times 0.03 \text{ rad} = 1.5 \text{ cm}$$

where $f = 50$ cm is the focal length of the lens and we have used the result of (a). The width of the central maximum is $2x_1$, or 3 cm.

32-9 Resolving power

Consider, in Figure 32-18, two *incoherent* slit sources S_1 and S_2, separated by a small angular distance α and emitting light waves that come to a focus on a screen CD after going through an aperture of width b. Assuming that the distances between the sources and the aperture and between the aperture and the screen are effectively infinite, we expect to find on the screen two Fraunhofer diffraction patterns with centers separated by the same angle α. By contrast to the laws of geometrical optics according to which there should appear two points of light on the screen, one at P_1 and the other at P_2, we find instead around each of these two points P_1 and P_2 an illuminated region associated with the diffraction patterns produced by each of the sources. If the angular separation α between the sources is very small, it may not be possible to ascertain the existence of two distinct sources. The purpose of this section is to see under what circumstances it is possible to distinguish clearly between two sources very close together.

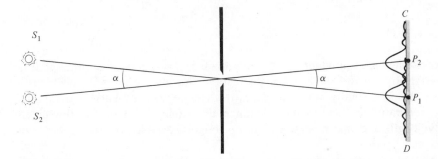

Figure 32-18

Figure 32-19 shows the various classes of diffraction patterns that can arise from two sources separated by a very small angle α. In each case, the dashed lines represent the separate diffraction patterns of the two sources, and the solid line, which represents their sum, is the observed intensity. Recall that the sources are assumed to be incoherent. The angle θ_1, as defined in (32-23), stands for the angular distance from the center of the principal maximum of a source to the point for which the intensity vanishes. For the sake of simplicity let us assume that the maximum intensities of the two sources are the same, and that they are monochromatic and radiating at the same wavelength λ. Figure 32-19a shows the case when the angle α between the sources is much larger than θ_1. Here the two maxima are clearly separated and we have no trouble concluding that two radiators are emitting. Similarly, Figure 32-19b shows the case in which $\alpha = 1.5\theta_1$ and, again, even though there is some overlap of the principal diffraction maxima, it is still easy to see that there exist two distinct sources. Finally, Figures 32-19c and

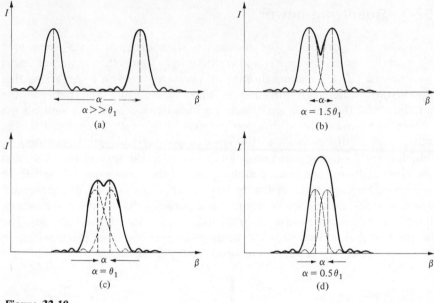

Figure 32-19

d show the cases in which $\alpha = \theta_1$ and $\alpha = 0.5\theta_1$, respectively. Although it is possible by careful observation to resolve the two sources in the former case, this is certainly not so in the latter. Hence we see that if two sources are separated by an angular distance less than θ_1, then for the given wavelength and aperture size it is no longer possible to resolve the two images.

A study of the patterns shown in Figure 32-19 leads one to the conclusion that the situation in Figure 32-19c represents the smallest separation angle α for which the images can still be resolved. It was decided, somewhat arbitrarily perhaps, by Lord Rayleigh that this angle α_1, which is given by

$$\alpha_1 = \theta_1 = \frac{\lambda}{b} \qquad (32\text{-}24)$$

where the second equality follows by use of (32-23) with $\theta_1 \ll 1$, should be the *criterion* for the resolvability of the two diffraction patterns. This choice for the angle of separation α_1 between the two images is known as *Rayleigh's criterion*, and defines the *resolving power* for the aperture b. In practical terms, what this means is that if the separation angle α between two sources is less than the value α_1 given in (32-24) then they cannot be resolved by use of an aperture of the given size b. Conversely, if the separation angle between two sources is greater than λ/b, then they are resolvable.

In the above discussion we have made use of (32-23) and thus have assumed that the sources and the aperture were long slits. For most cases of practical interest we are concerned with point sources, such as two stars or the opposite edges of an amoeba, and with circular apertures, such as the objective lens of a telescope or a microscope. To apply Rayleigh's criterion to these cases requires a knowledge of the diffraction pattern produced by a circular aperture. The results of this calculation—which was first carried out by Sir George Airy (1801–1892)—are surprisingly simple and are given by

$$\alpha_1 = 1.22 \frac{\lambda}{b} \tag{32-25}$$

That is, the minimum angle of separation α_1 between two point sources that radiate at the wavelength λ and which can be still resolved by use of a circular aperture of diameter b is given by $1.22\lambda/b$. So except for the factor 1.22, this formula for the resolving power of a circular aperture is the same as (32-24) for a slit.

Example 32-8 A "spy satellite" orbits the earth at a height of 100 miles. What is the minimum diameter b of the objective lens in a telescope that must be used to resolve the marked yard lines on a football field? Assume $\lambda = 5500$ Å and that the limitation is due to diffraction.

Solution The angular separation α between two lines separated by a distance of 5 yards and as seen 100 miles away is

$$\alpha_1 = \frac{5 \text{ yards}}{100 \text{ miles}} = \frac{5 \text{ yards}}{100 \text{ miles} \times 1760 \text{ yards/mile}}$$

$$= 2.8 \times 10^{-5} \text{ rad}$$

Substituting this value into (32-25) and solving for b we find that

$$b = \frac{1.22\lambda}{\alpha_1} = \frac{1.22 \times 5.5 \times 10^{-7} \text{ m}}{2.8 \times 10^{-5} \text{ rad}}$$

$$= 2.4 \text{ cm}$$

Example 32-9 Assuming that the pupil of the human eye has a diameter of 4.0 mm, what is the minimum separation distance y between a person and two small, red objects ($\lambda = 6500$ Å) separated from each other by a distance of 20 cm. (See Figure 32-20, which is *not* to scale!)

Solution According to the given data, the minimum angle α that can still be resolved is 20 cm/y. Equating this to $1.22\lambda/b$ in accordance with (32-25) and solving for y, we obtain

$$y = \frac{20 \text{ cm}}{1.22\lambda/b} = \frac{(20 \text{ cm}) \times (4.0 \times 10^{-1} \text{ cm})}{1.22 \times 6.5 \times 10^{-5} \text{ cm}} = 1.0 \text{ km}$$

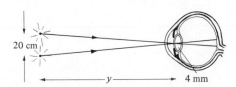

Figure 32-20

32-10   Interference by thin films

A source of illumination that is *not* a point source, and which may therefore be thought of as consisting of a large number of *incoherent* point sources, is known as an *extended* or a *diffuse* source. If light from such a source is reflected from a thin, transparent film—a thin layer of oil or the membrane of a soap bubble, for example—then under certain conditions interference effects will be observed. These effects manifest themselves frequently by the appearance of bright-colored bands across the surface of the film. The purpose of this section is to show how this phenomenon can be understood in terms of an interference effect.

Consider, in Figure 32-21, a thin film of material of thickness d and of refractive index n and illuminated by an extended monochromatic source of wavelength λ. To see how it is possible for the observer O to see interference effects, let us focus on a typical ray AB, which is emitted from a point A on the source and in a direction so that after reflection from the film it enters the observer's eye. Since the source is extended, light rays from various other points of the source will enter the observer's eye as well, but for the moment let us consider only that associated with the ray AB. As shown in the figure, not all of the light in the incident ray is reflected at the point B on the upper surface of the film. Some of it is refracted and enters the film along the direction BC. On arriving at point C, part of this refracted ray BC is reflected upward along CD, and the remainder is refracted out of the underside of the film. The ray CD, on striking the upper surface at the point D, will also be partially refracted out of the film and in a direction parallel to the ray BO. Finally, since both of the rays BO and DO, which enter the observer's eye, originally started out as a single ray at the point B, it follows that they are coherent and that a definite phase relationship exists between them. Hence, when these rays come together on a screen (the retina of the observer's eye, in this case), interference effects may be observed.

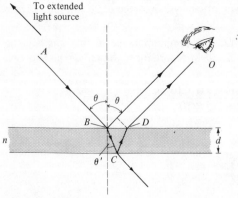

Figure 32-21

To simplify the analysis, let us confine ourselves in all of the following to observations near normal incidence. This means that $\theta \cong n\theta' \cong 0$ so that the distances $\overline{BC}$ and $\overline{CD}$ have a length d and the distance $\overline{BD}$ is negligible compared to the thickness d of the film. It follows that under this circumstance the path difference Δ between the rays BO and $BCDO$, which start out in phase at B, is

$$\Delta = 2d$$

We might expect, therefore, that constructive or destructive interference would be observed if this path difference $2d$ were an integral or a half-integral multiple of a wavelength, respectively. However, this is *not* so. Instead, as will be established below, maxima and minima in intensity occur in accordance with the formulas:

$$2nd = (m + \tfrac{1}{2})\lambda \qquad \text{(maxima)}$$
$$2nd = m\lambda \qquad \text{(minima)}$$

(32-26)

for m a nonnegative integer. These formulas differ in two ways from what one might expect based on our previous studies of interference: (1) by the appearance of the index of refraction n of the film on the left-hand side; and (2) by the interchange of the roles of maxima and minima in terms of the number of wavelengths in the path difference. The quantity $2nd \equiv n\Delta$ is also known as the *optical path difference*.

The necessity for the factor n in (32-26) can be seen in the following way. The wavelength λ in these formulas refers to the free-space wavelength of the light and not to its wavelength λ/n in the medium; see (29-24). Since the path difference of the interfering rays in Figure 32-21 is the extra distance, $\overline{BC} + \overline{CD} = 2d$, traveled by the ray $BCDO$ through the film, it follows that the wavelength λ/n in *the medium* is the one relevant for determining whether or not interference effects take place. Thus the reason for the factor n in (32-26) is so that the wavelength λ in these formulas refers to the free-space value.

With regard to the reversals of the roles of maxima and minima in (32-26), we may argue as follows. According to the first of the Fresnel equations in (30-17) the amplitude E_0'' of an electromagnetic wave reflected at normal incidence, at the interface between two media, will undergo a reversal of direction if $n_1 < n_2$. In other words, light incident from the lower index side of the interface between two media will undergo a phase change of $180°$ on reflection. According to the general relation between phase difference δ and path difference Δ in (32-7), a phase difference of $180°$, or π radians, corresponds to a path difference of a half of a wavelength. Therefore, if the path difference $2d$ in Figure 32-21 is an integral number of wavelengths, the interference is destructive. Correspondingly, constructive interference will be associated with a path difference $2d$ equal to an odd half integral multiple of a wavelength λ/n. It is for this reason then that the roles of maxima and minima are reversed in (32-26). Note that even if the film in Figure 32-21 is in a medium of refractive index n' $(>n)$, the relations in (32-26) are still applicable. According to (30-17), for this case a phase change of 180° occurs for the ray reflected at the lower surface, whereas at the upper surface the incident and the reflected rays remain in phase.

32-11   Observations of thin-film interference

In this section we apply (32-26) to several physical situations.

Consider first, in Figure 32-22a, light from a *monochromatic* extended source normally incident on two glass plates, which touch at one end and are separated by a small spacer, such as a piece of paper, at the other. In effect, we have a thin film of air whose thickness varies along the plates. It follows from (32-26) that the observer of the reflected rays will see a series of alternating dark and bright bands across the plate. See Figure 32-22b. Along the edge where the plates are in contact, $d = 0$. Hence, regardless of λ, (32-26) implies that destructive interference will be observed here. Further, since $n = 1$ for air, it follows, by setting $m = 0$ in (32-26), that for $d = \lambda/4$ a bright band will be seen on the plate. As shown in Figure 32-22b, bright bands will appear at film thicknesses of $3\lambda/4, 5\lambda/4, \ldots$, and these will be separated by dark bands with centers corresponding to film thicknesses of $d = \lambda/2, \lambda, 3\lambda/2, \ldots$.

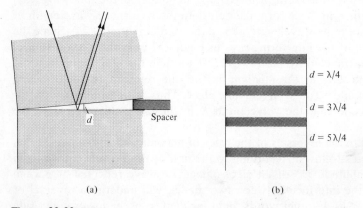

(a)   (b)

Figure 32-22

If the source is not monochromatic, then bands of light of other wavelengths may also appear on the plate. For example, if the source is sunlight, with its full spectrum of colors, the observer will see a sequence of bright bands with continuously varying colors from red to violet.

A second apparatus for observing interference in an air film is shown in Figure 32-23. Light from an extended source S (not shown) is normally incident on the flat side of a plano-convex lens B whose spherical face lies on a flat glass plate A. The air film between A and B then produces circular

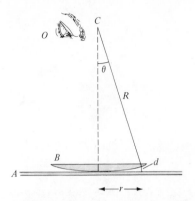

Figure 32-23

interference fringes, which will be seen by the observer O. These particular fringes are known as *Newton's rings* and were first studied in detail by Isaac Newton. It is ironic that these rings, which were really the first experimental evidence for the wave nature of light, should be credited to the creator and defender of the corpuscular theory.

To obtain a quantitative relation between the radius and wavelength of a given ring, let us consider the apparatus in more detail. If r is the radius of a given ring of order m, and d the corresponding thickness of the air film, then for small angles

$$d = R - R \cos \theta \cong R - R \left(1 - \frac{\theta^2}{2} \right) = \frac{R\theta^2}{2}$$

$$= \frac{r^2}{2R}$$

since $\cos \theta \cong 1 - (\theta^2/2)$ and, according to the figure, $\theta \cong \tan \theta = r/R$. Substituting this form for the thickness of the air film into (32-26) and using the value $n = 1$ for air, we find for the radius r_m of the $(m + 1)$th bright ring

$$r_m = \left[R\lambda \left(m + \frac{1}{2} \right) \right]^{1/2} \qquad m = 0, 1, 2, \dots \qquad \textbf{(32-27)}$$

Because of the fact that there is a phase change associated with the reflection from the lower glass plate, the spot in the middle of the plate is dark. That is, destructive interference takes place at the central point. This feature is clearly illustrated in Figure 32-24, which is an actual photograph of the fringes formed by this method.

Finally, let us consider in Figure 32-25 a precision instrument known as the *Michelson interferometer*, which also produces interference fringes when exposed to an extended source. Light from a diffuse source S on arriving at a half-silvered plate P_1 is broken up into a reflected beam (1) and a transmitted beam (2) of approximately equal intensity. Beam 1 travels upward toward a highly polished mirror M_1, from which it is reflected back through P_1 into the eye of the observer at O. The second beam is directed

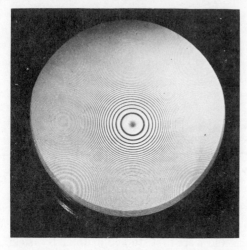

Figure 32-24 *Newton's rings produced by use of a monochromatic source. (Courtesy Bausch and Lomb, Scientific Optical Products Division.)*

first through a compensating plate P_2 so if $l_1 = l_2$, then the optical paths for the two beams will be the same. It is then reflected from a second mirror M_2, back through P_2, and finally into the observer's eye after being reflected by P_1. The observer O thus sees two images of the extended source S; one of these comes from its reflection in M_1 and the other in M_2. Just as for the situation shown in Figure 32-21, the rays that appear to come from these two images will interfere constructively or destructively depending on the number of wavelengths in the optical path difference between rays (1) and (2). For example, if the mirrors M_1 and M_2 are not precisely perpendicular to each other, then in effect we have the situation in Figure 32-22 and,

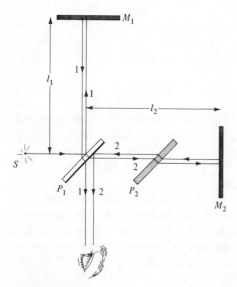

Figure 32-25

consistent with experiment, the observer sees a set of parallel, bright fringes. If the two mirrors are precisely perpendicular, then the observer sees a series of concentric circular fringes, reminiscent of Newton's rings.

The interferometer has played, and continues to play, an important part in precision measurements of various physical quantities. It was originally designed by Michelson in 1887 to measure the velocity of the earth through the "aether." The negative outcome of this Michelson–Morley experiment, as it is known, showed that the velocity of light is independent of the relative motion between the source and the observer. The significance of this fact was not fully appreciated until later when Einstein enunciated the special theory of relativity. Subsequently, the interferometer was used to measure the diameters of selected *red-giant* stars. More recently, the interferometer has been used to define the SI unit of length of the meter as the distance equal to 1,650,763.73 wavelengths of the bright orange-red line in the spectrum of krypton-86. The precision possible with this instrument is truly remarkable.

Example 32-10   Newton's rings are observed by use of light of wavelength 6000 Å and a plano-convex lens with a radius of curvature of 25 cm. What is the radius of the first and the twentieth ring?

Solution   Making use of the given values $R = 25$ cm and $\lambda = 6.0 \times 10^{-5}$ cm, we find by setting $m = 0$ in (32-27) that the radius r_1 of the first ring is

$$r_1 = \left[R\lambda \left(m + \frac{1}{2} \right) \right]^{1/2} = \left[25 \text{ cm} \times 6.0 \times 10^{-5} \text{ cm} \times \frac{1}{2} \right]^{1/2}$$
$$= 0.27 \text{ mm}$$

while that for the twentieth ring is found by setting $m = 19$

$$r_{20} = [25 \text{ cm} \times 6.0 \times 10^{-5} \text{ cm} \times 19.5]^{1/2}$$
$$= 1.7 \text{ mm}$$

Example 32-11   In Figure 32-23 suppose that plate A has a refractive index n_A, plate B a refractive index n_B, and the intervening space is filled up with a fluid of refractive index n_C. If

$$n_A < n_C < n_B \quad \text{or} \quad n_A > n_C > n_B$$

what is the formula analogous to (32-27) for the radius of the bright rings?

Solution   The derivation would go through just as above, but now there is *no* relative phase change on reflection. That is, at each reflection both rays suffer either no phase change or else both undergo a phase change of 180°. In either case the relative phase change is zero. Thus in place of the first of (32-26), the condition for a maximum becomes

$$2n_C d = m\lambda$$

Proceeding as in the derivation of (32-27), we conclude that rings occur now at the values of r given by

$$r_m = \left[\frac{R\lambda m}{n_C}\right]^{1/2} \qquad m = 0, 1, 2, \ldots$$

In particular, this means that a bright spot appears at the center of the plate under these circumstances!

32-12   Summary of important formulas

If, as in Figure 32-13, a plane monochromatic wave is incident on a barrier out of which have been cut N very narrow slits, the intensity $I(\theta)$ observed on a distant screen is

$$I(\theta) = I_0 \frac{\sin^2[(N\pi d \sin \theta)/\lambda]}{\sin^2[(\pi d \sin \theta)/\lambda]} \qquad (32\text{-}17)$$

where d is the spacing between the slits, λ the wavelength, and I_0 the intensity that would be observed if only one slit were present. For the special case $N = 2$ this reduces to

$$I(\theta) = 4I_0 \cos^2\left[\frac{\pi d \sin \theta}{\lambda}\right] \qquad (32\text{-}11)$$

In both (32-17) and (32-11) it is assumed that the width of each slit is very small compared to λ.

If plane monochromatic light of wavelength λ is incident on a slit of width b, and if we define the angle θ as in Figure 32-16, then the observed intensity $I(\theta)$ at the point θ is

$$I(\theta) = I_m \frac{\sin^2 \beta}{\beta^2} \qquad (32\text{-}20)$$

where the parameter β is defined by

$$\beta = \frac{\pi b \sin \theta}{\lambda} \qquad (32\text{-}21)$$

and where I_m is the intensity of the central maximum (Figure 32-17).

If two distant and *incoherent* point objects, such as two stars, are viewed through a circular aperture of size b, and if the minimum angle between the sources exceeds the angle α_1, given by

$$\alpha_1 = 1.22 \frac{\lambda}{b} \qquad (32\text{-}25)$$

then the sources can be resolved.

If monochromatic light from a diffuse or extended source is reflected by a thin film of refractive index n and of thickness d, then at normal incidence there will be maxima and minima in the observed intensity for

$$2nd = (m + \tfrac{1}{2})\lambda \qquad \text{(maxima)}$$
$$2nd = m\lambda \qquad \text{(minima)}$$

(32-26)

for m a nonnegative integer.

QUESTIONS

1. Define or describe briefly what is meant by the following: (a) interference; (b) Huygens construction; (c) incoherent sources; (d) destructive interference; and (e) Fresnel diffraction.

2. The light emitted from two very close stars is brought to a focus on a screen by a lens. Will an interference pattern appear on the screen? Explain.

3. It is often stated that a diffraction pattern is the result of the interference of the radiation emitted by a very large number of coherent sources. Discuss the meaning of this statement in terms of Huygens principle.

4. Explain why two coherent beams of light of the same amplitude must be traveling in parallel—or nearly parallel—paths in order for complete destructive interference to be observed.

5. In Figure 32-13, why is the path difference from A to P the same as that from D to P regardless of the detailed structure of the lens. (*Hint*: According to Fermat's principle, all rays that emanate from a point source spend the same amount of time traveling to a point of focus after going through a lens.)

6. An interference pattern is formed by two coherent light beams, one of which has twice the intensity of the other. In what way does this pattern differ from the one that would be formed if the beams were of the same strength?

7. Explain, in terms of a microscopic picture, why the light rays emitted from different points of an ordinary source are incoherent.

8. A series of equidistant interference maxima are observed in a two-slit experiment, such as in Figure 32-9. If the separation distance d between the slits is doubled, what happens to the spacing between the fringes on the screen? If light of a smaller wavelength is used what happens to the distance between the fringes?

9. Suppose that the angle α of the Fresnel double mirror in Figure 32-12a is halved. What happens to the distance between the virtual sources S_1 and S_2? What happens to the separation distance between neighboring fringes on the screen?

10. Explain why the vertex angle C of the Fresnel biprism in Figure 32-12b must be very nearly 180° in order for fringes to be produced.

11. A diffraction grating has a certain number of lines per unit length. How do you find, from this datum, the values for the parameters d and N to be used in connection with (32-17)? If you cannot obtain values for both d and N, what additional data do you require?

12. Explain why it is desirable to have the rulings of a diffraction grating as close together as possible. Why is it desirable also to have a large value for the parameter N?

13. How do you account for the existence of an upper limit to the order m of a principal maximum that can be

obtained by use of a diffraction grating? What happens to this upper limit as the wavelength is increased? What happens to it as the number of lines per unit length is decreased?

14. Suppose that you want to make a diffraction grating for infrared radiation of wavelength 10^{-4} meter. What grating spacing would be appropriate? What would be the appropriate spacing for microwaves with a wavelength of 10 cm?

15. Explain what is meant by Fresnel diffraction and contrast it with Fraunhofer diffraction.

16. A person looks at light through a very narrow space between two of his fingers. Does he see a Fresnel or a Fraunhofer diffraction pattern? Explain.

17. What happens to the maximum intensity of a Fraunhofer diffraction pattern as:
    (a) The wavelength is increased?
    (b) The slit width is increased?

18. Describe the nature of the single-slit Fraunhofer diffraction pattern that is obtained by use of a slit of width b equal to the wavelength of the light used.

19. What property of the lenses in telescopes is crucial in enabling us to distinguish between two very close stars? Explain why increasing the magnification of a telescope alone will not improve our ability to resolve two very close celestial objects.

20. It was discovered by Davisson and Germer that electrons of mass m and traveling at a velocity v will, under suitable conditions, behave as a *matter wave* with a wavelength

$$\lambda = \frac{h}{mv}$$

where h is Planck's constant ($= 6.63 \times 10^{-34}$ J-s). Explain, in qualitative terms, why an electron microscope—that is, a microscope that uses a stream of electrons instead of light rays—might enable us to see structures that cannot be resolved by ordinary microscopes.

21. The light from two very close celestial objects is viewed by use of a blue filter. Would the resolution be better or worse if a red filter were used instead?

22. Light from a diffuse source is incident on a Newton's rings apparatus as in Figure 32-23. Assuming that the plates A and B in the figure are both made of glass, compare the rings seen by reflection by the observer at O with those that are observed by transmission by an observer who looks upward from below plate A.

23. Suppose that a thin film of air, as in Figure 32-22, is exposed to an extended source of coherent light such as that supplied by a laser. In what way, if any, will the observed interference pattern differ from that obtained by using a monochromatic, but incoherent, extended source?

24. A soap film on a loop of wire is held so that the normal to the plane of the loop is horizontal. On being exposed to an extended light source, it is found that when viewed by reflected light, the top of the film is dark and various horizontal colored fringes appear below. Verify this experiment by use of materials directly available to you and explain the phenomenon.

PROBLEMS

1. What is the electric field amplitude E_0 associated with a monochromatic electromagnetic wave with an average energy flux of 2.0×10^{-2} W/m²?

2. Two point sources are a distance 2.0×10^{-7} meter apart and radiate coherently and in phase at $\lambda = 5500$ Å. With respect to an observer along the line joining them:
   (a) What is the phase difference between them?
   (b) What is the path difference between them?

3. Repeat both parts of Problem 2, but assume that the sources have a phase difference $\alpha_1 - \alpha_2 = 90°$. See (32-4).

4. Two sources S_1 and S_2 are separated by a distance Δ and located in the x-y plane at the points $(\pm\Delta/2, 0)$. Assuming that the sources radiate coherently and in phase at the wavelength λ, show that the condition for a maximum in intensity at the point (x, y) is

$$\sqrt{\left(x + \frac{\Delta}{2}\right)^2 + y^2} - \sqrt{\left(x - \frac{\Delta}{2}\right)^2 + y^2} = m\lambda$$

where m is an integer. Prove that this curve is a hyperbola.

5. If, in the double-slit arrangement in Figure 32-9, the fringes that appear on the screen are a distance of 0.5° apart, find the distance d between the slits if the source emits the bright line in the spectrum of sodium with a wavelength of 5892 Å.

6. Suppose that in Problem 5 the interfering light beams are brought to a focus on a screen by a lens of focal length 50 cm. What is the distance

on the screen between neighboring maxima in the center of the pattern?

7. Blue light, of wavelength 4700 Å, is used to illuminate two slits separated by a distance of 0.3 mm. How far from the slits must a screen be placed so that the fringes are 1 mm apart?

8. A double slit is illuminated by light of wavelength 6000 Å and it is found that the fringes on a screen 90 cm away are separated by a distance of 5.0 mm. A second source is now used with the same equipment and it is found that for it the fringes are separated by 5.7 mm. (a) What is the separation distance between the slits? (b) What is the wavelength of the second source?

*9. The radiation from two very narrow line sources S_1 and S_2 is projected onto a screen by a thin lens of focal length f; see Figure 32-26. Let I_1 be the average intensity on the screen at P when the radiation from S_2 is blocked off, and I_2 the corresponding intensity if S_1 is blocked.

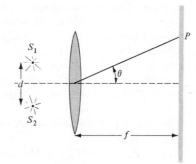

Figure 32-26

(a) Show that if S_1 and S_2 are coherent, then when both are radiating in phase the observed intensity $I(\theta)$ at P is

$$I(\theta) = I_1 + I_2$$
$$+ 2\sqrt{I_1 I_2} \cos \left[\frac{2\pi d \sin \theta}{\lambda} \right]$$

(b) Show that for $I_1 = I_2 = I_0$ this agrees with (32-11).

10. Suppose the ratio I_1/I_2 of the intensities of the sources in Figure 32-26 is 2.5.

(a) What is the ratio of the maximum to the minimum intensity observed on the screen?

(b) If $\lambda = 4500$ Å and $d = 0.01$ mm, what is the angular separation between the maxima?

(c) If $f = 40$ cm for the situation in (b), what is the spacing between maxima on the screen?

*11. Suppose that in a double-slit apparatus a very thin, transparent film of thickness t and refractive index n is placed in front of one of the slits. Show that the path difference Δ between the two beams is now given by

$$\Delta = d \sin \theta + t(n - 1)$$

and write down the conditions for maxima and minima in this case.

12. In a two-slit experiment fringes are observed on a screen by use of light of wavelength 6500 Å. If a very thin, transparent film of refractive index 1.45 is placed in front of one of the slits, it is observed that the fringes are displaced upward by a distance of about three fringe widths; that is, the central fringe is now located at the point where formerly the maximum associated with a path difference of 3λ was located.

(a) By how much has the film increased the optical path?

(b) Making use of the result of Problem 11, find the thickness of the film.

13. Using the facts that

$$e^{\pm iy} = \cos y \pm i \sin y$$

and

$$1 + r + r^2 + \cdots + r^{N-1} = \frac{1 - r^N}{1 - r}$$

derive (32-8).

14. Show that for small values of θ the distance x_m of the mth maximum from the point P_0 on the screen in Figure 32-10 is

$$x_m = \frac{m\lambda D}{d}$$

and use this to derive (32-14).

15. By making use of the techniques of the calculus prove the following limits (N and m are integers):

(a) $\lim\limits_{x \to 0} \dfrac{\sin \pi x}{\pi x} = 1$

(b) $\lim\limits_{x \to 0} \dfrac{\sin N\pi x}{\sin \pi x} = N$

(c) $\lim\limits_{x \to m} \dfrac{\sin N\pi x}{\sin \pi x} = (-1)^{m(N-1)} N$

*16. Show that the distance d between the two virtual images in the Fresnel biprism in Figure 32-12b is

$$d = 2a\alpha(n - 1)$$

where n is the refractive index and where a and α are defined in the figure, with α a very small angle. Calculate the distance between neighboring maxima on a screen 100 cm from the biprism for the choices $n = 1.5$, $a = 15$ cm, $\alpha = 0.01$ rad, and $\lambda = 5000$ Å.

*17. Consider, in Figure 32-27, a plane wave incident at an angle i with respect to the normal onto a barrier, out of which have been cut two thin slits separated by a distance d. Show that the intensity $I(\theta)$ at the point P on the screen described by the angle θ is still given by (32-11),

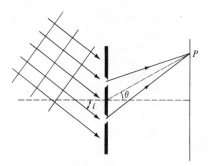

Figure 32-27

but with the replacement

$$d \sin \theta \to d(\sin \theta + \sin i)$$

18. Show that if monochromatic light of wavelength λ is incident at an angle i (with respect to the normal) on a diffraction grating, then the principal maxima occur at those values of θ given by

$$d(\sin i + \sin \theta) = m\lambda$$

$$m = 0, 1, 2, \ldots$$

(*Hint*: Use the method suggested in Problem 17.)

19. A diffraction grating has 2000 lines/cm and is illuminated with red light of wavelength 6600 Å.
    (a) What is the grating spacing?
    (b) At what angles do the first two principal maxima occur?
    (c) How many orders are observed?

20. A certain grating has 5000 lines/cm. How many orders are produced at a wavelength of (a) 4000 Å? (b) 5000 Å? (c) 7000 Å?

21. Yellow light from a sodium lamp of wavelength 5890 Å is incident normally on a diffraction grating, and five orders are observed. Assuming that the fifth maximum occurs at 88°, calculate the grating spacing.

22. Light of the wavelengths 6000 Å and 6300 Å falls normally on a grating that has 4000 lines/cm, and the diffracted light is brought to a focus

on a screen by a thin lens of focal length 100 cm.
    (a) What is the angular separation between the two beams in the first order?
    (b) What is the separation distance on the screen between these two maxima?
    (c) Repeat (b) for the third-order maxima.

23. An intense line is observed at an angle of 50° with respect to the normal of a diffraction grating of 6000 lines/cm. What are the possible wavelengths of this line?

24. Make a plot of the formula for I in (32-17) as a function of the variable

$$\gamma = \frac{\pi d \sin \theta}{\lambda}$$

for the following cases: (a) $N = 2$; (b) $N = 3$; and (c) $N = 100$.

*25. Consider the intensity formula for a diffraction grating in (32-17) as a function of the variable γ defined by

$$\gamma = \frac{\pi d \sin \theta}{\lambda}$$

(a) By using the results of Problem 15 show that the principal maxima are at the points

$$\gamma = m\pi \qquad (m = 0, 1, 2, \ldots)$$

and that the value of I here is

$$I = I_0 N^2$$

(b) Show that the zeros of I between the central and the first principal maximum occur at the points

$$\gamma = \frac{\pi}{N}, \frac{2\pi}{N}, \frac{3\pi}{N}, \ldots, \frac{(N-1)\pi}{N}$$

(c) Show that the secondary maxima are given by the roots of the equation

$$N \tan \gamma = \tan N\gamma$$

26. Light of wavelength 5400 Å is incident normally on a long slit of width 0.8 mm. If a lens of focal length 90 cm is placed behind the slit and a screen is placed in its focal plane:
   (a) What is the distance from the center of the principal maximum to the first minimum?
   (b) How far is it from the first minimum to the second minimum?

27. (a) Show by differentiation that the maximum for the intensity $I(\theta)$ in (32-20) occurs at the values of β satisfying

$$\tan \beta = \beta$$

   (b) Show that if β_i is the value for β at the ith maximum, then the intensity I_i at this maximum is

$$I_i = I_m \cos^2 \beta_i$$

28. For the physical situation in Problem 26, at what angle θ has the intensity dropped to half its value at the central maximum?

29. Make a plot, on the same graph, of the intensity distribution $I(\theta)$ as a function of θ produced by the Fraunhofer diffraction from a slit of width b if: **(a)** $\lambda = b$; **(b)** $\lambda = 0.1b$; and **(c)** $\lambda = 0.01b$. Interpret your results physically.

30. What is the angular diameter, as measured by θ_0, of the central maximum for each of the slit widths in Problem 29?

31. Show that if light is incident on a slit at an angle i with respect to its normal, then (32-20) for $I(\theta)$ is still applicable, except that the parameter β is now given by

$$\beta = \frac{\pi b}{\lambda}(\sin \theta + \sin i)$$

32. Light of wavelength 6200 Å is incident normally on a slit and projected onto a screen by a lens of focal length 75 cm. If the distance between the first two minima is 2.3 mm:
   (a) How wide is the slit?
   (b) How far away from the central maximum has the intensity dropped to half its central value?

33. Figure 32-28 shows the production of Fraunhofer diffraction due to plane waves of wavelength λ incident normally on two slits, each of width b and separated by a distance d. The intensity $I(\theta)$ at a point on the screen characterized by the angle θ is

$$I(\theta) = I_m \cos^2 \gamma \frac{\sin^2 \beta}{\beta^2}$$

where

$$\gamma = \frac{\pi d \sin \theta}{\lambda} \qquad \beta = \frac{\pi b \sin \theta}{\lambda}$$

   (a) Show that for $b \ll \lambda$ this reduces to our previous results obtained for Young's experiment.

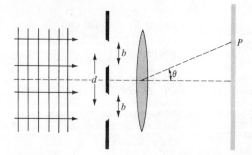

Figure 32-28

(b) For the special case $d \ll \lambda$, what should $I(\theta)$ reduce to? Show that it does.

(c) Show that for $d = b$, $I(\theta)$ reduces to the expected formula for Fraunhofer diffraction from a single slit of width $2b$.

34. For the graph in Figure 32-19c show that the minimum between the two peaks is approximately equal to $0.8I_m$, where I_m is the maximum intensity at a peak.

35. What is the minimum distance apart that two objects on the surface of the moon can be so that they can still be seen by the unaided eye? Assume that the pupil of the eye has a diameter of 4 mm, that the effective wavelength is 5500 Å, and that the distance to the moon is 3.8×10^5 km.

36. At what distance from an automobile can the unaided eye resolve its two headlights, assuming that they are 1.0 meter apart? Assume the data of Problem 35.

37. Two sunspots are a distance 8000 km apart. Assuming that the distance to the sun is 1.5×10^8 km and that the effective wavelength of light is 5500 Å, what is the diameter of the objective lens of a telescope that will just resolve the two spots?

38. What is the minimum angular separation between two distant stars that can just be resolved by the 200-inch telescope at Mount Palomar? Assume that the effective wavelength is 5500 Å.

39. A thin film of transparent material has a refractive index of 1.5 and a thickness of 10^{-4} cm. If it is illuminated with an extended source, what are the wavelengths of the fringes that appear in the reflected light? Assume normal incidence.

40. Suppose that in the apparatus in Figure 32-22 the glass plates are 10 cm long and the spacer is of such a thickness that the angle α between the plates is 2.0×10^{-4} rad. If the light is of wavelength 5000 Å and is viewed normally, calculate:

(a) The number of fringes that appear across the plate.

(b) The separation distance between the first two fringes.

41. Repeat Problem 40, but suppose that this time the space between the plates contains water of refractive index 1.33.

42. Suppose that in Figure 32-22 the upper plate has a refractive index of 1.4 and the lower one a refractive index of 1.6. Will there be a bright or a dark fringe along the line of contact between the two plates if: **(a)** The space between the plates contains air? **(b)** The space between the plates contains a material of refractive index 1.5? Justify your answer in each case.

43. A thin layer of oil of refractive index 1.4 lies on a glass plate of refractive index 1.5. What is the minimum thickness of the layer if it reflects very strongly light of wavelength 5000 Å incident from above? If a layer of oil of the same thickness lies between two glass plates, what wavelength will be strongly reflected? Assume normal incidence.

44. Generalize the results in (32-26) to the case where the angle of incidence θ is not zero (see Figure 32-21). Show in particular that if the film has a refractive index n and is in air, then for a maximum

$$2d \sqrt{n^2 - \sin^2 \theta} = \left(m + \frac{1}{2} \right) \lambda$$

45. In a Newton's rings experiment, the radius of the fifth ring is found to be 0.2 cm for light of wavelength 6500 Å. **(a)** What is the radius of curvature of the lens? **(b)** What is the radius of the twentieth ring?

46. If the space between the plates in a Newton's rings experiment is filled with a material of refractive index n_0, show that the radii of the rings decrease by the factor $n_0^{-1/2}$.

47. An experiment with Newton's rings is carried out by use of two plano-convex lenses (see Figure 32-29). If the radii of curvature of the two lenses are R_1 and R_2, respectively, show that the radii of the maxima are given by

$$r_m = \left\{ \lambda \left(m + \frac{1}{2} \right) \left(\frac{R_1 R_2}{R_1 + R_2} \right) \right\}^{1/2}$$

$$m = 0, 1, 2, \ldots$$

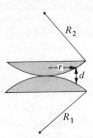

Figure 32-29

33 Polarization

Colors are the deeds and suffering of light.
                                    GOETHE

33-1   Introduction

The phenomena of diffraction and interference are very generally associated with both longitudinal waves as well as transverse waves. As we saw in Chapter 18, for a longitudinal wave, such as a sound wave, the oscillations of the constituents of the medium are along the direction of propagation of the wave so that, as a rule, these waves are symmetric about this direction. By contrast, for a transverse wave the underlying vibrations are at right angles to the direction of propagation. Here, as exemplified by the traveling wave on the string in Figure 18-6, there may be an asymmetry about the direction of propagation of the wave. The term *polarization* is used to characterize a wave of this type, which is asymmetric about its direction of travel; the wave itself is said to be *polarized*. Polarization is a phenomenon associated exclusively with transverse waves; longitudinal waves do not exhibit polarization.

In Chapter 29 we saw that the electric and the magnetic vectors **E** and **B** of an electromagnetic wave are invariably perpendicular to the direction of travel of the wave. It follows that electromagnetic waves in general, and light waves in particular, are transverse and should in principle exhibit polarization effects. This is indeed the case. The purpose of this chapter is to confirm this characteristic behavior of light by describing a number of methods that can be used to produce and to detect polarized light. Specifically we shall consider in detail the production of polarized light by (1) reflection and refraction; (2) selective absorption; and (3) double refraction or birefringence. This listing is by no means exhaustive; polarized light can be produced in other ways as well.

33-2 Linearly polarized light

As a preliminary to a discussion of practical methods for producing and observing polarized light, in this section and the next we shall describe the various states of polarization that are observed for light.

Consider, in Figure 33-1, an observer O viewing the radiation emitted by an accelerated charged particle. Assume for simplicity that the observer is far enough away from the particle so that the approaching wavefronts are planes and also that the acceleration **a** of the particle is perpendicular to the observation direction. According to (29-30), the electric vector **E** in the radiation that reaches the observer is proportional to, and oriented along, the direction of the acceleration **a**. Hence, as shown, if **a** is along the vertical, then the direction of the electric field vector in the approaching wavefronts will also lie along this direction. The orientation of the associated **B**-field is directed as shown and may be determined by the condition that the Poynting vector, $\mathbf{E} \times \mathbf{B}/\mu_0$, points along the direction of propagation of the wave. Furthermore, if the particle oscillates up and down at a certain frequency, then the electric vector in the observed radiation will similarly oscillate up and down with the same frequency. However, and this is an important point, the direction—although not the sense—of the electric field is unchanging and will in this case lie along the vertical throughout the particle's oscillations. Thus there is an asymmetry about the direction of propagation of the wave; in other words, the wave is *polarized*.

Any electromagnetic wave, such as this one, for which the direction, although not necessarily the sense, of the electric field is the same all along

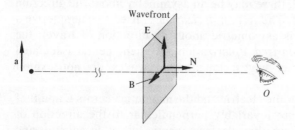

Figure 33-1

the wave is said to be *plane-polarized* or *linearly polarized* along that direction. For the situation in Figure 33-1, for example, this means that the light is linearly polarized along the direction of acceleration of the particle. Since the direction of the electric field **E** and that along which the wave propagates determine a unique direction for the associated **B**-field, in describing polarized light we shall from now on omit all reference to the latter field. Figure 33-2a shows the electric field vectors in the wavefront of an approaching wave that is plane-polarized along the vertical direction. The arrowheads at both ends of the lines represent the **E**-field and emphasize the fact that the sense of this field has no significance in this context. Figure 33-2b shows the corresponding case of light that is plane-polarized along the horizontal direction.

Figure 33-3 shows pictorially several light rays traveling to the right. The dots at various points along the ray in part (a) signify that the electric field vectors for this ray are perpendicular to the plane of the drawing. Hence this ray represents light linearly polarized perpendicular to the plane of the page. Correspondingly, the ray shown in Figure 33-3b represents light linearly polarized in the plane of the drawing. Since the electric field must be perpendicular to the light ray itself, the direction of polarization is along the vertical direction. Figure 33-3c shows how to represent the light resulting from the superposition of these two polarized light rays.

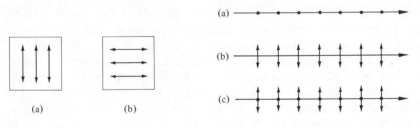

(a)

(b)

Figure 33-2   *Figure 33-3*

Let us consider now the radiation emitted by a collection of charged particles that separately accelerate in various directions. The observer, this time, will see radiation emitted by the totality of these particles and an instantaneous picture of the electric field vectors in a wavefront will now appear as in Figure 33-4a. In this case we say that the light is *unpolarized*. The distinction between unpolarized light and linearly polarized light should be carefully noted. For polarized light the *direction* of the electric vector in the wavefronts is constant in time at any fixed observation point, whereas for unpolarized light the direction of the electric vector changes, in a more-or-less random way, in the course of time.

There is an important relation between unpolarized and linearly polarized light, which may be established in the following way. Consider the instantaneous electric fields in the wavefront in Figure 33-4a and imagine taking their components along two mutually perpendicular directions. On adding

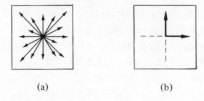

(a)   (b)

Figure 33-4

these together, we obtain, as shown in Figure 33-4b, the components of the resultant vector along these two directions. Since this process may be repeated at any subsequent time, it follows that the originally unpolarized beam can be thought of as the superposition of two overlapping beams, each linearly polarized along one of the two mutually orthogonal directions. Note that this resolution of unpolarized light into two uncorrelated and linearly polarized components may be carried out along *any* two mutually perpendicular directions in the wavefront. It is for this reason that unpolarized light rays are represented, as in Figure 33-3c, by a ray with electric vectors oriented along two mutually perpendicular directions.

33-3   Circular and elliptical polarization

In order to describe states of polarization other than the above, a more quantitative characterization of unpolarized and polarized light is necessary. To this end, suppose that a monochromatic electromagnetic wave of wavelength λ is propagating along the z-axis of a certain coordinate system. As for the unpolarized wave in Figure 33-4, let us resolve the associated electric field in the wave into a vertical field $\mathbf{E}_1$ along the x-axis and a horizontal field $\mathbf{E}_2$ along the y-axis (see Figure 33-5). The total field is $(\mathbf{E}_1 + \mathbf{E}_2)$, where according to (29-20) through (29-23)

$$\mathbf{E}_1 = \mathbf{i}E_0^{(1)} \sin\left[\frac{2\pi}{\lambda}(z - ct)\right]$$

$$\mathbf{E}_2 = \mathbf{j}E_0^{(2)} \sin\left[\frac{2\pi}{\lambda}(z - ct) + \alpha\right]$$

(33-1)

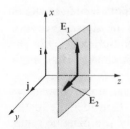

Figure 33-5

Here, $\mathbf{i}$ and $\mathbf{j}$ are unit vectors along the x- and y-axes, respectively, $E_0^{(1)}$ and $E_0^{(2)}$ are the amplitudes of the two component waves and α is the relative phase between them. The totality of all possible states of polarization of a light wave can be described completely in terms of these amplitudes $E_0^{(1)}$ and $E_0^{(2)}$ and the relative phase α.

If the two component waves in (33-1) are independent of each other, then the phase difference α between them must be *random*. In this case, the amplitudes $E_0^{(1)}$ and $E_0^{(2)}$ are generally equal to each other, and we say that the total wave described by $(\mathbf{E}_1 + \mathbf{E}_2)$ is *unpolarized*. If α is random but $E_0^{(1)} \neq E_0^{(2)}$, then the wave is said to be *partially linearly polarized*. Generally speaking, unless special pains are taken, the light emitted by a source will be unpolarized, since by symmetry $E_0^{(1)}$ and $E_0^{(2)}$ must be equal.

If the phase difference α between the electric fields in (33-1) vanishes, then the associated wave is plane-polarized. For in this case the total field $(\mathbf{E}_1 + \mathbf{E}_2)$ is

$$\mathbf{E}_1 + \mathbf{E}_2 = [\mathbf{i}E_0^{(1)} + \mathbf{j}E_0^{(2)}] \sin\left[\frac{2\pi}{\lambda}(z - ct)\right]$$

and since the direction of this field is constant in both space and time, it represents light that is linearly polarized along a certain direction. The angle θ between this polarization direction and the x-axis is easily found to be

$$\tan \theta = \frac{E_0^{(2)}}{E_0^{(1)}} \tag{33-2}$$

Besides unpolarized and linearly polarized light, (33-1) can also be used to characterize *circularly* and *ellipically* polarized light. If, for example, $E_0^{(1)} = E_0^{(2)} = E_0$ and $\alpha = -\pi/2$, then the sum of the fields in (33-1) is

$$\mathbf{E} = \mathbf{E}_1 + \mathbf{E}_2 = E_0\left\{\mathbf{i} \sin\left[\frac{2\pi}{\lambda}(z - ct)\right] - \mathbf{j} \cos\left[\frac{2\pi}{\lambda}(z - ct)\right]\right\} \tag{33-3}$$

since $\sin(\theta - \pi/2) = -\cos\theta$. To see the significance of this field, suppose that at a fixed instant t_0, the phase $2\pi(z - ct_0)/\lambda$ vanishes. It follows by substitution that at $t = t_0$, $\mathbf{E} = -\mathbf{j}E_0$. Similarly, a quarter of a period or $\lambda/4c$ later, the electric field will have the value $-\mathbf{i}E_0$ and at time $(t_0 + \lambda/2c)$ its value will be $+\mathbf{j}E_0$. Hence, as shown in Figure 33-6a, at a fixed point z, during the course of time this electric vector rotates clockwise in the x-y plane. Its tip therefore traces out in a period λ/c a circle of radius E_0 with uniform angular velocity. The radiation associated with this field is said to be *right circularly* polarized. Correspondingly, *left circularly polarized* light results if the tip of the electric vector traces out a circle in the opposite sense. The choice $E_0^{(1)} = E_0^{(2)} = E_0$; $\alpha = +\pi/2$ in (33-1) describes this type of polarized light. The term *circularly polarized* light, without qualification, is used very generally to describe either type. Note that in these definitions the direction—clockwise or counterclockwise—in which the tip of the electric vector travels is with reference to the observer toward whom the wave is advancing.

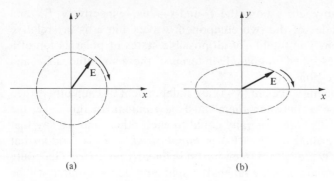

Figure 33-6

If the tip of the electric vector traces out an ellipse, then the associated light ray is said to be *elliptically polarized*; see Figure 33-6b. It is left as an exercise to confirm by use of arguments similar to those above that elliptically polarized light results if (1) $\alpha = \pm\pi/2$, but $E_0^{(1)} \neq E_0^{(2)}$; or (2) the phase angle α is not an integral multiple (including zero) of $\pi/2$. For example, the analogue of (33-3) for the case $\alpha = -\pi/2$ and $E_0^{(1)} \neq E_0^{(2)}$ is

$$\mathbf{E} = \mathbf{i}E_0^{(1)} \sin\left[\frac{2\pi}{\lambda}(z - ct)\right] - \mathbf{j}E_0^{(2)} \cos\left[\frac{2\pi}{\lambda}(z - ct)\right]$$

so that now at successive quarter-periods the electric vector assumes the values $-\mathbf{j}E_0^{(2)}$, $-\mathbf{i}E_0^{(1)}$, $+\mathbf{j}E_0^{(2)}$, and $+\mathbf{i}E_0^{(1)}$, respectively. Hence, assuming that $E_0^{(1)} > E_0^{(2)}$, the electric vector traces out clockwise an ellipse of semimajor and semiminor axes $E_0^{(1)}$ and $E_0^{(2)}$, respectively.

33-4 Polarization of light by reflection

For most of the remainder of this chapter we shall be concerned with methods for producing and detecting the various states of polarization defined above. One of the easiest methods to describe involves reflecting an unpolarized light beam at the interface between two dielectric media. This is the subject of this section.

Consider, in Figure 33-7a, a beam of *unpolarized* light incident, at an angle θ_1, onto the interface between, say, air and a medium of refractive index n. Part of the incident light will be reflected at the same angle θ_1 and the remainder will enter the medium at an angle θ_2 given by Snell's law:

$$\sin \theta_1 = n \sin \theta_2 \tag{33-4}$$

Now, although the amplitudes of the electric vectors—parallel and perpendicular to the plane of incidence—in the unpolarized incident ray are equal, this will *not*, in general, be true for the reflected and the refracted rays. The precise form of these amplitudes may be obtained by use of Fresnel's

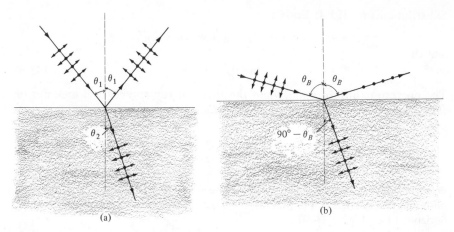

Figure 33-7

equations in (30-19) and (30-20). The first of these in (30-19) gives the amplitudes for the case of light polarized in the plane of incidence, and the second in (30-20) if the electric vector is perpendicular to the plane of incidence. It follows from these relations that the intensity of the reflected and the refracted light beams depends on the state of polarization of the incident ray as well as on the refractive index n and the angle of incidence θ_1. In general, for a given refractive index n the reflected and the refracted beams will be partially polarized, depending on the angle of incidence θ_1.

Of particular interest in this connection is a certain angle of incidence for which the reflected beam is totally linearly polarized along a direction perpendicular to the plane of incidence; see Figure 33-7b. The angle θ_B at which this occurs is known as *Brewster's angle* and may be thought of as that angle of incidence for which the angle between the reflected and the refracted rays is 90°. Reference to the first of Fresnel's equations in (30-19) shows that since tan 90° is infinite, at this angle θ_B, no light whose electric vector lies in the plane of incidence is reflected. Hence, if θ_1 has the value θ_B, so that $\theta_1 + \theta_2 = 90°$, then the electric vector in the reflected light must be perpendicular to the plane of incidence. Note that the transmitted beam need *not* be completely polarized even when the angle of incidence is θ_B. Reference to (30-20) shows that even at Brewster's angle, light which is plane-polarized perpendicular to the plane of incidence is partially reflected and partially transmitted. The above state of pure linear polarization of the reflected beam results only because there is no reflection of that part of the incident light which is polarized in the plane of incidence.

To obtain the relation between Brewster's angle θ_B and the refractive index n of the medium it is necessary to make use of Snell's law in (33-4). By definition of θ_B, the angle of refraction θ_2 associated with the angle of incidence $\theta_1 = \theta_B$ is

$$\theta_2 = 90° - \theta_B$$

Substitution into (33-4) leads to

$$\sin \theta_B = n \sin \theta_2 = n \sin (90° - \theta_B) = n \cos \theta_B$$

and thus

$$\tan \theta_B = n \qquad\qquad (33\text{-}5)$$

The corresponding formula when the incident ray travels in a medium of refractive index n_1 and strikes an interface with a medium of index n_2 is shown in the problems to be

$$\tan \theta_B = \frac{n_2}{n_1} \qquad\qquad (33\text{-}6)$$

Example 33-1   At what angle β above the horizon is the sun so that a person observing its rays reflected in water ($n = 1.33$) finds them linearly polarized along the horizontal (see Figure 33-8)?

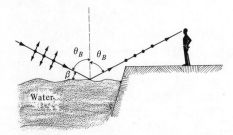

Figure 33-8

Solution   In order for the reflected rays to be linearly polarized, the angle of incidence must be Brewster's angle. Substituting the value $n = 1.33$ into (33-5), we find that

$$\tan \theta_B = 1.33$$

and thus $\theta_B \cong 53°$. The sun's angle β above the horizon is the complement of this angle. Hence

$$\beta = 37°$$

Example 33-2 *Linearly polarized* light is incident at Brewster's angle on the surface of a medium. What can you say about the refracted and reflected beams if for the incident beam:
   (a) The direction of polarization is parallel to the plane of incidence?
   (b) The direction of polarization is perpendicular to the plane of incidence?

Solution
   (a) At Brewster's angle the parallel component is completely refracted. Thus no light is reflected at all in this case.
   (b) Here as may be verified by use of (30-20) some of the incident light is reflected and some is refracted. Both the reflected and the refracted beams will be polarized perpendicular to the plane of incidence just as was the incident beam. The direction of polarization of a light beam is, of course, not altered on being reflected or refracted.

33-5 Dichroic crystals

Suppose that a *linearly polarized* light beam is incident on the interface of a refractive medium at Brewster's angle. If the direction of polarization lies in the plane of incidence, then according to the results of Section 33-4, none of the incident light is reflected; it is entirely refracted into the second medium. Moreover, if the reflecting surface is rotated about the direction of the incident ray as an axis and in a way so that the angle of incidence remains θ_B throughout, a steady variation in the intensity of the reflected light is observed. This intensity vanishes when the direction of polarization lies in the plane of incidence and is a maximum when it is at right angles to this plane. The results of Section 33-4 therefore can be used not only to devise methods for producing linearly polarized light but also for detecting the existence of- and the state of polarization of such light rays.

In practice it is found much more convenient not to use these properties of reflected light, but rather to produce and to detect polarized light by other means. Generally speaking, these alternate and more practical methods involve the usage of certain anisotropic crystals whose optical properties depend on the orientation of the electric vector in the light beam being transmitted. There are two types of crystals that we shall consider in this connection. The first type is known as *birefringent* or *doubly refracting* crystals, examples of which are calcite and quartz. These crystals will be discussed in Section 33-7. The purpose of this and the next section is to describe a second type, known as *dichroic crystals*. An example of such a crystal is the mineral tourmaline.

A dichroic crystal is one in which there exists a certain direction in the crystal, known as the *axis of easy transmission*, which has the following property: If a light beam goes through the crystal, then the component of the beam polarized along this axis is transmitted without essential modification. By contrast, a light beam that is polarized perpendicular to this direction undergoes absorption inside the crystal. Note that the axis of easy transmission is a *direction* in the crystal and not an axis through it. Regardless of the path followed by a light beam in traversing a crystal, at any point of this path that component of the light polarized perpendicular to the line through this point and parallel to the axis of easy transmission will be absorbed.

Consider, in Figure 33-9, a beam of unpolarized light incident normally on the face of a dichroic crystal whose axis of easy transmission is parallel to that face. If the incident light is resolved into one polarization component along, and the other at right angles to, the axis of easy transmission, it follows that only the former is transmitted. The component of the beam with the electric field vector perpendicular to the axis of easy transmission is absorbed. Therefore, assuming that the crystal is thick enough so that this perpendicular component is completely absorbed, it follows that the light transmitted through this dichroic crystal will emerge as linearly polarized light with the electric field along the axis of easy transmission.

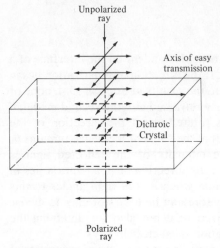

Figure 33-9

To illustrate the kind of physical processes involved in generating polarized light by use of dichroic crystals, let us consider briefly a certain substance developed by Edwin H. Land in 1938 and known as *Polaroid*. Polaroid normally comes in the form of thin plastic sheets and is made of a substance which contains long hydrocarbon chains to which are attached iodine atoms. In the manufacturing process, these thin sheets are stretched in a way so that these long molecular chains tend to line up in a parallel array along the direction of stretch. Associated with this regular structure are free electrons, which are not attached to any particular molecule and thus are free to travel, but only along the direction of alignment of the hydrocarbon chains. In particular they are *not* free to travel at right angles to this direction. If, now, as in Figure 33-10, a beam of unpolarized light is normally incident on such a sheet, these electrons can absorb energy only from that component of the radiation whose electric field is along the direction of the molecules. The electrons cannot move perpendicular to these chains, and hence cannot absorb the other component of the radiation. It follows that the component of the radiation polarized along the direction of the

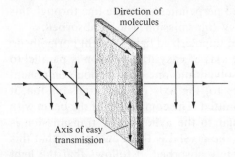

Figure 33-10

molecules will be absorbed and thus not transmitted. Therefore, as shown in Figure 33-10, the emerging light is linearly polarized in a direction perpendicular to that along which the molecules are aligned. The axis of easy transmission must therefore be perpendicular to this direction of the molecules.

33-6 Malus' law

Consider, in Figure 33-11, an observer O looking into a unpolarized light beam, which has been transmitted through two parallel Polaroid sheets with axes of easy transmission mutually perpendicular. The light transmitted through the first Polaroid will emerge linearly polarized along the vertical direction. Since the axis of the second Polaroid is horizontal, while the light incident on it is polarized along the vertical, it follows that the beam will be completely absorbed. Thus, the observer O will see no light! Experiment shows that he will actually see a small amount of blue light if it is present in the incident beam. This feature follows from the fact that Polaroid is not completely effective at all wavelengths.

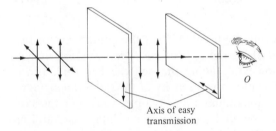

Axis of easy
transmission

Figure 33-11

Figure 33-12 shows a similar experiment, but this time with the axes of the parallel sheets of Polaroid at an angle θ relative to each other. Let us assume again that unpolarized light is incident from the left, and that the first Polaroid sheet, which is known as the *polarizer* when used in this way, has its axis of easy transmission at the angle θ with respect to the vertical. If the

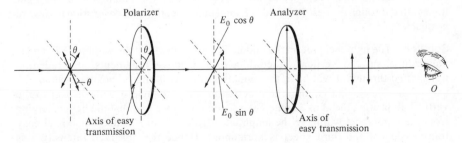

Figure 33-12

incident light is resolved into a component polarized along the axis of the polarizer, and a second at right angles to this axis, only the former will be transmitted. Hence the light emerging from the polarizer is polarized along a line that makes an angle θ with the vertical. Its intensity—neglecting any reflections or absorption—will be exactly half the incident intensity. The amplitude E_0 of the transmitted beam is the same as the amplitude of the corresponding component of the incident beam.

Let us now consider what happens to the beam on going through the second Polaroid sheet, or *analyzer*. The incident polarized light of amplitude E_0 may be thought of as the superposition of two polarized beams: one of amplitude $E_0 \cos \theta$ and polarized along the vertical; and the second of amplitude $E_0 \sin \theta$ polarized along the horizontal. Since the latter will be absorbed by the analyzer, it follows, as shown in the figure, that the transmitted beam will be polarized along the vertical and have the amplitude $E_0 \cos \theta$. According to (29-25) the intensity I of the transmitted beam is proportional to the square of this amplitude $E_0 \cos \theta$. For the same reason, the intensity I_0 of the radiation incident on the analyzer is proportional to $|E_0|^2$. Hence follows the basic relation

$$I = I_0 \cos^2 \theta \qquad (33\text{-}7)$$

which is known as *Malus' law*.

In deriving (33-7) we have explicitly assumed that the component of the incident beam that is polarized along the axis of easy transmission undergoes no reflection or absorption while going through the Polaroid. If we interpret the quantity I_0 in (33-7) not as the intensity of the radiation incident on the analyzer but rather as the maximum intensity observed when the axis of the polarizer and analyzer are parallel, then (33-7) is valid in general. For example, it is applicable if in place of two Polaroid sheets, we use two reflecting surfaces and reflect the light at Brewster's angle in each case. The term "Malus' law" is usually reserved for (33-7) when I_0 has this more general meaning.

Example 33-3 If a third Polaroid sheet with its axis at an angle θ with respect to the vertical is inserted between the two Polaroid sheets in Figure 33-11, what intensity will the observer O see now? Assume that the amplitude of the light transmitted through the first Polaroid is E_0 and that reflection and absorption losses are negligible.

Solution The light incident on the middle Polaroid sheet has the amplitude E_0 and is polarized along the vertical. According to the above arguments, the radiation transmitted through it will have the amplitude $E_0 \cos \theta$ and be polarized along the axis of this sheet, that is, along a direction making an angle θ with respect to the vertical. Repeating the above argument a second time, we find that the light emerging from the third sheet will have the amplitude $(E_0 \cos \theta) \sin \theta$ (since the axis of easy transmission of the third sheet is horizontal). Hence the observed intensity I is proportional to $[\cos \theta \sin \theta]^2 = (\sin^2 2\theta)/4$.

By contrast to this very striking result, whose validity is easily confirmed experimentally, if the added Polaroid sheet is placed in front of or behind the two Polaroid sheets in Figure 33-11 the observer will see no light for any value of the angle θ. Can you explain this physically?

Example 33-4   Suppose a beam of circularly polarized light is normally incident on a thin Polaroid sheet.

(a) What is the nature of the transmitted light?

(b) What happens to the observed intensity if the Polaroid sheet is rotated about the direction of the beam?

Solution

(a) Circularly polarized light, as we have seen, is the superposition of two light beams that are polarized along two orthogonal directions and have a phase difference of 90° between them. It follows that, just as for linearly polarized light, if such a beam goes through a Polaroid sheet, one of its two components will be absorbed and the other transmitted. Hence the emerging beam will be linearly polarized in a direction parallel to the axis of easy transmission of the sheet.

(b) Since the electric field vectors in two components of a circularly polarized beam have the same amplitudes and the phase difference of $\pm 90°$, it follows that the transmitted intensity is the same regardless of the orientation of the axis of easy transmission. Hence no variation in intensity would be observed as the Polaroid sheet is rotated. This is to be contrasted with the case of elliptically polarized light, for which there would be a variation in intensity as the sheet is rotated. Why?

33-7   Double refraction

The optical properties of a transparent, homogeneous, and isotropic medium, such as glass, are determined exclusively by its refractive index n. Thus a knowledge of n and its possible variation with wavelength makes possible—by use of Snell's law in (30-9) and Fresnel's equations in (30-19) and (30-20)—the calculation of the intensities and the directions of travel of light reflected and refracted at the surface of such a medium. Moreover, the velocity of light, c/n in such a medium, has the same value regardless of the direction of propagation. Of particular significance in this connection is the fact that this speed is the same for all states of polarization of the light beam.

To understand the physical basis for this independence of the speed of light in an isotropic medium to the direction of the electric field, let us recall (30-7):

$$n = \sqrt{\kappa} \tag{33-8}$$

which relates the refractive index of the medium to its dielectric constant. For a given electric field, as we saw in Chapter 22, the constituent dipoles of the medium tend to line up *along* the direction of the field. The extent to which this alignment takes place is measured by the strength of the dielectric constant κ in accordance with (22-29). Since, for a homogeneous and isotropic medium, κ is the same everywhere regardless of the direction or the strength of **E**, it follows from (33-8) that the velocity of a light wave in such a medium is also independent of the electric field.

There is another class of materials of considerable interest, which are also macroscopically homogeneous, but are not optically isotropic. These anisotropic crystalline materials are called *birefringent* or *doubly refracting*. Mainly for reasons of simplicity, in the following we shall be concerned only with the particular class of these materials known as *uniaxial* crystals. Some examples of uniaxial crystals are calcite ($CaCO_3$) and quartz (SiO_2). The significance of the term *uniaxial* will be clarified below.

Consider a light wave traversing a birefringent crystal. Experiment shows that its speed depends, in general, on both the direction of propagation of the wave and on its state of polarization. Most striking is the fact that on entering the crystal, an initially unpolarized light beam will split into two linearly polarized and distinct beams. One of these is called the *ordinary ray* or *o-ray* and the other the *extraordinary ray* or *e-ray*. The behavior of the *o*-ray is essentially the same as that of a light wave in an isotropic medium. For example, on entering the birefringent crystal the *o*-ray is refracted in accordance with Snell's law. Also the refractive index n_o of the *o*-ray as well as its speed c/n_o in the crystal is the same regardless of the direction of travel of the *o*-ray. By contrast, the *e*-ray behaves in a most unusual way. Its velocity of propagation is different for different directions in the crystal! Moreover, its direction of travel, on being refracted into the crystal, is *not* generally consistent with Snell's law. It is very convenient nevertheless to associate a certain refractive index n_e with the *e*-ray. If $n_e > n_o$ the crystal is said to be *positive*, whereas if $n_e < n_o$ the crystal is said to be *negative*. As will be seen in Section 33-8, where a precise definition of n_e is given, for a positive crystal, such as quartz, c/n_e is the smallest value for the velocity of the *e*-ray in the crystal, while for a negative one, such as calcite, c/n_e is the maximum speed of the *e*-ray. The speed of the *o*-ray in a positive crystal always exceeds that of the *e*-ray, whereas for a negative crystal the speed of the *e*-ray is never less than that of the *o*-ray. Depending on its direction of travel, the velocity of the *e*-ray in a positive crystal varies from a maximum value of c/n_o to a minimum of c/n_e, and for a negative crystal the velocity ranges from a minimum of c/n_o to a maximum of c/n_e. Note that the velocity of the *o*-ray is the same in all directions in the crystal.

Table 33-1 lists values for n_o and n_e for a number of positive (p) and

Table 33-1   Indices of refraction of birefringent crystals at 5893 Å

Substance	n_o	n_e
(n) Calcite ($CaCO_3$)	1.658	1.486
(n) Sodium nitrate ($NaNO_3$)	1.587	1.336
(p) Zinc sulfide (ZnS)	2.356	2.378
(p) Quartz (SiO_2)	1.544	1.553
(p) Ice (H_2O)	1.309	1.313

negative (n) crystals. Since for ice the difference $(n_e - n_o)$ is very small, care must be exercised to see double refraction by use of this substance.

Figure 33-13 shows an experiment that demonstrates double refraction. An unpolarized light beam is normally incident on one side of a doubly refracting medium with parallel opposite faces. At the point of entry, the beam splits into two rays: an o-ray and an e-ray. The former obeys the laws of geometrical optics and thus is transmitted without deviation. By contrast, the e-ray is bent away from the normal when entering the crystal and toward the normal on leaving. It thus emerges in a direction parallel to and displaced from the incident beam. Of considerable interest is the fact that the emerging rays are linearly polarized along mutually perpendicular directions. This feature, which will be examined in more detail in Section 33-8, is the reason that doubly refracting crystals are so useful in a study of polarized light.

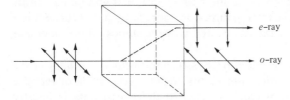

e-ray

o-ray

Figure 33-13

33-8 Polarization in birefringent crystals

To obtain a physical understanding of this distinctive behavior of doubly refracting crystals it is convenient to make use of the same physical ideas used to explain the optical behavior of isotropic materials. Since for isotropic substances, the constituent dipoles always tend to line up along the direction of the external field, the induced dipole moment per unit volume **P** is invariably parallel to **E**. By contrast, in a doubly refracting crystal, **P** does *not* generally lie along the direction of the electric field. Hence some modification of (22-29) and (22-30) is required in order to describe this unique behavior.

More detailed studies show that regardless of the complexity of a nonabsorbing crystal, there are always three mutually perpendicular directions in a crystal along which **P** will be parallel to the electric field. These three directions are known as the *principal axes of the crystal* and a distinct dielectric constant is associated with each one. Let us define a Cartesian coordinate system with axes along the principal axes of the crystal and let κ_1, κ_2, and κ_3 be the respective dielectric constants associated with the x-, y-, and z-axes; see Figure 33-14. If, for example, **E** is an electric field along the x-axis, the associated dipole moment **P** will be given by (22-29) as $\epsilon_0(\kappa_1 - 1)$**E** with dielectric constant κ_1. Similarly, if **E** is along the y-axis, the appropriate dielectric constant is κ_2. The significance of the fact that these dielectric

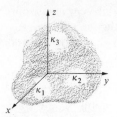

Figure 33-14

constants are different is that the respective indices of refraction n_1, n_2, and n_3 (in accordance with (33-8)) will also be different. Thus if a linearly polarized electromagnetic wave with its electric vector along the x-axis travels through the crystal, its speed will be $c/n_1 = c/\sqrt{\kappa_1}$. Similarly, if the wave is polarized along the y-axis or the z-axis, its speed will be $c/\sqrt{\kappa_2}$ or $c/\sqrt{\kappa_3}$, respectively. For polarization directions not along one of these three principal axes the velocity of the associated wave assumes an intermediate value.

The important feature to be noted here is that the *velocity of light in the crystal is determined by the direction of the electric field in the wave*; the propagation direction of the wave is related to the speed only indirectly through its relation to the direction of polarization. It is for this reason that on entering a birefringent crystal, an originally unpolarized beam will split into an *o*-ray and an *e*-ray, each characterized by its distinct state of polarization.

A crystal for which no two of the principal dielectric constants κ_1, κ_2, and κ_3 are equal is called a *biaxial crystal*. Feldspar, gypsum, and mica are examples of such crystals. The name *biaxial* has to do with the fact that for these crystals there are two distinct directions in the crystal along which the *o*-ray and the *e*-ray can travel at the same speed. These directions, which are known as the *optic axes* of the crystal, do *not* lie along any of the three principal axes.

In the following we shall confine our attention to a simpler class of crystals known as *uniaxial*, of which the substances listed in Table 33-1 are members. A uniaxial crystal is one for which two of the dielectric constants are the same. As the name implies, a uniaxial crystal has only one *optic axis* and this lies along that principal axis with the "unequal" dielectric constant. As for biaxial crystals, along the direction of the optic axis the speeds of the *o*-ray and the *e*-ray are precisely the same. Just as for the axis of easy transmission in dichroic crystals, the optic axis is a *direction* through the crystal and not an axis. Only along the unique *direction* of the optic axis is there no distinction between the *o*-ray and the *e*-ray.

Continuing our analysis of uniaxial crystals, let us assume to be specific that $\kappa_1 = \kappa_2 \equiv \kappa_0 \neq \kappa_3$ in Figure 33-14. The fact that the optic axis then lies along the z-axis in the crystal can be seen in the following way. The electric field associated with a wave traveling along the z-axis must lie in the x-y

plane. Hence, since $\kappa_1 = \kappa_2 = \kappa_0$, the velocity of this wave is $c/\sqrt{\kappa_0}$ and is independent of the state of polarization. We define the ordinary refractive index n_o of this crystal so that c/n_o is the velocity of a light wave traveling along the optic axis. According to (33-8), we have thus

$$n_o = \sqrt{\kappa_0} \qquad (33\text{-}9)$$

Consider now an unpolarized electromagnetic wave traveling along the x-axis of this same crystal. This wave may be thought of as the superposition of two components: one linearly polarized along the y-axis and the other along the z-axis. The dielectric constant associated with the former is κ_0, so this component travels at the velocity c/n_o in accordance with (33-9). It is the ordinary ray, or o-ray. The direction of polarization of this o-ray is *perpendicular to the plane defined by the propagation direction and the optic axis.* By contrast, the other component, which is polarized along the z-axis, travels at the velocity corresponding to the dielectric constant κ_3. It is the extraordinary ray, or e-ray, and the associated refractive index n_e is

$$n_e = \sqrt{\kappa_3} \qquad (33\text{-}10)$$

The direction of polarization of the e-ray is *in the plane determined by the optic axis and the direction of travel of the ray.* This latter feature, and the corresponding one stated above for the polarization direction of the o-ray, are very generally true; their validity is not restricted, as here, to light traveling along the principal axes of the crystal.

It is important to note that regardless of the direction of travel of an initially unpolarized light beam through a uniaxial crystal, there must be associated with it a component of the electric field along the x-axis or the y-axis. Hence any ray traveling through a uniaxial crystal will have an o-ray component whose velocity c/n_o is independent of the direction of travel. Unless the direction of the beam is along the optic axis itself, an e-ray will also be associated with it. However, the velocity of the e-ray varies with the direction in the crystal.

To determine the velocity of the e-ray along an arbitrary direction in the crystal it is necessary to carry out studies analogous to those described in Chapter 29. The results may be summarized in the following way. Imagine a pulse of light generated at some point O in a uniaxial crystal, and let u_x, u_y, and u_z be the components along the principal axes of the velocity $\mathbf{u}$ of one of these rays. Since the o-ray goes out with the same speed c/n_o in all directions, it follows that 1 second later the wavefront associated with the ordinary ray will be at the distance $|\mathbf{u}|$ from O, where

$$u_x{}^2 + u_y{}^2 + u_z{}^2 = \frac{c^2}{n_o{}^2} \qquad (o\text{-ray}) \qquad (33\text{-}11)$$

If we think of a set of "velocity axes" with directions parallel to the principal axes of the crystal, (33-11) represents a sphere of radius c/n_o centered at the origin of this system. The velocity of an o-ray in any direction is the vector along that direction from the origin to the point of intersection with the surface. Correspondingly, the surface associated with the e-ray is the ellipsoid of revolution

$$\frac{u_x^2}{c^2/n_e^2} + \frac{u_y^2}{c^2/n_e^2} + \frac{u_z^2}{c^2/n_o^2} = 1 \quad (e\text{-ray}) \tag{33-12}$$

and again the velocity of an e-ray in any direction is given by a vector along that direction from the origin to the point of its intersection with the ellipsoid. For an e-ray traveling along the z-axis, for example, we find by setting $u_x = u_y = 0$ in (33-12) that $u_z = c/n_0$, thus confirming that along the optic axis the e-ray and the o-ray travel at the same speed. Similarly, for an e-ray traveling along the x-axis, (33-12) predicts a speed c/n_e. Moreover, by use of (33-12) it is also possible to calculate the velocity of an e-ray in a direction other than a principal axis. The result, as shown in the problems, is that the speed u of an e-ray traveling along a direction making an angle θ with optic axis is

$$u = \frac{c}{[n_o^2 \cos^2 \theta + n_e^2 \sin^2 \theta]^{1/2}} \tag{33-13}$$

Figure 33-15 shows a cross section (of a plane containing the optic axis) through the surfaces defined by (33-11) and (33-12). In each case the optic axis is the line AB. Figure 33-15a shows the case of a positive crystal for which the e-ray ellipsoid is inside the o-ray sphere. Note the directions of polarization associated with a typical o-ray and e-ray. Figure 33-15b shows the corresponding situation for a negative crystal.

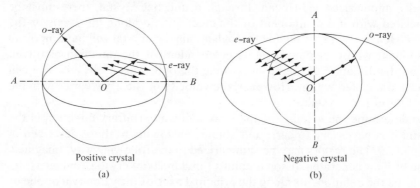

Positive crystal   Negative crystal

(a)   (b)

Figure 33-15

Figure 33-16 shows the formation of wavefronts in a negative uniaxial crystal as a result of unpolarized light being normally incident on a crystal face. Figures 33-16a and b consider the case when the optic axis is, respectively, parallel and perpendicular to the refracting surface. Figure 33-16c shows the corresponding case when the optic axis is perpendicular to the plane of the drawing. Finally, in Figure 33-16d we consider the more general case, where the optic axis makes an arbitrary angle with the surface.

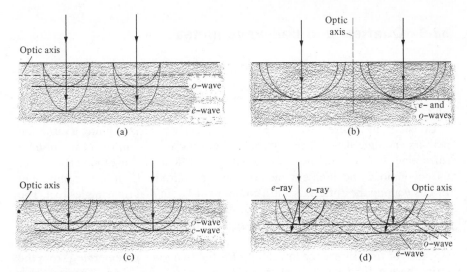

Figure 33-16

As shown by the arrows to the respective *e*- and *o*-wavefronts, on entering the crystal the two rays split up. This is the situation envisioned in Figure 33-13.

Example 33-5 Calculate the velocity in a calcite crystal of a light ray traveling at an angle of 45° with respect to the optic axis for:
(a) The *o*-ray.
(b) The *e*-ray.

Solution
(a) The velocity **u** of the *o*-ray is the same in all directions. Hence, by use of Table 33-1,

$$u = \frac{c}{n_o} = \frac{3.0 \times 10^8 \text{ m/s}}{1.658} = 1.8 \times 10^8 \text{ m/s}$$

(b) Because of the symmetry under rotations about the optic axis, we can assume that the *e*-ray lies in the *x*-*z* plane and makes equal angles with the *x*- and *z*-axes. Hence, setting $u_y = 0$ and $u_x = u_z$ in (33-12), we find that

$$u_x = \frac{c}{\sqrt{n_e^2 + n_o^2}} = \frac{3.0 \times 10^8 \text{ m/s}}{[(1.486)^2 + (1.658)^2]^{1/2}} = 1.35 \times 10^8 \text{ m/s}$$

Hence the speed *u* is

$$u = [u_x^2 + u_z^2]^{1/2} = \sqrt{2}u_x = 1.9 \times 10^8 \text{ m/s}$$

The same result can also be obtained by substitution into (33-13).

33-9   Quarter- and half-wave plates

One application of doubly refracting crystals is to the production and detection of circularly and elliptically polarized light. The purpose of this section is to describe a particular optical device known as a *quarter-wave plate*, which is very useful for this purpose.

Consider, in Figure 33-17, a light wave linearly polarized along a direction making an angle θ with the vertical and normally incident on a thin, doubly refracting sheet of thickness d and with the optic axis along the vertical. Let us decompose the incident electric field amplitude E_0 into a component $E_0 \cos \theta$ along the optic axis and a component $E_0 \sin \theta$ at right angles to it. The component along the optic axis is destined to become the amplitude of the *e*-ray and travels through the crystal at the speed c/n_e, and the other component is the amplitude of the *o*-ray which travels at the speed c/n_o. Because of this difference in velocity for the two rays, on emerging from the strip, there will be a certain phase difference δ between them. The emerging ray will in general no longer be plane-polarized as it was prior to entering the strip.

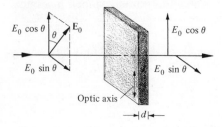

Figure 33-17

Now assuming that the incident wave is monochromatic with free-space wavelength λ, it follows from (29-24) that the wavelength of the *o*-ray in the medium is λ/n_o while that for the *e*-ray is λ/n_e. The phase difference δ that arises between these rays as a result of this wavelength difference may be obtained, by use of (32-7), as

$$\delta = \frac{2\pi}{\lambda/n_o} d - \frac{2\pi}{\lambda/n_e} d$$

so that

$$\delta = \frac{2\pi}{\lambda} d(n_o - n_e) \tag{33-14}$$

Thus, assuming no absorption, the light ray that emerges from the biefringent plate in Figure 33-17 consists of two linearly polarized components, of the respective amplitudes $E_0 \cos \theta$ and $E_0 \sin \theta$, and with the phase difference δ. Comparing this with the definitions of linearly, circularly, and elliptically polarized light in Section 33-3, we may draw the following conclusions:

1. If the thickness d of the plates is such that $\delta = \pm\pi/2$, and $\theta = 45°$ (so the amplitudes $E_0 \cos\theta$ and $E_0 \sin\theta$ are equal), then the emerging beam in Figure 33-17 is circularly polarized.
2. If the thickness d is such that $\delta = \pm\pi$, then the emerging beam is linearly polarized in a direction making an angle 2θ with respect to the incident direction; that is, the direction of polarization is rotated by an angle 2θ in this case.
3. If the thickness d is such that δ has a value other than an odd or even integral multiple of $\pi/2$, then the emerging beam is elliptically polarized.

The details of the arguments justifying these conclusions are found in the problems.

A *quarter-wave plate* is defined to be one for which the thickness d in (33-14) corresponds to a phase difference δ of $\pm 90°$. The importance of the quarter-wave plate is that it can be used as a tool to convert plane-polarized light to circularly polarized light, and conversely. If the direction between the optic axis of the quarter-wave plate and the direction of polarization of the incident radiation is $45°$, the emerging ray will be circularly polarized. At other angles the emerging beam will be elliptically polarized.

A related optical device is the *half-wave plate*. It differs from a quarter-wave plate in that its thickness d is such that the phase difference δ between the emerging rays is $180°$. The thickness of certain commercially available cellophane happens to be such that for certain wavelengths it is a reasonable approximation to a half-wave plate.

It should be noted that the phase difference δ between the emerging rays in Figure 33-17 depends on wavelength. Hence, in general, a given *quarter-wave plate* is useful only for an appropriately chosen color. For a given birefringent material, quarter-wave plates designed for different colors will have different thicknesses, according to (33-14).

Example 33-6 Describe the nature of the light transmitted through a quarter-wave plate if the incident light is circularly polarized.

Solution In accordance with (33-3), circularly polarized light is the superposition of two waves of equal amplitude, linearly polarized along perpendicular directions and with a phase difference of $90°$. As the light passes through the quarter-wave plate, this phase difference becomes zero or $180°$ since the effect of the plate is to change one of the trigonometric functions in (33-3) to the other or to its negative. In either event, the emerging beam comes out linearly polarized along a direction making an angle of $45°$ with the optic axis. If viewed through a sheet of Polaroid, the emerging light will vary in intensity from zero to some maximum as the quarter-wave plate is rotated about the direction of the beam. This procedure is therefore a definite test for circularly polarized light.

Example 33-7   Consider, in Figure 33-18, an optical system consisting of a half-wave plate between two sheets of Polaroid whose axes of easy transmission are perpendicular. Assuming that the incident beam is unpolarized, what variation in intensity is observed by the observer O as the half-wave plate is rotated about the axis of the beam?

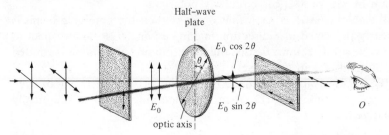

Figure 33-18

Solution   The light ray emerging from the first Polaroid sheet is linearly polarized along the vertical. Thus if it were not for the *half-wave* plate, the observer at O would see nothing. However, as shown in the figure, the half-wave plate—assuming that its optic axis makes an angle θ with the vertical—has the effect of rotating the electric vector by an angle 2θ. Hence the beam emerging from the half-wave plate may be thought of as the superposition of two linearly polarized waves: one of amplitude $E_0 \cos 2\theta$ and polarized along the vertical, and the other at right angles with amplitude $E_0 \sin 2\theta$. The observer at O will see the horizontally polarized component with intensity proportional to $\sin^2 2\theta$.

As the half-wave plate is now rotated about the direction of the beam, the observed intensity will vary with θ as $\sin^2 2\theta$. It thus vanishes when the optic axis of the half-wave plate is parallel to the axis of easy transmission of the first Polaroid sheet, rises to a maximum when this angle is 45°, and vanishes when it is 90°, and so forth.

33-10   Summary of important formulas

If an electromagnetic wave of wavelength λ is traveling in free space, then the total electric field $(\mathbf{E}_1 + \mathbf{E}_2)$ in the advancing wavefront may be written

$$\mathbf{E}_1 = \mathbf{i}E_0^{(1)} \sin\left[\frac{2\pi}{\lambda}(z - ct)\right]$$

$$\mathbf{E}_2 = \mathbf{j}E_0^{(2)} \sin\left[\frac{2\pi}{\lambda}(z - ct) + \alpha\right]$$

(33-1)

where $E_0^{(1)}$ and $E_0^{(2)}$ are the amplitudes and α is a phase angle; see Figure 33-5. If α is a random variable, then the radiation is said to be *unpolarized*. If $\alpha = 0$ or π, or one of $E_0^{(1)}$ or $E_0^{(2)}$ vanishes, then it is said to be *plane* or *linearly polarized* along the direction of the total field. If $\alpha = \pm \pi/2$ and $E_0^{(1)} = E_0^{(2)}$, then we say that the wave is *circularly* polarized. For other values of α the light is characterized as being *elliptically* polarized.

Suppose that a beam of unpolarized light that is originally traveling in air strikes the interface of a medium of refractive index n. If the angle of incidence is *Brewster's angle* θ_B defined by

$$\tan \theta_B = n \qquad (33\text{-}5)$$

then all of the reflected light is plane-polarized perpendicular to the plane of incidence.

The intensity I of an originally unpolarized light beam after being transmitted through a polarizer and an analyzer with axes at an angle θ relative to each other is

$$I = I_0 \cos^2 \theta \qquad (33\text{-}7)$$

where I_0 is the maximum intensity observed for parallel axes. This relation is known as *Malus' law.*

If a linearly polarized wave is normally incident on a doubly refracting plate with optic axis as shown in Figure 33-17, the phase shift δ between the emerging components polarized along and perpendicular to the optic axis is

$$\delta = \frac{2\pi}{\lambda} d(n_o - n_e) \qquad (33\text{-}14)$$

Here, λ is the free-space wavelength, n_o and n_e are the ordinary and extraordinary refractive indices, and d is the thickness of the plate.

QUESTIONS

1. Define or describe briefly what is meant by the following terms: (a) polarized light; (b) birefringence; (c) Brewster's angle; (d) uniaxial crystal; and (e) dichroic crystal.

2. Define or describe briefly what is meant by the following terms: (a) optic axis; (b) axis of easy transmission; (c) principal axes; (d) quarter-wave plate; and (e) extraordinary ray.

3. Contrast and compare the following pairs of light rays: (a) linearly polarized–unpolarized; (b) circularly polarized–linearly polarized; and (c) circularly polarized–elliptically polarized.

4. If you look at the sky on a clear, sunny day through a Polaroid sheet, you will observe a variation in intensity as the Polaroid is rotated about an axis along the line of sight. What conclusions can you draw about the state of polarization of this light? How could you resolve any ambiguities?

5. Describe in physical terms the mechanism underlying the operation of a dichroic material such as Polaroid.

6. What physical characteristics of a doubly refracting crystal, such as quartz, bring about its unusual optical behavior? Does this mean that ordinary commercial cellophane, which is also birefringent, is crystalline? Explain.

7. Unpolarized light is normally incident on a thin sheet of material. Describe the state of polarization of the emerging beam if the sheet is: (a) Polaroid; (b) birefringent with the optic axis in the plane of the sheet; (c) a quarter-wave plate.

8. Would you expect that the axis of easy transmission for the lenses of Polaroid sunglasses to be vertical or horizontal when worn by a person? (*Hint*: How could you eliminate some of the rays reflected from water (refer to Figure 33-8)?)

9. Two parallel Polaroid strips have their axes at right angles so that no light comes through the combination. Suppose that a third Polaroid strip is placed between these two. Explain in detail why light may now get through this combination of three Polaroid strips.

10. *Unpolarized* light is transmitted through a quarter-wave plate. What is the state of polarization of the emerging beam? Would your answer also hold for a half-wave plate?

11. What is the final polarization state of an originally right circularly polarized light beam after going through a half-wave plate? Justify your answer.

12. Linearly polarized light is viewed through a quarter-wave plate. What is the nature of the observed beam? Does the intensity vary as the quarter-wave plate is rotated about the direction of the incident beam? Explain.

13. What are the largest and smallest values for the speed of the *e*-ray in (**a**) quartz? (**b**) calcite? Use the data in Table 33-1 and Figure 33-15.

14. Which travels faster in a positive crystal, the *o*-ray or the *e*-ray? In a negative crystal?

15. An *o*-ray travels along the *x*-axis of a coordinate system fixed relative to a uniaxial crystal whose optic axis is along the *y*-axis. What is its direction

of polarization? What would be the direction of polarization if it were an *e*-ray?

16. What is the final state of polarization of an initially plane-polarized light beam after being transmitted successively through a quarter-wave plate and then a half-wave plate? Does the angle between the two optic axes play a role in your answer? Explain.

17. In Figure 33-13 determine the general direction of the optic axis in the crystal. (*Hint*: Refer to Figure 33-16d.)

18. Elliptically polarized light is viewed through Polaroid. What is the state of polarization of the transmitted light? Is any variation in intensity observed as the Polaroid is rotated about the beam direction?

19. For the situation in Example 33-4 suppose that a quarter-wave plate is placed in front of the Polaroid. Describe the nature of the variation in intensity that is now observed as the Polaroid sheet alone is rotated about the beam axis. If the quarter-wave plate and the Polaroid sheet are rotated in unison, what is the observed variation in intensity?

20. Repeat Question 19, but suppose this time that the incident light is elliptically polarized.

21. Describe in what ways two quarter-wave plates, one made of calcite and one made of quartz, differ. What is the distinction between the states of polarization of linearly polarized light transmitted through each plate?

22. Describe the significance and usefulness of the quarter-wave and half-wave plates for producing and analyzing polarized light.

PROBLEMS

Unless a statement is made explicitly to the contrary, assume in the following that the component of a light beam with the E-vector along the axis of easy transmission undergoes no losses on traversing a sheet of Polaroid.

1. Calculate Brewster's angle for heavy flint glass, which has a refractive index of 1.65 for the sodium line at $\lambda = 5893$ Å.

2. At an angle of incidence of 63°, it is found that the light reflected from a certain transparent substance is plane-polarized perpendicular to the plane of incidence. What is the refractive index for the given substance?

3. An unpolarized light beam traveling in a medium of refractive index n_1 strikes the interface with a second medium of index n_2. Show that Brewster's angle θ_B for this case is

$$\tan \theta_B = \frac{n_2}{n_1}$$

4. If θ_c is the critical angle for a certain refracting medium when in air, and if θ_B is the value for Brewster's angle for the same medium in air, show that

$$\cot \theta_B = \sin \theta_c$$

5. A monochromatic unpolarized light beam of intensity of 0.2 W/m^2 is normally incident onto a Polaroid sheet. (a) What is the intensity of the transmitted light? (b) What is the intensity observed if this beam goes through a second Polaroid sheet with its axis of easy transmission at an angle of 30° with respect to the first?

6. Let I_0 be the intensity observed for a light beam transmitted through two Polaroid sheets with parallel axes of easy transmission. If a third sheet is put between the original two in such a way that the observed intensity is reduced to $0.3I_0$, at what angle is the axis of easy transmission of the third sheet relative to that of the others?

7. A monochromatic wave of wavelength λ travels along the positive z-axis of a certain coordinate sys-

tem. Identify the state of polarization in each case if the electric field in the wave is given by:

(a) $\mathbf{E} = \mathbf{i}E_0 \sin \dfrac{2\pi}{\lambda}(z - ct)$

(b) $\mathbf{E} = \mathbf{i}E_0 \sin \dfrac{2\pi}{\lambda}(z - ct)$

$\qquad + 2\mathbf{j}E_0 \sin \dfrac{2\pi}{\lambda}(z - ct)$

(c) $\mathbf{E} = \mathbf{i}E_0 \sin \dfrac{2\pi}{\lambda}(z - ct)$

$\qquad + \mathbf{j}E_0 \cos \dfrac{2\pi}{\lambda}(z - ct)$

(d) $\mathbf{E} = \mathbf{i}E_0 \sin \dfrac{2\pi}{\lambda}(z - ct)$

$\qquad - 2\mathbf{j}E_0 \cos \dfrac{2\pi}{\lambda}(z - ct)$

(e) $\mathbf{E} = \mathbf{i}E_0 \sin \dfrac{2\pi}{\lambda}(z - ct)$

$\qquad + \mathbf{j}E_0 \sin \left[\dfrac{2\pi}{\lambda}(z - ct) + \dfrac{\pi}{3}\right]$

8. Consider a rotation of the coordinate system in Problem 7 by an angle α about the z-axis. The unit vecors $\mathbf{i}'$ and $\mathbf{j}'$ along the *new* axes are related to the old ones by

$$\mathbf{i}' = \mathbf{i} \cos \alpha + \mathbf{j} \sin \alpha$$
$$\mathbf{j}' = -\mathbf{i} \sin \alpha + \mathbf{j} \cos \alpha$$

(a) Express the electric fields in (a) and (b) of Problem 7 in terms of these unit vectors.

(b) Prove that the field in (c) of Problem 7 represents circularly polarized light in terms of the new axes for any choice for the angle α.

9. An unpolarized beam of light is incident on three parallel sheets of Polaroid, each of whose axes is at an angle of 45° with respect to that of its predecessor.

(a) What fraction of the incident radiation intensity comes through?

(b) If the middle Polaroid sheet is removed, what fraction of the incident intensity comes through?

(c) Compare your answers to (a) and (b), and explain them in physical terms.

10. Three strips of Polaroid are parallel and lined up so that the maximum intensity I_0 is transmitted.

(a) If the middle one is rotated by 30°, what is the intensity now?

(b) If in the initial situation, the last one is rotated by 30°, what intensity gets through?

(c) What is the intensity if the middle and the last Polaroid strip are rotated simultaneously by 45° in the same direction?

11. By starting with (33-3) and an analogous formula for left circularly polarized light (with the same amplitude and phase) prove that a plane-polarized light beam may be thought of as the superposition of two circularly polarized light beams of the same amplitude and phase, one polarized to the left and the other to the right.

12. Imagine viewing, through a Polaroid sheet, an incident light beam whose electric vector is characterized by the electric field in Problem 7b. What is the ratio of the minimum to the maximum transmitted intensity observed as the Polaroid is rotated about the beam axis?

13. Repeat Problem 12, but this time suppose the electric field in the incident beam is that given in Problem 7c. Explain why your answer *must* differ from that in Problem 12.

*14. Suppose that in Figure 33-16c the unpolarized light beam were incident at an angle θ. Assuming the optic axis of the crystal to be perpendicular to the plane of incidence:

(a) Why is Snell's law applicable to both the o-ray and the e-ray in this case?

(b) Assuming further that the crystal is calcite, and that $\theta = 30°$, use the data in Table 33-1 to calculate the angular separation between the two rays in the crystal.

15. Making use of (33-13), calculate by use of the data in Table 33-1 the velocity of an e-ray traveling at an angle of 30° with respect to the optic axis in (a) calcite and (b) quartz.

16. Show by use of (33-12) that the speed u of an e-ray which travels in a uniaxial crystal along a direction making an angle θ with the optic axis is

$$u = \frac{c}{\sqrt{n_o^2 \cos^2 \theta + n_e^2 \sin^2 \theta}}$$

17. Making use of the result of Problem 16, find the maximum and minimum values for the velocity of an e-ray in (a) calcite and (b) quartz.

18. Figure 33-19 shows an unpolarized light beam normally incident on what is known as a "Rochon" prism made of quartz. As shown, the light on entering travels along the optic axis of the first prism and then is doubly refracted at the interface with the second prism, whose optic axis is perpendicular to the plane of incidence.

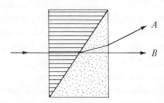

Figure 33-19

(a) Why is ray B undeviated?

(b) Which of the two rays is the e-ray?

(c) Describe the state of polarization of the two emerging rays.

*19. In Figure 33-19, if the incident ray strikes the interface between the two prisms at an angle of incidence of 30°, calculate the angular separation between the two emerging rays. Assume that both prisms are made of quartz and use the data in Table 33-1.

*20. Figure 33-20 shows an unpolarized light beam normally incident on a "Wollaston" prism. As shown, the incident beam on entering the prism first travels perpendicularly to the optic axis and then is doubly refracted at the interface with a second prism whose optic axis is perpendicular to the plane of incidence. Assuming that both prisms are made of quartz, and using the data in Table 33-1:

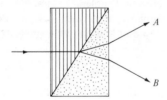

Figure 33-20

(a) Which of A and B is the o-ray and which is the e-ray?

(b) What is the direction of polarization of each of these rays?

(c) Assuming an angle of incidence of 30° at the interface, calculate the angular separation of the o-ray from the incident direction.

21. Repeat Problem 20, but suppose that this time the prisms are made of calcite.

22. (a) Show that the minimum thickness d_0 of a quarter-wave plate is given by

$$d_0 = \frac{\lambda}{4|n_e - n_o|}$$

where λ is the free-space wavelength of the light being used and n_e and n_o the two refractive indices.

(b) What other values can the thickness d have so that it is still a quarter-wave plate?

(c) Calculate the smallest thickness of a quarter-wave plate made of calcite for yellow light of wavelength 5893 Å.

23. Repeat all parts of Problem 22 for a half-wave plate.

24. By use of the data in Table 33-1 calculate the ratio of the thicknesses of two quarter-wave plates, one made of calcite and the other made of quartz. Assume the minimum thickness in each case.

25. A light beam, plane-polarized along the vertical, is transmitted through a quarter-wave plate with its optic axis at an angle θ with respect to the vertical. Assuming that $0 < \theta < 45°$:

(a) Show that the light is elliptically polarized.

(b) What is the spatial orientation of the major axis of the ellipse?

(c) Show that the ratio of the semimajor axis to the semiminor axis of the ellipse is $\cot \theta$.

26. Prove that if linearly polarized light is normally incident on a half-wave plate, and if the polarization direction makes an angle θ with the optic axis of the plate, then the plane of polarization of the transmitted light makes an angle 2θ with respect to direction of the incident polarization.

*27. Prove the following:

(a) If transmitted through Polaroid, circularly polarized light comes out plane-polarized.

(b) If transmitted through Polaroid,

elliptically   polarized   light
comes out plane-polarized.

(c) If the Polaroid in (a) is rotated
about the direction of the beam,
no variation in intensity is detected.

(d) If the Polaroid in (b) is rotated
about the direction of the beam,
there is a variation in intensity
but some light is always transmitted.

28. A quarter-wave plate is designed to
be used at a wavelength of 4000 Å.
Suppose that in an experiment
plane-polarized light of wavelength
5000 Å is transmitted through it. If θ
is the angle between the optic axis
of the plate and the polarization
direction of the incident beam, describe the polarization state of the
emerging light if: (a) $\theta = 0$; (b) $\theta =$
90°; (c) $\theta = 45°$. Is there any value
for θ for which the light is circularly polarized?

29. For the physical situation in Figure
33-18, suppose that the axes of both
Polaroid sheets are vertical. What is
the variation of the transmitted intensity observed in terms of θ as

the half-wave plate is rotated about
the direction of the beam?

30. If the axis of the polaroid strip next
to the observer in Figure 33-18 is
rotated so that its optic axis makes
an angle α with respect to the
vertical (the other one still being
along the vertical), find the variation
in the observed intensity as a
function of the angles θ and α as the
half-wave plate is rotated about the
direction of the incident beam.

*31. You have available only a Polaroid
strip and a quarter-wave plate with
which to analyze the state of polarization of a given beam of light.
Assuming that the light beam is
either plane-polarized, circularly
polarized, or elliptically polarized,
set up a procedure to determine
which it is.

*32. Repeat Problem 31, but this time
suppose that the light beam can be a
mixture of any two of these three
states of polarization.

*33. Repeat Problem 31, but this time
suppose that the incident beam
could in addition be unpolarized.

34 Epilogue

An intelligence knowing, at a given instant of time, all forces acting in nature, as well as the momentary positions and velocities of all things of which the universe consists, would be able to comprehend the motions of the largest bodies of the world and those of the lightest atoms ... provided his intellect were sufficiently powerful to subject all data to analysis; to him nothing would be uncertain; both present and future would be in his eyes.

PIERRE S. DE LAPLACE (1749–1827)

34-1 Introduction

It is not very difficult for us today to see why the late nineteenth-century physicists might have been overly proud of the achievements in their discipline during the preceding two centuries. With Newton's mechanics and Maxwell's theory of electromagnetism firmly established, it appeared to them to be only a matter of time until all physical happenings in the world could be described in full detail. They knew, for example, that mechanics, as encompassed in Newton's laws of motion and of gravity, correctly described not only the large-scale motions of the solar system and of the galaxy, but those of bodies of ordinary size as well. Moreover, as they had seen demonstrated by developments in the kinetic theory of gases and statistical mechanics, these laws appeared to be applicable on a microscopic level as

1043

well. Similarly, thanks to Maxwell they had acquired insights into electromagnetic phenomena—which by this time included optics as a special case—that no one could have conceived of only 50 years earlier.

As viewed by these late nineteenth-century physicists, then, these basic physical laws—or laws of *classical physics*, as they are known today—were logically consistent, complete, and fully capable of accounting in detail for all phenomena in the physical universe. Indeed, many expressed sorrow and concern for the plight of the twentieth-century physicist, whose central problem, they feared, would amount principally to ascertaining the "next decimal place." As Maxwell himself expressed it at the inauguration of the Devonshire (Cavendish) Laboratory at Cambridge in 1871:

> ... The opinion seemed to have got abroad that in a few years all the great physical constants will have been approximately estimated and that the only occupation which will then be left to men of science will be to carry on these measurements to another place of decimals.

As subsequent events have shown, this sort of complacency has no place in scientific research and thought. Indeed, the carrying out of measurements to the "next decimal place" plays the extremely important role in scientific research of helping to ascertain the limits of validity of physical laws. Except for the much rarer discovery of an altogether unforeseen phenomenon, this is often the only means available for uncovering new physical laws. In any event, largely in the pursuit of this secondary objective, during the 30-year period beginning about 1895 many physicists were pursuing studies to see to what extent the laws of classical physics were applicable at the microscopic level. Much to their surprise, they found that these laws, which had worked so well for macroscopic systems, were simply not applicable, without modification, to individual atoms and molecules. That is, they found that the application of the laws of Newtonian mechanics, and Maxwell's theory to atoms and molecules led in some cases to both quantitative as well as qualitative contradictions with experiment.

Out of this period of intellectual ferment there emerged ultimately two entirely new physical theories. These are (1) the *special*[1] *theory of relativity*, which was proposed by Albert Einstein in 1905; and (2) the theory of *quantum mechanics*, which was proposed by L. de Broglie, E. Schrödinger, W. Heisenberg, and others during the period 1924-1926.

With the theory of relativity, as we have seen earlier, Einstein extended the domain of validity of classical physics to high-speed phenomena. He argued that the laws of physics must be invariant, not under Galilean transformations as had been assumed in Newtonian mechanics, but rather under Lorentz transformations. As we saw in Sections 3-12 and 9-10, these

---

[1]The general theory of relativity, which was also proposed by Einstein in 1916, plays an important role only for phenomena that take place near very massive bodies, such as very large stars. It is normally of no significance in considerations of terrestrial, microscopic phenomena, and hence will not be considered further here.

ideas led to certain modifications of a number of physical quantities, including, for example, the mass, the energy, and the momentum of a particle. It is significant, however, that all the basic features of Newton's laws of motion, such as their deterministic nature and their associated laws of energy and momentum conservation, were retained.

The second theory developed during this period is quantum mechanics. In simplest terms this theory may be viewed as the generalization of Newton's laws required to describe systems with masses and dimensions of the order of those of atoms and molecules. The predictions of quantum mechanics for macroscopic bodies are precisely the same as those of Newtonian mechanics. Hence, quantum mechanics is often characterized as the generalization of Newtonian mechanics required to describe the behavior of microscopic systems. However, the changes imposed by the theory of quantum mechanics on basic classical concepts are much more severe, from a philosophical point of view, than those imposed by the theory of relativity. In particular, the deterministic nature of Newtonian mechanics, as expressed in the quotation from Laplace at the beginning of this chapter, had to be forfeited. Quantum mechanics is not a deterministic theory in the sense we are used to in classical mechanics.

34-2 Some experimental difficulties

By way of an introduction to some of the difficulties encountered in applying the laws of classical physics to microscopic phenomena, in this section we shall list certain predictions made by the classical laws. With each of these predictions we shall also state an experimental result that is incompatible with it. The remaining sections of this chapter are then devoted to a discussion of some of the assumptions—which eventually led to the discovery of quantum mechanics—that were made in order to account for these experimental results. This list is representative, but in no sense all-inclusive.

    1. Classical prediction: *An accelerating charged particle radiates electromagnetic energy.*

This prediction is in conflict with the self-evident fact that atoms are stable. An atom, as we know, consists of a positively charged nucleus, about which orbit an appropriate number of negatively charged electrons, so that overall the atom is electrically neutral. In order for this picture to be viable, it is necessary, according to classical ideas, for these electrons to accelerate as they orbit about the nucleus. But this means (Section 29-9) that they must radiate electromagnetic energy and thus gradually spiral into the nucleus. It follows therefore that the very existence of atoms is in contradiction with one or more facets of classical physics.

2. **Classical prediction:**   *The energy of a mechanical system can assume any value from a continuous range of values.*

This contradicts the fact that the energy of bound microscopic systems, such as atoms, molecules, and nuclei generally can assume only certain discrete values; other values are forbidden. The allowable energy values for these systems are known as *energy levels* or *energy eigenvalues*.

3. **Classical prediction**:   *An electromagnetic wave is a wave and not a stream of corpuscles such as envisioned by Newton.*

This conflicts with the fact that although, under many conditions, electromagnetic radiation is indeed a wave phenomenon, there exist others for which it is best described as consisting of particulate bundles of energy now known as *photons*. In other words, electromagnetic radiation has a dual nature: under some circumstances it acts as a wave and under others as a stream of particles. The relation between the parameters characterizing these two aspects is given in (30-1), which expresses the energy E and the momentum p of the photons in the wave in terms of its wavelength or frequency:

$$E = h\nu \qquad p = \frac{h}{\lambda} \tag{30-1}$$

4. **Classical prediction**:   *Particles, such as electrons or atoms, do not exhibit wavelike characteristics.* In particular, if a stream of electrons is incident on a slit system, such as that in Figure 34-1, neither diffraction nor interference phenomena will be observed on the screen.

This is in conflict with the experimental fact that if a beam of electrons is incident from the left on the two-slit system in Figure 34-1, then the electrons striking the screen will, under appropriate conditions, exhibit diffraction and interference. Specifically, the observed intensity with both slits open will not, in general, be the same as the sum of the intensities observed if only one slit is open at a time. This is a characteristic feature of

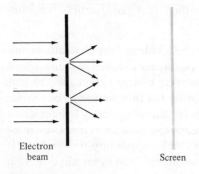

Electron
beam   Screen

Figure 34-1

an interference effect and demonstrates that matter—the electron beam in this case—has wavelike properties. Hence, just as there are particulate aspects associated with electromagnetic waves, experiments of this type show that there are also wavelike aspects associated with matter. The very existence of these *matter waves*, as they are known, makes it unambiguous that Newtonian mechanics is not always applicable at the microscopic level.

34-3 Thermal radiation

Experiment shows that if a noninflammable body, such as an iron poker, is heated, it will emit radiation whose quantity and quality generally depend on the temperature. At low temperatures, the wavelengths of the emitted rays are mainly in the infrared region of the electromagnetic spectrum (Section 29-7) and thus not directly observable by the unaided eye. However, as the temperature is raised, the body begins to glow with a dull red color which gradually turns a bright red. Eventually, when its temperature is sufficiently high, the radiated light appears to be white and the body is then said to be "white hot." The glow of the tungsten filament in a light bulb is an example of such a white-hot radiator.

A spectroscopic investigation of this type of radiation shows that it consists of a mixture of light of various wavelengths distributed continuously throughout the entire visible spectrum. That is, all colors, from red at one end to deep violet at the other, are present in the radiation. Moreover, the use of infrared and ultraviolet sensors shows that the emitted radiation is not confined to visible wavelengths alone but comes with varying intensities from neighboring parts of the spectrum as well. Thus, at low temperatures when the radiating body appears dull red, the most intense visible radiation it emits is at wavelengths of the order of 6500 Å; but there is a significant amount radiated at the invisible infrared wavelengths as well. As the temperature is raised, the maximum in intensity moves continuously to shorter and shorter wavelengths, as a result of which ultraviolet light becomes a more and more significant component of this radiation.

As a general rule, the amount and quality of the radiation emitted by a body depend on the nature of the body itself. However, under conditions of thermal equilibrium—that is, when the energy radiated per unit area per unit time by the body is equal to that absorbed from external sources—the nature and quality of this radiation must be the same for all bodies. For, how else could a state of thermal equilibrium exist? It follows that under conditions of thermal equilibrium, *all* bodies at temperature T will radiate electromagnetic energy with the same intensity at corresponding wavelengths. This energy radiated under conditions of thermal equilibrium is known as *blackbody radiation* or *isothermal radiation*. It depends only on the temperature and not on the nature of the material doing the radiating.

A convenient way to measure the distribution in wavelength of blackbody radiation is by use of a hollow body or cavity whose walls are maintained at some fixed temperature T. The radiation emitted by the interior walls of the cavity will be trapped inside, and eventually the enclosed radiation will come into thermal equilibrium with the walls. The characteristic features of blackbody radiation can then be measured by making a small opening in the enclosure and observing the emerging radiation. Figure 34-2 is a plot of the logarithm of the blackbody radiation intensity I or the energy radiated per unit time per unit area in a wavelength interval $d\lambda$ as a function of wavelength λ for various temperatures. The total energy radiated per unit area per unit time at all wavelengths is proportional to the area under the appropriate curve. Note that, consistent with our expectations, the higher the temperature the more energy is radiated.

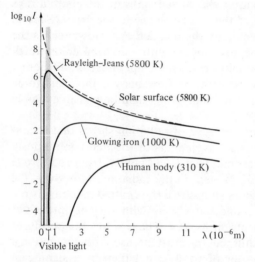

Figure 34-2

Isothermal radiation of this type plays an extremely important role in the physical world. It characterizes, for example, the heat radiation, generally at infrared wavelengths, radiated and absorbed by our own bodies at 310 K. Similarly the radiation emitted from the surface of the sun, which is at a temperature of approximately 5800 K, is also of this type. Indeed, the surface temperature of many stars is often inferred by studying the nature of the radiation they emit. Evidently, the continuous isothermal radiation emitted by a body is a very important physical quantity and one with which any acceptable physical theory must be able to come to grips.

34-4 Planck's proposal

If we were to apply the ideas of classical physics to the problem of blackbody radiation we would find a surprisingly simple and unambiguous answer. This is the *Rayleigh–Jeans* formula:

$$I = \frac{2\pi ckT}{\lambda^4} \qquad (34\text{-}1)$$

where $I \equiv I(\lambda, T)$ is the energy radiated per unit area, per unit time, and per unit wavelength interval at the wavelength λ when the body is in thermal equilibrium at the absolute temperature T. The constants k and c here are respectively, Boltzmann's constant and the speed of light, and have the values $k = 1.38 \times 10^{-23}$ J/K and $c = 3.0 \times 10^8$ m/s.

Unfortunately, this very simple Rayleigh-Jeans formula in (34-1) has a very serious drawback. For although it is in excellent, actually perfect, agreement with experiment at long wavelengths, it is in violent disagreement at short wavelengths. Indeed, because of the λ^{-4} singularity at $\lambda = 0$, the integrated spectrum—that is, the total energy radiated at all wavelengths—is infinite! This implies that at least one of the laws of classical physics that was used in deriving (34-1) and was assumed to be applicable at all wavelengths must be seriously in error. There is no ambiguity. As evidenced by the classical (dashed) curve in Figure 34-2, the classical formula disagrees with experimental facts.

In 1900 Max Planck (1858–1947) proposed a new formula for this important physical quantity I. Planck's formula is

$$I = \frac{2\pi hc^2}{\lambda^5}\left\{ \frac{1}{\exp\left[hc/\lambda kT\right] - 1} \right\} \qquad (34\text{-}2)$$

where k, c, λ, and T are as defined above and $h \cong 6.63 \times 10^{-34}$ J-s is Planck's constant. The various curves in Figure 34-2 are actually plots of this formula. For long wavelengths, it is easy to show by expanding the exponential that (34-2) reduces to the Rayleigh–Jeans formula in (34-1). By contrast to the incorrect Rayleigh–Jeans result at short wavelengths, however, Planck's formula for I is consistent with experiment at all wavelengths. Its prediction of an exponential decrease with decreasing wavelengths is fully in accord with observations at all wavelengths.

Now how did Planck arrive at this formula? Evidently an assumption that goes beyond the laws of classical physics is required. Indeed, to obtain (34-2), Planck was forced to make the hypothesis that only discrete units of radiant energy could be emitted by a body in thermal equilibrium. In other words, he found it necessary to assume that there is a smallest unit or *quantum* of electromagnetic energy that the constituent "oscillators" or atoms of a radiating body can emit. The result in (34-2) then followed provided that this quantum of energy, ϵ, is related to the frequency ν of the oscillators by

$$\epsilon = h\nu \qquad (34\text{-}3)$$

where h is Planck's constant.

This hypothesis in (34-3) that the energy emitted by the constituents of a body in thermal equilibrium can occur only in discrete bundles or quanta is completely foreign to classical physics. It states, in effect, that there is a smallest unit of energy, $h\nu$, and that all energy exchange can occur only in integral multiples of this unit. However, as strange as this hypothesis may sound at first, no one can quarrel with the success of its prediction, at least that in (34-2).

34-5   Quantization of the electromagnetic field

In a classic paper written in 1905, Einstein first suggested extending Planck's ideas on quantization to include the electromagnetic field itself. During the same year that he enunciated the theory of relativity he made this proposal in connection with an attempt to explain a phenomenon then under active experimental study and known as the *photoelectric effect*. In essence, he put forward the idea that an electromagnetic wave of frequency ν could, under some circumstances, be considered as a stream of corpuscles, or *photons* as we call them today, each with energy ϵ given in (34-3). Let us consider this matter briefly.

Experiment shows that if an electromagnetic wave of an appropriate wavelength is incident on certain substances, for example, zinc, potassium, or copper, electrons are emitted from its surface. These ejected electrons are called *photoelectrons*, and the phenomenon itself is known as the photoelectric effect. Figure 34-3 shows schematically an experimental setup for measuring the characteristics of these emitted electrons. An evacuated tube A contains a sample B and a collector C, which are maintained at some potential difference by a battery of emf $\mathscr{E}$. Under ordinary circumstances, with no radiation incident on B, the galvanometer reading is zero, since no current can flow around an open circuit. However, if light of an appropriate frequency is shined on the sample then, as shown in the figure, a current is found to flow through the galvanometer. The direction of this current corresponds to a stream of electrons going from B to the collector C inside the tube.

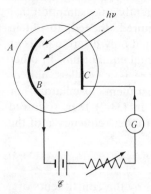

Figure 34-3

Studies of this effect usually involve measuring the current through the galvanometer as a function of the frequency and intensity of the incident radiation. These studies show that the photoelectric effect has a number of features that are unusual from the classical physicist's point of view. Very striking among these are:

1. Regardless of the intensity of the incident radiation, no electrons are emitted if the frequency is below a threshold value ν_c, which in general is different for different materials.
2. If the frequency of the incident radiation exceeds the threshold value ν_c for the given material, then electrons are emitted. The number of these electrons is proportional to the intensity of the incident light.
3. The distribution of the kinetic energies of the ejected electrons does not depend on the intensity of the incident radiation.
4. The maximum kinetic energy K_{max} of the emitted electrons is greater, the greater is the frequency ν of the incident radiation.

Each one of these properties of the photoelectric current is difficult to understand in terms of the concepts of classical physics. For example, the classical physicist would expect that the kinetic energies of the electrons would depend on the intensity of the incident radiation and certainly not its frequency. (See, for example, (29-25) in this connection.) The fact that below a threshold frequency ν_c there is no photoelectric current regardless of the incident intensity would be equally incomprehensible to him.

In a classic paper published in 1905, Einstein showed that these anomalous features could be accounted for by extending Planck's ideas of quantization to the electromagnetic field itself. He postulated that an electromagnetic wave of frequency ν consists of a stream of corpuscles—or *photons* as they are called today—each with energy E given by

$$E = h\nu \tag{34-4}$$

with h the same Planck's constant as in (34-3). Assuming, further, that a given electron in the metal is not likely to absorb more than one of these photons, he argued that the maximum energy acquired by any electron must be $h\nu$. Therefore, if this energy $h\nu$ is not large enough for an electron to overcome the potential barrier V_0 at the surface of the metal, no electrons will be emitted. However, if the energy $h\nu$ of the incident photons exceeds eV_0, then it *is* energetically possible for the electrons to leave the surface. By conservation of energy, the maximum kinetic energy of these ejected electrons must be the difference $(h\nu - eV_0)$. Hence

$$K_{max} = h\nu - eV_0 \tag{34-5}$$

This is Einstein's famous photoelectric-effect equation. The quantity eV_0 here is also known as the *photoelectric work function*. For a typical metal, such as zinc, it has a value of the order of 3.0 eV.

The value of K_{max} associated with a fixed frequency can be conveniently measured by reversing the polarity of the battery in Figure 34-3 and determining the minimum potential difference between B and C for which no current flows. Figure 34-4 shows a plot of K_{max} against ν as obtained in this way. The fact that the curve is a straight line is a beautiful confirmation of Einstein's proposal and its consequence in (34-5). The intercept along the horizontal axis is the threshold frequency ν_c and is generally different for different materials. It is related to the potential barrier V_0 by $\nu_c = eV_0/h$. Note that the slope of the curve is the same for all materials and has the value h.

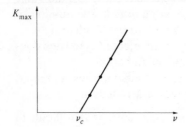

Figure 34-4

34-6 The Compton effect

A second, and equally impressive, experimental justification for the photon concept was discovered by Arthur H. Compton (1892–1962) about two decades later in 1923. Compton was interested in the distribution of the X rays scattered by the electrons in a carbon target. To this end he carried out experiments that measured the wavelength λ' of the emerging X rays as a function of the scattering angle θ and the wavelength λ of the incident X ray; see Figure 34-5. Assuming, with Einstein, that the incident X rays may be thought of as a stream of particles, each of energy $h\nu = hc/\lambda$, and with momentum $h\nu/c = h/\lambda$, and that energy and momentum are conserved during the collision, he deduced the relationship

$$\lambda' - \lambda = \frac{h}{mc}(1 - \cos\theta) \tag{34-6}$$

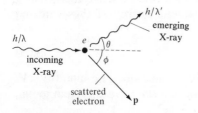

Figure 34-5

where m is the mass of an electron. And this is precisely the relation between λ' and θ that he measured in his experiment. The quantity h/mc in this relation has dimensions of length and is known as the *electron-Compton* wavelength. Its numerical value is approximately 2.4×10^{-12} meter. The term *Compton scattering* is used very generally today to describe processes that involve the scattering of photons from charged particles.

To derive (34-6), let us suppose that the velocity of the electron is sufficiently small that its momentum and kinetic energy are mv and $mv^2/2$, respectively. (See Problem 12 for the corresponding analysis for relativistic electrons.) Assuming, then, that before the collision the electron is at rest, and afterward it has a velocity v, it follows from the law of energy conservation that

$$\frac{hc}{\lambda} = \frac{hc}{\lambda'} + \frac{1}{2} mv^2 \qquad (34\text{-}7)$$

since, for example, $h\nu = hc/\lambda$ is the initial energy of the photon. Moreover, since the momentum of a photon of energy $h\nu$ is $h\nu/c = h/\lambda$, it follows from the law of conservation of linear momentum that

$$\frac{h}{\lambda} = \frac{h}{\lambda'} \cos\theta + mv \cos\phi$$

$$\frac{h}{\lambda'} \sin\theta = mv \sin\phi \qquad (34\text{-}8)$$

where θ and ϕ are as defined in Figure 34-5. Finally eliminating the electron parameters v and ϕ from among (34-7) and (34-8), we find that the wavelength λ' of the scattered photon is indeed given by the very simple formula in (34-6).

34-7 Atomic spectra

Imagine an originally evacuated glass tube filled with an element in a gaseous state; for example, hydrogen, helium, or neon. If the gas is heated, say by passing an electric discharge through it, the tube will glow with a color characteristic of the given element. A spectroscopic analysis of this light shows that its spectrum consists of a sequence of discrete lines, each with its distinctive wavelength or color. Figure 34-6 shows schematically a possible experimental setup. Light emitted by the gas in the tube T is passed through a spectroscopic slit S and then viewed directly with the eye or on a viewing screen V after transmission through a diffraction grating D. Each line on the screen is an image of the spectroscopic slit associated with a particular wavelength. The sequence of lines obtained this way is called the *emission spectrum*, or the *line spectrum*, or simply the *spectrum* of the element in the tube. Experiment shows that no two elements have the same spectra; the emission spectrum of each element is a unique property

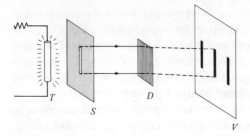

Figure 34-6

characteristic of the atoms of that element only. Figure 34-7 shows a photograph of the spectrum of copper.

Let us consider, for example, the spectrum of hydrogen. A spectroscopic analysis of this element shows the existence of four prominent lines in the visible region. These are a red-orange line at 6563 Å, a bluish-green line at 4861 Å, a blue-violet line at 4341 Å, and a deep violet line at 4101 Å. These four lines are members of a sequence of lines in the hydrogen spectrum known as the *Balmer series.* A second important series in the hydrogen spectrum, which occurs in the ultraviolet, is known as the *Lyman series.* Some members of this series are associated with the wavelengths 1216 Å, 1026 Å, and 973 Å. In a third series, known as the *Paschen series,* the lines occur in the infrared. Two examples are the lines at 18,756 Å and 12,821 Å.

Similarly, the emission spectrum of helium has seven prominent lines in the visible part of the spectrum. These include a red-orange line at 6678 Å, a yellow line at 5875 Å, and on to the violet end of the spectrum through the sequence of lines at 5015 Å, 4921 Å, 4713 Å, 4471 Å, and 4026 Å.

As a general rule, any element when in the gaseous state and heated sufficiently will emit radiation consisting of a collection of spectral lines characteristic of that element. Indeed, a practical way for ascertaining the existence of a certain element in a chemical substance is by heating it and then analyzing the spectrum of its vapor. Thus, the element sodium is known to emit two very closely spaced and intense yellow lines; one at 5890 Å and the other at 5896 Å. Hence, if these two lines are detected in the spectrum of some substance it is safe to conclude that the element sodium must be present in it. No other element emits these two lines.

Very closely related to the emission spectrum of an element is its *absorption spectrum.* This is obtained by making a spectroscopic analysis of continuous or thermal radiation after it has passed through a vapor of the

Figure 34-7 *A photograph showing some of the spectral lines of copper. (Courtesy Bausch and Lomb, Analytical Systems Division.)*

element in question. A possible experimental setup would be similar to that in Figure 34-6, but with the tube T replaced by a continuous source plus an intervening tube containing the vapor of the element. This time, as illustrated in Figure 34-8, the pattern on the screen is a continuous spectrum characteristic of thermal radiation, but cut by a series of dark lines. These lines constitute the absorption spectrum of the element. Assuming that the spectroscopic analysis is made by use of a diffracting grating, we may associate with each of these lines a wavelength in accordance with the basic formula in (32-18). We find in this way that each of the lines in the absorption spectrum coincides with some line in the emission spectrum. However, except under extreme conditions of gas temperature—for example, that of the gases surrounding the sun—*not* all lines of the emission spectrum have their counterparts in the absorption spectrum.

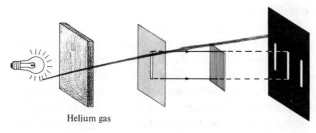

Helium gas

Figure 34-8

If, for example, a laboratory experiment is carried out to produce the absorption spectrum of hydrogen, in general only the Lyman series in the ultraviolet region is obtained. It is extremely difficult to produce the Balmer or the Paschen series in an absorption spectrum. To do so would require raising the gas temperature to prohibitively high values.

However, the fact that the Balmer and the Paschen lines do indeed exist in the absorption spectrum of hydrogen can be established by making a spectroscopic analysis of the radiation emitted by a star—for example, our own sun. Indeed, William H. Wollaston and Joseph Fraunhofer first discovered the existence of absorption spectra during the early part of the nineteenth century by making a spectral analysis of the light emitted by the sun. Fraunhofer is credited with the discovery of nearly 750 absorption lines in the solar spectrum. He was also the first to note that in many cases the dark lines in the absorption spectrum appear in the same positions as do certain emission lines of selected elements. Finally, in 1859 Gustav R. Kirchhoff established the significance of the Fraunhofer lines—as the lines in the solar absorption spectrum are known—by producing the absorption spectra of various elements in the laboratory. Today it is generally agreed that the surface of the sun is at a temperature of nearly 5800 K and radiates mainly isothermal radiation at that temperature. Prior to reaching the earth, however, this continuous spectrum of radiation passes through the hot gases surrounding the sun. The dark lines observed in the solar spectrum arise because of the presence of a variety of elements in these vapors. Nearly 65 elements have been detected on the sun in this way. This method is also used to determine the chemical composition of other stars.

Although in this discussion we have considered only the spectra of atoms, it should be noted that other microscopic systems, including notably molecules and nuclei, also display discrete spectra. It is apparent, therefore, that these various spectra represent an important clue to the structure of all microscopic systems. And as such, the existence and quantitative features of these spectra should at least be consistent with the laws of physics. Yet, despite much effort, no one has ever succeeded in explaining even the existence of these spectra by use only of classical physics.

34-8   The Bohr model

The first successful quantitative explanation of atomic spectra was developed by Niels Bohr (1885–1963) in 1913. To achieve this, Bohr found it necessary to introduce a number of *ad hoc* assumptions, each one of which was in contradiction with, or went beyond some facet of, classical physics. Nevertheless, his model was extremely successful in accounting in detail for many of the significant spectroscopic features of the hydrogen atom, and thus it is of more than passing interest. Unfortunately, this *Bohr model*, as it is known, is applicable only to hydrogen, or hydrogenlike ions, such as He^+, Li^{++}, Be^{+++}, and so forth. It is not useful for describing the spectra of other atoms. Ultimately it was abandoned, but not until the discovery of quantum mechanics more than a decade later.

Because of the revolutionary nature of the postulates underlying Bohr's model and also because it leads to basically the same hydrogen spectrum as does the correct theory, let us consider this model briefly. For hydrogen Bohr's postulates may be stated as follows:

1. The electron travels about the proton in Keplerian orbits by virtue of the Coulomb force of attraction between them.
2. Not all orbits are allowed. The allowable ones are those for which the angular momentum L of the electron about the proton is an integral multiple of $h/2\pi$, where h is Planck's constant. The allowable orbits are therefore those for which the angular momentum L has one of the values:

$$L = \frac{nh}{2\pi} \qquad (n = 1, 2, \ldots) \tag{34-9}$$

The quantity n is often referred to as a *quantum number.*

3. While in one of these allowed orbits the atom is stable. That is, the accelerating electron does *not* radiate electromagnetic energy, and its energy E_n in the orbit characterized by any integer n, is constant in time.

4. Radiation is emitted from the atom only when the electron makes a transition from one allowed orbit, say of energy E_n to another of lower energy $E_{n'}$. The frequency ν of the radiation emitted in such a transition is given by

$$h\nu = E_n - E_{n'} \qquad \textbf{(34-10)}$$

Note the similarity between (34-10) and the Planck–Einstein energy quantization condition in (34-3) and (34-4).

In order to apply this theory to the calculation of the emission spectrum for hydrogen, let us consider an electron of mass m and charge $-e$ traveling at a uniform velocity $\mathbf{v}$ in a circular orbit of radius a about a proton of charge $+e$; see Figure 34-9. According to Coulomb's law, the electron experiences an attractive force $e^2/a^2 4\pi\epsilon_0$, and also undergoes a radially inward centripetal acceleration of magnitude v^2/a. Hence, by Newton's second law,

$$\frac{mv^2}{a} = \frac{e^2}{4\pi\epsilon_0}\frac{1}{a^2} \qquad \textbf{(34-11)}$$

Since the angular momentum of the electron in this circular orbit is mva, it follows from Bohr's second postulate that for an allowable orbit the parameters v and a must be related by

$$mva = \frac{nh}{2\pi} \qquad \textbf{(34-12)}$$

Finally, from the fact that the total energy E_n of the electron when in this orbit is the sum of its kinetic and potential energies, we have

$$E_n = \frac{1}{2}mv^2 - \frac{e^2}{4\pi\epsilon_0 a}$$

In order now to obtain an explicit formula for the electron energy in an allowable orbit, let us solve (34-11) and (34-12) for v and a and substitute the result into this formula for E_n. We find in this way Bohr's prediction for the energy levels for the hydrogen atom:

$$E_n = -\frac{1}{2}\frac{e^2}{4\pi\epsilon_0 a_0}\frac{1}{n^2} \qquad (n = 1, 2, \ldots) \qquad \textbf{(34-13)}$$

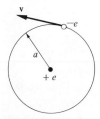

Figure 34-9

where a_0 is an important parameter known as the *Bohr radius*, and is defined by

$$a_0 = \frac{h^2 \epsilon_0}{\pi m e^2} \qquad (34\text{-}14)$$

Its numerical value is approximately 5.3×10^{-11} meter. According to Bohr's theory, then, the totality of values for the energy of an electron in a hydrogen atom are those given by (34-13). No other values for the energy are possible.

It is often very convenient and useful to represent the result in (34-13) on what is known as an *energy-level diagram*. As shown in Figure 34-10, such a diagram consists of an energy axis running vertically upward and a sequence of horizontal lines that intersect this axis at the allowable values for the energy. The state of lowest energy, or *ground state*, of the atom is represented by the lowest horizontal line on this diagram. As can be readily established by the substitution of the value $n = 1$ into (34-13), the energy E_1 of the ground state of hydrogen is -13.6 eV; the first *excited state* at $n = 2$ occurs at -13.6 eV$/2^2 \cong -3.4$ eV, and so forth for the other excited states.

The vertical arrows in the energy-level diagram in Figure 34-10 refer to the possible transitions that the electron can make in going from a given excited state to another one of lower energy. The frequency ν associated with the light emitted in any one of these transitions can be easily calculated by use

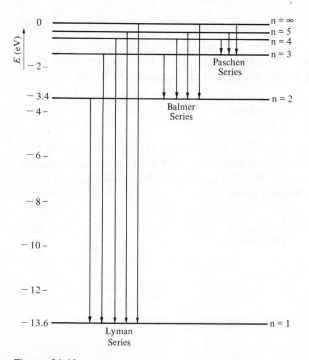

Figure 34-10

of (34-10) and (34-13). The result is given by the formula

$$\nu = \frac{1}{8h\pi\epsilon_0}\frac{e^2}{a_0}\left(\frac{1}{n'^2}-\frac{1}{n^2}\right)$$

where the quantum number n characterizes the initial state and n' that of the final state. As shown in the figure, for $n' = 1$ the frequencies for $n = 2, 3, \ldots$ correspond to the Lyman series. Similarly, for $n' = 2$ the transitions $n = 3, 4, \ldots$ represent the Balmer series, while for $n' = 3$ the lines for $n = 4, 5, \ldots$ are the Paschen series. In each case the spectral lines form an infinite sequence whose "series limit" is determined by setting $n \to \infty$. For example, since $\lambda = c/\nu$, it follows that the wavelengths of the lines of the Lyman series are given by

$$\frac{1}{\lambda} = \frac{1}{8\pi ch\epsilon_0}\frac{e^2}{a_0}\left(1-\frac{1}{n^2}\right) \qquad n = 2, 3, 4, \ldots \qquad \text{(Lyman)}$$

and, consistent with experiment, this leads to wavelengths of $1216\,\text{Å}$, $1026\,\text{Å}$, $973\,\text{Å}, \ldots, 912\,\text{Å}$ for $n = 2, 3, 4, \ldots, \infty$, respectively. Similarly, for $n' = 2$, we obtain the wavelengths of the lines in the Balmer series

$$\frac{1}{\lambda} = \frac{1}{8\pi ch\epsilon_0}\frac{e^2}{a_0}\left(\frac{1}{4}-\frac{1}{n^2}\right) \qquad n = 3, 4, \ldots \qquad \text{(Balmer)}$$

and this formula correctly predicts the observed wavelengths of $6563\,\text{Å}$, $4861\,\text{Å}$, $4341\,\text{Å}, \ldots, 3650\,\text{Å}$ for $n = 3, 4, 5, \ldots, \infty$, respectively.

Thus by use of his model Bohr produced a quantitative way for describing the important features of the emission spectrum for hydrogen. Also, from the assumption that at low temperatures hydrogen atoms are mostly in the ground state and that on being exposed to radiation they preferentially absorb photons with energy $h\nu$ given by (34-10), it follows that the absorption spectrum should consist only of the Lyman series. This is also consistent with experiment. Thus, provided that we grant the validity of Bohr's postulates, the beginnings of a theory of atomic spectra emerges. Unfortunately, however, Bohr's theory leads to incorrect predictions for the spectra of atoms other than hydrogen and hydrogenlike atoms. Thus to obtain a realistic understanding of the spectra of atoms, as well as those of molecules and nuclei, it was necessary to await the discovery and development of quantum mechanics.

34-9   Matter waves

In 1924 still another hypothesis was proposed: this one by L. de Broglie (1892–   ). At the time of its being put forward this hypothesis appeared to be even more revolutionary than that of Planck and Einstein on the wave-particle duality of photons and those of Bohr on the existence of energy levels in atoms. Although it is not related to experiment as directly as are Bohr's postulates, de Broglie's hypothesis played a significant role in the subsequent development of quantum mechanics.

In brief, what de Broglie suggested was that the Planck–Einstein wave-particle duality of the electromagnetic field was also shared by electrons. That is, he hypothesized that under some circumstances electrons behave as ordinary particles while under others they take on the characteristics of waves. Moreover, he proposed arguments which suggested that the wavelength λ associated with matter waves should be related to the momentum $p = mv$ of the electron by

$$\lambda = \frac{h}{p} \tag{34-15}$$

with h Planck's constant. In essence, de Broglie's argument for this relation consisted of the requirement that (34-9) be satisfied and that the circumference of the electron's orbit in the hydrogen atom be an integral number of wavelengths. (See Problem 23.)

An immediate and very important consequence of de Broglie's proposal was that it led other physicists, notably Erwin Schrödinger (1887–1961), to think in terms of matter waves and to attempt to quantify this concept by setting up an appropriate equation for such a wave. Ultimately, this effort led to the development of the theory of quantum mechanics, a key element of which is a wave equation for matter waves known as the *Schrödinger equation*. The role that this equation plays in quantum mechanics is very similar to that played by Newton's laws of motion in classical mechanics. Some knowledge of this wave equation of Schrödinger is essential for an understanding of much of the physics developed during the twentieth century.

A second important consequence of de Broglie's hypothesis was to motivate physicists to generate and to look for matter waves in the laboratory. In 1928 C. J. Davisson and L. H. Germer of the United States and G. P. Thompson of Scotland succeeded in this effort. In the Davisson–Germer experiment electrons with energies of the order of 100 eV were incident on a nickel crystal. An analysis of the scattered electrons showed that their spatial distribution could best be described as the diffraction of plane waves by the regularly spaced atomic planes in the crystal. Moreover, the wavelength λ associated with these waves turned out to be precisely that given by the de Broglie relation in (34-15). The experiment of G. P. Thompson was similar, but this time the electron diffraction patterns resulted from an electron beam going through very thin gold foils. Figure 34-11 illustrates the types of diffraction patterns found in this way. The existence of matter waves, as strange as this might appear to the classical physicist, is definitely an experimental reality of the physical world. The fact that particles other than electrons also exhibit wavelike behavior was subsequently confirmed in 1931 by Estermann, Frish, and Stern. They succeeded in demonstrating the diffraction of helium atoms by scattering a monoenergetic beam of these atoms from a lithium fluoride crystal.

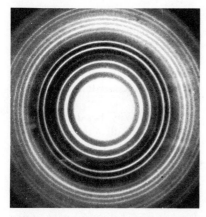

Figure 34-11 *Diffraction pattern produced by directing an electron beam through polycrystalline aluminum. (Courtesy Project Physics, Holt, Rinehart and Winston, New York.)*

It is important to recognize that the de Broglie relation in (34-15) is not in conflict with normal observations of macroscopic physical systems. As we saw in Chapter 32, the fact that light is a wave motion manifests itself only when it interacts with systems with dimensions of the order of its wavelength. For the case of a macroscopic body with momentum $p = mv \sim$ 1 kg-m/s^2, the wavelength $\lambda = h/mv$ of the associated matter wave is, according to (34-15), of the order of 6×10^{-34} meter. Hence the wave aspects of this system are completely masked in this case. However, for the case of an electron with energy of 100 eV, as in the Davisson–Germer experiment, we find that $p \sim 5 \times 10^{-24}$ kg-m/s, so this time $\lambda \cong 10^{-10}$ meter. Since this is of the order of the interatomic spacing in the nickel crystal, diffraction effects should be, and are indeed, observed.

QUESTIONS

1. Define or describe briefly what is meant by the following terms: (a) blackbody radiation; (b) photons; (c) matter wave; (d) absorption spectrum; and (e) photoelectric effect.
2. Why do you suppose the term "blackbody" radiation is used to describe the isothermal radiation emitted by a body in equilibrium at a temperature T?
3. Describe in operational terms the meaning of the following statements: (a) The surface of the sun has a temperature of 5800 K. (b) Barnard's star has a surface temperature of 3000 K.
4. List the assumptions which Planck introduced to obtain the correct blackbody spectrum? How did his result compare with the classical Rayleigh–Jeans formula?
5. What is the photoelectric effect? What new physical assumptions did Einstein introduce to obtain his basic result in (34-5)?
6. Compare and contrast the assumptions made by Einstein in connection with the photoelectric effect and those made by Planck in connection with his work on blackbody radiation.
7. Can a stream of photons undergo diffraction or interference? Explain.
8. What is Compton scattering? What was the significance of Compton's original experiments in 1923?
9. What is meant by the *emission spectrum* of an element? How is it

related to the absorption spectrum of that element?

10. Describe in physical terms why the emission spectrum of an element will in general appear to exhibit more lines than the corresponding absorption spectrum. What is the nature of the lines missing from the absorption spectrum?

11. Explain in physical terms why the solar spectrum contains only absorption lines. Why might one expect to see emission lines from the sun during a complete solar eclipse?

12. It has been recently discovered that we (and, presumably, the entire universe) are immersed in "three-degree radiation." Explain what you think this means and how this fact

could be discovered experimentally. (See Problem 7.)

13. List the basic postulates of the Bohr model and for each one state that feature of classical physics with which it is inconsistent.

14. In what way or ways did the Bohr model fail? Why, besides its historical value, is it still of interest today?

15. Describe an experiment that confirms the existence of matter waves. Explain why matter waves manifest themselves only at the microscopic level.

16. How do you account for the fact that the existence of matter waves remained undetected until relatively recently?

PROBLEMS

1. Calculate the rate of emission per unit area per unit wavelength interval at the wavelength of 6000 Å from a blackbody of temperature: (a) 6000 K; (b) 1000 K; (c) 300 K.

2. By integrating (34-2) over all wavelengths derive the *Stefan–Boltzmann* law

$$\mathscr{F} = \sigma T^4$$

for the total $\mathscr{F}$ rate at which energy is radiated per unit area from a blackbody. Express the Stefan–Boltzmann constant σ, whose value is 5.7×10^{-8} W/m²-K⁴, as a definite integral. (*Hint*: Change dummy variables to remove the T dependence from the integrand for $\mathscr{F}$.)

3. Assuming that you radiate as a blackbody at 310 K, calculate your rate of radiation per unit area by use of the Stefan–Boltzmann law.

4. Assuming that the sun radiates 4.0×10^{26} watts and is a blackbody at 5800 K, calculate its surface area and radius. Compare with the known value of 7.0×10^8 meters for the latter and account for any differences.

5. Show that the intensity I in (34-2) is a maximum at that value for wavelength λ_m which satisfies

$$\lambda_m T = \text{constant}$$

where the constant has the value 0.2898×10^{-2} m-K.

6. By use of the results of Problem 5 find the value λ_m at which I in (34-2) is a maximum for (a) the surface of the sun at 5800 K and (b) the human body at 310 K. State in each case in what part of the electromagnetic spectrum the peak in the intensity occurs.

7. It has been recently discovered that we are embedded in 2.7-K blackbody radiation (often called the *three-degree radiation*). By use of the result of Problem 5, calculate the value λ_m at which this radiation has its peak. Does the fact that the earth's atmosphere is opaque for wavelengths just below $\cong 0.5$ cm affect our ability to measure the full spectrum of this radiation?

8. In a certain experiment, it is found that the lowest wavelength of light

that will produce photoelectrons from a cesium sample is 7.0×10^{-7} meter. Calculate (a) the associated value for the threshold frequency ν_c and (b) the value of the photoelectric work function for the given sample.

9. The photoelectric work function for a certain sample of copper is 3.0 eV.
   (a) What is ν_c for this sample?
   (b) If light of wavelength 3000 Å is incident on the sample, what is the maximum kinetic energy of the ejected electrons?

10. Monochromatic light of wavelength 6500 Å and of intensity 2.0 W/m^2 is normally incident on a surface. Calculate:
    (a) The energy of the photons in the beam.
    (b) The number of photons incident on the surface per unit area per unit time.

11. Assuming that the total rate of radiation of 4.0×10^{26} watts by the sun is at a wavelength of 5500 Å, calculate the number of photons the sun emits per second.

12. Consider the Compton scattering event in Figure 34-5.
    (a) Show that the conservation-of-energy relation is very generally given by

$$\frac{hc}{\lambda} + mc^2 = \frac{hc}{\lambda'} + c[p^2 + m^2c^2]^{1/2}$$

   where m is the mass of the electron.
   (b) Write down the corresponding relations implied by the law of momentum conservation in terms of λ', p, ϕ, and θ.
   (c) Eliminate the parameters p and ϕ from these relations and thus confirm the validity of (34-6).

13. Suppose that a beam of 2.0×10^5 eV photons is scattered by the electrons in a carbon target.
    (a) What is the wavelength associated with these photons?
    (b) What is the wavelength of those

photons scattered through an angle of 90°?
   (c) What is the energy of the scattered photons that emerge at an angle of 60° relative to the incident direction?

14. Calculate numerical values for:
    (a) The electron Compton wavelength.
    (b) The proton Compton wavelength.
    (c) The pion Compton wavelength ($m_\pi = 274m$, with m the electron mass).

15. Calculate the frequency of the line with the longest wavelength in the Balmer series of hydrogen. What is the wavelength of this line?

16. Calculate the frequency and the wavelength of the lines of shortest and longest wavelength in the Lyman series of hydrogen.

17. Prove that the energy levels $E_n(Z)$ of a hydrogenlike atom whose nucleus carries a charge Ze is

$$E_n(Z) = Z^2 E_n$$

with E_n the corresponding energy of ordinary hydrogen. (*Hint:* Repeat the same steps used to derive (34-13) but using the form of the Coulomb force appropriate to this case.)

18. The *binding energy* of an electron in an atom is the minimum energy required to remove it from the atom. Calculate the binding energy of the electron in: (a) hydrogen; (b) He$^+$; and (c) Li^{++}. (*Hint:* Use the result of Problem 17.)

19. Prove that the speed v_n of an electron in hydrogen in a state corresponding to the quantum number n is

$$v_n = \frac{e^2}{2\epsilon_0 h n}$$

What is the radius of the electron's orbit?

20. Calculate, by use of the result of Problem 19, the time required for an

electron to make a complete orbit when it is in an allowable state.

21. Calculate the de Broglie wavelength associated with a man of mass 80 kg and traveling at a speed of 5 m/s. Would you expect to see diffraction effects in this case? Explain.

22. What is the de Broglie wavelength associated with:
    (a) A 2.0 keV electron?
    (b) A proton of the same energy?

23. Prove that the Bohr condition in (34-9) is equivalent to the statement that the circumference of the electron orbit in hydrogen is an integral number of de Broglie wavelengths.

APPENDIX A   Differentiation formulas

The purpose of this appendix is to list for reference purposes a number of formulas that are useful in calculating derivatives.

Let $x(t)$, $x_1(t)$, and $x_2(t)$ represent functions of the independent variable t, and let α be a constant independent of time and n a positive integer. Then

$$\frac{d}{dt}(\alpha t^n) = n\alpha t^{n-1} \tag{A-1}$$

$$\frac{d}{dt}[x_1(t) + \alpha x_2(t)] = \frac{dx_1}{dt} + \alpha\frac{dx_2}{dt} \tag{A-2}$$

$$\frac{d}{dt}[x_1(t)x_2(t)] = x_1\frac{dx_2}{dt} + x_2\frac{dx_1}{dt} \tag{A-3}$$

$$\frac{d}{dt}f(x(t)) = \frac{df}{dx}\frac{dx}{dt} \tag{A-4}$$

where, in (A-4), f is a function of the independent variable t only through its explicit dependency on the function $x(t)$. The relation in (A-4) is known as the *chain rule*.

To give some idea as to how these results are obtained, let us derive (A-1). Applying the definition of a derivative

$$\frac{dx}{dt} = \lim_{\Delta t \to 0}\frac{x(t + \Delta t) - x(t)}{\Delta t}$$

to the function $x = \alpha t^n$, we obtain

$$\frac{d}{dt}(\alpha t^n) = \lim_{\Delta t \to 0}\frac{\alpha(t + \Delta t)^n - \alpha t^n}{\Delta t}$$

i

By use of the binomial expansion

$$(a + b)^n = a^n + na^{n-1}b + \frac{n(n-1)}{2!} a^{n-2}b^2 + \cdots + b^n$$

and the fact that n is a positive integer, this may be simplified to the desired result:

$$\frac{d(\alpha t^n)}{dt} = \lim_{\Delta t \to 0} \alpha \frac{t^n + nt^{n-1}\Delta t + \cdots + (\Delta t)^n - t^n}{\Delta t}$$

$$= \alpha nt^{n-1}$$

More generally, it can be shown that (A-1) is valid for n any positive or negative number.

The validity of (A-2), (A-3), and (A-4) and related formulas may be established in a similar way.

APPENDIX C

The derivatives of the sine and the cosine

The purpose of this appendix is to derive the rules for differentiating the sine and the cosine functions. To this end, let us first establish the limit

$$\lim_{\theta \to 0} \frac{\sin \theta}{\theta} = 1 \qquad \text{(C-1)}$$

where the angle θ is expressed in radians. The fact that for sufficiently small angles, $\theta \cong \sin \theta$ is easily established by reference to a table of trigonometric functions. Expressing θ in radians we find the values $\sin 0.10 = 0.09983$, $\sin 0.05 = 0.04998$, $\sin 0.01 = 0.01000$, and, more generally, for $\theta \leqslant 0.10$ rad ($\cong 6°$) we find that θ and $\sin \theta$ differ from each other by less than 1 percent.

To establish the validity of (C-1), consider in Figure C-1 a circle of radius R centered at A, and let AB and AC be two radii subtending an angle θ. Construct from the point B the line BE tangent to the circle at B and the line BD to be perpendicular to AC. The angle between BD and BE is also θ. By

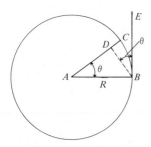

Figure C-1

definition of a radian, the length of the arc BC is $R\theta$ and thus

$$\theta = \frac{\text{arc } BC}{R}$$

Further, in the right triangle ABD,

$$\sin \theta = \frac{BD}{R}$$

and thus by division we find that

$$\frac{\sin \theta}{\theta} = \frac{BD}{\text{arc } BC}$$

In the limit as $\theta \to 0$, since BD becomes parallel to BE, it follows that the length of BD must become more and more equal to the arc BC. The validity of (C-1) is thus established.

We shall now use (C-1) to calculate the derivative of $\sin \theta$. According to the definition of a derivative, we have

$$\frac{d}{d\theta} \sin \theta = \lim_{\Delta\theta \to 0} \frac{\sin (\theta + \Delta\theta) - \sin \theta}{\Delta\theta}$$

$$= \lim_{\Delta\theta \to 0} \frac{\sin \theta \cos \Delta\theta + \sin \Delta\theta \cos \theta - \sin \theta}{\Delta\theta} \qquad \text{(C-2)}$$

where the second equality follows by use of the identity

$$\sin (A + B) = \sin A \cos B + \cos A \sin B$$

Since $\cos 0 = 1$, it follows that for small $\Delta\theta$, $\cos \Delta\theta \cong 1$. Hence (C-2) becomes

$$\frac{d}{d\theta} \sin \theta = \lim_{\Delta\theta \to 0} \frac{\cos \theta \sin \Delta\theta}{\Delta\theta}$$

$$= \cos \theta \qquad \text{(C-3)}$$

where the second equality follows by use of (C-1).

Similarly, the derivative of the cosine is found to be

$$\frac{d}{d\theta} \cos \theta = \lim_{\Delta\theta \to 0} \frac{\cos (\theta + \Delta\theta) - \cos \theta}{\Delta\theta}$$

$$= \lim_{\Delta\theta \to 0} \frac{\cos \theta \cos \Delta\theta - \sin \theta \sin \Delta\theta - \cos \theta}{\Delta\theta}$$

$$= \lim_{\Delta\theta \to 0} \frac{- \sin \theta \sin \Delta\theta}{\Delta\theta} \qquad \text{(C-4)}$$

$$= - \sin \theta$$

where the second equality follows from the identity

$$\cos (A + B) = \cos A \cos B - \sin A \sin B$$

and the fourth from (C-1).

Applying the chain rule (A-4), and (C-3), (C-4), we find that if h is a function of t, then

$$\frac{d}{dt} \sin [h(t)] = \cos [h(t)] \frac{dh}{dt} \tag{C-5}$$

$$\frac{d}{dt} \cos [h(t)] = - \sin [h(t)] \frac{dh}{dt} \tag{C-6}$$

For the special case, $h(t) = \omega t + \alpha$, with ω and α constants, these formulas reduce to (6-17) and (6-18), respectively.

APPENDIX G Table of trigonometric functions

Degrees	Radians	Sine	Cosine	Tangent	Cotangent		
0	0	0	1.0000	0	——	1.5708	90
1	.0175	.0175	.9998	.0175	57.290	1.5533	89
2	.0349	.0349	.9994	.0349	28.636	1.5359	88
3	.0524	.0523	.9986	.0524	19.081	1.5184	87
4	.0698	.0698	.9976	.0699	14.301	1.5010	86
5	.0873	.0872	.9962	.0875	11.430	1.4835	85
6	.1047	.1045	.9945	.1051	9.5144	1.4661	84
7	.1222	.1219	.9925	.1228	8.1443	1.4486	83
8	.1396	.1392	.9903	.1405	7.1154	1.4312	82
9	.1571	.1564	.9877	.1584	6.3138	1.4137	81
10	.1745	.1736	.9848	.1763	5.6713	1.3963	80
11	.1920	.1908	.9816	.1944	5.1446	1.3788	79
12	.2094	.2079	.9781	.2126	4.7046	1.3614	78
13	.2269	.2250	.9744	.2309	4.3315	1.3439	77
14	.2443	.2419	.9703	.2493	4.0108	1.3265	76
		Cosine	Sine	Cotangent	Tangent	Radians	Degrees

Degrees	Radians	Sine	Cosine	Tangent	Cotangent		
15	.2618	.2588	.9659	.2679	3.7321	1.3090	75
16	.2793	.2756	.9613	.2867	3.4874	1.2915	74
17	.2967	.2924	.9563	.3057	3.2709	1.2741	73
18	.3142	.3090	.9511	.3249	3.0777	1.2566	72
19	.3316	.3256	.9455	.3443	2.9042	1.2392	71
20	.3491	.3420	.9397	.3640	2.7475	1.2217	70
21	.3665	.3584	.9336	.3839	2.6051	1.2043	69
22	.3840	.3746	.9272	.4040	2.4751	1.1868	68
23	.4014	.3907	.9205	.4245	2.3559	1.1694	67
24	.4189	.4067	.9135	.4452	2.2460	1.1519	66
25	.4363	.4226	.9063	.4663	2.1445	1.1345	65
26	.4538	.4384	.8988	.4877	2.0503	1.1170	64
27	.4712	.4540	.8910	.5095	1.9626	1.0996	63
28	.4887	.4695	.8829	.5317	1.8807	1.0821	62
29	.5061	.4848	.8746	.5543	1.8040	1.0647	61
30	.5236	.5000	.8660	.5774	1.7321	1.0472	60
31	.5411	.5150	.8572	.6009	1.6643	1.0297	59
32	.5585	.5299	.8480	.6249	1.6003	1.0123	58
33	.5760	.5446	.8387	.6494	1.5399	.9948	57
34	.5934	.5592	.8290	.6745	1.4826	.9774	56
35	.6109	.5736	.8192	.7002	1.4281	.9599	55
36	.6283	.5878	.8090	.7265	1.3764	.9425	54
37	.6458	.6018	.7986	.7536	1.3270	.9250	53
38	.6632	.6157	.7880	.7813	1.2799	.9076	52
39	.6807	.6293	.7771	.8098	1.2349	.8901	51
40	.6981	.6428	.7660	.8391	1.1918	.8727	50
41	.7156	.6561	.7547	.8693	1.1504	.8552	49
42	.7330	.6691	.7431	.9004	1.1106	.8378	48
43	.7505	.6820	.7314	.9325	1.0724	.8203	47
44	.7679	.6947	.7193	.9657	1.0355	.8029	46
45	.7854	.7071	.7071	1.0000	1.0000	.7854	45
		Cosine	Sine	Cotangent	Tangent	Radians	Degrees

APPENDIX H   Table of e^x and e^{-x}

x	e^x	e^{-x}	x	e^x	e^{-x}
0.0	1.0000	1.0000	3.0	20.086	.04979
0.1	1.1052	.90484	3.1	22.198	.04505
0.2	1.2214	.81873	3.2	24.533	.04076
0.3	1.3499	.74082	3.3	27.113	.03688
0.4	1.4918	.67032	3.4	29.964	.03337
0.5	1.6487	.60653	3.5	33.115	.03020
0.6	1.8221	.54881	3.6	36.598	.02732
0.7	2.0138	.49659	3.7	40.447	.02472
0.8	2.2255	.44933	3.8	44.701	.02237
0.9	2.4596	.40657	3.9	49.402	.02024
1.0	2.7183	.36788	4.0	54.598	.01832
1.1	3.0042	.33287	4.1	60.340	.01657
1.2	3.3201	.30119	4.2	66.686	.01500
1.3	3.6693	.27253	4.3	73.700	.01357
1.4	4.0552	.24660	4.4	81.451	.01228
1.5	4.4817	.22313	4.5	90.017	.01111
1.6	4.9530	.20190	4.6	99.484	.01005
1.7	5.4739	.18268	4.7	109.95	.00910
1.8	6.0496	.16530	4.8	121.51	.00823
1.9	6.6859	.14957	4.9	134.29	.00745
2.0	7.3891	.13534	5.0	148.41	.00674
2.1	8.1662	.12246	5.5	244.69	.00409
2.2	9.0250	.11080	6.0	403.43	.00248
2.3	9.9742	.10026	6.5	665.14	.00150
2.4	11.023	.09072	7.0	1096.6	.00091
2.5	12.182	.08208	7.5	1808.0	.00055
2.6	13.464	.07427	8.0	2981.0	.00034
2.7	14.880	.06721	8.5	4914.8	.00020
2.8	16.445	.06081	9.0	8103.1	.00012
2.9	18.174	.05502	9.5	13360	.00007
3.0	20.086	.04979	10.0	22026	.00005

APPENDIX J   Experimental values of basic physical constants[a]

Speed of light	c	$2.997924562(11) \times 10^8$ m/s
Gravitational constant	G	$6.6732(31) \times 10^{-11}$ N-m^2/kg^2
Avogadro's number	N_0	$6.022169(40) \times 10^{23}$ atoms/mole
Boltzmann's constant	k	$1.380622(59) \times 10^{-23}$ J/K
Gas constant	R	$8.31434(35)$ J/mole-K
Triple point of water	T_t	273.16 K (exact)
Electron mass	m_e	$9.109558(54) \times 10^{-31}$ kg
Proton mass	m_p	$1.672614(11) \times 10^{-27}$ kg
Quantum of electric charge	e	$1.6021917(70) \times 10^{-19}$ C
Permittivity of free space	ϵ_0	$8.8541853(59) \times 10^{-12}$ F/m
Permeability of free space	μ_0	$4\pi \times 10^{-7}$ H/m (exact)
Planck's constant	h	$6.626196(50) \times 10^{-34}$ J-s
Bohr radius	a_0	$5.2917715(81) \times 10^{-11}$ m
Electron Compton wavelength	$\dfrac{h}{m_e c}$	$2.4263096(74) \times 10^{-12}$ m

[a]The values in this table (except for c) are adapted from, B. N. Taylor, W. H. Parker, and D. N. Langenberg, *Fundamental Constants and Quantum Electrodynamics*, New York, Academic Press, 1969. The value for c is taken from the article by K. M. Evenson *et al.*, *Phys. Rev. Lett.* **29**, 1346 (1972). The numbers in parentheses refer to the error in the last digits of the listed values.

APPENDIX K   Units for physical quantities

Quantity	Symbol	Derived Unit	SI Unit
Acceleration	**a**	m/s^2	m/s^2
Angle	θ	radian (rad)	—
Angular acceleration	α	rad/s^2	s^{-2}
Angular momentum	**L**	J-s	$kg\text{-}m^2/s$
Angular velocity	ω	rad/s	s^{-1}
Area	S, A	m^2	m^2
Capacitance	C	farad(F)	$A^2\text{-}s^4/kg\text{-}m^2$
Charge	q, e	coulomb(C)	A-s
Charge density			
Volume	ρ	C/m^3	$A\text{-}s/m^3$
Surface	σ	C/m^2	$A\text{-}s/m^2$
Linear	λ	C/m	$A\text{-}s/m$
Coefficient of linear expansion	α	K^{-1}	K^{-1}
Conductivity	σ	$(\Omega\text{-}m)^{-1}$	$A^2\text{-}s^3/kg\text{-}m^3$
Current	i, I	ampere(A)	A
Current density	**J**	A/m^2	A/m^2
Electric dipole moment	**p**	C-m	A-s-m
Electric displacement	**D**	C/m^2	$A\text{-}s/m^2$
Electric field	**E**	V/m	$kg\text{-}m/A\text{-}s^3$
Electric flux	Φ, Φ_E	V-m	$kg\text{-}m^3/A\text{-}s^3$
Electromotive force	$\mathscr{E}$	volt(V)	$kg\text{-}m^2/A\text{-}s^3$
Energy	E, T, V	joule(J)	$kg\text{-}m^2/s^2$

Quantity	Symbol	Derived Unit	SI Unit
Entropy	S	J/K	kg-m^2/s^2-K
Flow rate	V_R	m^3/s	m^3/s
Force	$\mathbf{F}$	newton(N)	kg-m/s^2
Frequency	ν	hertz(Hz)	s^{-1}
Gravitational field strength	$\mathbf{g}$	m/s^2	m/s^2
Heat	Q	J	kg-m^2/s^2
Inductance	L	henry(H)	kg-m^2/A^2-s^2
Length	l, d, L	m	m
Magnetization	$\mathbf{M}$	A/m	A/m
Magnetic dipole moment	$\boldsymbol{\mu}$	N-m/T	A-m^2
Magnetic field	$\mathbf{H}$	A/m	A/m
Magnetic flux	Φ_B	weber(Wb)	kg-m^2/A-s^2
Magnetic induction	$\mathbf{B}$	tesla	kg/A-s^2
		(T = Wb/m^2)	
Mass	m, M	kg	kg
Mass density	ρ	kg/m^3	kg/m^3
Molar specific heat	C	J/mole-K	kg-m^2/s^2-mole-K
Moment of inertia	I	kg-m^2	kg-m^2
Momentum	$\mathbf{p}$	N-s	kg-m/s
Particle density	n	molecules/m^3	m^{-3}
Permeability of free space	μ_0	H/m	kg-m/A^2-s^2
Permittivity of free space	ϵ_0	F/m	A^2-s^4/kg-m^3
Polarization	$\mathbf{P}$	C/m^2	A-s/m^2
Potential	V	volt(V)	kg-m^2/A-s^3
Power	P	watt(W)	kg-m^2/s^3
Poynting vector	$\mathbf{N}$	W/m^2	kg/s^3
Pressure	P	N/m^2	kg/m-s^2
Resistance	R	ohm(Ω)	kg-m^2/A^2-s^3
Resistivity	ρ	Ω-m	kg-m^3/A^2-s^3
Specific heat	c	J/kg-K	m^2/s^2-K
Spring constant	k	N/m	kg/s^2
Temperature	T	kelvin(K)	K
Time	t	second	s
Torque	τ	N-m	kg-m^2/s^2
Velocity (speed)	$\mathbf{v}$	m/s	m/s
Volume	V	m^3	m^3
Wave number	k	m^{-1}	m^{-1}
Wavelength	λ	m	m
Work	W	J	kg-m^2/s^2

APPENDIX L Conversion to the Gaussian system of units

The form of the equations of electromagnetism used in this text is applicable provided all quantities are expressed in SI (mks) units. To convert any of these equations to the corresponding Gaussian form, replace each of the SI symbols for the quantities listed below under "SI" by the corresponding one listed under "Gaussian." For example, under this replacement the relation $C = \varepsilon_o A/d$ becomes $4\pi\varepsilon_o C = \varepsilon_o A/d$, so that the corresponding Gaussian form is $C = A/4\pi d$. Note that in this process the symbols for mass, length, time, and other mechanical quantities are not changed and that any noncancelling factors of $(\varepsilon_o\mu_o)^{-\frac{1}{2}}$ are to be replaced by the speed of light c.

Physical quantity	SI	Gaussian
Capacitance	C	$4\pi\varepsilon_o C$
Conductivity	σ	$4\pi\varepsilon_o\sigma$
Charge (as well as charge density, current, current density and polarization)	$q[\rho,\ i,\ \mathbf{J},\ \mathbf{P}]$	$\sqrt{4\pi\varepsilon_o}\ q[\rho,\ i,\ \mathbf{J},\ \mathbf{P}]$
Dielectric constant	κ	$\varepsilon_o\kappa$
Displacement vector	$\mathbf{D}$	$\sqrt{\varepsilon_o/4\pi}\ \mathbf{D}$
Electric field (and potential and flux)	$\mathbf{E}[V,\ \Phi]$	$(4\pi\varepsilon_o)^{-\frac{1}{2}}\ \mathbf{E}[V,\ \Phi]$
Inductance	L	$L/4\pi\varepsilon_o$
Magnetic field	$\mathbf{H}$	$(4\pi\mu_o)^{-\frac{1}{2}}\ \mathbf{H}$
Magnetic induction (and flux)	$\mathbf{B}[\Phi_m]$	$\sqrt{\mu_o/4\pi}\ \mathbf{B}[\Phi_m]$
Magnetization	$\mathbf{M}$	$\sqrt{4\pi/\mu_o}\ \mathbf{M}$
Relative permeability	κ_m	$\mu_o\kappa_m$
Resistance	R	$R/4\pi\varepsilon_o$

The table below lists side by side equivalent amounts of various physical quantities in SI and Gaussian units. The symbol α represents the ratio $c/10^8$m/s, which, according to (30-3), has the value 2.997925. For many purposes the approximation $\alpha = 3$ is sufficient.

Physical quantity	Symbol	SI	Gaussian
Capacitance	C	1 F	$\alpha^2 \times 10^{11}$cm
Charge	q	1 C	$\alpha \times 10^9$ statcoul
Charge density	ρ	1 C/m³	$\alpha \times 10^3$ statcoul/cm³
Conductivity	σ	1 $(\Omega\text{-m})^{-1}$	$\alpha^2 \times 10^9$ s^{-1}
Current	i	1 A	$\alpha \times 10^9$ statamps
Current density	J	1 A/m²	$\alpha \times 10^5$ statamp/cm²
Displacement vector	D	1 C/m²	$4\pi\alpha \times 10^5$ statvolt/cm
Electric field	E	1 V/m	$\alpha^{-1} \times 10^{-4}$ statvolt/cm
Energy (work)	$U(W)$	1 J	10^7 erg
Force	F	1 N	10^5 dynes
Length	l, d	1 m	10^2 cm
Magnetic field	H	1 A/m	$4\pi \times 10^{-3}$ oersted
Magnetic flux	Φ_m	1 Wb	10^8 maxwells
Magnetic induction	B	1 T	10^4 gauss
Magnetization	M	1 A/m	$(4\pi)^{-1} \times 10^{-3}$ oersted
Mass	m	1 kg	10^3g
Polarization	P	1 C/m²	$\alpha \times 10^5$ statcoul/cm²
Potential	V	1 V	$\alpha^{-1} \times 10^{-2}$ statvolt
Power	P	1 W	10^7 erg/s
Resistance	R	1 Ω	$\alpha^{-2} \times 10^{-11}$ s/cm
Time	t	1 s	1 s

THE PERIODIC TABLE OF THE ELEMENTS

The number directly above the chemical symbol for each element is its atomic number Z, and the number below is its atomic weight in amu. For unstable elements, the atomic weight of the most stable isotope is given in parentheses.

1a	2a	3b	4b	5b	6b	7b	8	8	8	1b	2b	3a	4a	5a	6a	7a	0
1 H 1.01																	2 He 4.00
3 Li 6.94	4 Be 9.01											5 B 10.81	6 C 12.01	7 N 14.01	8 O 16.00	9 F 19.00	10 Ne 20.18
11 Na 22.99	12 Mg 24.31											13 Al 26.98	14 Si 28.09	15 P 30.97	16 S 32.06	17 Cl 35.45	18 Ar 39.95
19 K 39.10	20 Ca 40.08	21 Sc 44.96	22 Ti 47.90	23 V 50.94	24 Cr 52.00	25 Mn 54.94	26 Fe 55.85	27 Co 58.93	28 Ni 58.71	29 Cu 63.55	30 Zn 65.37	31 Ga 69.72	32 Ge 72.59	33 As 74.92	34 Se 78.96	35 Br 79.90	36 Kr 83.80
37 Rb 85.47	38 Sr 87.62	39 Y 88.91	40 Zr 91.22	41 Nb 92.91	42 Mo 95.94	43 Tc (99)	44 Ru 101.07	45 Rh 102.91	46 Pd 106.4	47 Ag 107.87	48 Cd 112.	49 In 114.82	50 Sn 118.69	51 Sb 121.75	52 Te 127.60	53 I 126.90	54 Xe 131.30
55 Cs 132.91	56 Ba 137.34	57–71 *	72 Hf 178.49	73 Ta 180.95	74 W 183.85	75 Re 186.2	76 Os 190.2	77 Ir 192.2	78 Pt 195.09	79 Au 196.97	80 Hg 200.59	81 Tl 204.37	82 Pb 207.19	83 Bi 208.98	84 Po (210)	85 At (210)	86 Rn (222)
87 Fr (223)	88 Ra (226)	89–103 **	104 Ku (260)														

* Lanthanide Series	57 La 138.91	58 Ce 140.12	59 Pr 140.91	60 Nd 144.24	61 Pm (147)	62 Sm 150.35	63 Eu 151.96	64 Gd 157.25	65 Tb 158.92	66 Dy 162.50	67 Ho 164.93	68 Er 167.26	69 Tm 168.93	70 Yb 173.04	71 Lu 174.97
** Actinide Series	89 Ac (227)	90 Th 232.04	91 Pa (231)	92 U 238.03	93 Np (237)	94 Pu (242)	95 Am (243)	96 Cm (247)	97 Bk (247)	98 Cf (251)	99 Es (254)	100 Fm (253)	101 Md (256)	102 No (254)	103 Lw (257)

Solutions to Odd-Numbered Problems

Chapter 19

1. (a) $9.63 \times 10^4 C$;
(b) $-9.63 \times 10^4 C$

3. 58 N; repulsive

5. 1.2×10^2 N; attractive

7. $1.44 \times 10^{-37} C$

9. $F_y = \dfrac{qQ}{4\pi\epsilon_0} y [x^2 + y^2 + z^2]^{-3/2}$;

$F_z = \dfrac{z}{y} F_y$

11. (a) 4.8×10^{-2} N; (b) 6.2×10^{-2} N;
(c) 6.5×10^{-2} N

13. (a) $0.91 \, q^2/4\pi\epsilon_0 a^2$; directed toward center of square;
(b) same as (a)

15. (a) zero; (b) $ybQq/\pi\epsilon_0[b^2 - y^2]^2$

17. $Qqa/2\pi\epsilon_0[a^2 + y^2 + z^2]^{3/2}$

19. along the axis and away from the circle if $qQ > 0$

21. (a) $dF_H = Q\lambda a^2 d\theta/4\pi\epsilon_0[a^2 + b^2]^{3/2}$;
$dF_V = bdF_H/a$;
(b) $Q\lambda ab/2\epsilon_0[a^2 + b^2]^{3/2}$; directed vertically up for $Q\lambda > 0$

Chapter 20

1. 4.5×10^8 N/C; 4.5×10^6 N/C;
4.5×10^2 N/C; directed radially outward

3. (a) zero; (b) 2.4×10^4 N/C in the $-x$ direction; (c) 2.8×10^4 N/C in the $+x$ direction

5. (a) $p/9\pi\epsilon_0 a^3$ in $+x$ direction;
(b) same as (a)

7. —

9. $1.8 \times 10^{-9} C/m^2$

11. (a) 1.0×10^{-7} N/C; (b) 9.4×10^5 m

13. $\dfrac{\lambda}{2\pi\epsilon_0}\left[\dfrac{1}{y} - \dfrac{1}{(y^2 + l^2)^{1/2}}\right]$ parallel to line charge, toward $-\lambda$ part

15. 9.5×10^{-11} C/m

17. $\lambda/2\pi\epsilon_0 a$, perpendicular to the diameter joining the endpoints

19. (a) zero

21. $\dfrac{\sigma b}{2\epsilon_0}\left[\dfrac{1}{(a^2 + b^2)^{1/2}} - \dfrac{1}{(R_0^2 + b^2)^{1/2}}\right]$; along axis

23. —

25. (b) $\dfrac{q}{2\epsilon_0}\left[1-\dfrac{a}{(R_0^2+a^2)^{1/2}}\right]$;

    (c) $q/2\epsilon_0$
27. (a) zero; (b) $-2.5\ \mu C$; (c) $+2.5\ \mu C$
29. (a) $-r\rho_0/2\epsilon_0$; (b) $-\rho_0 a^2/2\epsilon_0 r$;

    (c) zero
31. (b) $\rho_0 r(a^2-r^2)/4\epsilon_0 a^2$; radially

    outward
33. —
35. (a) $Ze/4\pi\epsilon_0 r^2$;

    (b) $7.5\times10^{44}Zer/4\pi\epsilon_0 A$
37. (a) $\rho_0 r/3\epsilon_0$; (b) $\rho_0 a^3/3\epsilon_0 r^2$;

    (c) $\rho_0(a^3+b^3-r^3)/3\epsilon_0 r^2$;

    (d) $\rho_0(a^3+b^3-c^3)/3\epsilon_0 r^2$
39. —
41. (a) $-q/4\pi a^2$; $q/4\pi b^2$;

    (b) $-q/4\pi a^2$; zero
43. $q/4\pi\epsilon_0 r^2$ for $r\leq a$;

    zero for $a\leq r\leq b$;

    $(Q+q)/4\pi\epsilon_0 r^2$ for $r\geq b$

Chapter 21

1. (a) -6.0×10^{-4} J; (b) 6.0×10^{-4} J;

   (c) zero; (d) -5.2×10^{-4} J
3. (a) 5.7×10^4 V/m; (b) planes parallel

   to the charged plane;

   (c) -1.1×10^4 V
5. (a) 1.0×10^4 V/m; (b) -6.0×10^{-4} J
7. (a) $\dfrac{q}{4\pi\epsilon_0}\left[\dfrac{3d-4a}{a(d-a)}\right]$; (b) $3d/4$
9. (a) zero; (b) zero; (c) zero
11. 2.3×10^{-2} J; -1.3×10^{-2} J;

    1.1×10^{-2} J
13. (a) $\dfrac{q}{\pi\epsilon_0}(y^2+a^2/2)^{-1/2}$;

    (b) $\dfrac{qy}{\pi\epsilon_0}(y^2+a^2/2)^{-3/2}$; directed

    vertically away from square
15. $2\alpha(x\mathbf{i}+y\mathbf{j}-2z\mathbf{k})$
17. —
19. (a) zero;

    (b) $\dfrac{q_0\sigma}{2\epsilon_0}[R-(R^2+z^2)^{1/2}+\sqrt{z^2}]$
21. —
23. (a) $\dfrac{Ze}{4\pi\epsilon_0 r}$; (b) $6.8\times$

    $10^{54}\dfrac{Ze}{A}(1.8\times10^{-30}A^{2/3}-r^2/2)$

    (meters); (c) 3.4×10^{-4} J

25. (b) $-\sigma a/\epsilon_0$; (c) $-\sigma(d-b)/\epsilon_0$
27. (b) $\dfrac{Q_0}{4\pi\epsilon_0 a}$
29. $\dfrac{q^2}{4\pi\epsilon_0 a}[4+\sqrt{2}]$
31. -4.4×10^{-18} J
33. $\dfrac{-3q^2}{2\pi\epsilon_0 a}$; $\dfrac{3q^2}{4\pi\epsilon_0 a}$
35. (b) 2.9×10^{21} V/m
37. 4.6×10^{-13} J; 1.5×10^{-10} J
39. (a) -1.1×10^{-3} V; (b) 26 V
41. (a) -4.7×10^{-2} V; (b) 2.4×10^2 V/m
43. (a) $\dfrac{\rho_0}{3\epsilon_0}\left[\dfrac{a^2}{b}(b-a)+\dfrac{(a^2-r^2)}{2}\right]$

    (b) $\dfrac{\rho_0 a^3}{3\epsilon_0}\left(\dfrac{1}{r}-\dfrac{1}{b}\right)$; (c) $-\dfrac{\rho_0 a^3}{3b^2}$
45. $\dfrac{\rho_0}{2\epsilon_0}(b^2-a^2)$
47. —

Chapter 22

1. 1.2×10^{-5} C
3. (a) $1.2\times10^{-5}\ \mu F$;

   (b) 1.2×10^{-7} C/m²
5. (a) $1.1\times10^{-3}\ \mu F$; (b) $1.1\times10^{-2}\ \mu C$;

   (c) $8.8\times10^{-2}\ \mu C/m^2$;

   $-8.9\times10^{-2}\ \mu C/m^2$
7. (a) 1.3×10^4 V; (b) 6.3×10^6 V/m
9. 0.16 F
11. (a) $8.0\ \mu F$; (b) $1.9\ \mu F$
13. $1.6\ \mu F$
15. $3.2\ \mu F$
17. (a) zero on C_5, 15 μC on each of

    the others; (b) $3.0\ \mu F$
19. (a) 6 μF; 1.5 μF; (b) 1 μF; 2 μF;

    4.5 μF; 9 μF;

    (c) 12 μF; 0.75 μF; 1.2 μF; 3.0 μF;

    2.3 μF; 4.0 μF; 5.0 μF; 1.8 μF;

    7.5 μF
21. (a) 5.6×10^{-5} J; 1.9×10^{-5} J;

    (b) 7.5 μC; 23 μC; (c) 5.6×10^{-5} J
23. $\dfrac{\epsilon_0 A\kappa_1\kappa_2}{a\kappa_2+b\kappa_1}$
25. —
27. —
29. (b) $\pm Q(\kappa-1)/\kappa$; (c) no
31. $4\pi\epsilon_0\left[\dfrac{1}{a}-\dfrac{1}{b}+\dfrac{(1-\kappa)}{\kappa}\left(\dfrac{1}{c}-\dfrac{1}{d}\right)\right]^{-1}$;

    yes

33. —

35. (a) $-\dfrac{q}{4\pi a^2}\left(\dfrac{\kappa-1}{\kappa}\right)$; (b) $\dfrac{q}{4\pi b^2}\left(\dfrac{\kappa-1}{\kappa}\right)$

37. (c) $-\epsilon_0 A V_0^2 dy/y^2$

39. (b) $\dfrac{\epsilon_0 V^2}{2d}[A + ay(\kappa - 1)]$

41. —

43. —

Chaper 23

1. (a) 0.20 A; (b) 4.0×10^4 A/m^2

3. 2.5×10^5 A/m^2; 1.3×10^5 A/m^2

5. (a) 3.3 A; (b) 200 C;
(c) 1.9×10^{22} electrons

7. $i_{Ag}/i_{Al} = 1.8$

9. 160 Ω

11. (a) 100 Ω; (b) 1.0 A

13. (a) 10 i; (b) 50 i

15. (a) 0.40 Ω; (b) 4.0×10^5 Ω

17. (a) 2.5 μs; (b) 4.0 A; (c) 1.0×10^{-3} J;
(d) $800 \exp\{-2t/2.5 \,\mu s\}$ (W)

19. (a) 0.50 A; (b) 91 μC; 460 μC;
(c) 0.41 A; 0.041 A

21. (a) $5 \times 10^{-4}[1 - \exp(-10^3 t)]$ (C)
(b) $0.5 \exp(-10^3 t)$ (A)
(c) 2.5×10^{-2} J

23. (b) $C\epsilon^2/2$

25. $\dfrac{\epsilon}{d}(1 - e^{-t/RC})$; $C = \epsilon_0 A/d$

27. (c) $\dfrac{Q_0}{RC_1}\exp[-t(C_1 + C_2)/RC_1 C_2]$

29. (a) 2.6×10^{-8} Ω-m;
(b) 3.7×10^{-8} Ω-m

31. —

33. 677 Ω

Chapter 24

1. (a) 23 Ω; (b) 1.7 Ω

3. 2R/3; 1:1:2

5. —

7. (a) 240 Ω; (b) 0.50 A; (c) 60 W

9. 0.68

11. —

13. —

15. (b) 4.6 W by 10-volt battery; 12 W
by 20-volt battery

17. (a) 0.50 A; (b) $i_{25} = 1.4$ A (up);
$i_{50} = 1.9$ A (right); $i_{10} = 0.11$ A (left);
(c) no

19. —

21. (a) 1.0×10^{-5} s; (b) 30 A;
(c) 300 μC

23. (a) 5.0 A; (b) zero on the 3 μF
capacitor; 100 μC on the 2 μF
capacitor; zero

25. (a) 3.8 A; (b) same as Problem 23b

27. (a) 0.050 A   (left);
(b) 0.30 A   (counterclockwise);
(c) 0.35 A   (clockwise).

29. ϵ/R_1 through R_1; $\epsilon/2R_2$ through R_2;
$\epsilon/2R_3$ through R_3

31. (b) $i_1 = i_3 = 5.0$ A; $i_2 = 0$

33. —

Chapter 25

1. (a) 1.5×10^{-3} N   (west); (b) zero;
(c) 1.1×10^{-3} N   (vertically up)

3. (a) 1.1×10^{-16} N   (east);
(b) 1.2×10^{14} m/s^2

5. positive z-axis

7. (a) zero; (b) $\mu_0 idl/4\pi a^2$ (along $-x$
axis); (c) $\mu_0 idl/4\pi a^2$ (along $-y$
axis)
(d) $\mu_0 idl/8\pi a^2$ (in x-y plane
midway between the y and $-x$
axis)

9. —

11. —

13. 3.6×10^{-5} T

15. —

17. $\dfrac{\mu_0 i}{2\pi}\left\{\left[\dfrac{y-2a}{x^2+(y-2a)^2} - \dfrac{y}{x^2+y^2}\right]\mathbf{i}\right.$
$\left. + \left[\dfrac{x}{x^2+y^2} - \dfrac{x}{x^2+(y-2a)^2}\right]\mathbf{j}\right\}$

19. 40 A; 1.6×10^6 A

21. $\mu_0 i\alpha/4\pi a$ (down)

23. $\mu_0 i/4a$

25. —

27. —

29. 1.6×10^3 turns/m

31. zero

33. —

35. (a) $-\mu_0 Ni$; (b) $+\mu_0 Nih/l$
(c) zero; (d) zero

37. (a) $\mu_0 iR/(a^2 + R^2)^{1/2}$

39. (a) $\mu_0 ir/2\pi a^2$; (b) $\mu_0 i/2\pi r$;
(c) $\dfrac{\mu_0 i}{2\pi r}\left[\dfrac{c^2 - r^2}{c^2 - b^2}\right]$; (d) zero

Chapter 26

1. (a) 210 m; (b) 1.3×10^{-3} s;
   (c) 4.8×10^{3} rad/s

3. (a) $\dfrac{R_\alpha}{R_p} = 2$; (b) $\dfrac{\omega_\alpha}{\omega_p} = \dfrac{1}{2}$;
   (c) $\dfrac{E_\alpha}{E_p} = 4$

5. 12 cm below slit; $E({}^7\text{Li}) : E({}^6\text{Li}) =$ 7 : 6

7. (a) 7.2×10^{7} rad/s; (b) 69 orbits;
   (c) 3.0×10^{-6} s

9. 1.2×10^{4} MeV; no

11. (a) $m\dfrac{d\mathbf{v}}{dt} = q\epsilon\dfrac{\mathbf{v} \times \mathbf{r}}{\mathbf{r}^3}$

13. (a) 9.6×10^{16} rad/s
    (b) 4.6×10^{-10} m

15. —

17. (a) iaB directed perpendicular to the wire and in the plane of the coil

19. —

21. 9.7×10^{-5} N between longer wires; 2.4×10^{-6} N between shorter wires

23. $\dfrac{\mu_0 Iic}{2\pi}\left[\dfrac{1}{a} - \dfrac{1}{b}\right]$;   (left); yes

25. —

27. (a) 0.63 A-m^2; (b) 4.0×10^{-2} A-m^2;
    (c) 0.31 A-m^2

29. 9.4×10^{-3} N-m

31. —

33. $il^2 B / 4\pi$

35. (a) 1.3×10^{-5} m/s;
    (b) 2.5×10^{-10} m^3/C
    (c) 2.5×10^{28} charge carriers/m^3

9. (a) 1.0 V/m   (up);
   (b) 0.20 V (upper end)

11. 5.6×10^{-4} V

13. (a) 17 Ω; (b) 3.8×10^{-3} W;
    (c) 7.5×10^{-4} N; (d) 3.8×10^{-3} W

15. (b) $Blv/R \cos \alpha$; clockwise as viewed from above

17. (b) $\omega B l^2 / 2$; end at P at higher potential

19. —

21. $\dfrac{\mu_0 iv}{2\pi} \ln\left[1 + \dfrac{l}{a}\cos \alpha\right]$

23. (a) zero; (b) $\frac{1}{2}\omega B l^2 \sin \omega t$ (upward in one, down in the other);
    (c) $\omega l^2 B \sin \omega t$

25. —

27. (a) counterclockwise;
    (b) 6.3×10^{-4} V; (c) 6.3 μA

29. (a) $\mu_0 n\pi a^2 (i_0 + \alpha t)$; (b) $\mu_0 n\pi a^2 \alpha$;
    (c) $\mu_0 n\pi a^2 \alpha / R$; opposite to solenoid current

31. (a) $(\mu_0 I\alpha / 2\pi) \ln(1 + b/a)$;
    clockwise
    (b) same as (a)

33. (a) $\mu_0 i_0 (b - \sqrt{b^2 - a^2}) \sin \omega t$
    (b) $\omega \mu_0 i_0 (b - \sqrt{b^2 - a^2}) \cos \omega t$

35. —

37. (a) $q\alpha r / 2$; directed to the right and tangent to a concentric circle; (b) 8.8×10^{8} m/s^2; opposite to the direction in (a)

39. $m\dfrac{d\mathbf{v}}{dt} = \dfrac{q}{2}\mathbf{r} \times \dfrac{d\mathbf{B}}{dt} + q\mathbf{v} \times \mathbf{B}$

Chapter 27

1. (a) 2.0 V/m; along $+y$ direction
   (b) zero; (c) 1.0 V/m along $+z$ direction

3. (a) 1.6×10^{-5} T (perpendicular to plane of wires)
   (b) 4.8×10^{-4} V/m (perpendicular to the wires and in their plane);
   (c) 3.2×10^{-4} V/m; (along the wires)

5. $\dfrac{\mu_0 ivd}{2\pi x(d - x)}$; directed to the right

7. $(\mu_0 \rho_0 a^3 / 3r^3)\mathbf{v} \times \mathbf{r}$ (outside)
   $(\mu_0 \rho_0 / 3)\mathbf{v} \times \mathbf{r}$ (inside)

Chapter 28

1. (a) 6.3×10^{-3} H/m;
   (b) 2.4×10^{-6} Wb; (c) 4.8×10^{-3} Wb

3. (a) 66 μH; (b) 6.6×10^{-7} Wb

5. —

7. (a) 10 mH; (b) 5.0×10^{-6} Wb;
   0.10 Wb

9. $\dfrac{\mu_0 li}{\pi} \ln\left(\dfrac{d - a}{a}\right)$

11. (a) 1.0×10^{-4} s; (b) 6.3×10^{-2} A;
    (c) 10 V

13. —

15. —

17. —
19. —
21. —
23. (a) $ABCDEFA$; (b) $ABEFA$
25. (a) 1.0 A; (b) 0.83 A; (c) 42 μC
27. (a) 0.50 A; (b) 0.67 A; (c) 4.0 A;
     (d) 4.7 A
29. (a) 1.6×10^4 rad/s; (b) 4.0×10^{-4} s;
     (c) 6.3×10^{-6} J
31. —
33. —
35. —
37. —
39. —

Chapter 29

1. 5.5×10^{14} Hz; 1.1×10^7 m^{-1}
3. (a) 5.0×10^{14} Hz;
     (b) 3.1×10^{15} rad/s; (c) 1.0×10^7 m^{-1}
5. (a) 95 V/m; (b) 3.2×10^{-7} T; (c) no
7. Saturn; 16 W; Jupiter; 0.86 W
9. 4.1×10^{-7} m; 5.5×10^{-7} m
11. (b) $\dfrac{\epsilon_0 \mathscr{E}}{RCd} e^{-t/RC}$
13. yes
15. —
17. —
19. 7.5×10^8 m^2
21. zero; 46°
23. —
25. 1.2×10^8 m/s; 2.9×10^{-7} m
27. 3.7×10^{-7} m
29. —
31. $\dfrac{2q^2 a_0^2}{3\pi\epsilon_0 c^3}$
33. (a) qvB/m
35. 7.4×10^{-24} J

Chapter 30

1. (a) 500 s; (b) 3.1×10^3 s;
     2.1×10^3 s
3. 1.5; 3000 Å
5. 2α
7. —
9. (a) 42°; (b) 70°; (c) no emerging
     beam
11. (a) 26°; (b) 78°

13. 1.53
15. (a) 16°; 19°; (b) 3.2°
17. $\sqrt{2}$
19. 31°
21. —
23. 96%; reflected
25. 5.8
27. —
29. —
31. —

Chapter 31

1. 90 cm
3. (a) 30 cm behind mirror;
     (b) 3.0; (c) virtual and erect
5. (a) 0.75 m in front of mirror;
     (b) -7.5×10^{-6}; (c) 1.5×10^{-4} m
7. (a) 13 cm behind mirror; virtual
     (b) 0.67; (c) 6.7 m; erect
9. (a) $D = s^2/(s + R/2)$; (b) zero
11. —
13. —
15. (a) 18 cm behind mirror; 3.0 cm
     high; erect; (b) 15 cm behind
     mirror; 3.8 cm high; erect
17. -1
19. (a) at center of bowl; (b) erect
21. a distance $\dfrac{(d + h)(4d + 3h)}{h}$ above
     mirror
23. (a) 8.8 R; virtual image in water
     (b) 8.8 R; real image in glass
25. (a) 90 cm to right of left surface
     (b) real; inverted; 4.0 cm;
     (c) 10 cm to right of right surface;
     (d) inverted; real; -1
27. 1.6 R to left of left surface of the
     sphere
29. (a) 20 cm; (b) -20 cm;
     (c) converging
31. —
33. (a) 60 cm; (b) 20 cm
35. —
37. (a) away from lens; (b) 12 cm
39. —
41. 10 cm to left of the second lens;
     virtual; inverted; 20 cm high
43. 12 cm to the right

45. —
47. —
49. —

Chapter 32

1. 3.9 V/m
3. (a) 1.2π rad, or 0.23π rad;
   (b) 2.0×10^{-7} m
5. 6.8×10^{-5} m
7. 64 cm
9. —
11. $\Delta = m\lambda$ for a maximum;
    $\Delta = (m + 1/2)\lambda$ for a minimum;
    $m = 0, 1, 2, \ldots$
13. —
15. —
17. —
19. (a) 5.0×10^{-4} cm; (b) $7.6°$; $15°$;
    (c) 7
21. 2.9×10^{-4} cm
23. $\dfrac{13,000}{m}$ Å; $m = 1, 2, 3, \ldots$,
25. —
27. —
29. —
31. —
33. (b) $\mathrm{Im}(\sin^2 \beta / \beta^2)$
35. 6.4×10^4 m
37. 1.3 cm
39. $\dfrac{3.0 \times 10^4}{m + \frac{1}{2}}$ Å; $m = 0, 1, 2, \ldots$
41. (a) 107; (b) 0.94 mm
43. 1.8×10^{-5} cm; 10,000 Å
45. (a) 1.4 m; (b) 4.2 mm
47. —

Chapter 33

1. $59°$
3. —

5. (a) 0.1 W/m^2; (b) 0.075 W/m^2
7. (a) linearly polarized; (b) linearly
   polarized; (c) left circularly
   polarized; (d) elliptically polarized;
   (e) elliptically polarized
9. (a) 1/8; (b) zero
11. —
13. 1
15. (a) 1.86×10^8 m/s; (b) 1.94×10^8 m/s;
17. (a) 0.67C; 0.60C; (b) 0.65C; 0.64C
19. 5.2×10^{-3} rad
21. (a) A is e-ray; B is o-ray
    (b) A (B) is linearly polarized
    perpendicular to (in) plane of
    diagram; (c) $5.6°$
23. (a) $d_0 = \dfrac{\lambda}{2|n_e - n_o|}$; (b) $3d_0$; $5d_0$; $\ldots$
    (c) 17,000 Å
25. (b) at angle θ to the vertical
27. —
29. $\cos^2 2\theta$
31. —
33. —

Chapter 34

1. (a) 9.0×10^{13} W/m^3;
   (b) 1.8×10^5 W/m^3;
   (c) 8.1×10^{-20} W/m^2
3. 530 W/m^2
5. —
7. 1.1×10^{-3} m; yes
9. (a) 7.2×10^{14} Hz; (b) 1.1 eV
11. 1.1×10^{45} photons/s
13. (a) 6.2×10^{-12} m; (b) 8.6×10^{-12} m;
    (c) 1.7×10^5 eV
15. 4.56×10^{14} Hz; 6.58×10^{-7} m
17. —
19. $n^2 a_0$
21. 1.7×10^{-36} m; no; λ is too small

Index

Index